T0313213

Statistical Analysis of Financial Data

CHAPMAN & HALL/CRC
Texts in Statistical Science Series

Joseph K. Blitzstein, *Harvard University, USA*
Julian J. Faraway, *University of Bath, UK*
Martin Tanner, *Northwestern University, USA*
Jim Zidek, *University of British Columbia, Canada*

Recently Published Titles

Theory of Spatial Statistics
A Concise Introduction
M.N.M van Lieshout

Bayesian Statistical Methods
Brian J. Reich and Sujit K. Ghosh

Sampling
Design and Analysis, Second Edition
Sharon L. Lohr

The Analysis of Time Series
An Introduction with R, Seventh Edition
Chris Chatfield and Haipeng Xing

Time Series
A Data Analysis Approach Using R
Robert H. Shumway and David S. Stoffer

Practical Multivariate Analysis, Sixth Edition
Abdelmonem Afifi, Susanne May, Robin A. Donatello, and Virginia A. Clark

Time Series: A First Course with Bootstrap Starter
Tucker S. McElroy and Dimitris N. Politis

Probability and Bayesian Modeling
Jim Albert and Jingchen Hu

Surrogates
Gaussian Process Modeling, Design, and Optimization for the Applied Sciences
Robert B. Gramacy

Statistical Analysis of Financial Data
With Examples in R
James E. Gentle

For more information about this series, please visit:
https://www.crcpress.com/Chapman–HallCRC-Texts-in-Statistical-Science/
book-series/CHTEXSTASCI

Statistical Analysis of Financial Data

With Examples in R

James E. Gentle

CRC Press
Taylor & Francis Group
Boca Raton London New York

CRC Press is an imprint of the
Taylor & Francis Group, an **informa** business

A CHAPMAN & HALL BOOK

First edition published 2020
by CRC Press
6000 Broken Sound Parkway NW, Suite 300, Boca Raton, FL 33487-2742

and by CRC Press
2 Park Square, Milton Park, Abingdon, Oxon, OX14 4RN

Library of Congress Cataloging-in-Publication Data

Library of Congress Control Number:2019956114

ISBN: 978-1-138-59949-9 (hbk)
ISBN: 978-0-429-48560-2 (ebk)

To María

Preface

This book is intended for persons with interest in analyzing financial data and with at least some knowledge of mathematics and statistics. No prior knowledge of finance is required, although a reader with experience in trading and in working with financial data may understand some of the discussion more readily. The reader with prior knowledge of finance may also come to appreciate some aspects of financial data from fresh perspectives that the book may provide. Financial data have many interesting properties, and a statistician may enjoy studying and analyzing the data just because of the challenges it presents.

There are many texts covering essentially the same material. This book differs from most academic texts because its perspective is that of a real-world trader who happens to be fascinated by data for its own sake. The emphasis in this book is on financial *data*. While some stale datasets at the book's website are provided for examples of analysis, the book shows how and where to get current financial data, and how to model and analyze it.

Understanding financial data may increase one's success in the markets, but the book does not offer investment advice.

The organization and development of the book are data driven. The book begins with a general description of the data-generating processes that yield financial data. In the first chapter, many sets of data are explored statistically with little discussion of the statistical methods themselves, how these exploratory analyses were performed, or how the data were obtained. The emphasis is on the *data-generating processes of finance:* types of assets and markets, and how they function.

The first chapter may seem overly long, but I feel that it is important to have a general knowledge of the financial data-generating processes. A financial data analyst must not only know relevant statistical methods of analysis, the analyst must also know, for example, the difference between a seasoned corporate and a T-Bill, and must understand why returns in a short index ETF are positively correlated with the VIX.

While reading Chapter 1 and viewing graphs and other analyses of the various datasets, the reader should ask, "Where could I get those or similar datasets, and how could I perform similar analyses?" For instance, "How could I obtain daily excess returns of the SPY ETF, as is used in the market model on page 61?".

These questions are addressed in an appendix to Chapter 1. Appendix A1, beginning on page 139, discusses computer methods used to obtain real financial data such as adjusted closing stock prices or T-Bill rates from the web, and to mung, plot, and analyze it.

The software used is R. Unless data can be obtained and put in a usable form, there can be no analysis. In the exercises for the appendix, the reader is invited to perform similar exploratory analyses of other financial data.

In the later chapters and associated exercises, the emphasis is on the *methods of analysis* of financial data. Specific datasets are used for illustration, but the reader is invited to perform similar analyses of other financial data.

I have expended considerable effort in attempting to make the Index complete and useful. The entries are a mix of terms relating to financial markets and terms relating to data science, statistics, and computer usage.

Financial Data

This book is about financial data and methods for analyzing it. Statisticians are fascinated by interesting data. Financial data is endlessly fascinating; it does not follow simple models. It is not predictable. There are no physical laws that govern it. It is "big data". Perhaps best of all, an endless supply of it is available freely for the taking.

Access to financial data, and the financial markets themselves, was very different when I first began participating in the market over fifty years ago. There was considerably more trading friction then; commissions were *very* significant. The options markets for the retail investor/trader were almost nonexistent; there were no listed options (those came in 1973). Hedging opportunities were considerably more limited. There were no ETFs. (An early form was released in 1989, but died a year later; finally, the first Spider came out in 1993.) Most mutual funds were "actively managed", and charged high fees.

It has been an interesting ride. The euphoria of the US markets of the '60s was followed by a long period of doldrums, with the exception of the great year of 1975. The Dow's first run on 1,000 fizzled to close 1968 at 943. Although it ended over 1,000 in two years (1972 and 1976), it was not until 1982 that it closed at and stayed over 1,000. Now, since 1999, it's been in five figures except for two crashes. The 80s and 90s went in one direction: up! The "dot.com crash" was a severe hiccup, when the Dow fell back to four figures. Not just the dot.coms, but almost everything else got smashed. Then, back on track until the crash caused by the financial crisis, when the Dow again fell to four figures. In early 2018, at historically low volatility, it broke through 26,000, and then within two years, 29,000. No one could understand the volatility, but lots of traders (especially the "smart money") made money trading volatility, until suddenly they lost money on vol trades, in many cases a *lot* of money (especially the "smart money"). Also, no one understood the

Christmas Eve crash of 2018 (although many analysts "explained" it), but the brave ones made lots of money in the new year.

My interest in playing with financial data as a statistician is much more recent than my participation in the market, and was not motivated by my trading. The only kind of "formal" analysis that I do for my own portfolios is to run a market model (equation (1.35)) weekly, for which I entered data manually and used Fortran until sometime in the 1990s; then I used spreadsheet programs until about 2000; then I used R. I don't enter data manually anymore. Until a few years ago, I used R directly, but now I often use a simple Shiny app I wrote to enter my time intervals and so on. I get most price data directly from Yahoo Finance using `quantmod` (see page 171), but data on options are still problematic. The data-generating process itself is interesting and fun to observe, and that's why I do it.

Like anyone else, I would like to believe that the process is rational, but like any other trader, I know that it is not. That just makes analysis of financial data even more interesting.

Outline

Chapter 1 is about exploratory data analysis (EDA) of financial data. It is less quantitative than the remaining chapters. The chapter ends with a summary of general "stylized facts" uncovered by the EDA in the chapter. Chapter 1 began as a rather brief and breezy overview of financial data, but grew in length as terms and topics came up in developing the later chapters.

I think that a data analyst, in addition to knowing some general characteristics of the data, should have at least a general understanding of the data-generating process, and a purpose of Chapter 1 is to provide that background knowledge. Chapter 1 introduces terms and concepts that relate to financial data and the markets that generate that data. (Among other guidelines for the content of this chapter, I have attempted to include most of the terms one is likely to hear on the CNBC daily episodes or on other financial programs in the mass media.) When I use data from Moody's Seasoned Baa Corporate Bonds in examples or exercises later in the book, I want the reader to know what "Moody's" is, what "seasoned" means, and what "Baa corporate bonds" are, and I do not want to have to intrude on the point of the examples in later chapters to explain those terms there. Those terms are defined or explained in Chapter 1.

The exercises for Chapter 1 are generally conceptual, and involve few computations, unlike the exercises for the appendix to Chapter 1 and later chapters.

The full R code for most of the graphs and computations in Chapter 1 is available at the website for the book. The appendix to Chapter 1 discusses the R code and describes how the data were obtained. The main reason for placing the appendix at this point is that for all of the remaining chapters, R will be used extensively without much comment on the language itself, and many of the exercises in those chapters will require R, and will require internet access to real financial data. The exercises for the appendix involve the use of R, in some cases just to replicate illustrations in Chapter 1.

Instead of stale data on the website for the book or some opaque proprietary database, I want the reader to be able to get and analyze real, current data.

The remaining chapters are about statistical methods. The methods could be applied in other areas of application, but the motivation comes from financial applications.

Chapter 2 harks back to the exploratory data analyses of Chapter 1 and discusses general nonparametric and graphical methods for exploring data.

Chapter 3 covers random variables and probability distributions. Although the chapter does not address statistical analysis *per se*, these mathematical concepts underlie all of statistical inference. Distributional issues particularly relevant for financial data, such as heavy-tailed distributions and tail properties, are emphasized.

Chapter 3 also describes methods for computer generation of random numbers that simulate realizations from probability distributions. Some of the basic ideas of simulation are presented in Section 3.3, and Monte Carlo methods are used in later chapters and in the exercises.

Chapter 4 discusses the role that probability distributions play in statistical inference. It begins with a discussion of statistical models and how to fit them using data. The criteria for fitting models involve some form of optimization ("least" squares, "maximum" likelihood, etc.); therefore, Chapter 4 includes a small diversion into the general methods of optimization. The chapter continues with basic concepts of statistical inference: estimation, hypothesis testing, and prediction. Specific approaches, such as use of the bootstrap, and relevant applications, such as estimation of VaR, are described. Analysis of models of relationships among variables, particularly regression models, is discussed and illustrated.

In view of the recent browbeating by some statisticians about the use of the word "significant", I feel compelled to mention here that I use the term extensively; see the notes to Chapter 4, beginning on page 469.

Chapter 5 provides a brief introduction to the standard time series models, and a discussion of why these models do not work very well in practice. Time series models that account for certain types of heteroscedasticity (GARCH) are discussed, and methods of identifying and dealing with unit roots in autoregressive models are developed. Chapter 5 also addresses topics in vector autoregressive processes, in particular, cointegration of multiple series.

A couple of major topics that are not addressed are the analysis of fixed assets, such as bonds, and the pricing of derivative assets using continuous-time diffusion models. These topics are mentioned from time to time, however.

For any of the topics that are covered, there are many additional details that could be discussed. Some of these are alluded to in the "Notes and Further Reading" sections.

There are of course many smaller topics, such as processing streaming data and high-frequency trading, and the resulting market dynamics, that cannot be discussed because of lack of space.

Software and Programming

The software I use is R. Although I often refer to R in this book, and I give some examples of R code and require R in many exercises, the reader can use other software packages.

A reader with interest but even with no experience with R can quickly pick up enough R to produce simple plots and perform simple analyses. The best way to do this is to look at a few code fragments, execute the code, and then make small changes to it and observe the effects of those changes. The appendix to Chapter 1 displays several examples of R code, and the code to produce all of the graphs and computations shown in Chapter 1 is available at the website for the book.

If the objective is to be able to use R to perform some specific task, like making a graph, the objective can be achieved quickly by finding some R code that performs that kind of task, and then using it with the necessary modifications. (This is not "programming".)

R is a rich programming language. If the objective is to learn *to program* in R, it is my oft-stated belief that "the way to learn to program is to get started and then to program". That's the way I learned to program; what more can I say? That applies to other things also: the way to learn to type is to get started (get a keyboard and find out where the characters are) and then to type; the way to learn to swim is to get started (find some water that's not too deep) and then to swim.

Although I have programmed in many languages from Ada to APL, I don't believe it's desirable to be a programmer of all languages but master of none (to rephrase an old adage). I'd prefer to master just one (or three).

Prerequisites

The prerequisites for this text are minimal. Obviously some background in mathematics, including matrix algebra, is necessary. There are several books that can provide this background. I occasionally make reference to my own books on these topics. That is not necessarily because they are the best for that purpose; it is because I know where the material is in those books.

Some background in statistics or data analysis and some level of scientific computer literacy are also required. I assume that the reader is generally familiar with the basic concepts of probability such as *random variable, distribution, expectation, variance,* and *correlation.* For more advanced concepts and theory, I refer the reader to one of my exhausting, unfinished labors of love, *Theory of Statistics,* at

```
mason.gmu.edu/~jgentle/books/MathStat.pdf
```

Mention of rather advanced mathematical topics is made occasionally in the text. If the reader does not know much about these topics, the material in this book should still be understandable, but if the reader is familiar with these topics, the mention of the topics should add to that reader's appreciation of the material.

No prior knowledge of finance is assumed, although a reader with some background in finance may understand some of the discussion more quickly.

In several places, I refer to computer programming, particularly in R. Some of the exercises require some simple programming, but most exercises requiring use of the computer do not involve programming.

Note on Examples and Exercises

The book uses real financial data in the examples, and it requires the reader to access real data to do the exercises. Some stale datasets are also available at the website for the book. The time periods for the data are generally the first couple of decades of the twenty-first century.

The book identifies internet repositories for getting *real* data and *interesting* data. For the exercises, the reader or the instructor of a course using the book is encouraged to substitute for "2017", "2018", or any other bygone time period, a more interesting time period.

In addition to real data, the book discusses methods of simulating artificial data following various models, and how to use simulated data in understanding and comparing statistical methods. Some exercises ask the reader to assess the performance of a statistical technique in various settings using simulated data.

Data preparation and cleansing are discussed, and some exercises require a certain amount of data wrangling.

The exercises in each chapter are not necessarily ordered to correspond to the ordering of topics within the chapter. Each exercise has a heading to indicate the topic of the exercise, but the reader is encouraged to read or skim the full chapter before attempting the exercises. The exercises vary considerably in difficulty and length. Some are quite long and involved.

Supplementary Material

The website for the book is

`mason.gmu.edu/~jgentle/books/StatFinBk/`

The website has a file of hints, comments, and/or solutions to selected exercises. Supplemental material at this site also includes R code used in producing examples in the text. Although I emphasize real, live data, the website also has a few financial datasets from the past.

The website has a list of known errors in the text that will be updated as errors are identified.

A complete solutions manual is available to qualified instructors on the book's webpage at

`www.crcpress.com`

Because adjusted asset prices change over time, the reader must be prepared to accept slight differences in the results shown from the results arising from data that the reader may access at a later time.

Acknowledgments

First, I thank John Chambers, Robert Gentleman, and Ross Ihaka for their foundational work on R. I thank the R Core Team and the many package developers and those who maintain the packages for continuing to make R more useful.

Jim Shine has read much of the book and has worked many of the exercises. I thank Jim for many helpful comments.

I thank the anonymous reviewers for helpful comments and suggestions.

I thank my editor John Kimmel. It has been a real pleasure working with John again, as it was working with him on previous books.

I thank my wife, María, to whom this book is dedicated, for everything.

I used TEX via LATEX 2_ε to write the book. I did all of the typing, programming, etc., myself, so all mistakes are mine. I would appreciate receiving notification of errors or suggestions for improvement.

Fairfax County, Virginia

James E. Gentle
December 27, 2019

Contents

Preface   vii

1 The Nature of Financial Data   1
   1.1 Financial Time Series   5
        1.1.1 Autocorrelations   7
        1.1.2 Stationarity   7
        1.1.3 Time Scales and Data Aggregation   8
   1.2 Financial Assets and Markets   12
        1.2.1 Markets and Regulatory Agencies   16
        1.2.2 Interest   20
        1.2.3 Returns on Assets   29
        1.2.4 Stock Prices; Fair Market Value   32
        1.2.5 Splits, Dividends, and Return of Capital   45
        1.2.6 Indexes and "the Market"   48
        1.2.7 Derivative Assets   63
        1.2.8 Short Positions   65
        1.2.9 Portfolios of Assets: Diversification and Hedging . . .   66
   1.3 Frequency Distributions of Returns   76
        1.3.1 Location and Scale   79
        1.3.2 Skewness   81
        1.3.3 Kurtosis 81
        1.3.4 Multivariate Data   82
        1.3.5 The Normal Distribution   87
        1.3.6 Q-Q Plots   91
        1.3.7 Outliers 93
        1.3.8 Other Statistical Measures   94
   1.4 Volatility 98
        1.4.1 The Time Series of Returns   99
        1.4.2 Measuring Volatility: Historical and Implied   102
        1.4.3 Volatility Indexes: The VIX   108
        1.4.4 The Curve of Implied Volatility   112
        1.4.5 Risk Assessment and Management   113
   1.5 Market Dynamics   120
   1.6 Stylized Facts about Financial Data   129
   Notes and Further Reading   130
   Exercises and Questions for Review   132

Appendix A1: Accessing and Analyzing Financial Data in R . . . 139
   A1.1   R Basics 140
   A1.2   Data Repositories and Inputting Data into R 158
   A1.3   Time Series and Financial Data in R 172
   A1.4   Data Cleansing 183
   Notes, Comments, and Further Reading on R 187
   Exercises in R 191

2 Exploratory Financial Data Analysis   **205**
  2.1 Data Reduction   207
      2.1.1 Simple Summary Statistics   207
      2.1.2 Centering and Standardizing Data 208
      2.1.3 Simple Summary Statistics for Multivariate Data . . . 208
      2.1.4 Transformations   209
      2.1.5 Identifying Outlying Observations 210
  2.2 The Empirical Cumulative Distribution Function 211
  2.3 Nonparametric Probability Density Estimation 217
      2.3.1 Binned Data   217
      2.3.2 Kernel Density Estimator 219
      2.3.3 Multivariate Kernel Density Estimator 220
  2.4 Graphical Methods in Exploratory Analysis 221
      2.4.1 Time Series Plots 222
      2.4.2 Histograms   222
      2.4.3 Boxplots 224
      2.4.4 Density Plots 226
      2.4.5 Bivariate Data   227
      2.4.6 Q-Q Plots 229
      2.4.7 Graphics in R 234
  Notes and Further Reading   238
  Exercises   239

3 Probability Distributions in Models of Observable Events **245**
  3.1 Random Variables and Probability Distributions 247
      3.1.1 Discrete Random Variables 248
      3.1.2 Continuous Random Variables   252
      3.1.3 Linear Combinations of Random Variables;
              Expectations and Quantiles 256
      3.1.4 Survival and Hazard Functions 257
      3.1.5 Multivariate Distributions 258
      3.1.6 Measures of Association in Multivariate Distributions 261
      3.1.7 Copulas 264
      3.1.8 Transformations of Multivariate Random Variables . . 267
      3.1.9 Distributions of Order Statistics   269
      3.1.10 Asymptotic Distributions; The Central Limit Theorem 270
      3.1.11 The Tails of Probability Distributions 273

3.1.12 Sequences of Random Variables; Stochastic Processes 278
3.1.13 Diffusion of Stock Prices and Pricing of Options . . . 279
3.2 Some Useful Probability Distributions 282
3.2.1 Discrete Distributions 283
3.2.2 Continuous Distributions 285
3.2.3 Multivariate Distributions 294
3.2.4 General Families of Distributions Useful in Modeling . 295
3.2.5 Constructing Multivariate Distributions 309
3.2.6 Modeling of Data-Generating Processes 311
3.2.7 R Functions for Probability Distributions 311
3.3 Simulating Observations of a Random Variable 314
3.3.1 Uniform Random Numbers 315
3.3.2 Generating Nonuniform Random Numbers 317
3.3.3 Simulating Data in R 321
Notes and Further Reading 323
Exercises 325

4 **Statistical Models and Methods of Inference** **335**
4.1 Models 336
4.1.1 Fitting Statistical Models 340
4.1.2 Measuring and Partitioning Observed Variation 340
4.1.3 Linear Models 343
4.1.4 Nonlinear Variance-Stabilizing Transformations 344
4.1.5 Parametric and Nonparametric Models 345
4.1.6 Bayesian Models 346
4.1.7 Models for Time Series 346
4.2 Criteria and Methods for Statistical Modeling 347
4.2.1 Estimators and Their Properties 347
4.2.2 Methods of Statistical Modeling 349
4.3 Optimization in Statistical Modeling; Least Squares
and Maximum Likelihood 358
4.3.1 The General Optimization Problem 358
4.3.2 Least Squares 363
4.3.3 Maximum Likelihood 371
4.3.4 R Functions for Optimization 375
4.4 Statistical Inference 376
4.4.1 Confidence Intervals 379
4.4.2 Testing Statistical Hypotheses 381
4.4.3 Prediction 385
4.4.4 Inference in Bayesian Models 386
4.4.5 Resampling Methods; The Bootstrap 393
4.4.6 Robust Statistical Methods 396
4.4.7 Estimation of the Tail Index 399
4.4.8 Estimation of VaR and Expected Shortfall 404
4.5 Models of Relationships among Variables 408

    4.5.1   Principal Components   409
    4.5.2   Regression Models   413
    4.5.3   Linear Regression Models   418
    4.5.4   Linear Regression Models: The Regressors   422
    4.5.5   Linear Regression Models: Individual Observations and
            Residuals   428
    4.5.6   Linear Regression Models: An Example   435
    4.5.7   Nonlinear Models   449
    4.5.8   Specifying Models in R   454
  4.6   Assessing the Adequacy of Models   455
    4.6.1   Goodness-of-Fit Tests; Tests for Normality   456
    4.6.2   Cross-Validation   463
    4.6.3   Model Selection and Model Complexity   467
  Notes and Further Reading   469
  Exercises   472

5 **Discrete Time Series Models and Analysis**   **487**
  5.1   Basic Linear Operations   495
    5.1.1   The Backshift Operator   495
    5.1.2   The Difference Operator   497
    5.1.3   The Integration Operator   500
    5.1.4   Summation of an Infinite Geometric Series   500
    5.1.5   Linear Difference Equations   501
    5.1.6   Trends and Detrending   505
    5.1.7   Cycles and Seasonal Adjustment   508
  5.2   Analysis of Discrete Time Series Models   510
    5.2.1   Stationarity   514
    5.2.2   Sample Autocovariance and Autocorrelation Functions;
            Stationarity and Estimation   518
    5.2.3   Statistical Inference in Stationary Time Series   523
  5.3   Autoregressive and Moving Average Models   528
    5.3.1   Moving Average Models; MA(q)   529
    5.3.2   Autoregressive Models; AR(p)   534
    5.3.3   The Partial Autocorrelation Function, PACF   547
    5.3.4   ARMA and ARIMA Models   549
    5.3.5   Simulation of ARMA and ARIMA Models   555
    5.3.6   Statistical Inference in ARMA and ARIMA Models .   556
    5.3.7   Selection of Orders in ARIMA Models   560
    5.3.8   Forecasting in ARIMA Models   561
    5.3.9   Analysis of ARMA and ARIMA Models in R   561
    5.3.10 Robustness of ARMA Procedures; Innovations with
            Heavy Tails   566
    5.3.11 Financial Data   568
    5.3.12 Linear Regression with ARMA Errors   571
  5.4   Conditional Heteroscedasticity   575

        5.4.1   ARCH Models   576
        5.4.2   GARCH Models and Extensions   580
    5.5   Unit Roots and Cointegration   584
        5.5.1   Spurious Correlations; The Distribution of the
                Correlation Coefficient   584
        5.5.2   Unit Roots   592
        5.5.3   Cointegrated Processes   599
    Notes and Further Reading   603
    Exercises   604

References   **615**

Index   **623**

1

The Nature of Financial Data

In this chapter, I discuss the operation of financial markets and the nature of financial data. The emphasis is on *data*, especially data on returns of financial assets. Most of the data discussed are taken from various US markets in the late twentieth and early twenty-first centuries. The emphasis, however, is not on the specific data that I used, or data that I have put on the website for the book, but rather on current data that you can get from standard sources on the internet.

The objectives of the chapter are to

- provide a general description of financial assets, the markets and mechanisms for trading them, and the kinds of *data* we use to understand them;

- describe and illustrate graphical and computational statistical methods for exploratory analysis of financial *data*;

- display and analyze historical financial *data*, and identify and describe key properties of various types of financial *data*, such as their volatility and frequency distributions;

- discuss the basics of risk and return.

The statistical methods of this chapter are exploratory data analysis, or EDA. We address EDA methods in Chapter 2.

I introduce and describe a number of concepts and terms used in finance ("PE", "book value", "GAAP", "SKEW", "EBITDA", "call option", and so on). I will generally italicize these terms at the point of first usage, and I list most of them in the Index. Many of these terms will appear in later parts of the book.

The computations and the graphics for the examples were produced using R. There is, however, no discussion in this chapter of computer code or of internet data sources. In the appendix to this chapter beginning on page 139, I describe some basics of R and then how to use R to obtain financial data from the internet and bring it into R for analysis. In that appendix, I refer to specific datasets used in the chapter. The R code for all of the examples in this chapter is available at the website for the book.

Types of Financial Data

Financial data come in a variety of forms. The data may be prices at which transactions occur, they may be quoted rates at which interest accrues, they may be reported earnings over a specified time, and so on. Financial data often are derived quantities computed from observed and/or reported quantities, such as the ratio of the price of a share of stock and the share of annual earnings represented by a share of stock.

Financial data are associated either with specific points in time or with specific intervals of time. The data themselves may thus be prices at specified times in some specified currency, they may be some kind of average of many prices at a specified time (such as an index), they may be percentages (such as interest rates at specified times or rates of change of prices during specified time intervals), they may be nonnegative integers (such as the number of shares traded during specified time intervals), or they may be ratios of instantaneous quantities and quantities accumulated over specified intervals of time (such as price-to-earnings ratios).

When the data are prices of assets or of commodities, we often refer generically to the unit of currency as the *numeraire*. Returns and derivative assets are valued in the same numeraire. Most of the financial data in this book are price data measured in United States dollars (USD).

Our interest in data is usually to understand the underlying phenomenon that gave rise to the data, that is, the *data-generating process*. In the case of financial data, the data-generating process consists of the various dynamics of "the market".

The purpose of the statistical methods discussed in this book is to gain a better understanding of data-generating processes, particularly financial data-generating processes.

An analysis of financial data often begins with a plot or graph in which the horizontal axis represents time. Although the data may be associated with specific points in time, as indicated in the plot on the left in Figure 1.1, we often connect the points with line segments as in the plot on the right in Figure 1.1. This is merely a visual convenience, and does not indicate anything about the quantity between two points at which it is observed or computed. The actual data is a discrete, finite set.

Models and Data; Random Variables and Realizations: Notation

We will often discuss *random variables* and we will discuss data, or *realizations* of random variables. Random variables are mathematical abstractions useful in statistical models. Data are observed values. We often use upper-case letters to denote random variables, X, Y, X_i, and so on. We use corresponding lower-case letters to denote observed values or realizations of the random variables, x, y, x_i, and so on.

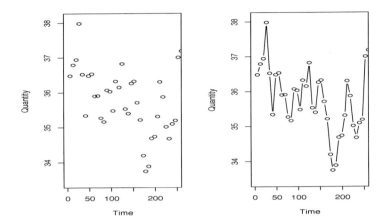

FIGURE 1.1
Financial Data

I prefer to be very precise in making distinctions between random variables and their realizations (in notation, the conventional distinction is upper-case letters for random variables and lower case for realizations), but the resulting notation often is rather cumbersome. Generally, therefore, in this book I will use simple notation and may use either upper- or lower-case whether I refer to a random variable or a realization of one.

We form various functions of both random variables and data that have similar interpretations in the context of the model and the corresponding dataset. These functions are means, variances, covariances, and so on. Sometimes, but not always, in the case of observed data, we will refer to these computed values as "sample means", "sample variances", and so on.

An important concept involving a random variable is its *expected value* or *expectation*, for which I use the symbol $E(\cdot)$. It is assumed that the reader is familiar with this terms. (It will be formally defined in Section 3.1.)

The expected value of a random variable is its *mean*, which we often denote by μ, and for a random variable X, we may write

$$E(X) = \mu_X. \tag{1.1}$$

Another important concept involving a random variable is its *variance*, often denoted by σ^2 and for which I use the symbol $V(\cdot)$. The variance is defined as the expected value of the squared difference of the random variable from its mean, and for the random variable X, we may write

$$V(X) = E\left((X - \mu_X)^2\right) = \sigma_X^2. \tag{1.2}$$

The square root of the variance is called the *standard deviation*.

For two random variables, X and Y, we define the *covariance*, denoted by $\mathrm{Cov}(\cdot, \cdot)$, and defined as

$$\mathrm{Cov}(X, Y) = \mathrm{E}\left((X - \mu_X)(Y - \mu_Y)\right). \tag{1.3}$$

A related concept is the *correlation*, denoted by $\mathrm{Cor}(\cdot, \cdot)$, and defined as

$$\mathrm{Cor}(X, Y) = \frac{\mathrm{Cov}(X, Y)}{\sqrt{\mathrm{V}(X)\mathrm{V}(Y)}}. \tag{1.4}$$

We often denote the correlation between the random variables X and Y as ρ_{XY}. The correlation is more easily interpreted because it is always between -1 and 1. As mentioned, it is assumed that the reader is familiar with these terms, but we mention them here to establish notation that we will use later. These terms will be formally defined in Section 3.1. The sample versions are defined in Section 1.3.

For an ordered set of random variables $X_1, \ldots, X_d$, I denote the d-vector consisting of this ordered d-tuple as X, that is, with no special font to distinguish a vector from a scalar. I write this as $X = (X_1, \ldots, X_d)$. A vector is not a "row vector"; it has the same arithmetic properties as a column in a matrix. The mean of the d-vector X is the d-vector $\mathrm{E}(X) = \mu_X$. (The notation is the same as for scalars; we must depend on the context.)

The *variance-covariance matrix* of X, which is often denoted as Σ_X, following the equations above, is

$$\mathrm{V}(X) = \mathrm{E}\left((X - \mu_X)(X - \mu_X)^{\mathrm{T}}\right) = \Sigma_X, \tag{1.5}$$

where the notation w^{T} means the transpose of the vector w. (The same kind of notation is used to denote the transpose of a matrix. The expression in the expectation function $(X - \mu_X)(X - \mu_X)^{\mathrm{T}}$ is the *outer product* of the residual vector with itself.)

Linear Combinations of Random Variables

In statistical analysis of financial data, we often form linear combinations of the variables. Here, we establish some notation and state some facts from elementary probability theory. We will develop these and other facts in Section 3.1.8.

Suppose we have the random vector $X = (X_1, \ldots, X_d)$ with $\mathrm{E}(X) = \mu_X$ and $\mathrm{V}(X) = \Sigma_X$. Now, consider the constants $w_1, \ldots, w_d$ and write $w = (w_1, \ldots, w_d)$ to denote the d-vector. The *dot product* $w^{\mathrm{T}}X$ is the *linear combination*

$$w^{\mathrm{T}}X = w_1 X_1 + \cdots + w_d X_d. \tag{1.6}$$

Using the properties of the expectation operator, it is easy to see that

$$\mathrm{E}\left(w^{\mathrm{T}}X\right) = w^{\mathrm{T}}\mu_X \tag{1.7}$$

and

$$\mathrm{V}\left(w^{\mathrm{T}}X\right) = w^{\mathrm{T}}\Sigma_X w. \tag{1.8}$$

We will develop these facts more carefully in Chapter 3, but we will use them in various applications in this chapter.

1.1 Financial Time Series

Data collected sequentially in time, as those shown in Figure 1.1, are *time series data*, and the dataset itself is also called a *time series*. If there is only one observation at each time point, the time series is *univariate*; if there are more than one at each time point, the time series is *multivariate*.

One of the first features to note about financial data-generating processes is that they are time series that may have nonzero autocorrelations. Observations are not independent of each other. Financial time series differ from other common time series, however, in at least two ways. Firstly, the time intervals between successive observations in financial time series are usually not of constant length. Secondly, the data in a financial time series are often subject to revision or adjustment.

Notation and General Characteristics of Time Series Data

We usually denote an element of a time series with a symbol for the variable and a subscript for the time, X_t, for example. We may denote the whole time series in set notation as $\{X_t\}$, but in this case, the elements of the set have a strict order. For many other types of data, we use similar notation, symbols for the variables and a subscript that just serves as an index to distinguish one observation of the variables from another, X_i or (Y_i, Z_i), for example. For such variables, we often assume they are *exchangeable*; that is, in computations for an analysis, we can process the data in the order $X_1, X_2, \ldots$ or any other order such as $\ldots, X_2, X_1$.

In time series, the order is important; the data are not exchangeable. In fact, the order itself is often the primary feature to be analyzed.

Because the quantity of interest, X, is a function of time, we sometimes use the notation $X(t)$ instead of X_t.

For many statistical analyses, we assume that the data are *independently and identically distributed* (iid). We obviously do not make this assumption in the analysis of a time series. The properties of a given observation may be dependent on the properties of the preceding observations; hence, the data are possibly neither independently nor identically distributed.

In a sense, a time series is just one observation on a multivariate random variable, and a specific value, such as x_t, is just one element of a multivariate observation with special properties. The value at the point t_0, that is, x_{t_0}, may be dependent on other values x_t for which $t < t_0$. The present may be dependent on the past, but not on the future. (In a more precise mathematical exposition, we would refer to this as a filtration of the underlying σ-fields. This setup is the reason that I will often use the terms "conditional expectation", "conditional variance", and so on, and from time to time I may use notation such as $\mathrm{E}(X_t|\mathcal{F}_{t-1})$. I will hold the mathematical theory to a minimum, but I will always attempt to use precise terminology. I will discuss conditional distributions and conditional expectation further on page 260.)

In many statistical models for time series, time is assumed to be a discrete quantity measured on a linear scale; that is, it is "equally spaced". Although, as we will see, most financial data are not observed at equally-spaced intervals, the assumption may nevertheless allow for adequate models. Time, of course, is a continuous quantity (or at least, we consider it to be measurable at any degree of granularity).

In the simple discrete time series in which we assume equal spacing of the data over time, we can measure time with integers, and we can denote the "next time" after a given time t as $t + 1$. Considering the times to be discrete and equally spaced, we can express the dependence on the past in general as

$$x_t = f(\ldots, x_{t-2}, x_{t-1}) + \epsilon_t, \tag{1.9}$$

where f is some function, possibly unknown, and ϵ_t is an unobservable random variable. A model such as equation (1.9) is used often in statistical applications to represent dependence of one observable variable on other variables. The random variable ϵ_t is a "residual", an "error term", or especially in time series applications, an "innovation". We generally assume that the error term, as a random variable, has the following properties.

- For each t, the *expected value* of the error term is zero, that is, $\mathrm{E}(\epsilon_t) = 0$.

- For each t, the *variance* of the error term is constant, that is, $\mathrm{V}(\epsilon_t) = \sigma^2$, where σ is some positive, but perhaps unknown, number.

- For each t and each $h > 0$, the *correlation* of two error terms is zero, that is, $\mathrm{Cor}(\epsilon_t, \epsilon_{t+h}) = 0$.

We also often assume that the elements of the ordered set $\{\epsilon_t\}$ are iid. Random variables that are iid have the same variance and have zero correlation, as in the assumptions above. Another common assumption is that the error terms have a normal distribution. (In that case, the zero correlation implies independence.) As we will see, however, these assumptions often do not hold for financial data.

Equation (1.9) should be thought of as a "model". In this case, the model does not include any specification of the time period that it covers, that is, it does not specify the values of t for which it applies, other than the fact that t is an integer (that is, the time series is "discrete"). Restrictions of course could be placed on the range of values of t in the model, but often none are. This means, in a broad sense, that

$$t = \ldots, -2, -1, 0, 1, 2, \ldots$$

I will cover time series in more detail in Chapter 5, and will consider some specific forms of the model (1.9). In the remainder of this section, I will mention some of the basic issues in dealing with time series data, especially financial data.

1.1.1 Autocorrelations

One of the most important characteristics of a time series are the correlations among pairs of terms in the series X_t and the lagged value X_{t-h}. We call these correlations, $\text{Cor}(X_t, X_{t-h})$, *autocorrelations*. We also refer to nonzero correlations or autocorrelations informally as "dependencies".

The autocorrelations may be zero, positive, or negative. They are generally different for different values of h, and even for a given value of h, they may or may not be constant for all values of t.

We stated that a common assumption in the model (1.9) is that the error terms ϵ_t are iid. This would of course imply that the autocorrelations of the error terms are 0 for any t and $h \neq 0$. (This, however, does not imply that the autocorrelations of the observable X_t are 0.)

For a financial time series $\{X_t\}$, it seems reasonable that $\text{Cor}(X_t, X_{t-h})$ may be larger in absolute value for small values of the lag h than for larger values of h. For h sufficiently large, it is plausible that the autocorrelation for any financial quantity, say a stock price, is zero.

An interesting question in financial data has to do with *long-range* dependencies. Some empirical studies indicate that there are nonzero autocorrelations at larger values of h than general economic considerations may lead one to expect.

1.1.2 Stationarity

"Stationarity", in general, refers to the constancy in time of a data-generating process. Constancy in time does not mean that the observed values are constant; it means that salient properties of the *process*, such as (marginal) variance or correlations, are constant.

In terms of autocorrelations, a type of stationarity is that for given h, $\text{Cor}(X_t, X_{t-h})$ is constant for all values of t. This type of stationarity implies that the variance $V(X_t)$ is constant for all t. Intuitively, if this is the case, our statistical analysis is simpler because we can aggregate data over time.

For this type of stationarity, the *autocorrelation function* (ACF) is a very useful tool. Because of this stationarity, the ACF is a function only of the lag h:

$$\rho(h) = \text{Cor}(X_t, X_{t-h}). \tag{1.10}$$

We will give precise technical definitions of types of stationarity in Chapter 5, but for now we may note how unlikely it would be for a data-generating financial process, such as trading stocks, to be constant over time. Most financial processes change with exogenous changes in the general economy and even, possibly, with changes in the weather.

1.1.3   Time Scales and Data Aggregation

In many simple time series models, it is assumed that the data are collected at discrete and equally-spaced intervals. When the time intervals are equal, the analysis of the data and the interpretation of the results are usually simpler. For such data, we can denote the time as $t, t+1, t+2$, and so on; and the actual time, no matter what index is used to measure it, maps simply to the index for data, x_t.

For financial time series, this is almost never the case.

The simple time series exhibited in Figure 1.1 are weekly unadjusted closing prices, $x_1, x_2, \ldots$, of the common stock of Intel Corporation, traded on the Nasdaq market as INTC, for the period from January 1, 2017 through September 30, 2017. The closing price at the end of the first week, January 6, is x_1. In the following, I will continue to use the INTC prices to illustrate characteristics of stock prices and by extension, of financial data in general.

A "weekly close" is the closing price on the last day of a week in which the market is open. There are 35 observations shown in Figure 1.1 because there were 35 weeks in that period.

The time interval between two successive observations is 7 days, except for observations 15, 16, and 17. Because April 14 was the Good Friday holiday, x_{15} was the closing price at the end of the trading day of Thursday, April 13, and so there were 6 days of real time between $t = 15$ and $t = 16$, and 8 days of real time between $t = 16$ and $t = 17$. There were only 4 days of trading between $t = 15$ and $t = 16$. Likewise for example, because of the Martin Luther King holiday on January 16, there were only 4 days of trading between $t = 2$ and $t = 3$. A careful analysis of a financial time series may need to take into consideration these kinds of differences from simple time series with equally-spaced time.

In Figure 1.2 we also show the daily and monthly closing prices of INTC for the same period. For the daily closing prices, shown in the graph in the upper left side of Figure 1.2, there are 188 observations because there were 188 days in which the market was open during that period. Also, as we remarked concerning Figure 1.1, although we are not using any data between day t and day $t + 1$, we connect the discrete data with a line segment, as if the prices were continuous variables. As with the data shown in Figure 1.1, the actual time interval between two observations in the time series varies. The variation in the length of the intervals in the daily data is especially relatively large due to weekends and holidays.

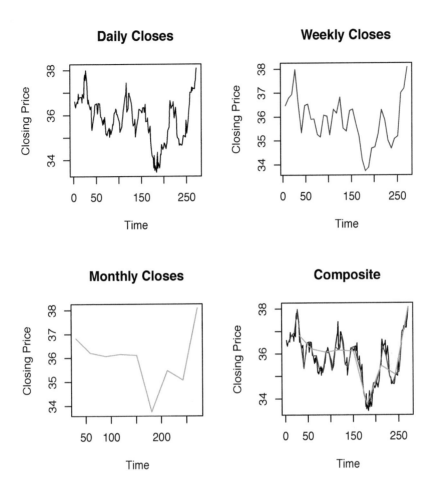

FIGURE 1.2
Closing Prices (Unadjusted) of INTC for the Period January 1, 2017 through
September 30, 2017; *Source:* Yahoo Finance

The graph in the upper right side of Figure 1.2 is a plot of the same data as
shown in Figure 1.1; that is, the weekly closes of INTC. Again, the horizontal
axis is marked in days since January 1, 2017. Although continuous line segments are drawn for visual convenience, there are only 35 actual observations
used to make the plot.

The plot in the lower left side of Figure 1.2 shows the monthly closes of
INTC over this same period. There are only 8 observations. The time intervals between two successive observations are different because the lengths of

months are different. In addition, for that reason and also because of holidays, the number of trading days between successive observations may be different.

The plots in Figure 1.2 represent data from the same data-generating process collected at time intervals of different lengths. The different datasets have different *sampling frequencies.* Time series data collected at a higher frequency may display characteristics that are different from properties of data from the same process collected at a lower frequency.

The Irregularity of Time

The "trading day" depends on the market where trading occurs. Although small amounts of INTC stock are traded in several markets around the world, the primary market for INTC is the Nasdaq National Market, which is headquartered in New York. The trading day for Nasdaq and also for the New York Stock Exchange is from 9:30am to 4:00pm New York time. Thus the closing time is 4:00pm New York time. (There is an "after hours" market lasting four hours after the market close and a "pre-market" starting five and one-half hours before the market open that allows trading that follows a unique set of rules.)

As we saw above, even for "daily" stock price data, say $y_1, y_2, \ldots$, the time intervals are not equal because the United States markets are closed on weekends and on nine official holidays per year. For the daily data in the upper left side of Figure 1.2, y_1 for example is the price on the first trading day, Tuesday, January 3; y_4 is three days later, January 6 (the same as x_1 in the weekly data); and y_5 is six days later, January 9, skipping the weekend. The differences are small enough that they can usually be ignored, but we should be aware of them. The effect of these time differences, whether from a weekend or from a holiday, is called a "weekend effect." (In Exercise A1.10, you are asked to investigate these differences using real data.)

These irregular intervals are apparent in Table 1.1.

TABLE 1.1

The Irregularity of Time; January, 2017

Day of year	1	2	3	4	5	6	7	8	9	10	11	12	13	14	15	16	17	...
Daily data			1	2	3	4			5	6	7	8	9				10	...
Weekly data				1								2						...

Occasionally a catastrophic event will result in market closure. For example, following the terrorist attacks before the market open on September 11, 2001 (a Tuesday), the US markets suspended trading for that day and was closed for 6 days, until the following Monday.

Even if trading occurred every day, the daily closing prices would not be determined at equal times from one day to the next, because the last trade of a given stock occurs at different times from day to day.

The successive trades over nonuniform intervals that nonetheless establish prices are called "nonsynchronous trading". There have been many studies of the effect of nonsynchronous trading and of the "weekend effect" on stock prices. These studies (and theories based on them) are generally inconclusive.

In most plots of daily price data, the time axis is marked as if the prices were observed at equal intervals. The horizontal axis in the plot in the upper left of Figure 1.2, however, is marked proportionally to real time, but the gaps for the weekends and holidays are hardly noticeable in the graph.

Measuring Time

Time is a continuous quantity. We use various units to measure it, seconds, hours, days, and so on. We associate an observation in a time series with a specific time, which we may measure as a date, specified according to some formal calendar or time format, or as a number of time units from some fixed starting time.

In the examples in the preceding section, we have used various time units, days, weeks, and months. These time measures present problems of two types. First of all, months and years are not linear units of time. Months have unequal relationships to days and to weeks. Years, likewise, have unequal relationships to days, weeks, and months.

Another type of problem with these units is that the numbers of opportunities for financial transactions are not constant within any of them. The realtime interval between two weekly closings is usually 7 days but sometimes 6 or 8 days, and there are usually 5 trading days between one weekly close and the next one, but, as we mentioned, with the INTC data in some cases there were only 4 trading days within the week.

Even taking a day as the basic unit of time, we have a problem of matching trading days with real time because of weekends and holidays.

Another way of measuring time is to ignore any relationship to real time. A trading day, without any regard to weekends or holidays, is a unit of time, and for many financial time series, such as stock prices, it is an appropriate unit of time, even though it is not of uniform length in real time.

Tick Time

We can carry this approach to measuring time even further. We may measure time by *ticks*; that is, by whenever a transaction occurs. This unit of time, which is different for each financial quantity whether it is the price of INTC or the price of IBM, for example, has no linear connection with real time.

There are some interesting and anomalous characteristics of data collected at shorter intervals of time, say every minute. At this frequency, the data exhibit some unexpected properties, which we refer to generally as "market microstructure". This effect of frequency has been known for some time. In the beginning, one minute was a very short period, and the number of trades in a single stock in one minute was generally small. Nonetheless, the higher frequency in collection of the data reveals basic differences in market behavior.

Since the 1990s many financial institutions have placed larger and larger numbers of orders to buy or sell the same stock. A stock may be bought and less than a second later sold. The decision to buy or sell is based on an "algorithm", and all orders are limit orders (see page 35). A fast rate of buying and selling is called "high-frequency trading". (There is no specific rate that qualifies as "high-frequency".) In the analysis of high-frequency trading, the time intervals are usually just taken to be the lengths of time between successive trades of the same stock; hence, the time intervals are unequal. In this chapter, and indeed in this book, we will generally be interested in analyzing daily, weekly, or similar data, rather than in the characteristics of high-frequency trading or of the market microstructure.

Aggregating Time

One of the most interesting properties of financial data is the proportional change from one period to another. The nature of the frequency distribution of these changes depends on the length of the time interval over which the change is measured. The statistical distribution is different for changes measured over shorter periods, say days, from changes over longer periods, say months. At even shorter periods of time, say seconds, the differences become very pronounced.

Depending on our objectives in analysis of financial data, it is sometimes desirable to use data over longer periods of time. This is a type of aggregation of data, but it is not the same type of aggregation as means or medians that are used in statistical data analysis when the time characteristics are ignored.

1.2   Financial Assets and Markets

Finance is generally concerned with prices of assets, particularly with changes in those prices, and with interest rates.

Types of Assets

An asset is anything that can be traded. We will also refer to a collection of assets as an asset, although the collection may not be tradable as a single collection. Common types of financial assets are cash, bonds, shares of stock, shares in master limited partnerships (MLPs), shares in real estate investment trusts (REITs, pronounced "reets"), contracts, and options, or portfolios consisting of those types of assets. Bonds are sometimes referred to as "fixed-income assets"; shares of stock, as "equity assets"; contracts, as "futures"; and options, as "derivative assets". A special type of contracts is called "swaps". A swap is an agreement between two parties to exchange sequences of cash flows for a set period of time.

Bonds are contracts representing a loan, its carrying interest, and often its amortization schedule. Shares of stock are documents (perhaps virtual documents) representing ownership and the rights of that ownership. In terms of total value of assets, the worth of all existing bonds in any economy generally far exceeds that of all existing equity assets.

REITs and MLPs are both "pass-through" entities under the US tax code. Most corporate earnings are taxed twice, once when the earnings are booked and again when distributed as dividends. However REITs and MLPs avoid this double taxation since earnings are not taxed at the corporate level.

REITs own assets based on real estate and they may also generate income by real estate management. A REIT may act as a holding company for mortgage debt and earn interest income, or it may be actively involved in managing real properties and generate income from rent. Most MLPs are formed to own or manage in the energy and natural resources.

The US tax laws mandate that REITs pay out 90% of earnings in the form of dividends to their shareholders, but MLPs are not required to distribute any part of their earnings.

There are a number of products assembled from other assets and sold by financial companies. The number of these products such as stock or bond mutual funds, hedge funds, and ETFs is greater than the number of underlying assets.

A *financial instrument* is the asset itself or some document representing ownership of that asset. We also refer to a financial instrument as a *security*, or as a *fund*. The simplest financial instrument is a stock share. That is the kind we will discuss most often, and we will usually refer to it as a "stock". Other financial instruments are composed of various kinds of assets and/or loans or other commitments. Financial services companies make a substantial portion of their revenue by making and selling financial instruments or by extracting various types of "management fees". Some of these instruments are very simple and are just composed of bonds or shares of stock. Other financial instruments include borrowed funds (which must be repaid) and other future obligations. These future obligations are called "leverage", and the funds composed of them are called leveraged financial instruments.

Foreign Exchange

Trades in currencies are often conducted as hedges against changes in currency exchange rates when future expenditures must be made in a different currency. Much of the trading activity in the foreign exchange market ("FX" or "forex"), however, is conducted based on speculation that the currency being bought will rise in value versus the base currency. Of course there are two sides to every trade.

The exchange rates of some currencies versus others do vary considerably, and the return on some FX activity can be far more profitable than trades in stocks. Other pairs of currencies exhibit less exchange rate volatility, however.

Figure 1.3 shows the daily rate of the US dollar versus the euro from the date when the euro officially replaced the former European Currency Unit. (The graph is the number of US dollars per 1 euro. In this form, the rates are similar to the prices of a stock or other asset valued in US dollars. The directions of quoted exchange rates follow tradition; dollars per British pound, for example, but Japanese yen per dollar.)

FIGURE 1.3
The US Dollar versus the Euro, January 1999 to December 2017; *Source:* Quandl

Financial assets in a cryptocurrency, such as Bitcoin or Ethereum, are similar to cash in a local currency in some ways. Most assets in cryptocurrencies, however, are held, rather than as a store of value, in anticipation of changes in their value.

Types of Owners

Owners of financial assets may be individuals or they may be "institutions". *Institutional investors* include endowment funds, pension funds, mutual funds, hedge funds, commercial banks, and insurance companies. (These terms have common meanings, but their legal meanings are defined by governmental regulatory agencies.) The assets of institutional investors are often ultimately owned by individuals, but financial decisions by institutional investors generally do not directly involve the individual investors. A large proportion of the shares of large publicly-traded corporations are beneficially owned by institutional investors. (For example, approximately two-thirds of the shares of INTC throughout the year 2017 were owned by institutional investors.)

In this book I will often refer to persons or institutions buying or selling assets. I will use the terms "trader" and "investor" synonymously to refer to such a buyer or seller. (In some financial literature, a distinction is made between these two terms based on the intent of the transaction. This distinction is not relevant for our discussions.)

Returns

Financial assets produce returns, that is, income in the form of interest or dividends, or change in value of the asset itself (hence, a return may be negative). Returns over a given period are generally computed as a proportion or percentage of the price of the asset at the beginning of the period.

We refer to the income (interest or dividends) plus the relative change in price as the *total return* on the asset. The interest or dividends are referred to as the *yield*. Often, the unqualified term "return" refers just to the part of the return that is due to price fluctuations.

Prices and returns are stated or computed in the same numeraire. If the currency itself is in play, there may be an additional return due to currency exchanges.

Risk

Risk refers generally to variation in the value of an asset or in the returns on the asset. Some assets are more risky than others. INTC is a riskier asset than US dollars (USD) held in cash, for example. Cash in USD is not without risk, however. Inflation and other things may affect its real value.

The most common use of the term "risk" is generally for the variation in returns, rather than for variation in asset values. In financial analysis, a useful concept is that of an asset with a fixed value and a fixed rate of return. Such an asset is called *risk-free*; other assets are called *risky*. Although risk-free assets do not exist, we often consider US treasuries to be risk-free.

1.2.1   Markets and Regulatory Agencies

We will use the term "market" in different ways. In one sense, "market" refers to the set of trading dynamics that allows variations in the prices of assets. In another type of usage of the term, "market" refers to an organization that provides a mechanism for trading financial instruments. Additionally, "market" refers to some overall measure of the total value of the assets traded within a "market" (in the previous sense).

Markets and Exchanges

Financial instruments are bought and sold through various types of markets. The most common type of market is an *exchange*. Exchanges were once buildings or large rooms ("trading floors") where all trades were made, but now an exchange may be "electronic", meaning traders do not need to be on a common trading floor, but rather are connected electronically. The more prominent exchanges include the *New York Stock Exchange* (*NYSE*), the *Nasdaq National Market* (*Nasdaq*), the *London Stock Exchange* (*LSE*) and the *Tokyo Stock Exchange* (*TSE*). Exchanges are regulated by various government agencies (see below), and also the exchanges impose their own rules on trading members and on securities that are traded on the exchange.

An *alternative trading system* (ATS) is a venue set up by a group of traders for matching the buy and sell orders of its members. An ATS is generally regulated differently from an exchange. Most alternative trading systems are registered as *broker-dealers*. An important alternative trading system is the "over the counter" (OTC) market. The OTC Bulletin Board (OTCBB) and OTC Link are electronic quotation systems that display quotes and other information for inter-dealer transactions.

Alternative trading systems are often used by institutional investors (that is, "big players") to find counterparties for transactions. Because ATS transactions do not appear on national exchange order books, institutional investors building a large position in an equity may use an ATS to prevent other investors from buying in advance. ATSs used for these purposes are referred to as *dark pools*. In most countries, government agencies are increasing their oversight of ATSs.

Some markets specialize in a particular type of instrument. The two largest US markets, the NYSE and the Nasdaq, deal primarily in stocks, for example. The *Chicago Board Options Exchange* (*CBOE*) is the world's largest market for stock options. The CBOE was only established in 1973, but has grown rapidly, and now also handles trades of other options, such as on exchange-traded funds, interest rates and volatility indices. The *Chicago Mercantile Exchange* (*CME*), as well as other markets in the CME Group, the *Chicago Board of Trade* (*CBOT*), and the *New York Mercantile Exchange* (*NYMEX*) together with its *COMEX* division, offers futures and options based on interest rates, equity indexes, foreign exchange, energy, agricultural commodities, metals, weather and real estate. These futures and options can provide insurance or "hedges" against variation in the prices of the underlying assets. We discuss options in Section 1.2.7 and hedging in Section 1.2.9.

Markets make money by charging fees to companies for their assets to be traded in the market, by charging fees for traders or brokers to be approved to make transactions in the market, and finally by charging a very small fee on all transactions.

The markets also serve as regulators of financial transactions of the instruments traded on them. Each market may have minimum requirements for various measures of economic status for any company whose securities are traded in that market. The market may take various measures to ensure orderly trading, such as providing *market makers* with obligations under certain conditions to take opposing sides of proposed trades. The market may also *halt trading* in a particular security when prices show an excessive trend or variation.

Regulatory Agencies

Most national governments have one or more agencies that regulate the trading of securities. In the US, the *Securities and Exchange Commission (SEC)*, created in 1934, has the mission "to protect investors, maintain fair, orderly, and efficient markets, and facilitate capital formation". In addition to regulating formation and trading of stock of public corporations, the SEC oversees all option contracts on stocks and stock indexes. The SEC monitors the NYSE, the Nasdaq, and other US stock markets.

The SEC also regulates the activities of various participants in the market. One regulation, for example, limits the amounts that a broker or dealer can lend to a client to use in buying stock. Stocks bought with funds lent by a broker or dealer are said to be bought *on margin*. As prices fluctuate, the values in the client's account may fail to meet the margin requirements. In this case, the client is issued a *margin call*, and the client must deposit additional funds or liquidate some positions.

The SEC also regulates various financial products, such as open-end and closed-end *mutual funds*, *hedge funds*, and *exchange-traded funds (ETFs)*. These are collections of bonds, stock shares, options, and other derivatives whose value is determined by the total market value of the assets in the collection. Ownership in the fund is divided into shares that are sold or traded according to regulations set out by the SEC.

The differences in these fund products are determined by the differences in the regulations of the SEC. The differences in mutual funds and hedge funds, for example, have nothing to do with the difference in "mutual" and "hedge", but rather in the SEC regulations governing these two asset classes. A mutual fund, for example, can be sold to almost anyone as long as the buyer is given a copy of the prospectus, which describes the general properties of the fund. A hedge fund, on the other hand, can be sold only to *accredited investors*. "Accredited investor", who may be an individual or an institution, is a term defined by the US Congress. Accredited investors include *high net worth individuals* (*HNWIs*). There are, of course, other regulatory differences. The prices at which open-end mutual funds are traded are determined at the end of each trading day by computing the net asset value of the collection of assets in the fund.

A mutual fund may be "actively managed", that is, the fund manager may choose the individual stocks and the amount of each to include in the fund (subject to some restrictions, possibly), or the fund may consist of a fixed set of stocks in a relatively fixed proportion. Mutual funds of the latter type often consist of stocks that make up a particular index in the proportion that they are included in the index. (We discuss market indexes beginning on page 49.) Such mutual funds are called "index funds", and are currently by far the largest category of mutual funds.

An ETF trades on the open market and its price is determined by similar market dynamics that determine the price of other equity assets traded in the open market. If the securities composing an ETF pay dividends, the ETF typically distributes the dividends directly to the holders of ETF shares in the same way that an ordinary stock may distribute dividends (that is, there is a "declaration" and an ex-dividend trading date). (There are some subtle differences in ETF distributions and distributions by the component instruments that depend on the nature of the holdings. For example, the distribution of an MLP in the US is taxed very differently from the distribution of an ETF, even if the ETF owns only MLPs. Of course, expenses are skimmed prior to distribution in either case.)

Special kinds of debt obligations may also be traded on exchanges. An *exchange-traded note* (*ETN*) is an unsecured and unsubordinated debt obligation issued by a financial institution that is traded in a stock market. It is a debt obligation of the institution that issues it, usually a commercial bank, but it has no principal protection. An ETN is generally designed to track some market index, and its value is contractually tied to that index. An ETN, however, usually does not hold any assets whose prices are included in the index; instead, the assets of an ETN are generally swaps or derivatives.

The *Commodity Futures Trading Commission* (*CFTC*) has similar regulatory oversight of options contracts on foreign exchange ("FX" or "forex") and of futures contracts. The CFTC monitors the CBOE, the CBOT, the CME, and other options and futures markets. In addition to these government agencies, in the US the *Financial Industry Regulatory Authority* (*FINRA*), created in 2007, is a non-government body devoted to investor safety and market reliability through voluntary regulation. *The National Futures Association* (*NFA*) is a similar non-government body that watches over options and futures contracts on forex and commodities.

General economic conditions are influenced through two types of policies that relate to two different aspects: *fiscal* (taxation and spending) and *monetary* (money and government bonds). Fiscal policies are generally determined by the executive and/or legislative branch of the government, while monetary policies are usually set by a banking agency of the government, often called a *central bank*. In the US, the *Federal Reserve System*, which was created by Congress in 1933, operates as the central bank.

Various international consortia often make formal agreements about regulations that are to be adopted and enforced by the individual national governmental agencies. One of the most influential of these associations is called the *Group of Ten*, which was established in 1962 by ten countries to cooperate in providing financial stability for the member nations. (The Group of Ten is not to be confused with the Group of Seven, "G7" or "G8", depending on whether Russia is included.) The Group of Ten created the *Basel Committee* to improve banking supervision worldwide, and to serve as a forum for cooperation on banking supervisory matters among the member countries, of which there are now over forty. The Basel Committee has promulgated accords, called Basel I (in 1988), Basel II (in 2006) and Basel III (in 2010, with subsequent modifications), that set standards for balance sheets of banks and that prescribe operational procedures to ensure those standards are met.

Other financial transactions are overseen by various agencies, both government and non-government. Various sectors of the financial industry have regulatory bodies. For example, the *Financial Accounting Standards Board* (*FASB*) is a non-government regulatory agency in the US that promulgates certain accounting principles, and the *CFA Institute* (formerly the *Chartered Financial Analyst Institute*) is a non-government agency that, among other things, sets standards for how investment firms calculate and report their investment results.

Financial transactions in all of the European Union member countries are subject to various regulations agreed on by the member countries. The group of all European national markets or the set of regulations governing transactions in those markets is called the "Euromarket".

Some regulatory agencies, whether government or private, make money by skimming fees from financial transactions that they oversee. The SEC, for example, currently collects a fee of 0.002% of all stock and ETF transactions on the US exchanges, and FINRA collects a fee of 0.01% of stock and ETF transactions. These are not insubstantial sums. The US Congress has not yet (as of 2019) imposed a tax on stock, ETF, or options transactions, but such a tax is occasionally included in bills that come before the House.

Rating Agencies

There are companies that assign credit ratings to financial instruments and/or to companies or government entities as a whole. The companies issuing the ratings are called *credit rating agencies*. (Similar companies that rate the credit worthiness of individuals are called *credit bureaus*.)

The three largest credit rating agencies are Moody's Investors Service, Standard & Poor's (S&P), and Fitch Ratings. These three together control 95% of the global market for ratings.

The most common types of ratings given by credit rating agencies are for individual corporate bonds or for other debt instruments.

The ratings are based on probabilities or on probabilistic expectations. These, of course, are abstract mathematical entities, but in order for them to be useful, they must correspond to empirical frequency distributions. Ratings assigned by credit rating agencies do not have a strong historical correspondence to observed reality. There are many reasons for this, one of which is that the provider of the rated instrument pays the credit rating agency to rate it.

One of the biggest financial disasters of all time, the financial crisis of 2008, was exacerbated by the failure of the ratings assigned to various financial instruments to reflect the true (or realized) risk of the instruments. One reason for this, in addition to the biases implied by the fundamental business structure of rating agencies, is that the producers of financial products for sale are always more clever than the raters of the products, the regulators, or even the buyers of those products, simply because they have relatively more stake in the game. In the years prior to 2008, government policies encouraged lenders to make loans to economically-disadvantaged persons. The lenders obediently lent, and subsequently bundled these loans into financial products that concealed the overall risk in an instrument that consisted of many highly-risky components. It appeared bo be "diversified", but it was not. The simplest of these bundled products are called *collateralized debt obligations (CDOs)*, but there are many derivative products based on CDOs. CDOs and other collections of loans are also often subdivided into separate collections with varying risk, interest, and maturity profiles. These separate collections are called *tranches*. It is possible that the credit rating agencies just did not understand these products.

The Dodd-Frank Wall Street Reform and Consumer Protection Act of the US Congress of 2010 imposed certain regulations of credit rating agencies designed to make less likely failure to rate risky products correctly.

1.2.2   Interest

A *loan* is a financial transaction in which one party provides a counterparty with a financial instrument, usually cash, with the understanding that the counterparty will return the full value of the financial instrument at some specified future time and also will pay the first party for the use of the financial instrument during that time. The amount of the loan is called the *principal* and the payment for its use is called *interest*. The principal may be returned, or *amortized*, incrementally over the course of the loan.

Bonds, including government-issued "bills" and "notes", are essentially loans made by the *bearer* of the bond to the *issuer* of the bond. The bond has a *term* of duration at the end of which, the *due date*, the issuer is obligated to return the original principal of the loan, called the *par value* or synonymously *face value* of the bond. In some cases, the issuer pays to the bearer a fixed amount of interest at regular times. A bond with regular payouts is called a *coupon bond.* (The name is a carry-over from the early twentieth century, when bonds were physical pieces of paper with attachments ("coupons") that were exchanged for cash interest.) The interest is generally based on a specified percentage of the par value of the bond. The *nominal interest rate* is the percentage of the par value that is paid per year, regardless of the times when the payments are made. The interest for one year may all be paid at the end of the year, or it may be paid in regular increments at intervals through the year. For this reason, we may refer to the total interest paid during the year as a percentage of the par value as the *nominal annual percentage rate*, or *nominal annual interest rate*. (These terms are not used consistently, however.) The interest rate of a particular bond depends on rates of other bonds, "prevailing interest rates", as well as on the perceived probability that the issuer will repay the full par value (or the alternative, that the issuer will *default*).

Bonds are negotiable instruments; that is, they can be traded in an open market. The price at which a bond is traded may be greater or less than the par value of the bond. This is because prevailing interest rates may change and/or the perception of the probability of default may change. The duration is also a consideration, because the longer the time to the due date, the larger is the probability of variation in the value of the bond. Another thing that can have a significant effect on the price of bonds, more than most other financial assets, is the relative value of the currency in which the bond is denominated compared to other world currencies, the *foreign exchange rate*.

The issuer continues to pay the same amount of interest on a coupon bond at the fixed time intervals, so long as the issuer does not default. Because the price of the bond fluctuates with market conditions, however, the effective interest rate or *yield* also fluctuates.

Interest Rates

In some cases, the interest rate is set by the issuer of a loan. In other cases, a bond with a fixed par value and a fixed duration is offered for sale in the open market. Thus, the actual interest rate depends on what the purchaser is willing to pay. Variations in the rate continue as further trades occur in the "secondary market".

Financial assets generate value continually, and so depository institutions such as banks and other holders of assets that borrow or lend assets even temporarily (overnight, for example) pay or charge interest for those borrowed or lent assets. Interest rates for these funds are generally set by central banks of the national or supranational governing body. In the US, the *Federal Open Market Committee* (*FOMC*), which is a component of the Federal Reserve System, sets a "target" interest rate for these funds. This interest rate is called the *fed fund rate* or the *fed discount rate*. A bank with surplus balances in its reserve account may lend to other banks in need of larger balances. The rate that the borrowing institution pays to the lending institution is determined between the two banks. The weighted average rate for all of these negotiated rates is called the *effective fed funds rate*. The effective fed funds rate is determined by the market, but is influenced by the Federal Reserve through open market operations to reach the fed funds rate target.

Two other central banks whose rates are very important in global economics are the *European Central Bank* (*ECB*), which is the central bank for the Euromarket, and the *Bank of Japan* (*BOJ*).

Another standard interest rate is a *prime rate*, which is the rate at which banks will lend money to their most-favored customers. Obviously, banks do not move in lockstep, and furthermore the rate at which they "will lend money" is not an observable phenomenon. *The Wall Street Journal* publishes a rate that is widely accepted as the "prime rate." Many commercial interest rates are tied to *The Wall Street Journal* Prime Rate. Brokerage firms, for example, often set their interest rates on margin funds as some fixed amount above the prime rate reported by *The Wall Street Journal*. There are other standard interest rates, such as for interbank loans. One of the most important of these has been the *London Interbank Offered Rate* (*LIBOR*).

Interest rates, of course, are generally stated as percentages. Because the changes in interest rates are often relatively small, a different unit is often used to refer to rates, especially to changes in rates. A *basis point* is 0.01 percentage point. A significant change in the fed discount rate, for example, would be 50 basis points, or half of 1%.

In finance, the term "risk" generally refers to fluctuation in asset values or fluctuation in the flow of funds. Because of inflation and other economic processes, all interest rates carry risk. As in many other areas of finance, however, we conceive of an ideal. We often speak of a *risk-free interest rate*. In computations, we commonly use the current rate on a three-month US Treasury Bill (T-Bill) as the "risk-free interest rate".

Figure 1.4 shows the interest rates (by month, in the secondary market) over a multi-decade period. The maximum rate, 16.3%, occurred in the mid-1980s. The minimum rate, 0.1%, occurred in the 1930s and again in 2011.

US Treasuries

The US Treasury issues paper with various terms. These fixed-rate assets are called "Treasuries" generically. In general, a Treasury with a term of less than a year is called a "bill", with a term longer than a year and up to 10 years is called a "note", and with a term of more than 10 years is called a "bond".

The interest rates on Treasuries depend on supply and demand. The interest rate often depends on the term of the loan. In general, the longer the term, the higher the interest rate. This is not always the case, however, and analysis of the variation in yields of loan instruments with different terms may provide some information about the overall economy and possibly directions in which prices of stocks or other assets may move.

Some sample historical rates are shown in Table 1.2. We see the general

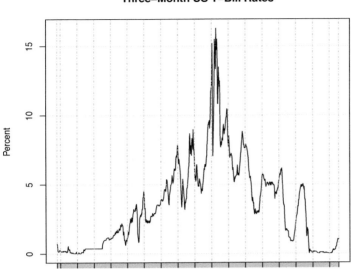

FIGURE 1.4
"Risk-Free" Interest Rates, January 1934 to September 2017; *Source:* FRED

decrease in rates from the early 1980s. We also observe in Table 1.2 patterns of rates as a function of the term of the instrument; rates for longer terms are generally higher.

TABLE 1.2
Historical Interest Rates of Treasury Bills, Notes, and Bonds; *Source:* FRED

	Term			
Date	three-month	two-year	five-year	thirty-year
January 1982	12.92	14.57	14.65	14.22
June 1989	8.43	8.41	8.29	8.27
January 1997	5.17	6.01	6.33	6.83
January 2008	2.82	2.48	2.98	4.33
January 2017	0.52	1.21	1.92	3.02

The yield as a function of term is called the *yield curve*. A simple yield curve consists of linear interpolants between each pair of successive observed points, and this is the kind of yield curve we will use. Instead of a broken-line curve, a smoothed curve can be fitted to the observed values by various methods, such as by splines (see Fisher, Nychka, and Zervos, 1995) or by use of a parametric linear combination of exponential discount functions (see Nelson and Siegel, 1987, or Svensson, 1994). We will not describe these methods here. For a given class of corporate bonds we may have many more observations from which to construct a yield curve.

The underlying dynamics are referred to as the *term structure*. The most frequently reported yield curve compares the three-month, two-year, five-year, and thirty-year US Treasury debt. (The thirty-year Treasury constant maturity series was discontinued in 2002, but reintroduced in 2006.)

The structure shown for the last three dates (1997, 2008, and 2017) in Table 1.2 is a "normal" yield curve (with a small blip in 2008). The structure in June, 1989, is an inverted yield curve, and that in January, 1982 is a "humped" curve. The early 1980s was a period of very high inflation in the US, and the high interest rates reflect that fact. Figure 1.5 helps to visualize these curves.

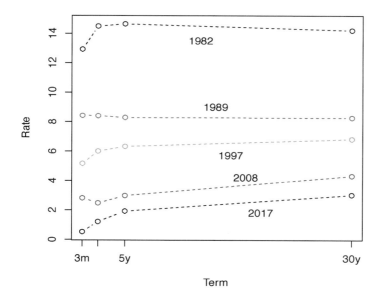

FIGURE 1.5
Yield Curves (*from* Table 1.2)

The structure of the yield curve is one of the most reliable leading indicators of economic *recessions* and of market *corrections*. A recession is generally defined as two consecutive quarters of negative economic growth as measured by the gross domestic product, or GDP. A market corrections is generally defined as a drop in the relevant market index by 10% from its previous high.

In an overwhelming majority of recessions in the US over the past century, the yield was inverted sometime during the preceding twelve months, and conversely in almost all instances of inverted yield curves, the economy entered a recession within twelve months. The inverted yield curve in 1989 shown in Figure 1.5, for example, was followed by a recession in 1990. Most periods with inverted yield curves are followed by, or are coincident with, market corrections. There are far more corrections, however, that are not preceded by a period with an inverted yield curve (or, indeed, by any identifiable indicator). (A very interesting yield curve occurred on June 5, 2019. You are asked to get the data and plot it in Exercise A1.3(c)iii.)

The fed fund rate, set by the FOMC, is one determinant of the rates on Treasuries, especially the shorter term bills. This is because the bids for Treasuries depend on the cost of money, and for banks, this is determined by the fed fund rate.

Figure 1.6 shows more complete yield curves at the end of each quarter in 2017. (Note that the scale is different in this figure.) These are the yield curves for the eleven Treasuries with one-month, three-month, six-month, one-year, three-year, five-year, seven-year, ten-year, twenty-year, and thirty-year maturities. These yield curves were more typical of a period of low interest rates. The curves generally show a sharp increase in yields of bills with maturities up to one year. These are the yields most affected by policies of the FOMC, which after the 2008 financial crisis implemented a low-interest policy (even called "ZIRP", for zero interest rate policy). The FOMC began slowly raising the rates at the end of 2015. (On page 438, we show a graph of the effective fed funds rate in that period, and comment further on its changes.)

The profits of commercial banks are dependent to some extent on an increasing yield curve. Much of their profits come from the spread between short-term interest rates, which is what they pay for deposits, and long-term rates, which they receive for lending. A small or inverted spread tends to constrict credit, and hence to depress economic growth.

Corporate Bonds

Corporations issue fixed interest-bearing securities, called corporate bonds. There are various types of these instruments. As with Treasuries, their rates of interest depend on supply and demand, that is, what an investor is willing to pay for a bond with a given face value and maturity. What an investor is willing to pay for a corporate bond is also based on an additional consideration: the probability that the corporation may default on the bond.

State and local governments also issue bonds, called generically "municipal

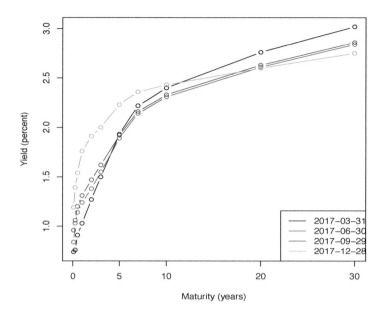

FIGURE 1.6
Yield Curves; *Source:* US Treasury Department

bonds" or "muni bonds". The prices and rates of these bonds are determined similarly to corporate bonds with a major difference. The income produced by most of these bonds is exempt from federal income tax and from income taxes within the corresponding local taxing districts.

The credit rating agencies assign corporate and municipal bonds to a few categories (usually seven) representing different levels of risk. The designations used by the agencies are different, but they basically consist of one to three letters, possibly with an appended "+" or "-", such as "AAA", "AA+", "B", "Caa", and so on. The top few categories are called "investment-grade" bonds, and the others are called "speculative" bonds. A widely-watched set of bonds are the Moody's Seasoned Aaa Corporate Bonds.

Interest Calculations and Present Values

The interest generated by an asset may be added to the principal of the asset and then interest paid on that interest. This is called *compounding*.

As mentioned above, interest is often paid in regular installments through the year. If the annual interest rate is r and interest is paid in n regular installments (coupons) per year, the rate for each is r/n. (This is the *coupon rate*, r_c.) If the interest is compounded (that is, the interest is added to the principal and interest accumulates on the previously accrued interest), at the end of the year a beginning principal of P plus its accrued interest will have a value of $P(1 + r/n)^n$. This leads to the well-known formula for the value of principal plus interest after t years at an annual rate of r compounded n times per year,

$$P(1 + r/n)^{nt}. \tag{1.11}$$

This amount is the *future value*, say A. The *present value* of A in the case of compounding of finite simple returns is

$$A(1 + r/n)^{-nt}. \tag{1.12}$$

Given a future value, an interest rate and type of compounding and a term, we can compute a present value. The rate in the factor (in this case r/n) is called the *forward rate*.

In a generalization of the factors in equations (1.11) and (1.12), we can have different rates in different periods, yielding a factor $(1+r_1)(1+r_2)\cdots(1+r_n)$. Of course, in any discussion of rates, we must be clear on the length of the period. In most rate quotes, the period is one year.

Bonds are generally traded on the secondary market, so instead of the principal, we refer to the par value, which remains constant, and the *price* or the value of the bond, which may fluctuate. If a bond that matures T years from now has a par value of A and it pays coupons of c n times per year, at the end of T years the total accumulated value will be $nTc+A$. The present value, which would be the current price in the secondary market, is the discounted value of $nTc + A$. Using the coupon rate r_c, the discounted value of A is just $A(1 + r_c)^{-nT}$. The discounted rates for the coupons, however, are different, because they are received at different times. The rate for the first is $(1+r_c)^{-1}$, for the second, $(1 + r_c)^{-2}$, and so on. Hence, the present value of all future coupon payments is

$$\sum_{t=1}^{nT} c(1 + r_c)^{-t}, \tag{1.13}$$

which can be expanded to

$$\frac{c}{r_c}\left(1 - (1 + r_c)^{-nT}\right).$$

(This follows from the relation $\sum_{t=1}^{k} b^{-t} = (1 - b^{k+1})/(1 - b)$, which, for any $b \neq 1$, we get by factoring the expression $1 - b^{k+1}$.)

Now, adding the discounted par value we have

$$\frac{c}{r_c}\left(1 - (1 + r_c)^{-nT}\right) + A(1 + r_c)^{-nT},$$

and simplifying this expression, we have the present value, which also is the fair current market price of the bond,

$$p = \frac{c}{r_c} + \left(A - \frac{c}{r_c}\right)(1 + r_c)^{-nT}. \tag{1.14}$$

The *yield to maturity* of an asset with a fixed principal that pays interest installments is the compounded yield that discounts the principal value plus all its payouts, to yield the current price of the asset.

For a given current price p, given coupon payout c, given par value A, the *coupon yield to maturity* in T years is the value r_c that solves equation (1.14).

A coupon, representing a fixed payment at a fixed future time, can be a negotiable instrument; hence, can be "stripped" from the bond and sold separately. (The analogy was even more appropriate when coupons were actual paper attachments.) The principal portion is without coupons, and the coupons can be negotiated in various ways, including being retained by the issuing institution. Virtual stripping can be done in many ways, of course, and in the 1980s a number of investment firms offered various types of stripped US Treasury Bonds. (BNP Paribas offered COUGRs; Salomon Brothers sold CATS; Merrill Lynch sold TIGRS; and Lehman Brothers offered LIONs. Because of the acronyms these instruments were called feline securities.)

Zero coupon bonds are similar to coupon bonds, except coupon payments are not made. They have par values and nominal interest schedules. The interest schedule of a zero coupon bond is stated as a rate and an associated period at which the interest is compounded. This results in the discounted price of a zero coupon bond. If the par value is A and the rate is r_c compounded n times per year, the present price T years until maturity is

$$p = A(1 + r_c)^{-nT}.$$

In 1985 the US Treasury introduced Separate Trading of Registered Interest and Principal of Securities (STRIPS), which allowed the sale of the principal and coupon components of Treasury Bonds separately. All issues from the Treasury with a maturity of 10 years or longer are eligible for the STRIPS process.

Continuous Compounding

Interest is often compounded continuously, and in taking the limit in expression (1.11) as $n \to \infty$, we get the formula for principal plus interest that is compounded continuously for t years,

$$Pe^{rt}. \tag{1.15}$$

In financial applications, the *present value* or *discounted value* of any future value is often computed using a risk-free interest rate (that is, the rate of a three-month US T-Bill) with continuous compounding. The discounted value of a future value A, for t years hence at a risk-free rate of r, is

$$Ae^{-rt}. \tag{1.16}$$

Compare this with equation (1.12).

1.2.3 Returns on Assets

Financial assets change in value due to accrued interest, distributed realized earnings, or just due to changes in the market value of the asset. The change in market value, of course, can be positive or negative. We may refer to the change itself, positive or negative, as the return. We also call the proportional change the *return*. This latter is the meaning used in this book, although sometimes for emphasis, I may call this the *rate of return*. For most assets, analysis of the rate of return is often of more interest than analysis of the asset prices themselves.

If P_0 is the value of an asset at a time t_0 and at time t_1 the value is P_1, the simple proportional *single-period return* for $t = t_1 - t_0$ is

$$R = (P_1 - P_0)/P_0. \tag{1.17}$$

This is also just called the *simple return*, with a time period being understood.

From a beginning value of P_0, after one period, the ending value is

$$P_1 = P_0(1 + R). \tag{1.18}$$

If R is constant for n periods of length t, and the change in value at the end of each period is taken into account (compounded), from a beginning value of P_0, after n periods, the ending value is

$$P_n = P_0(1 + R)^n, \tag{1.19}$$

which is essentially the amount shown in expression (1.11).

In order to make the expressions above more precise, we need to specify the length of the time period. The standard way of doing this is to use a year as the base length of time. For converting simple rates to annual rates, compounding is generally assumed to occur once a year, at year end; hence, the *annualized simple return* given the simple return for a period of t years is just the return for the period divided by t (where t is usually a fraction of a year). Conversely, for example, to convert an annualized simple rate to a daily rate, we divide by 365. Standards are necessary, for example, to compare one financial institution's reports of investment performance with another company's reports. The CFA Institute, mentioned above, through its *Global Investment Performance Standards* (*GIPS*), describes how rates of return should be reported by financial institutions.

Continuous Compounding; Log Returns

In most financial analyses (as opposed to public reports), we generally assume that returns are compounded continuously. This simplifies the analysis for aggregating returns over time, but complicates aggregation of returns from multiple assets.

If r is the annual rate, with continuous compounding, P_0, after t years becomes

$$P_1 = P_0 e^{rt}. \tag{1.20}$$

This leads to
$$r = (\log(P_1) - \log(P_0))/t, \qquad (1.21)$$
which is called a *log return*.

Given the two values P_0 and P_1 and the time interval t measured in years, r in equation (1.21) is called the *annualized log return*.

For a given time interval, the *log return* is just
$$r = \log(P_2) - \log(P_1), \qquad (1.22)$$

or $\log(P_2/P_1)$. It is common to speak of the *daily log return* when t is the length of one day and P_1 and P_2 are closing prices of successive days. (Recall that for most assets, the interval in a "daily" event is 1 day less than 80% of the time; it is often 3 days, and sometimes 4 days or longer.) Likewise, we speak of *weekly log returns* and *monthly log returns*, and of course the lengths of the intervals of "weeks" or "months" vary.

Returns are computed using stock prices in the same way as they are computed from asset values. Using the unadjusted closing prices of INTC stock, the annualized log return from January 3, 2017, to September 30, 2017, for example, is

$$
\begin{aligned}
r &= (\log(38.08) - \log(36.60))/0.661 & (1.23) \\
&= 0.0600. & (1.24)
\end{aligned}
$$

(The unadjusted closing prices do not account for dividends, so the "total return" for that period is larger. I will discuss this again in Section 1.2.5.)

The log return and the simple return are not the same. The relationship for a single period, from equation (1.19), is

$$r = \log(1 + R), \qquad (1.25)$$

where r is the log return and R is the simple return. (When discussing log returns and simple returns in the same context, I will often distinguish them by use of lower case for log returns and upper case for simple returns. *I will not always make this distinction in the notation, however.*)

If the return is small, the log return and the simple return are close in value, as we see in Figure 1.7. The log return is always smaller; 5% simple return corresponds to 4.88% log return and a -5% simple return corresponds to -5.13% log return for example, as seen in the graph. Note that a very large negative log return can correspond to a loss of more than 100%.

Aggregating Log Returns over Time

Log returns have a very nice aggregational property. If P_1, P_2, P_3, P_4, P_5 is a sequence of prices at times t_1, t_2, t_3, t_4, t_5, the corresponding sequence of log

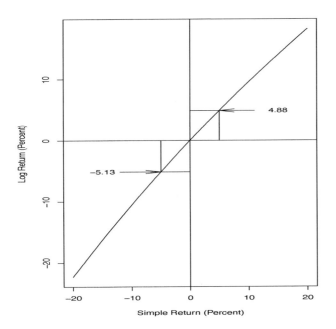

FIGURE 1.7
Log Returns and Simple Returns

returns at times t_2, t_3, t_4, t_5 is $r_2 = \log(P_2) - \log(P_1)$, $r_3 = \log(P_3) - \log(P_2)$, and so on. The log return from time t_1 to t_5 is $\log(P_5) - \log(P_1) = r_2 + r_3 + r_4 + r_5$. In general, given times $t_1, \ldots t_n$ with r_i being the log return at time t_i, the log return from t_1 to t_n is given by

$$r = \sum_{i=2}^{n} r_i. \tag{1.26}$$

Aggregating Simple Returns from Multiple Assets

Suppose we have two assets with values P_{10} and P_{20} at time t_0. If at time t_1 the prices are P_{11} and P_{21}, then the simple returns are $R_1 = P_{11}/P_{10} - 1$ and $R_2 = P_{21}/P_{20} - 1$. The "combined return" depends not just on the relative values from one time to the next, but also on the relative total values in the two time periods, $P_{10}/(P_{10} + P_{20})$ and $P_{11}/(P_{11} + P_{21})$. The simple return on the combined assets is

$$
\begin{aligned}
R &= (P_{11} + P_{21})/(P_{10} + P_{20}) - 1 \\
&= \frac{P_{10}}{P_{10} + P_{20}}\left(\frac{P_{11}}{P_{10}} - 1\right) + \frac{P_{20}}{P_{10} + P_{20}}\left(\frac{P_{21}}{P_{20}} - 1\right) \\
&= \frac{P_{10}}{P_{10} + P_{20}}R_1 + \frac{P_{20}}{P_{10} + P_{20}}R_2, \quad\quad\quad\quad (1.27)
\end{aligned}
$$

and of course this extends immediately to a portfolio of more than two assets. The log return of the total of two assets is often approximated as above.

Log Returns, Simple Returns, Interest Rates, and Excess Returns

Log returns usually follow more tractable mathematical models, and they are generally the type of returns used in measuring and analyzing volatility (see Section 1.4). Because simple returns over multiple assets can be aggregated easily, simple returns are generally the type of returns used in analyzing portfolios (see Section 1.2.9).

Many financial analyses are predicated on the concept of a basic return due to a general, passive economic change. The change may be due to a "risk-free" interest rate, or it may be the return of "the market". (We will discuss "the market" in Section 1.2.6.) The return of a given asset is then compared with the return of that reference benchmark. The *excess return* relative to the specified benchmark is just the return (log or simple) of the asset minus the return (usually of the same type) of the benchmark for the same period. Another way the term "excess return" is used is with specific reference to the difference of the return and the risk-free interest rate. Although the risk-free interest rate is a simple return, for assets where the return is measured as a log return, the corresponding excess may involve adjusting a log return with a simple return. Since log returns and simple returns are generally close, combining them in a measure or excess return yields very little distortion.

The idea of comparing the return of a given asset with the return of the market is the basis of the "capital asset pricing model" (CAPM).

1.2.4   Stock Prices; Fair Market Value

A major concern in financial analysis is to determine the "correct price" or the "value" of a given asset. We call this the "fair price". Ordinary prices are observable; fair prices are not.

Stock Prices and Market Value

An asset that is actively traded has a "price" at which it is traded in a given market at a given point in time. The "price" may be the last price at which the asset was traded, or it may be some intermediate value between the highest price for which a bid has been submitted and the lowest price at which the asset is offered for sale. If there are different prices in different markets, the force of trading will cause the prices to converge to a single value, subject to the market friction of transportation, registrations, regulations, or other logistics.

The *market value* of a publicly traded corporation is the stock price times the number of shares outstanding, that is, the number of shares that the corporation has issued and which are not owned by the corporation itself. The market value of a stock is also called the market capitalization of the stock. A related term is *enterprise value* or EV, which is the market value plus the net long-term debt of the corporation and the equity attributable to preferred stock or other minority interest, and less cash and cash equivalents. The enterprise value is a measure of the worth of a company in a merger of corporations or in a corporate takeover.

Fluctuations in Stock Prices and Trading Volume

The market price of any financial asset in any numeraire fluctuates from time to time. For bonds, as indicated above, the fluctuations may depend on prevailing interest rates and the perceived probability that the issuer will repay the principal.

For shares of stock, the reasons for fluctuations are more complicated. A share of stock represents ownership in a business concern. When a business is first incorporated, the number of shares, possibly of different types, is set. Over time the corporation may authorize additional shares. All shares authorized may not be issued or sold. Each of the shares that are actually issued, that is, the "shares outstanding" or the "float", represents a proportional ownership in the business. The total value of the corporation divided by the number of shares outstanding is the simplest measure of the value of an individual share. (This of course is a "kick the can" valuation; what is the value of the business?) A corporation can reduce the number of shares outstanding simply by purchasing shares on the open market and placing them in the corporate reserves. This is called a *stock buyback*. A stock buyback does not change the total value of the corporation, but it does affect per share measures such as the earnings per share.

The number of share traded, that is the "trading volume", varies daily. For most stocks, on an average day, the trading volume is 1% to 5% of the float. Often high volume is associated with large variations in price, although the opposite is also often the case, because of poor "price discovery". The direction of price change and the relative volume when the change is positive to when it is negative may be important indicators of future price movements.

During a given trading day, the price of a share of stock may fluctuate for no apparent reason. It is of interest to note the open, high, low, and closing price of the day. A dataset containing this information is called an open, high, low, and closing (OHLC) dataset, and such datasets are often displayed in "candlestick" plots. Figure 1.8 shows candlesticks of prices of stock of Intel

Corporation (INTC), for the first quarter of 2017. (The closing prices are the same as the prices in the upper left graph in Figure 1.2 on page 9.)

The number of shares traded on each day is shown in the bar graph in the lower panel of Figure 1.8. The different colors of the bars and of the boxes in the candlesticks indicate whether the price increased or decreased for the day.

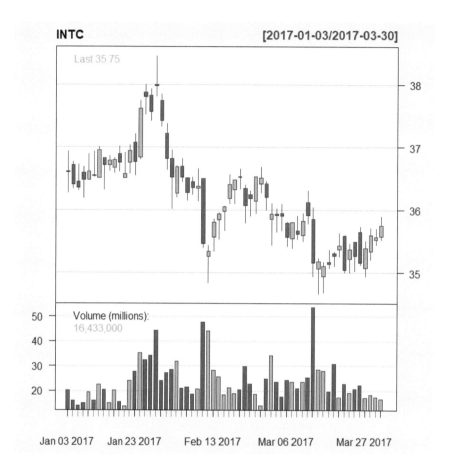

FIGURE 1.8
Daily Open, High, Low, and Closing (OHLC) Prices and Volume of INTC January 1, 2017 through March 31, 2017; *Source:* Yahoo Finance

Market Prices and Values

A great deal of effort is expended on attempts to determine the appropriate price or *fair market value* of a share of stock. Study of the profitability and general status of the underlying business in order to determine the worth of a share of stock is called *fundamental analysis*. The overall value of the company, how much the company is earning, how well its products or services are being received and how well these compare with those of competitor companies, the amount of debt the company carries, and other general considerations about the company are studied in fundamental analyses. Indications about what the price of the stock will be in the future may come from the price changes in the past. When the price of a stock has generally been increasing in the recent past, it is more likely that the price will continue to increase. Of course, it will not continue to increase forever. In *technical analysis*, patterns of price changes and the volume of the trading are studied and used to predict future price moves.

We might ask why was the price of one share of INTC stock $35.10 on March 15, 2017? The simple answer is that the last trade on the Nasdaq stock market on that day occurred at an average price per share of $35.10. That was the *market price* at that time. During that same day, however, some shares were traded as high as $35.17 and some as low as $34.68.

As with any trade, there were two sides to each trade on March 15, a buyer and a seller. From both sides, the price appeared fair at the time. Many potential buyers had specified *bid* prices, that is, prices at which they were willing to buy some amount of the stock. These are called *limit orders*, specifically, limit buy orders. Likewise, many potential sellers had specified *ask* prices, that is, prices at which they were willing to sell some amount. These are called limit sell orders. Each trade occurred in the open market. We conclude that the fair market value of the stock at the time of the last trade was $35.10.

The market price of a tradable asset can be specified as a range from the highest bid price to the lowest ask price, or it can be taken as the average of these two numbers. While this may seem simple enough, there are many complicating factors, the most important of which is the size of the bids, the size of the asks, and the amount of the asset to be traded. (If there is a bid of $34 for 100 shares and an ask of $36 for 5,000 shares, the market price of 1,000 shares, that is, the price at which 1,000 shares could be bought or the price at which 1,000 shares could be sold, is not necessarily between $34 and $36.)

National Best Bid and Offer

The best (lowest) available ask price and the best (highest) available bid price is called the *national best bid and offer* (NBBO). These are the prices quoted by brokerage firms. Regulation NMS of the SEC requires that brokers guarantee customers this price. (Since these prices can change constantly, it is not clear what the regulation means.)

The NBBO is updated throughout the day from all regulated exchanges. Prices on dark pools and other alternative trading systems may not appear in these results, however.

The open market makes it unlikely that some clueless buyer would pay much more than the market value of the stock, because for any stock there will almost always be potential sellers with ask prices at or just above the market value. Even if the buyer offered to buy the stock at a price above the lowest ask price, the buy order would be executed at the lowest ask price, unless another buy order came in at essentially the same time, in which case, the trade price would be at the next lower ask price. If the buyer does not specify a bid price, the buy order is a *market order*, and is executed at the lowest available ask price. Likewise, a market sell order is executed at the highest available bid price.

Another type of order is called a *stop order*, which is intended to "stop losses". A stop order to sell specifies a *stop price* at a price lower than the current price, and the stop order becomes a market order if the highest bid price drops to the stop price. For a security experiencing a rapid decline in price, the stop order may be executed substantially below the stop price, because of other sell orders. A variation on a stop order is a *stop-limit order*, which like a stop order specifies a stop price, but it also specifies a limit price (an ask price), and becomes a limit order if the highest bid price drops to the stop price. A stop order or stop-limit order can also be a buy order, which is similar to the sell order with the bid and ask prices reversed. (A stop price to buy can be set for a short position, which we will discuss later.)

For someone who does not participate actively in the market, there are two aspects of the market that may be surprising. One is how big it is. On an average trading day, well over 20 million shares of INTC are traded, for example. The other is how fast things happen. A market order to buy or sell INTC will be executed in less than a second.

Another thing that might be surprising to someone who does not partic-ipate in the market is the price fluctuations, or risk, of equity assets. Over the one-year period prior to the ending period in Figures 1.1 and 1.2, INTC ranged in price from approximately 33 to 38. On the other hand, for example, during that same period, the share price of CSX Corporation (CSX) ranged from approximately 26 to 53, and the share price of Transocean Ltd. (RIG) ranged from approximately 9 to 19. The shares of CSX generally increased in price, gaining over 100% in the 52 weeks. The shares of RIG decreased by over 50%. Other stocks experienced even more variation in price.

There are always potential buyers of INTC with specified bid prices and potential sellers with ask prices. Quotes of these prices are available at any time, although if the market is closed, these bid/ask prices may not be mean-ingful. During market hours, the *bid-ask spread*, that is, the difference in the bid and ask prices, for an actively traded stock will be of the order of 0.01%. Not all stocks can be as readily traded as INTC, however. For some stocks of smaller regional companies, there may be no active bid or ask. A quoted bid or ask may not be active; that is, there may be no one willing to buy at the quoted bid, and no one willing to sell at the quoted ask. Such prices are said to be "stale". For "thinly-traded" stocks, when there are both bid and ask prices, the bid-ask spread may be large, in excess of 1%. We refer to the bid-ask spreads and the efforts by brokers to match them as "trading friction". With internet brokers and online trading, the trading friction has become greatly reduced from former times.

Pricing Gaps

Anomalies occur of course. The price (and the market value) can change almost instantaneously when some major event is announced. There are many types of events that can be market movers. Sudden personnel changes at the executive level, announcements of earnings much greater or much less than expected, proposals of mergers, and other unexpected events can cause the stock price to change immediately. Many of these events are announced in the late afternoon after the market is closed, or in the morning before the market opens. The greatest changes often occur in the after hours market, and the traders in that market often realize larger gains or conversely suffer larger losses than those who trade only in the regular market.

Another thing that can move the price of a stock is the revelation of a large holding in the stock by some well-known trader or some large institution. This can allow the large trader to realize profits in the holding by selling after the revelation. (US law requires the revelation of holdings greater than certain thresholds, but the revelation is not simultaneous with the acquisition.) In addition to actual events, statements that are propagated through various media, whether or not they are true, can move the market or particularly the price of an individual security.

Initial Public Offering

For a newly issued stock, the fair market value is unknown. One of the most common and visible ways that a newly issued stock gets on the market is through an *initial public offering* (*IPO*) managed by an *underwriter*, which is generally an investment bank or a consortium of financial institutions. IPOs are tightly regulated in the US by the SEC. The underwriter must register and provide a prospective outlying a business plan and potential risks. The underwriter sets an ask price. As trading begins, there is a period of *price discovery* during which the price fluctuates as buyers and sellers set their bids and asks.

Stocks with a history of prices are also subject to fluctuations, some just due to general market variability and some due to major news events. The fluctuations following announcement of a major event potentially affecting the market value of a stock are also called *price discovery*.

An asset class that is even more difficult to price are cryptocurrencies. These are units in a decentralized digital ledger maintained by a *blockchain*. As with other non-revenue-producing assets, the value of cryptocurrencies is determined simply by supply and demand. The supply is known in the ledger, but the demand varies with traders' sentiments. A cryptocurrency is launched through a *initial coin offering* (*ICO*), which is not currently (2019) regulated. Following the ICO, the process of price discovery is rarely orderly because of the fundamental nature of the asset.

Seasoned Security

After an IPO or any direct placement, a security may be publicly and openly traded (that is, without restrictions except the usual regulatory restrictions). This is called the "secondary market".

A security that has been publicly traded in a secondary market long enough to eliminate any short-term effects from its initial placement is called a *seasoned security*. Some regulatory agencies specify the length of time of public trading for the term seasoned to be used. For example, the Euromarket requires a security to be publicly traded for 40 days in order for it to be called seasoned. The term is often used to refer to certain corporate bonds. Most stocks are seasoned, but because most corporate bonds have a finite maturity, some are not seasoned.

Law of One Price

Assets such as stock can often be traded in more than one market. Furthermore, equivalent assets can often be constructed through options or futures contracts using discounted present values. A fundamental principle of asset pricing is the "law of one price", which basically states that the prices of an asset in different markets and the prices of any equivalent asset are all the same. All prices must be adjusted for currency exchange rates and present values.

An *arbitrage* is a set of trades in the same or equivalent assets at different prices. Arbitrages can exist only temporarily because trading will adjust the prices. Generally, in financial analyses, we assume no opportunities for arbitrage exist; that is, we assume the law of one price.

Book Value

Because stock prices represent perceived value in shares of assets, we might compute the value of the underlying assets and divide by the number of shares outstanding. This is called the *book value*. The value of all of a company's assets include such tangible things as land, buildings, office equipment, and so on, as well as intangible items such as reputation and "good name". Book value, therefore, can be rather variable, depending on how the accounting department of a corporation assigns value to a diverse set of assets.

A company can affect the book value by stock buybacks. If shares are purchased on the open market at a price below the book value, the book value is increased; conversely, if shares are purchased at a price above the book value, the book value is decreased.

The reported book value per share of INTC on December 31, 2016, was

$14.19. (This figure is from the official 10-K filing by the company.) The actual closing price for INTC on December 31 was $36.27. Whether or not the book value had changed, the closing price on January 3 was $36.60.

In this case, the share price was well over twice the book value. On January 3, 2017, most stocks in the same industry sector as INTC (technology) had a much higher ratio of price to book. On the other hand, many other stocks had a much lower ratio.

Trend Lines; Moving Averages

A clearer picture of stock prices can often be obtained by smoothing out the data in some way, perhaps just by drawing a "trend line". A quantitative rule for fitting a trend line may be a simple least squares linear regression. A subjective procedure may be based on drawing a line that appears to go through the majority of the data, or it may be a line that subjectively bounds the data from above or below.

A simple objective smoothing line is a moving average (MA). For a given length of a window, this is the simple average of the prices over the days in the preceding window.

$$\mathrm{MA}_t(k) = \frac{1}{k} \sum_{i=1}^{k} p_{t-i}. \tag{1.28}$$

For daily stock prices, the previous 20, 50, or 200 days are commonly used. At the beginning of a sequence of MAs, the averages may be based just on the available data, so the first average is an average of one, the second, an average of two, and so on.

A variation of an MA is called an *exponentially weighted moving average* (EWMA). An EWMA is a weighted average of the current price (in financial applications, the previous day's price) and the previous EWMA. If at time t, the current price is p_t, and the proportion assigned to the current price is α, then

$$\mathrm{EWMA}_t = \alpha p_t + (1 - \alpha)\mathrm{EWMA}_{t-1}. \tag{1.29}$$

The recursive formula (1.29) does not indicate how to get a value of the EWMA to get started. In practice, this is usually done by taking as a starting point an MA for some chosen number of periods k. Although an EWMA at any point in time includes all previous prices since the inception of the average, it is often expressed in terms of a number of periods, say k, just as in ordinary MAs. The number of periods is defined as $2/\alpha - 1$. Other ways of sequentially weighting the prices are often used. A popular one is called "Wilder smoothing". Some MAs with different names are different only in how the "period" is used in forming the weight.

In Figure 1.9 we show the INTC data with two different simple MAs, and an EWMA.

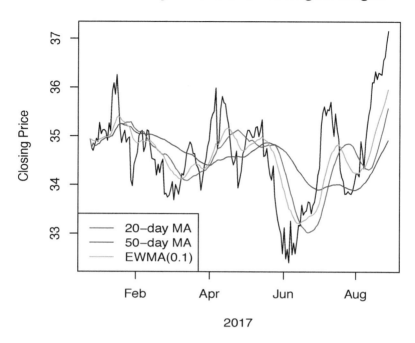

FIGURE 1.9
Moving Averages of INTC for the Period January 1, 2017 through September 30, 2017; *Source:* Yahoo Finance

As we see in the figure, the moving averages are considerably smoother than the raw data. The 50-day MA is smoother than the 20-day.

Moving averages are *smoothing models*; they smooth out the noise of a time series. A time series is converted into a smoother series. A time series $\ldots, p_{t-2}, p_{t-1}, p_t$ is smoothed into another time series $\ldots, S_{t-2}, S_{t-1}, S_t$,

$$\ldots, p_{t-2}, p_{t-1}, p_t$$
$$\ldots, S_{t-2}, S_{t-1}, S_t,$$

where the $\{S_t\}$ have smaller variation or else follow some time series model more closely. A smoothed series is also called a "filter". (In general, a *filter* is a function that maps one time series into another time series over a subset of the same time domain.)

The EWMA (1.29) follows the model

$$S_t = \alpha p_t + (1 - \alpha)S_{t-1}. \tag{1.30}$$

Smoothed series can be smoothed repeatedly. If the series in equation (1.30) is denoted by $\{S_t^{(1)}\}$ and α is denoted by α_1, then the kth *order exponential smooth* is defined as

$$S_t^{(k)} = \alpha_k S_{t-1}^{(k-1)} + (1 - \alpha_k) S_{t-1}^{(k)}. \tag{1.31}$$

Exponential smoothing is widely used in forecasting of economic time series. The so-called *Holt-Winters forecasting method* uses a third-order exponential smooth. The reason that the Holt-Winters filter uses a third-order smooth is to deal with seasonality in the time series.

On a given day, the price may be above or below any of its moving averages. Many analysts decide that the current market price of a stock is either greater than or less than the fair price based on the relationship of the current price to one or more of the moving averages.

The moving averages we have described above are averages of previous values, excluding the current value. This is appropriate in applications to stock prices as in Figure 1.9. The moving average on a given day does not include the price for that day. In other applications, however, moving averages may be used to smooth time series data by averaging the data on either side of a given point.

Another consideration for a moving average over a time series with a fixed starting point is what value to use for the moving average if not enough prior data are available. An obvious way of handling this situation for the first few items in the series is just to average all available previous data until enough data for the moving average are available.

Revenue and Earnings

Shares in stock represent ownership in equities that may produce earnings as a result of the business operations or other transactions made as part of the business.

The total amount of funds minus adjustments that a business receives during a given period is called the *revenue* for that period. Revenue is sometimes called "sales". The adjustments may be for returns or canceled sales or they may be for discounts given after the sales. Sales may be for merchandise or for services or other considerations. Revenue is also sometimes called the "top line".

A measure of the price of a share relative to its value is the ratio of the price to the sales or revenue attributable to one share. This is the *price-to-sales ratio* or *PSR*.

Revenue minus expenses is called *earnings*. The earnings are for a stated period, usually for a specific quarter or specific year. Earnings are also sometimes called "net income", "profit", or the "bottom line".

A widely-used measure of the value of a share is the amount of earnings attributable to one share. This is the *earnings per share* (*EPS*). The earnings may be computed over some recent period (*trailing earnings*) or they may be estimated for some future period (*forward earnings*). The standard period for earnings is one year, although most corporations compute and report earnings quarterly.

One way of using earnings to establish a value of a share of stocks is by computing the *price-to-earnings* (*PE*) ratio. As indicated above this ratio can be trailing or forward, although the user of the PE may neglect to state which meaning is being used.

A corporation can choose to disperse some of the earnings to the shareholders in the form of dividends, or else to reinvest the earnings in improving the overall prospects of the corporation. The proportion of earnings paid in dividends is called the *payout ratio*. The relative proportions allocated to dividends and to growth of the business itself depend on the nature of the business, and change over the life of a corporation. Newer corporations generally allocate less of the earnings to dividends than do more mature corporations.

A general method for assigning a value to any asset is to equate it to the total income expected to be generated by the asset over some time period, or even in perpetuity, discounted back to the present. This valuation method is called *discounted cash flow* or *DCF*. The cash income c_i at future times $t_1, t_2, \ldots$ are discounted to the present time t_0 as in equation (1.16) as $c_i e^{-r(t_i - t_0)}$. (Note that c_i is not necessarily proportional to a growing principal, as it would be in the case of compounded interest. An infinite series of these terms may therefore converge.)

Because dividends are the only tangible benefits to the stock owners who do not trade their shares, an obvious measure of the value of a share of stock is the present value of the total of all future dividends. Williams (1938) suggested this method of valuation of stocks, and gave a simple formula for the fair value of a stock based on that pricing mechanism. The *Gordon growth model* assumes a constant growth rate of dividends. The model yields a convergent infinite series that gives a "terminal value" for the stock. Obviously, the necessary assumptions about the future dividend stream and the time series of risk-free rates make this method of valuing stock essentially useless except as a theoretical exercise.

Based on the reported earnings of INTC for the calendar year 2016 and the closing price on January 3, 2017, the trailing PE for INTC on that date was 15.19. The earnings for the calendar year 2016 were known at that time. Was the price of 36.60 on January 3 justified? A lower price would result in a smaller PE, and a higher price would yield a larger PE. On January 3, 2017, most stocks in the same industry sector as INTC (technology) had higher PEs than 15.19.

A forward PE for INTC on January 3, based on the closing price of 36.60, was 11.32. A forward PE is an estimate or projection, and of course depends on who is making the projection. Various financial institutions use statements made by the corporation, both objective income statements and balance sheets and subjective statements ("conference calls"), to estimate earnings for the coming periods. (In this case, the forward PE for INTC was provided by Thomson Reuters.)

A company can affect its EPS and consequently the PE by stock buybacks. When the total number of shares outstanding is decreased, other things being equal, the EPS is increased and the PE is decreased.

Robert Shiller has suggested a "cyclically adjusted price to earnings (CAPE) ratio" that uses earnings per share, adjusted for inflation, over a 10-year period to smooth out fluctuations in corporate profits that occur over different periods of a business cycle. The ratio is the price divided by the average of ten years of earnings adjusted for inflation. (Adjustments for inflation are somewhat subjective, although we will not go into the details.) The CAPE ratio is also known as the Shiller PE ratio.

The ratio is more often used for indexes rather than for individual stocks. It is sometimes used to assess whether the market is undervalued or overvalued.

Earnings, on the surface, may seem to be a rather objective quantity, but there are many discretionary aspects to earnings. The FASB defines specific accounting principles, called *Generally Accepted Accounting Principles* (*GAAP*), that bring some consistency to financial statements. Earnings that are computed following these principles are called *GAAP earnings*. Even so there are various instances where discretionary spending or bookings can affect GAAP earnings, especially within a given time period. There are various types of non-GAAP earnings. One of the most common is *EBITDA* (earnings before interest, taxes, depreciation, and amortization).

Another meaningful valuation ratio is the enterprise value to the EBITDA, or EV/EBITDA. This more accurately reflects the liquidation value of the corporation to its free cash flow, and hence, is the primary measure in evaluating the worth of a company in a merger of corporations of in a corporate takeover.

Supply and Demand

A major dictum of economics is that prices are determined by supply and demand. The "supply" of a given stock depends on how many shares are available. "Demand" in the case of stocks is generally based on the perceived value rather than on economic need or on productive uses to which the good can be applied. Nevertheless, there is a certain but unquantifiable demand for many stocks. Aside from other considerations about the value of a stock such as book value or earnings per share, the number of people wanting to own the stock may affect its price.

A key assumption in many idealized financial analyses is that asset prices are in equilibrium; that is, supply equals demand for each asset.

Stocks are somewhat different from other economic goods, because they are often "sold short"; that is, a trader may borrow shares and sell the borrowed shares (see Section 1.2.8, beginning on page 65). The short seller must eventually buy back the shares so as to return what was borrowed, so there is a demand for a stock that has been sold short. Traders often look at the *short interest* in a stock, that is, the amount of the shares that have been sold short, to assess this component of the demand for a stock. The short interest of some "high-flying" stocks may be over 30% of the total outstanding shares of the stocks.

Corporate buybacks, which reduce the number of shares available, may also affect the price through the supply and demand mechanism.

Asset Pricing Models

Another approach to pricing assets that we alluded to above is to regress the return of an asset on the returns of other assets, each return being discounted by a risk-free return. This results in a measure of the returns relative to "the market". The model value of a return coupled with an observed value determines a model price for the asset. This "market model" can also be combined with other economic factors, and the resulting model is called a "factor model". There are various forms of this general kind of asset pricing model, such as the "capital asset pricing model" (CAPM) which is a regression relationship of the price of an asset and the prevailing prices of similar assets; and a model based on assuming no arbitrage; that is, following "arbitrage pricing theory" (APT).

We will illustrate a simple regression model on page 61, and discuss regression models more generally in Section 4.5.2.

Arbitrage opportunities depend on expected returns exceeding the risk-free rate. A common approach to financial modeling assumes that the rate of expected returns is equal to the risk-free rate. This assumption and approaches based on it are called *risk neutral*.

Behavioral Economics

While fundamental values of the underlying assets and possibly supply and demand may affect stock prices, the beliefs, whims, greed, and hopes of investors affect the prices of stocks significantly. As noted earlier, stock prices that have risen recently are more likely to continue to rise. In many cases, this seemingly occurs for no other reason than that the prices had recently been rising. This tendency for prices to continue to move in a given direction is called *momentum*.

Upward momentum is often sustained by traders' fear of missing out, or FOMO. When traders observe other traders making profits, their own decision-making processes are affected. Of course this works both ways, and when others' losses are observed, traders may make quick moves to mitigate their own losses. In traders' parlance, the "fear" in FOMO is actually "greed", and since greed is generally trumped by fear, traders often "take the stairs up and the elevator down." FOMO is the primary driver behind many participants in IPOs and most participants in ICOs.

Behavioral economics are a pervasive influence in asset prices. How people feel about the overall economy or about the overall political climate may affect stock prices.

1.2.5 Splits, Dividends, and Return of Capital

The number of shares outstanding for any given corporation is the result of a decision by the board of directors (and perhaps existing shareholders). If there are, say, 10 million outstanding shares worth S each, there could just as well be 20 million worth $S/2$ each. This is the idea of *stock splits*.

Because most US stocks trade for somewhere between $20 and $200 per share, a company whose stock is trading for $300 per share may feel that investors would be more comfortable dealing with stock around $100 per share. A three-for-one split would immediately reduce the price of a share to one-third of the previous price, while not affecting the total value of any investor's holding in the stock.

On the other hand, a company whose stock is trading for $5 per share may increase the value of a share by a *reverse split*. A one-for-three reverse split would reduce an investor's number of shares held to one third of the previous number while increasing the value of each by a factor of three.

Stock splits do not result in *income* to the owners of the stock. ("Income" is a technical term defined by government taxing agencies. It does not necessarily refer to funds received. A holder of a mutual fund, for example, may receive "income" and, consequently, a tax liability without receiving any funds or any increase in the value of mutual fund assets.)

A stock split is also called a *stock dividend*, especially if the split results in less than two shares for each current share.

Dividends paid in cash are simply called *dividends*, and a large number of stocks pay cash dividends. Because a cash dividend reduces the overall value of the corporation, a cash dividend reduces the share price by the amount of the dividend per share. Cash dividends result in *income* to the owners of the stock. Many corporations and brokerage firms offer *dividend reinvestment programs* ("DRIPs"), whereby dividends can be used to purchase shares or fractional shares at the price when the dividend is paid.

A corporation may also *return capital* to shareholders. Because a return of capital reduces the overall value of the corporation, return of capital reduces the share price by the amount returned per share. Funds considered to be return of capital are not *income* to the owners of the stock. Rather they reduce the *cost basis* of the shares. (What is considered "income" and what is considered "cost basis" are very important because of tax computations.)

Splits and Adjusted Prices

Whenever a stock split occurs, prices in the historical record of that stock must be adjusted in order for historical plots of prices, such as in Figures 1.1

and 1.2, to be meaningful. The adjustment for a stock split is simple. If a stock is split k for 1, all prices prior to the time of the split are adjusted by dividing by k.

Adjustments for splits must be made not just for graphs of historical prices to be meaningful, but adjustments are also necessary for real financial reasons involving conditional orders and various types of contracts that specify a share price. After a two-for-one split, for example, a call option with a strike of 90 is adjusted to have a strike of 45. (We will discuss options in Section 1.2.7, beginning on page 63.)

The *adjusted price* for a stock on any given date may be different at different future dates. Thus, in Table 1.3, at time t_1 the market price of a share of a particular stock was S_{t_1} and at time t_2 the market price was S_{t_2}. At a later time, t_3, there was a k for 1 split; hence, at that time, the adjusted price for time t_1 became S_{t_1}/k and the price for time t_2 became S_{t_2}/k, while the current market price was S_{t_3}. These were the adjusted prices until another adjustment. At a later time, t_4, there was an m for 1 split; hence, at that time, the adjusted price for time t_1 became $S_{t_1}/(km)$, the price for time t_2 became $S_{t_2}/(km)$, the price for time t_3 became S_{t_3}/m, while the current market price was S_{t_4}. The historical adjusted prices change with each split. The adjustments are the same if the splits are reverse splits; that is, when k is less than 1.

TABLE 1.3
Adjusted Prices Following Stock Splits

	time	t_1	t_2	t_3	t_4
prices	$\rightarrow$	S_{t_1}	S_{t_2}		
	split			k for 1	
adjusted prices	$\rightarrow$	S_{t_1}/k	S_{t_2}/k	S_{t_3}	
	split				m for 1
adjusted prices	$\rightarrow$	$S_{t_1}/(km)$	$S_{t_2}/(km)$	S_{t_3}/m	S_{t_4}

Dividends and Return of Capital

Other types of adjustments are made to reflect payment of dividends or return of capital. Both of these events are announced by the corporation ahead of time, and an *ex-dividend date* is set. All shareholders of record at market open on the *ex-dividend date* will receive the dividend or return of capital several days later. Except for certain extraordinary dividends and for return of capital, these adjustments do not affect option strike prices or prices specified in conditional orders.

The adjustment is straightforward. Suppose a corporation announces a dividend of d per share with an ex-dividend date of t_2. If the unadjusted price of the share at time $t_1 < t_2$ is S_{t_1}, we would expect the price at time t_2 to be $S_{t_1} - d$, within the normal variation due to trading; that is,

$$S_{t_2} = S_{t_1} - d.$$

The adjusted price at t_1, therefore is

$$\widetilde{S}_{t_1} = \left(1 - \frac{d}{S_{t_1}}\right) S_{t_1}. \tag{1.32}$$

All previous historical prices should be adjusted by this amount:

$$\widetilde{\widetilde{S}}_{t_k} = \left(1 - \frac{d}{S_{t_1}}\right) \widetilde{S}_{t_k}, \quad \text{for } t_k < t_2, \tag{1.33}$$

where $\widetilde{S}_{t_k}$ is the adjusted price at time t_2.

As an example, in July 2017, INTC declared a dividend of 0.273 per share, with an ex-dividend date of August 3, 2017. Therefore, from August 3 until the next adjustment event occurs, the historical adjusted prices prior to August 3, 2017, are the previous prices adjusted by a factor of $1 - 0.273/36.64$ or 0.9925491. In Table 1.4, we see, for example, on August 1 with an actual closing price of 36.35, the adjusted close is $(1 - 0.273/36.64)36.35$ or 36.07916.

TABLE 1.4
Adjusted Prices of INTC Following Dividends; *Source:* Yahoo Finance

	Open	High	Low	Close	Adjusted Close	
2017-08-01	35.928	36.703	35.837	36.35	36.079	
2017-08-02	36.603	36.945	36.331	36.64	36.367	
2017-08-03	36.550	36.590	36.150	36.49	36.490	0.273 dividend
2017-08-04	36.450	36.560	36.100	36.30	36.300	
2017-08-07	36.390	36.550	36.220	36.43	36.430	

These are the historical adjusted prices on August $3, 4, \ldots$ until the next adjustment event occurs. After another adjustment event occurs, the historical prices prior to that future time will be adjusted.

Carrying this example one step back, in April, 2017, INTC declared a dividend of 0.273 per share, with an ex-dividend date of May 3, 2017. Between May 3 and August 3 (which was the next taxable event), the historical prices would be adjusted by $(1 - 0.273/36.97)$. These adjusted prices would be adjusted again on August 3; hence, on August $3, 4, \ldots$ until the next adjustment event occurs, the historical adjusted price for May 1 is

$$(1 - 0.273/36.97)(1 - 0.273/36.64)36.31 = 35.77333,$$

TABLE 1.5

Adjusted Adjusted Prices of INTC Following Dividends; *Source:* Yahoo Finance

	Open	High	Low	Close	Adjusted Close	
2017-05-01	36.652	36.946	36.479	36.31	35.773	
2017-05-02	36.916	37.586	36.895	36.97	36.424	
2017-05-03	36.996	37.449	36.895	36.98	36.704	0.273 dividend
2017-05-04	37.268	37.389	36.915	36.85	36.575	
2017-05-05	37.137	37.207	36.774	36.82	36.546	

as shown in Table 1.5.

Return of capital also results in adjustments of prices. The adjustments are the same as for dividends. The only differences are in the tax consequences for the owner of the stock.

Total Rates of Return

The "total rate of return" is the proportional change in asset prices plus the proportional amount of distributed earnings received. Use of adjusted prices accounts for dividends, but there is a slight difference in the rate of return computed from adjusted prices and from unadjusted prices with the dividend added in. This is easily seen from simple arithmetic and equation (1.32):

$$\frac{S_{t_2} - \widetilde{S}_{t_1}}{\widetilde{S}_{t_1}} \neq \frac{S_{t_2} + d - S_{t_1}}{S_{t_1}}. \tag{1.34}$$

If the dividend is relatively small, the difference in the two expressions is small.

With return of capital resulting in adjustments of prices, there are also differences in the way the total rate of return could be computed. These small differences may make for differences in the advertising claims for certain financial products, but for our study of methods of statistical analysis, there is no practical difference, so long as we are aware of the different computations. (The CFA Institute, mentioned on page 19, sets guidelines for these computations by mutual funds; this affects advertising.)

1.2.6 Indexes and "the Market"

There are thousands of stocks actively traded in the US markets. On any given day, some go up in price and some go down. An *index* is some overall measure of stock prices at one point in time relative to the prices at another point in time. In addition to indexes that attempt to summarize stock prices, there are many other indexes that measure prices of other financial assets or other aspects of "the market". In this section we will focus on stock indexes.

Stock Indexes

A *stock index* is an average of the prices of a set of stocks, maybe the "whole market", maybe just the stocks of the largest companies, or perhaps just the stocks of some other group, such as the larger biotech companies. For any given set of stocks, however, it is not clear how to average the stock prices in a meaningful way.

In the 1890s, Charles Dow, of *The Wall Street Journal*, began to use the mean of the closing prices of just a few stocks as a measure of the overall price level of the stock market. Dow created different indexes for different sectors of the market. Currently, a widely followed index is the *Dow Jones Industrial Average* (DJIA, or just the "Dow"), which is a successor of one of Dow's original indexes. The DJIA currently is the sum of the stock prices of 30 large corporations traded on the NYSE or on Nasdaq divided by a fixed *divisor*. (INTC, for example, is one of the 30 stocks in the DJIA.) From time to time, the stock components of the Dow are changed. The divisor is also changed so that at the time of the replacement of one stock with another, the value of the index does not change.

As an "average", the DJIA is *price weighted*; that is, it is just the sum of the prices of the components divided by a fixed amount. The weight of a single component stock does not depend on how large the company is, only on what its share price is. For example, on January 3, 2017, the (unadjusted) closing price of INTC was 36.60. The closing price of Goldman Sachs (GS), another Dow component, was 241.57. The DJIA closed at 19,881.76 on that day. This figure is

$(36.60 + 241.57 + \text{sum of closing prices of 28 other stocks})/0.14523396877348,$

using the value of the divisor on that date. (The divisor changed on June 26, 2018.)

The effect of the price of GS on the Dow was over 6 times the effect of the price of INTC, even though the market value of all outstanding shares of INTC was approximately twice the market value of all outstanding shares of GS.

If a component stock of the DJIA undergoes a 2 for 1 split, its weight in the DJIA is reduced by a factor of 2. The divisor is changed at that time so that the computed value of the DJIA does not change. The relative weights of all of the component stocks change.

Two other widely followed indexes are the *Standard & Poor's 500 Index* (S&P 500) and the *Nasdaq Composite Index*. Both of these are *market-value weighted* indexes, that is, each component is weighted by market value of the stock (see page 32). A market-value weighted index is also called a *capitalization weighted* index. Stock splits do not change the total market value of a stock, so stock splits do not affect market-value weighted indexes. The S&P 500, consisting of approximately 500 large cap US stocks traded on the NYSE or on Nasdaq, is one of the most common benchmarks for the US stock market. (INTC, for example, was one of the 505 stocks included in the index during 2017 and 2018.) While the larger number of stocks in the S&P 500 compared to the DJIA may indicate broader coverage of the market, because of the market cap weighting, the largest 2% of the stocks in the S&P 500 (that is, 10 stocks) have in recent years accounted for over 18% of the value of the index.

In 2003, Standard & Poor's created the S&P 500 Equal Weight Index (EWI), which consists of the same stocks as in the S&P 500, but it weights them equally; that is, the values of all of the stocks in the index are the same regardless of the stock prices or the market values of the stocks. Both market-value weighted and equally weighted indexes must rebalance the relative amount of each entry as the price changes.

The Nasdaq Composite is a weighted average of the prices of approximately 3,000 stocks traded on Nasdaq. The Nasdaq Composite is also dominated by the 10 or 15 largest cap stocks. (INTC, for example, is one of the stocks in the Nasdaq Composite, and it carries much more weight than most of the other stocks in the index.)

The "major indexes" are the Dow Jones Industrial Average, the S&P 500 Index, and the Nasdaq Composite Index. Figure 1.10 shows the relative performance of these three indexes from 1987 through 2017. (The Dow Jones Industrial Average is often called the "Dow Jones" or just the "Dow", although there are other Dow Jones indexes, the most common of which are the Dow Jones Transportation Average and the Dow Jones Utility Average.)

Two things that stand out in Figure 1.10 are the strong correlations between the DJIA and the S&P 500 and the divergence and wild swings of the Nasdaq. The DJIA consists of just 30 stocks; yet it tracks the broader index of large cap stocks very well. Another thing that makes the agreement even more remarkable is the fact that the DJIA is price-weighted. As we mentioned above, this means that the relative influences on the index by the component companies are not consistent with what we might assume are the relative influences these companies may have on the overall market because of their size. On the other hand, a market-value weighted index such as the S&P 500 or the Nasdaq Composite may be dominated by just a few large-cap companies. In Exercise 4.20 of Chapter 4, you are asked to form an index based on principal component loadings of the Dow 30 stocks and consider the performance of that index. (Without constraints, such an index has short positions.)

The rapid rises of the Nasdaq in the 1990s and in the period after 2010 are both notable. The reason in both cases is the types of companies in the Nasdaq. There is a preponderance relative to the overall market of companies in the computer and information technology sector and in the biotech sector. Both of these areas have seen rapid developments in recent years, and companies involved in the area saw rapid growth in these two periods.

Even more noticeable than the rapid rises of the Nasdaq is its precipitous fall around 2000. This occurred during the "dot com" bust. There were many

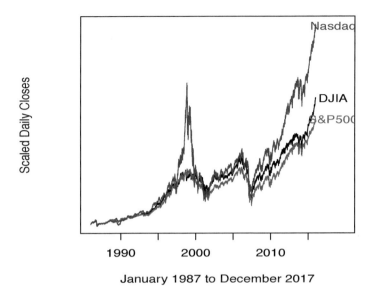

FIGURE 1.10
Daily Closes of Major Indexes for the Period 1987–2017; *Source:* Yahoo Finance

companies in the information technology space that had no earnings and no capital but which held forth the promise of rapid future growth. Even many companies in this sector that did have earnings and a reasonable balance sheet had PE ratios in the triple digits. The market sentiment that sustained these lofty prices began to fade in the late 1990s, and by 2000 had reached almost panic levels.

Many information technology companies went out of business around 2000; and the stocks of companies with good earnings and good prospects were also affected. Intel Corporation at the time was making leading-edge microprocessors and other chips. As a result, the stock of Intel had soared to price levels that were not sustainable, and it was hit very hard along with the stocks of companies whose assets consisted primarily of dreams. Figure 1.11 shows the performance of INTC from 1987 through 2017. It corresponds very closely to the Nasdaq Composite for the same period shown in Figure 1.10. Its rise since 2010 has not quite matched the increase in the Nasdaq Composite, however. Since 2000, the market sector of Intel has been called "old technology".

We notice in Figure 1.10 that between 1987 and 2017 all three indexes

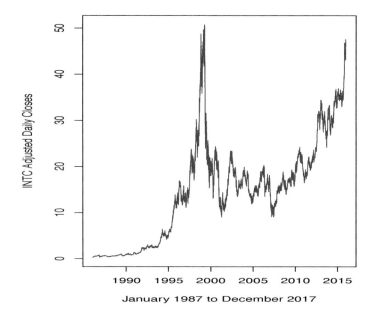

FIGURE 1.11
Adjusted Daily Closes of INTC for the Period 1987–2017; *Source:* Yahoo Finance

fell to multi-year lows on two separate occasions. The dot com bust around 2000 affected most stocks; not just the high-tech ones. Although there is no shortage of explanations for any market sell-off, in most cases, all we can say is "they happen". A simple explanation for the decline is that the stock prices were "too high". This explanation only kicks the can; the questions are what is too high, and why is it too high.

The other market low in the period shown in the graphs, around 2008, has more explainable causes. This was the so-called "Great Recession", whose causes have been traced to "greed" of both borrowers and lenders, to fraud on the part of lenders who repackaged loans and mortgages into under-secured financial products, and to stupidity on the part of the bankers and investors who bought these products.

In addition to the three major market indexes mentioned above, there are a number of other widely-followed indexes. Some of the more important ones are the Russell indexes. The Russell 3000 is a market-capitalization-weighted index of the 3,000 largest (more-or-less) US publicly-traded corporations. It is sometimes considered to be a good measure of the "whole market". The Russell 2000 is an index of the smallest 2,000 of the 3,000 in the full index.

It is often used as a measure of the stocks of "small-cap" companies. There is also the New York Stock Exchange (NYSE) Composite, which averages all stocks traded on the NYSE.

There are also a variety of proprietary indexes. Some that are widely used by academics are the various CRSP indexes built from the database compiled by the Center for Research in Security Prices at the University of Chicago. The CRSP database was first compiled in the 1960s and consisted of monthly prices of all NYSE stocks going back to 1926. The database has been extended to include daily prices, prices of stocks traded in other US markets, and other financial data. In the days before modern computing and communications, the CRSP data was widely used in academic research.

Adjustments of Stock Indexes

At any point in time, a stock index is computed based on the current prices of stocks. As we mentioned, the S&P 500 and the Nasdaq are self-adjusting for stock splits, including stock dividends, because they are market-weighted. The DJIA adjusts by changing the divisor. These indexes, however, are not adjusted for ordinary dividends or for return of capital. The values shown in Figure 1.10 are not adjusted. They do not adequately reflect the total return of the stock values that they represent.

We could form a series of adjusted values for the indexes just as we formed historical adjusted prices of stock in Section 1.2.5. This would mean, however, that the plots in Figure 1.10 would have to be changed every time one of the stocks in any of the indexes paid a dividend.

As we saw in the examples in Tables 1.4 and 1.5, the adjusted values for a given date change over time. This is the nature of the adjustments; they are specific to a specific time, the time they are made. They are effective until the next event that requires a price adjustment.

Another way of approaching the problem of accounting for total return is to form a separate series in which dividends or other returns are added in. There is no way of avoiding the problem that total return is specific to a specific time, however. In the case of adjusted closing prices in Section 1.2.5, the specific time is the immediate present. In the case of a separate augmented series, the specific time is some chosen time in the past.

Any data analyst, of course, could choose some specific time in the past and accumulate the dividends on all stocks in an index from that time forward (and many people and institutions have done this for the major indexes, starting at various times). One common total return index for the S&P 500 is called SP500TR. Its starting point is January 1988.

Figure 1.12 shows the relative performance of the S&P 500 total return (SP500TR) versus the unadjusted S&P 500 from January 1988 to December 2017.

S&P500 and S&P500 Total Return

Daily Values

S&P500 Total Return

S&P500

1990   2000   2010

January 1988 to December 2017

FIGURE 1.12
S&P 500 and S&P500 Total Return Daily Values for the Period January 1988
to December 2017; *Source:* Yahoo Finance

Logarithmic Scale

After viewing the graphs of the three indexes over the period of 30 years, we
will pause here to point out something about those graphs or graphs of other
assets that tend to grow at a fairly steady rate.

In graphical plots of asset prices or indexes that generally grow over time
at an amount roughly proportional to their size, the rate of change when the
values are small may be obscured by the changes when the values are larger.
In the plots in Figure 1.10 on page 51, for example, it may appear that the
indexes increased more rapidly after 2010 than they did in the period from
1990 to 2000 (ignoring the Nasdaq bubble).

For any quantity with a constant return, the values plotted on a linear
scale, the rate of change may not appear to be constant. For

$$y = b^x,$$

a plot of y versus x is exponential, but a plot of $\log(y)$ versus x is linear. For
this reason, plots such as those in Figure 1.10 are sometimes made with a
logarithmic scale on the vertical axis. Although in most plots in this book, I

will continue to use a linear scale, this may be a good time to present a quick illustration of the difference.

Figure 1.13 shows the same plots as shown in Figure 1.10, but on a logarithmic scale for the vertical axis.

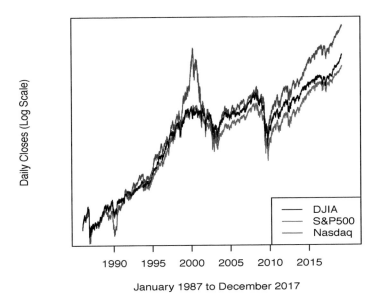

FIGURE 1.13
Daily Closes of Major Indexes on a Logarithm Scale; *Source:* Yahoo Finance

The general slopes of the plots in Figure 1.13 are linear, compared to the exponential slopes in Figure 1.10. The logarithmic scale sometimes makes the interpretations simpler.

Market Sectors

A "sector" of the market consists of stocks with some commonality, such as the nature of the business the corporation pursues, or the relative price movements of the stock relative to price movements of the market as a whole. The term may be used informally in this sense, and no particular set of stocks is identified with the sector. The two major stock index providers, S&P Global (formerly Standard and Poors) and MSCI, Inc. (formerly Morgan Stanley Capital International), have developed a standard set of twelve sectors, called the Global Industry Classification Standard or GICS. The sectors are further divided into industry groups, industries, and sub-industries. The GICS assigns a code from each grouping to every company publicly traded in the market. The GICS coding system is widely used in the financial industry.

Exchange-Traded Funds Based on Indexes or Market Sectors

An index is not a tradable asset, although there are tradable derivatives based on many indexes. There are also mutual funds and exchange-traded funds (ETFs) that track various indexes by forming portfolios of stocks that correspond to the particular indexes or the stocks of a market sector. An ETF that tracks an index is called an index ETF and one that tracks a sector is called a sector ETF. Shares of ETFs trade on the open market just like ordinary shares of stocks.

The individual securities in an ETF are owned by a financial institution that forms shares representing proportional beneficial interest in the portfolio of individual securities. The underwriting financial institution generates revenue by regularly extracting a proportionate amount of the total value of the portfolio. There are many financial institutions that have formed ETFs and listed them in the market. Some of the larger financial institutions are Black-Rock, which has several ETFs with the general name of iShares; State Street Global Advisors, which has several ETFs with the general name of SPDR; and ProShares.

There are several ETFs that track the major indexes. One of the largest is the SPDR SPY, which tracks the S&P 500. SPY, which was begun in January 1993, is the oldest of the ETFs. The dividend yield of SPY mirrors that of the stocks in the S&P 500. Figure 1.14 shows the relative performance of SPY and the S&P 500 from 1993 to 2017. In both cases, the prices shown are adjusted for dividends. (The S&P 500 shown is the SP500TR, which is also plotted in Figure 1.12.) The vertical axis is the difference of SP500TR and the adjusted SPY monthly closes as a percentage of the SP500TR. Over the twenty-five year period, the charges extracted by the sponsor reduced the net by about 3.5%. The fluctuations in the graph are due to trading variation in SPY and the timing of dividend payouts.

There are also EFTs that track many other indexes, or just market sectors. A sector ETF may weigh the stocks in the sector equally, that is, may maintain an almost-equal total value of all stocks in the sector, or it may weight the stocks equal to the stocks' relative market capitalizations.

Many sector ETFs are based on the GICS market sectors, but the stocks in the portfolio of an ETF may also be taken to define a "market sector" different from any of the GICS sectors. An ETF's portfolio can also consist of combinations of shares of stock, contracts on commodities, or commodities themselves (such as the SPDR GLD, which owns physical gold bullion).

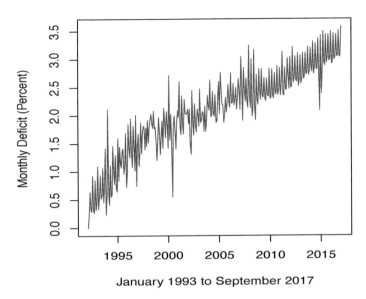

S&P500 Total Return Minus SPY (Scaled)

January 1993 to September 2017

FIGURE 1.14
Differences in Adjusted Daily Closes of the S&P 500 (Total Return) and of
SPY 1993 through 2017; *Source:* Yahoo Finance

Leveraged Instruments: Ultra and Inverse ETFs

ETFs built around one of the major indexes may seek to exceed the return
of the index, at least on a daily basis. Such ETFs are called "ultra ETFs".
The methods of attempting to achieve the excessive returns of an ultra ETF
are use of leverage, debt, options, and swaps and other derivatives; hence, an
ultra ETF differs in a fundamental way from an ETF that seeks to mirror the
performance of an index or to represent a sector of the market. An ETF of
the latter type generally owns stocks instead of swaps or derivatives.

ProShares' Ultra S&P500 (SSO) seeks daily returns (before fees and expenses!) that are twice those of the S&P 500 Index, and their UltraPro S&P500
(UPRO) seeks daily returns that are three times those of the S&P 500 Index.
The daily log returns of UPRO, which was begun in June 2009, compared
with the S&P 500 returns of the same days during four different periods are
shown in Figure 1.15. The returns of UPRO are in the same direction but
greater in absolute value in all cases.

There are also ETFs that seek to return the inverse (negative) of the return

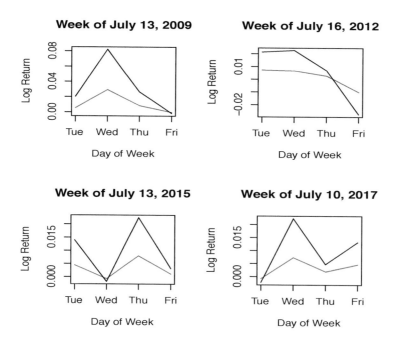

FIGURE 1.15
Daily Returns of UPRO (Unadjusted) and of the S&P 500 for Selected Weeks
Source: Yahoo Finance

of the index, again, on a daily basis. For example, ProShares' Short S&P 500
ETF (SH) seeks to mirror the inverse (negative) of the one-day performance
of the S&P 500. Their UltraShort S&P 500 ETF (SDS) seeks to mirror twice
the negative of the S&P 500 one-day returns. The methods of attempting to
achieve these returns are by use of short sales and by trading in derivatives.
The daily log returns of SDS, which was begun in July 2006, compared with the
S&P 500 returns of the same days during four different periods are shown in
Figure 1.16. The returns of SDS are greater in absolute value, but in different
directions in all cases.

ETFs that seek to mirror the negative of an index, a market sector, or a
commodity, are called "inverse ETFs" or "short ETFs". They are also called
"reverse ETFs". There are many such ETFs. For example, ProShares has an
inverse ETF called UltraShort Oil & Gas (DUG) that seeks daily investment
results (before fees and expenses!) that correspond to two times the inverse
(negative) of the daily performance of the Dow Jones US Oil & Gas Index. This
fund attempts to achieve this kind of return primarily by being short futures

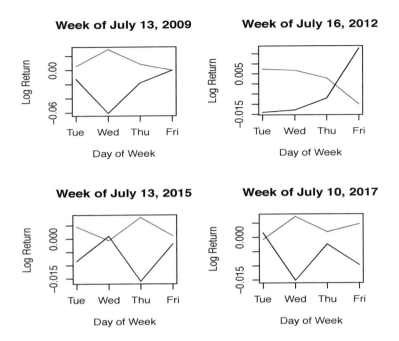

FIGURE 1.16
Daily Returns of SDS (Unadjusted) and of the S&P 500 for Selected Weeks;
Source: Yahoo Finance

contracts on the commodities themselves. (ProShares also has an ETF called DIG that attempts to mirror the returns of the Dow Jones US Oil & Gas Index.)

The nature of the assets owned in ultra and inverse ETFs make them very different from ETFs that own just stocks. The financial institutions that put together ultra and inverse ETFs tout them as short-term holdings, and emphasize that the objective of such funds is to produce a *daily* return that has a stated relationship to the *daily* return of its target.

There are hundreds of ETFs and their total trading volume (in dollars) is of the same order of magnitude as the total trading volume of all listed stocks. Table 1.6 shows a few ETFs, and their market focus. (Except for those following the major indexes, the ETFs shown in the table are not necessarily heavily-traded ones.)

Other Markets, Indexes, and Index Proxies

There are major markets for stocks and other securities in many countries. In Europe, there are active markets in London, Paris, and Frankfurt. In Asia, there are active markets in Tokyo, Hong Kong, and Shanghai. Each of these

TABLE 1.6

Some ETFs

QQQ	Nasdaq 100
DIA	Dow Jones Industrial Average
SPY	S&P 500
SSO	$2\times$ S&P 500
SDS	$-2\times$ S&P 500
EEM	MSCI Emerging Markets Index
GLD	gold bullion
GDX	stocks of gold-mining companies
AMLP	MLPs in the energy sector
DEMS	stocks expected to prosper under policies of the Democratic Party
GOP	stocks expected to prosper under policies of the Republican Party

markets has one or more widely-used indexes. Table 1.7 lists some of these markets and their indexes. In Exercise A1.25, you are asked to compare the historical performance of some of these indexes and the S&P 500 in a graph similar to that in Figure 1.10.

TABLE 1.7

International Markets and Indexes

London	FTSE-100 (capitalization-weighted)
Paris	CAC-40 (capitalization-weighted)
Frankfurt	DAX (capitalization-weighted)
Tokyo	Nikkei-225 (price-weighted)
Hong Kong	Hang Seng (capitalization-weighted)
Shanghai	SSE Composite (capitalization-weighted)

Indexes of course cannot be traded, but there are active options markets on some of these international indexes. In addition, there are many ETFs that track these indexes, or sectors within the indexes. There are ETFs that track combinations of these indexes, or other measures of international markets. There are also many ETFs that track smaller international markets. Various smaller markets are referred to as "emerging markets".

The reason there are so many ETFs is the income generated for the financial institution that assembles an ETF.

The Market; Excess Returns

The foregoing discussion makes clear that there is no single, monolithic "market". As we have mentioned, however, financial analyses of individual assets often focus on the relative return of the individual asset compared to some market benchmark. The benchmark may be one of the major indexes mentioned above, or it may be some other overall measure of the market or some sector of the market.

As mentioned earlier, we often hypothesize a risk-free asset, and then focus on the *excess return* of an asset over and above the return of the abstract risk-free asset. If $R_{F,t}$ is the risk-free return at time t, and $R_{i,t}$ is the return of an asset i at time t, the excess return of that asset is $R_{i,t} - R_{F,t}$, which may be negative. As mentioned earlier, generally, the returns are of the same type, simple or log, but occasionally a simple return (interest rate) is used to adjust a log return.

Next, we consider some "market" or group of assets. The "market" may be defined as an index, for example. Let $R_{M,t}$ be the return of this market. Then the excess return of the market is $R_{M,t} - R_{F,t}$. The simple linear regression of the excess return of asset i on the excess return of the market, called the *market model*, is

$$R_{i,t} - R_{F,t} = \alpha_i + \beta_i(R_{M,t} - R_{F,t}) + \epsilon_{i,t}. \tag{1.35}$$

Figure 1.17 shows a scatterplot of the annualized daily log returns of the S&P 500 market index $R_{M,t}$ and the annualized daily log returns of the common stock of Intel (unadjusted for dividends) $R_{i,t}$, both adjusted for the daily interest rate in the secondary market for 3-month US T-Bills $R_{F,t}$ for the year 2017. (Note that the scales of the axes are very different.) The least-squares regression line is also shown in Figure 1.17. We see that there is a positive linear relationship, but the linearity is not very strong. The adjusted R-squared value is 0.2056.

The "alpha" of the asset, α_i, represents excess growth (assuming it is positive), and is one of the talking points of mutual fund salespeople. The "beta" of the asset, β_i, represents the adjusted volatility of the asset relative to the adjusted volatility of the market.

Volatility; Data and Beta

Equations similar to (1.35) occur often in financial literature. Such equations may lead a financial analyst to believe that the market actually works that way. For any given asset and any given set of data, $\{R_{i,t}, R_{M,t}, R_{F,t} \,|\, t = 1, \ldots, T\}$, the regression equation (1.35) may or may not provide a good fit. The R-squared for the fitted line should be reasonably large before any serious consideration should be given to the alpha and beta. (Of course, "reasonably large" is subject to the analyst's interpretation.)

Beta is often used to assess the suitability of a particular stock relative to other normalized assets. Ignoring the fixed returns, and letting R_i be the

Annualized Excess Log Returns, 2017

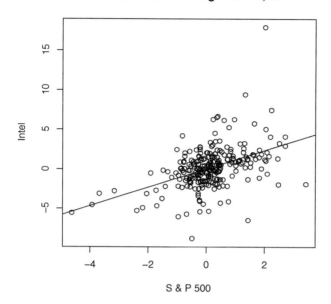

FIGURE 1.17
Scatterplot of INTC Returns and Those of S&P 500 for 2017, Both Adjusted for 3-Month T-Bill Interest; *Sources:* Yahoo Finance and FRED

returns of asset i and R_M be the returns of the market M, a frequently-used alternate formula for β_i is

$$\beta_i = \frac{\text{Cov}(R_i, R_M)}{\text{V}(R_M)}. \tag{1.36}$$

(This is not the same as a least-squares value for β_i in equation (1.35), but it is generally very close.)

A beta of 1 indicates that the asset's price moves in lock-step with an index of the particular market. A beta of less than 1 indicates that the security is less volatile than the market, and a beta greater than 1 indicates that the security's price is more volatile than the market. Of course, beta could also be negative for some asset. It is obviously negative for short positions.

The interpretation of *relative volatility* is applied to the absolute value of beta.

Mathematical expressions such as equations (1.35) and (1.36) convey a sense of exactitude and precision that do not exist in reality. Financial analysts will often cite a figure for beta or some other derived quantity as if it were a hard datum. Financial quantities, such as the returns in equation (1.35) or (1.36), are usually assigned values based on historical data, and the regression or the correlation depends on what historical period is used. Longer time periods provide better statistics under an assumption of constancy, but obviously suffer from greater effects of changes in the economic environment. The values also depend on another parameter of the data, the frequency. Higher frequency of measurements, such as daily returns versus weekly returns, yields more informative data, but the data contain more noise.

1.2.7 Derivative Assets

There are many kinds of financial transactions that can be made with any given financial asset. The main types of transaction of course are trades of the asset itself. The asset can also be hypothecated as collateral for some other transaction. A financial transaction may be made for a future transaction. (Committing to a future transaction is itself a transaction.)

The main type of financial assets we have discussed so far are stocks, and the main type of financial transactions on those assets that we have discussed are stock trades. There are many other types of financial transactions that can be made on the basis of the stocks, without trading the stocks. Buying other stocks using a loan made with the original stocks as collateral is an example. (This is buying "on margin", as mentioned above.)

Options on Stocks

Another type of transaction that can be made on an owned stock is a promise or agreement to sell that stock at some future time at some specified price, if the other party to the transaction wants to buy it at that price. The agreement is an *option*. Obviously, the owner of the stock would not make this agreement without some compensation, which of course would be paid by the other party to the transaction.

The contract representing this option is itself an asset, owned by the party that paid the owner of the stock. The option is a *derivative asset* or just a "derivative", because its value is derived from the value of an *underlying* asset.

Types of Stock Options

The type of derivative described above is a *call option*, because the holder of the option can "call in" the stock. Since the option involves stock, we also refer to it as a call stock option.

A similar stock option would be one in which the holder has the right to sell a stock at some future time at some specified price. The other party to the transaction has the obligation to buy it at that price if the holder of the option wishes to sell. This type of derivative is a *put option*, because the holder of the option can "put" the underlying stock to the other party.

Two additional terms common to both call and put stock options are *expiration* (or "expiry"), the date at which the option contract expires, and *strike price*, the price at which the stock could be bought or sold. Another term is *moneyness*, which at any give time, for a call option, is the price at the time minus the strike, and for a put option, is the strike minus the price at the time. If the moneyness is positive, the option is said to be *in the money*; otherwise, it is *out of the money* or *at the money*. If the price of the underlying is S, the *intrinsic value* of a call option with strike K is $(S - K)_+$, where

$$(x)_+ = \max(x, 0),$$

and the intrinsic value of a put with strike K is $(K - S)_+$. The *time value* of an option is the difference between the market value of the option and its intrinsic value. Except for deep out-of-the-money options, the time value is positive.

Options on specific stocks with fixed strikes and expirations are listed and traded on the CBOE and in other markets. A stock option is for a specific number of shares of the underlying; generally one option is for 100 shares. The number of units (usually stock shares) of the underlying per option is called the *multiplier*.

There are many variations of options. A basic difference is when an option can be *exercised* by the holder of the option. Most listed call and put stock options can be exercised at any time by the holder. Such options are called "American" options. Options that can only be exercised at expiry are called "European". Some with "in between" possibilities for exercise are called "Bermudan". Determining the fair market value of various types of options is a major concern in finance and we will mention some approaches in Section 3.1.13, but we will not consider the problem in any detail in this book.

There are listed call and put options on ETFs; in fact, some of the most actively traded options are on ETFs, especially on ETFs that track the major market indexes. There are no publicly traded derivatives on mutual funds and hedge funds (at least at this time).

Certain corporate actions can affect the notional price of each share of stock (see Section 1.2.5, beginning on page 45). The strike price of an option is adjusted to reflect some changes in the notional price of the underlying. An adjustment in price of the underlying due to stock splits or stock dividends (which are essentially the same as splits) is reflected in a one-to-one adjustment to the strike price of an option on that underlying. Ordinary cash dividends paid by the underlying, however, do not affect the strike prices of options.

There are also call and put options on things that have a value but that cannot actually be traded, such as an index. (The volume of trading in index options far exceeds the volume of trading in the options of any single stock.) Settlement of a stock option may involve trading of the stock; settlement of an option on a thing that cannot be traded requires a "cash settlement".

Prices of Options

Various models have been developed to relate the fair market price of an option to the corresponding strike price, the expiry, and the price of the underlying. Many of these models are based on stochastic differential equations that model a diffusion process. The most famous model of this type is the Black-Scholes-Merton model. Under certain assumptions that only partially correspond to observational reality, solutions to the differential equation yield a formula for the value of the option. The simplest formulas are called Black-Scholes formulas. The only quantity in the formula that is not observable is the volatility. These formulas themselves are fairly straightforward, but the development of the formulas is beyond the scope of this book.

There are various measures of the sensitivity of the price of an option to changes in the price of the underlying or changes in the length of time to expiry. These measures, which are generally partial derivatives of the price given by a formula, are called the "Greeks", because most of them are represented by the name of a Greek letter. For example, the *theta* is the sensitivity of the option to a small decrease in time to maturity; that is, the rate of decay of the time value of an option. The most commonly used Greek is the *delta*, which is the relative change in the price of an option to a change in the price of the underlying (see Exercise 1.14b). The delta of a call option is positive, and that of a put option is negative. The deltas of options with near-term expiries are generally larger in absolute values than options with longer terms. The deltas of options in the money approach ± 1, and options out of the money approach 0, as the time to expiration approaches 0. The *gamma* is the sensitivity of the delta to the change in the price of the underlying; that is, it is the second derivative of the option value to the price of the underlying.

1.2.8 Short Positions

In a *short sale* of stock, the seller borrows the stock from someone who owns it and sells it to a third party. (In practice, of course, the short seller only deals with a broker, and effectively just sells short to the market.) The short seller has a "short position" in the stock. We also use "short" as a verb: the stock is *shorted*.

At some point the short seller buys back the stock; that is, the seller "covers" the short. (Behind the scenes, the amount of stock that was sold short is bought on the open market, and that amount is returned to the original lender.)

Stocks that are viewed as "overpriced" are more likely to be shorted. The amount of short interest in a stock provides a lower barrier to the price declines, however, because the short seller must eventually cover the position. As mentioned previously, the short interest of some "high-flying" stocks may be over 30% of the total outstanding shares of the stocks.

Risks and Regulations of Shorts

The owner of an asset has a "long position" in the asset. The return on a clear long position is bounded below by -1 or -100%. The return on a short position has no lower bound, in theory.

For the duration of the short position in stock, the short seller must pay any cash dividends or return of capital issued on the shorted stock. (These go to the original holder of the stock.)

The short seller does not get the float of the proceeds of the sell, because those are held for security of the loan. Furthermore, the loan can be called at any time, at the discretion of the original holder. (In practice, other lenders can often be found. The available stock in the market that can be found for lending, that is, for shorting, may be very low. Such a stock is called "hard-to-borrow", and the broker may impose an interest on the market value of such stock.)

Because the market value of the stock could rise without limit, the potential loss to the short seller has no limit. We have used "risk" with a general meaning of variation or fluctuation, positive or negative, especially variation in returns generally. In the context of returns on short positions, "risk" often is used to refer to the unlimited negative return.

Short sales are tightly regulated by the SEC, by the brokerage firms, and by the markets (the exchanges). For income tax purposes in the US, all short sales are considered short-term, no matter the duration of the short position.

Shorts of Derivatives

A "sell to open" of an option establishes a short position in that asset. If the seller of a call option owns enough shares of the underlying to cover the option, the option is called a *covered call option* in the seller's account. Any portion of the call option not covered is called a *naked call option*. A naked call option has unlimited potential loss to the seller.

A put option is naked unless the seller has a short position in the underlying. In practice, most short put option positions are naked. A naked put option has limited potential loss to the seller because the price of the stock cannot go below zero.

The SEC and the markets regulate naked options only in a general way through margin requirements. Brokerage firms must approve each of their clients for different types of options trading, however.

A major practical difference in shorting options and shorting stocks is that the float of a short derivative sale is free (so long as an adequate margin position is maintained), whereas for a short stock sale it is not.

1.2.9   Portfolios of Assets: Diversification and Hedging

Financial assets are generally purchased in the hope that they will increase in price. Given a set of assets available for purchase, one may choose the subset of assets with the prospect for greatest increase in value. The simple solution to this clearly is the single asset with largest expected return.

The problem, of course, is that the return is not guaranteed. The fact that the return is variable, that is, it has risk, introduces another component to the objective. Not only do we want to maximize return, but we also want to minimize risk. With this objective, the solution may not be a single asset. This requires "diversification".

Portfolios; Returns and Risk

Suppose there are N individual assets available for trading. We approach the problem of selecting assets by defining *random variables*

$$R_1, R_2, \ldots, R_N$$

for the returns of the individual N available assets. We form a *portfolio* of the N assets in which at the beginning the proportional amount of the ith asset is w_i, where $\sum_i w_i = 1$. Hence, if the value of the portfolio at time t_0 is P_0, the amount of the ith asset at time t_0 is

$$P_{i,0} = w_i P_0. \tag{1.37}$$

A negative value of w_i indicates a short position in the ith asset.

The relative weights of the assets in a portfolio change over time. Portfolios are often *rebalanced* by buying and selling to maintain a target set of weights.

The question is how to choose the w_i so as to maximize expected return and to minimize risk. This is an *optimization problem* in which the *objective function* is some combination of total return and total risk, and the *decision variables* are the w_i. (Optimization problems arise in many areas of financial and statistical analyses. We will discuss optimization problems and methods for solving them in Section 4.3, beginning on page 358.)

We also should note that this approach entails some assumptions that may be unrealistic. One of the most obvious problems is the fact that economic conditions and the nature of the assets themselves are not constant.

We have used "risk" with a general meaning of variation or fluctuation. In the present context, we define *risk* as a specific measure of the variation. We define risk of the return to be the *standard deviation* of the random variable representing the return.

Next we want to combine individual returns and risks into the return and risk of a portfolio. How to combine individual returns depends on the meaning of "return", whether simple, log, multi-period, and so on. As we have seen, log returns can be aggregated easily over time (equation (1.26)), whereas aggregation of simple returns over multiple time periods depends on the compounding schedule, and in any event, is a power of a base return rate. Log returns, on the other hand, cannot easily be aggregated over multiple assets; log returns cannot be combined linearly (see Exercise 1.5). The single-period simple return of a portfolio, however, is a linear combination of individual simple returns, as we will now show. (That is why when we analyze returns on a portfolio, we generally use simple returns. Of course, the two types of returns are not very different.)

The combination of the individual risks into the risk of the portfolio is straightforward. It depends not only on the individual risks but also on the correlations between the returns.

Let us define $R_1, R_2, \ldots, R_N$ to be the single-period simple returns. Let P_0 be the total portfolio value at time t_0 and let P_1 be value time t_1. Let $P_{i,0}$ be the value of the ith asset at time t_0, and let $P_{i,1}$ be the value at time t_1.

If at time t_1 the value of the ith asset is $P_{i,1}$, then the value of that asset in the portfolio is

$$\frac{w_i P_0}{P_{i,0}} P_{i,1};$$

hence, the portfolio value at time t_1, P_1, is

$$P_1 = \sum_{i=1}^{N} \frac{w_i P_0}{P_{i,0}} P_{i,1}.$$

Using equation (1.18) on page 29, we can write $P_{i,1}/P_{i,0}$ as $(1 + R_i)$ so we have

$$
\begin{aligned}
P_1 &= \sum_{i=1}^{N} w_i P_0 \frac{P_{i,1}}{P_{i,0}} \\
&= \sum_{i=1}^{N} w_i P_0 (1 + R_i) \\
&= P_0 \left(1 + \sum_{i=1}^{N} w_i R_i \right),
\end{aligned}
\tag{1.38}
$$

because $\sum_{i=1}^{N} w_i = 1$. Thus, we have the return on the portfolio as the linear combination of the returns of the individual assets,

$$
\begin{aligned}
R_{\mathrm{p}} &= \sum_{i=1}^{N} w_i R_i \\
&= w^{\mathrm{T}} R,
\end{aligned}
\tag{1.39}
$$

where w is the vector of weights, $w = (w_1, w_2, \ldots, w_N)$, and R is the vector of returns, $R = (R_1, R_2, \ldots, R_N)$ and w^{T} represents the transpose of the vector w. (This is in the notation from page 4; note that I do not use a special font to denote a vector or matrix.)

This development is the same as that leading up to equation (1.27) on page 32, and it generalizes that result.

The expected return of the portfolio, from equation (1.7), is

$$\mathrm{E}(R_{\mathrm{P}}) = w^{\mathrm{T}} \mathrm{E}(R),
\tag{1.40}$$

which we sometimes write as μ_P.

If Σ is the conditional variance-covariance matrix of the N individual returns $R_1, R_2, \ldots, R_N$, that is, $\Sigma = V(R)$, then the variance of the portfolio return, from equation (1.8), is

$$V(R_p) = w^T \Sigma w, \tag{1.41}$$

which we sometimes write as σ_P^2. (Note the difference in the notation for the matrix Σ and the symbol for summation, Σ.)

The standard deviation of the portfolio, that is, the risk of the portfolio, is $\sqrt{V(R_p)}$, or σ_P.

In practice, the values in equations (1.40) and (1.41), that is, $E(R)$ and $V(R)$, are not known. Even if these were relatively constant values, they may not be easy to estimate. Nevertheless, these equations provide a basis for analysis.

An *optimal portfolio* has large $E(R_p)$ and small $V(R_p)$. These may be competing desiderata, however, and "large" and "small" are relative terms.

It is clear that the portfolio with maximum expected return has $w_i = 0$ for all except w_j, where R_j is the maximum of the vector R.

The portfolio can always be chosen so that the risk of the portfolio is no greater than the risk of any of the assets. (Just let $w_i = 0$ for all except w_j, where Σ_{jj} is the minimum along the diagonal of Σ, that is, the asset with the smallest risk.)

The portfolio with minimum risk, which is generally less than the risk of the single asset with the smallest risk, can be determined by differentiation since the risk function is twice-differentiable in the variables w_i. Constraints such as $\sum_{i=1}^{N} w_i = 1$ or $w_i \geq 0$ to prevent short selling can be handled by Lagrange multipliers (see page 362). General methods of constructing a portfolio with minimum risk subject to lower bounds on the total return are in the field of numerical optimization, specifically quadratic programming, which we will discuss on page 370.

From the theory of matrices, because the variance-covariance matrix Σ is nonnegative definite, we know that the quadratic form in equation (1.41) representing the variance has a minimum. A portfolio yielding that minimum is called the *minimum variance portfolio* (MVP). This is the portfolio with minimum risk, but the expected return of the MVP may not be optimal.

Another consideration is whether the portfolio can include short positions. Because of regulatory considerations or other issues, short positions may not be allowed. Such constraints may affect the optimal composition.

A Portfolio with Two Risky Assets; An Example

The issues concerning the returns of any portfolio can be addressed in a portfolio consisting of just two risky assets.

We consider two assets with simple returns R_1 and R_2, and a portfolio containing the assets in the relative amounts w_1 and w_2, where $w_1 + w_2 = 1$.

We assume the returns R_1 and R_2 are random variables with means μ_1 and μ_2, variances σ_1^2 and σ_2^2, and correlation ρ (hence, covariance $\rho\sigma_1\sigma_2$). The mean return of the portfolio, from equation (1.41), is

$$\mu_P = w_1\mu_1 + w_2\mu_2, \tag{1.42}$$

and the variance, from equation (1.41), is

$$\sigma_P^2 = w_1^2\sigma_1^2 + w_2^2\sigma_2^2 + 2w_1w_2\rho\sigma_1\sigma_2. \tag{1.43}$$

If $\mu_2 \geq \mu_1$, the mean return is maximized if $w_1 = 0$ and $w_2 = 1$. We see that for $\rho < 1$, the risk is minimized if

$$w_1 = \frac{\sigma_2^2 - \rho\sigma_1\sigma_2}{\sigma_1^2 + \sigma_2^2 - 2\rho\sigma_1\sigma_2}. \tag{1.44}$$

(You are to show this in Exercise 1.11a; differentiate and set to 0.) This value of w_1, together with $1 - w_1$, defines the minimum variance portfolio.

For any two assets, if we know μ_1, μ_2, σ_1^2, σ_2^2, and ρ for the returns of those assets, the analysis is straightforward. In practice, of course, these are unknown quantities, so we estimate these parameters using observed data and/or statistical models for the returns.

To proceed with an example, let us consider two assets with $\mu_1 = 0.000184$, $\mu_2 = 0.00178$, $\sigma_1^2 = 0.0000800$, $\sigma_2^2 = 0.000308$, and $\rho = 0.285$.

These values are the sample means, variances, and correlation for the daily returns of INTC and MSFT for the period January 1, 2017, through September 30, 2017, so this example is the analysis of a portfolio consisting of just INTC and MSFT, using observed historical returns.

We will consider only long positions, and so for each value of w_1 between 0 and 1, we compute the expected return and the risk of the corresponding portfolio. The parametric curve in w_1, with $0 \leq w_1 \leq 1$, is shown in Figure 1.18 (the curve with a red and a blue portion). This parametric curve represents the complete set of feasible points for (σ_P, μ_P) under these conditions.

The parametric curve in Figure 1.18, which is the parabolic curve lying on its side, is the *feasible set* for all combinations (σ_P, μ_P), given the values of the means, risks, and covariance. The minimum variance portfolio corresponds to the point on the curve with minimum risk, as shown.

For the parabolic curve in Figure 1.18, the correlation between the returns of the two assets is $\rho = 0.285$. As the correlation ρ increases toward 1, the curve representing the feasible set becomes flatter, and for $\rho = 1$ is the straight line shown in green in the figure. For a correlation of 1, it is not possible to reduce the risk by mixing the assets. The minimum variance portfolio in that case would consist of MSFT only, and it happens in that example that the MVP is also the one with maximum expected returns. (In Exercise A1.14, you

Return/Risk Set for Portfolio of INTC and MSFT

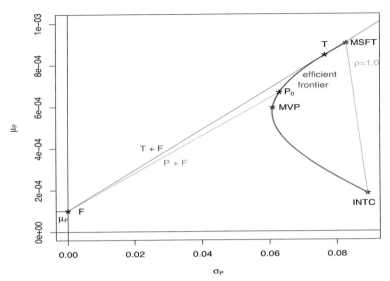

FIGURE 1.18
Expected Return versus Risk for Portfolio of INTC and MSFT

are asked to produce some similar curves with various values of the correlation coefficient.)

Notice that for the points in the feasible set with expected returns less than that of the MVP (the portion of the curve in red), there are other points with larger expected returns with the same risk (the portion of the curve in blue).

A portfolio corresponding to the feasible curve with expected returns greater than that of the MVP is called an *efficient portfolio,* and the portion of the feasible curve above the MVP point is called the *efficient frontier* (the blue curve in Figure 1.18).

A Portfolio with a Risk-Free Asset

We now introduce a "risk-free" asset into the portfolio of the two risky assets. The concept of a risk-free asset as a useful abstraction in financial analysis. It is an asset with a fixed return; that is, the standard deviation of the return is 0. For practical analyses, we commonly use the current rate on a three-month US Treasury Bill as the "risk-free interest rate".

Suppose that the daily return on an available risk-free asset is 0.0001. It is shown as F in Figure 1.18, and $\mu_F = 0.0001$.

Now, consider any efficient portfolio consisting of the two risky assets. A typical one is called "P_0" and shown in Figure 1.18. The line connecting P_0 and the risk-free return on the vertical axis represents the feasible points in a portfolio consisting of a mix of the portfolio P_0 and the risk-free asset. The risk of any such combined portfolio is less than the risk of portfolio P_0, but its return is also smaller.

Now, consider a different efficient portfolio. Identify the efficient portfolio such that the line representing the feasible points of the combination of that portfolio and the risk-free asset is tangent to the original feasible curve for the two risky assets. The portfolio at this point of tangency is called the *tangency portfolio*. The line connecting the risk-free asset return and the tangency portfolio is called the *capital market line*.

It is easily seen that, for any fixed risk, the return of the tangency portfolio is greater than that of any efficient portfolio (consisting of just the risky assets). The portion of the line extending beyond the point of tangency (which is hardly distinguishable from the feasible set for the two-asset portfolio in the figure) represents combined portfolios with short positions.

The slope of a line in the (σ_P, μ_P) plane connecting F to any efficient portfolio P_0 is an important characteristic of that portfolio. The slope of such a line is

$$\frac{\mu_{R_{P_0}} - \mu_F}{\sigma_{R_{P_0}}}. \tag{1.45}$$

This quantity is called *Sharpe ratio*. An efficient portfolio with a larger Sharpe ratio has a larger expected return than an efficient portfolio with a smaller Sharpe ratio. The tangency portfolio has the largest Sharpe ratio of any efficient portfolio. There are other, similar measures used to evaluate portfolios. A common one is the *Treynor ratio*, in which the portfolio beta is used instead of the risk.

Notice that the optimal portfolio consisting of the two risky assets plus the risk-free asset are all on the capital market line, and thus, all have the same mix of the two risky assets, namely, that of the tangency portfolio.

The concept of a risk-free asset with a fixed return leads to the concept of *excess returns*, which we encountered on page 32 and later on page 61 in discussion of the market model. The excess return of an asset i is the return R_i minus the return of a risk-free asset, R_F. Excess returns are central to the capital asset pricing model (CAPM) and to the capital market line in portfolio analysis.

The expected excess return is $E(R_i - R_F) = \mu_i - \mu_F$, in the notation used above.

The weights for the tangency portfolio are determined so that the line from the point $(0, \mu_F)$ in the (σ_P, μ_P) plane intercepts the efficient frontier in exactly one point. That is the point corresponding to the tangency portfolio, and the line is the capital market line. If $\rho < 1$, the weight for asset 1 in the tangency portfolio is

$$w_{T_1} = \frac{(\mu_1 - \mu_F)\sigma_1^2 - (\mu_2 - \mu_F)\rho\sigma_1\sigma_2}{(\mu_1 - \mu_F)\sigma_1^2 + (\mu_2 - \mu_F)\sigma_2^2 - (\mu_1 + \mu_2 - 2\mu_F)\rho\sigma_1\sigma_2}. \tag{1.46}$$

This is somewhat similar to the weights for the MVP in equation (1.44), except the volatilities are adjusted by the expected excess returns. Knowing the weights for the tangency portfolio, it is a trivial matter to work out the expected return μ_T and the risk σ_T of the tangency portfolio (see Exercise 1.11b). It is also instructive to consider the degenerate case in which $\rho = 1$, as you are asked to do in Exercise 1.11c.

These analyses extend to portfolios consisting of more than two assets. The relevant space is still the two-dimensional (σ_P, μ_P) plane, but the parametric curve is constructed as the maximum return of any portfolio with a given risk. The determination of the feasible set is a quadratic programming problem with various constraints on the w_is mainly depending on whether or not short selling is allowed.

Because the models for portfolio analysis involve properties of the assets such as expected return and risk, which are unknown, the analyses are subject to the errors and approximations incurred in supplying actual values for these properties. Several suggestions have been made for improving the statistical analyses, including, for example, use of the bootstrap. We will not pursue these more advanced topics here, however.

Hedges

For a given position in an asset, a *hedge* is another position whose price change is expected to be negatively correlated with the price change of the given position. Either position is a hedge against the other position.

Either position may be long or short. A simple example is two long positions, one in a given stock and another in put options on that stock. Another example is a long position in a given stock and a short position in call options on that stock (see also Exercise 1.12).

If the returns on two assets are positively correlated, a combination of opposing positions is called a *cross hedge*.

We also speak of an asset as a hedge against another asset if same-side positions in each constitute a hedge.

Another variation on the term "hedge" applies to a situation in which one position is not presently realized. Commodity futures contracts can be hedges against *future positions*. For example, an airline company may purchase futures on jet fuel, but the company has no present offsetting position in jet fuel.

Hedges reduce risk, but they also often reduce expected gain (because if one position increases in value, the other decreases). By purchasing futures on jet fuel, an airline controls its future expenses, but it may incur either a gain or a loss on the purchase, because the price of the jet fuel may be more or less in the future.

The idea behind a hedge can be seen very clearly in equation (1.43), expressing the variance of the total simple return of two assets with simple returns R_1 and R_2 and held in relative amounts w_1 and w_2 as

$$w_1^2 \mathrm{V}(R_1) + w_2^2 \mathrm{V}(R_2) + 2w_1 w_2 \mathrm{Cov}(R_1, R_2).$$

Even if the covariance is positive, the combination of the two assets may have smaller volatility than either asset alone, but hedging is most effective when $\text{Cov}(R_1, R_2)$ is negative. The relative amounts of w_1 and w_2 determine the amount of the reduction in risk. The ratio w_2/w_1 is called the *hedge ratio*.

A common strategy by stock market participants is "covered call writing"; that is, going short on call options on a stock with a long position in the stock sufficient to fulfill the calls.

As an example, consider the owner of 100 shares of Microsoft Corporation, MSFT, who on October 24, 2016, sells 1 April 2017 65 MSFT call option. At the time of the sale MSFT had just traded at 61.38. The option was sold at 1.35 ($135.00, because the multiplier is 100). The seller of the option is short the option and long the underlying. The short option position is fully offset by the long position in the underlying, so the call is covered. This is a hedge on the MSFT position because if the price of MSFT declines, the price of the option will likewise decline, so the net position does not change as much as it would with only one of the two assets. If the price of MSFT does not go beyond 65, the seller realizes a $135.00 gain at option expiration, minus brokerage costs.

Another common type of hedge by the holder of a long stock position is the purchase of a put option on the stock. The option is called a "married put". For example, the owner of the 100 shares of MSFT on October 24, 2016, could buy 1 April 52.5 MSFT put option for 1.15. (Again, the multiplier is 100.) The owner is now long both the option and the underlying. This is a hedge on the MSFT position because if the price of MSFT declines, the price of the option will increase, so the net position does not change as much as it would with only one of the two assets. The put option is similar to an insurance policy on the MSFT stock.

Notice that in both of these cases, if the price of MSFT would rise significantly, the hedger would have made more gain without the hedge.

Hedges are also easily formed by other types of option combinations. For example, a "call spread" is a long position in a call option at a strike price of K_1 and a short position of an equal number of call options in the same underlying, the same expiration, but at a strike price of K_2. The relationship between the strike prices K_1 and K_2 provides various possibilities for call spreads. A "put spread" is a similar combination of put options.

Another type of spread is a "calendar spread", which is either a long and a short call option, or a long and a short put option with the same strike price but with different expirations. In a long position in a calendar spread, the holder is long the longer-dated option and short the shorter-dated option, whether they are both calls or both puts. In a short position in a calendar spread, the holder is long the shorter-dated option and short the longer-dated option. Obviously, to establish a long position in a calendar spread requires a net expenditure because the premium on the longer-dated option is larger than the premium on the shorter-dated option.

Other common option combinations are "straddles" and "strangles". A straddle consists of same-side positions in equal numbers of puts and calls on the same underlying with the same strikes and same expirations. Two common modifications of a straddle are a "strip", which is the same as a straddle, except the number of puts is twice the number of calls, and a "strap", in which the number of calls is twice the number of puts. A strangle consists of same-side positions in equal numbers of puts and calls with higher strikes than the puts on the same underlying with the same expirations. A straddle or strangle position may be long or short. Both straddles and strangles are "market neutral" instruments because the profit/loss profile is the same if the underlying increases or decreases in price.

There are many other combinations of three or more options, but we will not discuss them here. Combinations, in general, are composed of positions that are hedges against one another.

Other types of hedges are for portfolios. For example, a long position in a portfolio of large capitalization stocks could be hedged by buying puts on an ETF that tracks the S&P 500, or by buying an ETF that tracks the inverse of the S&P 500, such as SH or SDS (see page 59).

As the asset mix in a portfolio changes due to purchases and sales, the hedging relationships may change. A portfolio management strategy will often involve *dynamic hedging*, which means transactions that preserve hedged relationships in the presence of other transactions.

In the use of options to hedge positions in the underlying assets, the relative values of the option and underlying positions may change. The relative change in the price of an option to a change in the price of the underlying is called the *delta* of the option. In general, the delta depends on the type of option (call or put), the price of the underlying, the strike price, and the time to expiry; and it will be different for each option. See Exercise 1.14.

Another use of the term "hedge" that is very common is in "hedge fund". A hedge fund is a type of fund defined by the regulations of the SEC that govern sales of the fund. The positions in a hedge fund may or may not be strongly hedged.

Pairs Trading

Two assets whose prices are correlated, either positively or negatively, may present opportunities for trading strategies that have expected positive returns, given certain assumptions.

The idea is to assume some simple model for the price movement of each asset. This may be just some kind of trend line, such as a moving average, which in general is a curved line, or a simple linear regression line. Once this trend line is fitted, using historical data, we assume that the near-term price movements will follow the trend line.

For two stocks whose prices are strongly related, we may assume that their prices at a given point in time should have the same relationship to their respective moving averages. If, however, one price is high relative to its trend line ("overpriced"), and the other is low relative to its trend line ("underpriced"), we may expect the one that is low will gain in price more rapidly than the one that is high. This is the idea behind *pairs trading.* It might be a good strategy to buy the one that is high and sell short the one that is low.

There are many sophisticated methods of identifying under- and overpriced stocks. ("Sophisticated" does not mean that the method works any better than a simple guess.)

1.3   Frequency Distributions of Returns

Given any set of data, one of the first ways of understanding the data is to study its frequency distribution; that is, how the different observations are spread over the range of values, and how frequently the observations take on specific values or occur within specific ranges.

Histograms

One of the easiest ways to get an overview of the frequency distribution is by means of a histogram.

A histogram is a graph with lines or bars whose sizes indicate the relative frequency of observed or expected values. A histogram of the daily unadjusted closing prices of INTC in the period January 1, 2017, through September 30, 2017, is shown in Figure 1.19, for example. (These are the same data as shown in the time series graph in the upper left side of Figure 1.2 on page 9.)

Histograms represent either frequencies (counts) or relative frequencies of bins of a given set of data, without regard to the order in which the data occurred. The bins in the histogram 1.19 represent values between 33.0 and 33.5, between 33.5 and 34.0, and so on. A histogram showing relative frequencies, as in the one in Figure 1.19, are based on relative frequencies that account for both the width and the length of the bins in such a way that the sum of the products of the relative frequencies and the areas of the bins is 1.

While the histogram in Figure 1.19 may be interesting, it misses the main point for these data. The data are a time series. The graphs in Figure 1.2 or in Figure 1.11 are more meaningful than the histogram because they show the data over time.

The frequency distribution of a set of returns, however, may be very informative. As we mentioned earlier, we are more often interested in the returns than in the stock prices themselves. For the remainder of this section, we will consider how stock returns are distributed.

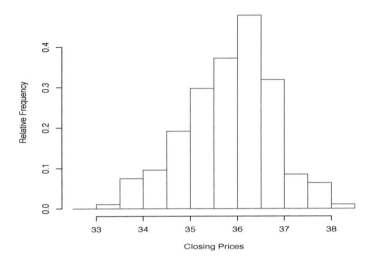

FIGURE 1.19
Frequency of Unadjusted Daily Closes INTC for the Period January 1, 2017 through September 30, 2017; *Source:* Yahoo Finance

Histograms of Returns

A histogram shows the frequency distribution of the returns without any consideration of the order in which they occurred. Figure 1.20 is a histogram of the daily log returns for INTC, computed from the same data as in Figure 1.19. The frequency distribution of the returns is of more interest than the distribution of prices.

A histogram gives us a simple view of the shape of a frequency distribution. It is formed by dividing the range of the data into bins, and then counting how many data points in the given sample fall into each of the bins. Of course we would get a slightly different histogram if we were to change the number of bins or shift them slightly in one direction or the other.

Jumps

One thing that is quickly apparent in looking at graphs of financial data is that there are occasional extreme changes. Although a histogram smooths out frequency distributions, it is obvious from the histogram in Figure 1.20 that the return is occasionally much more extreme than "average". We can also see this dramatically in the time series plot in Figure 1.29 on page 99.

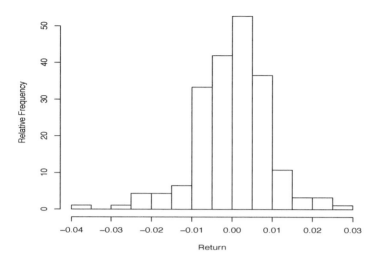

FIGURE 1.20
Frequency of Daily Simple Rates of Return for INTC for the Period January
1, 2017 through September 30, 2017; *Source:* Yahoo Finance

These extreme events are called "jumps" ("bear jump", "correction", or
"crash", if the asset price decreases; and "bull jump" or "gap up" if the price
increases). The jumps correspond to *outliers* in the statistical distribution of
returns.

Another Graphical View of the Distribution of Returns

An alternative view of the shape of a frequency distribution could be formed
by focusing on one point in the range and counting how many data points are
near that chosen point. How "near" is a similar decision to the question in a
histogram of how many bins, that is, how wide are the bins. This is a general
issue of a "smoothing parameter" that we will encounter in other applications
of statistical data analysis.

A simple way of proceeding following the approach suggested above would
be to choose some fixed width for the bins, form a bin centered at each point,
count the number of data points in that bin, and then scale appropriately
(similar in the scaling done in the histogram). This approach does not require
that the points that we choose to form the bin around be data points themselves; in fact, it may be better just to choose a large number of arbitrary, but
evenly spaced, points in the range of the data. Continuing with this approach,
we would get a picture of the overall density of the distribution, and we could
draw a curve through the individual values.

The approach outlined above is called "nonparametric kernel density estimation". There are many variations in this general approach and there are many theoretical results supporting specific choices. We will briefly consider some of the issues in Section 2.3 beginning on page 217.

Figure 1.21 is a graph of the density using a kernel density estimator for the same set of the daily log returns for INTC shown in the histogram in Figure 1.20. Different choices in the method would yield different graphs. It is interesting to note the small humps on the sides of the figure. The "tail properties" of distributions of rates of return will be of particular interest.

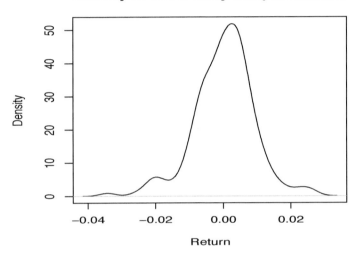

Density of INTC Daily Simple Returns

FIGURE 1.21
Density of Daily Simple Rates of Return for INTC for the Period January 1, 2017 through September 30, 2017; *Source:* Yahoo Finance

1.3.1 Location and Scale

Two of the most important characteristics of a frequency distribution of a sample of data are its location and shape. The location in general is determined by the "center" of the distribution, which can be identified as the mean of the sample or possibly its median or mode. The shape of a sample is more complicated to describe. The simplest descriptor of shape is the scale, which is a measure of how spread out the data are. Two other measures of shape are skewness and kurtosis, which we will discuss below.

The terms we define here, such as "mean" and "variance", refer to a given sample of data. To emphasize this, we sometimes use phrases such as "sample mean" and so on. There are corresponding terms for properties of a "population", which is a mathematical abstraction. We have used some of these terms already, but we will define them in Chapter 3. Although we assume the reader is generally familiar with these terms and we may use them prior to defining them, for completeness, we will give precise definitions at various points.

The *sample mean*, which is the *first sample moment*, is

$$\bar{r} = \sum_{i=1}^{n} r_i/n. \tag{1.47}$$

For the INTC daily returns over the period January 1 through September 30, 2017, $\bar{r} = 0.000251$.

The *second central sample moment*,

$$\sum_{i=1}^{n}(r_i - \bar{r})^2/n, \tag{1.48}$$

gives an indication of the spread of the sample in the vector r. A modification of the second central sample moment is the *sample variance*,

$$
\begin{aligned}
s^2 &= \frac{1}{n-1}\sum_{i=1}^{n}(r_i - \bar{r})^2 \\
&= \frac{1}{n-1}\left(\sum_{i=1}^{n} r_i^2 - n\bar{r}^2\right)
\end{aligned} \tag{1.49}
$$

The nonnegative square root of the sample variance is the *sample standard deviation*.

For a sample denoted by $r_1, \ldots, r_n$, we often use the notation "s_r^2" for the sample variance. We may also denote this as "$V(r)$".

The sample variance is usually the preferred measure because under an assumption of random sampling, it is an unbiased estimator of the population variance. The standard deviation, however, may be more relevant because it is in the same units as the data themselves. The standard deviation is a measure of the *scale* of data or variables. If all data are multiplied by a fixed amount, that is, the data are scaled by that amount, the standard deviation of the scaled data is the scale factor times the standard deviation of the original dataset.

Without some reference value, we cannot say that a given value of the variance or standard deviation is "large" or "small". For the INTC daily log returns, for example, we have $s^2 = 0.0000795$ and $s = 0.00887$, but whether these values indicate that INTC is a "risky" or "volatile" asset depends on the variance or standard deviation of returns of other assets.

While the first two moments provide important information about the sample, other moments are required to get a better picture of the shape of the distribution.

1.3.2 Skewness

The third central sample moment,

$$\sum_{i=1}^{n}(r_i - \bar{r})^3/n, \tag{1.50}$$

is a measure of how symmetric the frequency distribution is. Scaling by the third power of the standard deviation removes the units, and we define the sample skewness as the pure number

$$\frac{\sum_{i=1}^{n}(r_i - \bar{r})^3/n}{s^3}. \tag{1.51}$$

The third central moment indicates symmetry or lack thereof. A positive skewness indicates that the positive tail (on the right-hand side) is longer than the negative tail. A negative skewness indicates the opposite. (There are other, slightly different definitions of the sample skewness; see Joanes and Gill, 1998.)

For the INTC daily simple returns, the skewness is -0.3728. The histogram in Figure 1.20 seems to be more-or-less symmetric, but there does appear to be a very slight skewness to the left as the sample skewness indicates.

1.3.3 Kurtosis

The fourth central sample moment,

$$\sum_{i=1}^{n}(r_i - \bar{r})^4/n, \tag{1.52}$$

is a measure of how the distribution is spread out. Like the variance or the standard deviation, the fourth moment by itself does not tell us very much. Combined with the variance or standard deviation, however, the fourth central moment does tell us a lot about the shape of the distribution. While the second central moment is a measure of *how much* the distribution is "spread out", the fourth central moment indicates just *how* it is spread out.

Scaling by the square of the variance (the fourth power of the standard deviation) removes the units. The kurtosis, which is a pure number, defined as

$$\frac{\sum_{i=1}^{n}(r_i - \bar{r})^4/n}{s^4}, \tag{1.53}$$

indicates how peaked or flat the distribution is. (There are other, slightly different definitions of the sample kurtosis; see Joanes and Gill, 1998.)

For the INTC daily simple returns, the kurtosis is 4.627.

Because the exact kurtosis of an abstract normal distribution is 3, we define *excess kurtosis* as the kurtosis in the formula above minus 3. This allows simple comparisons with the normal distribution; a negative excess kurtosis means the sample is flatter than a normal distribution, and a positive excess kurtosis means the sample is more peaked than a normal distribution Many authors simply use "kurtosis" to mean "excess kurtosis". The excess kurtosis of the INTC daily simple returns is 1.627.

A histogram of a sample with a smaller kurtosis would be flatter than the histogram in Figure 1.20, and the tails would be less elongated. A histogram of a sample with a larger kurtosis would be more peaked, and the tails would be more elongated.

1.3.4 Multivariate Data

Multivariate data are data with multiple related variables or measurements. In the parlance of statistics, an "observation" consists of the observed values of each of the "variables", in a fixed order. In other disciplines, an observation may be called a "case", and the variables are called "features" and their values, "attributes". For variables x, y, and z, we may denote the ith observation as the vector

$$(x_i, y_i, z_i).$$

The means and variances and other moments of the individual variables have the same definitions and are of the same relevance as they are for univariate data. We also use similar notation; for example, $\bar{x}$ is the mean of x. We may use subscripts to denote other quantities; for example, s_x^2 may denote the variance of x. In multivariate data, however, we are often more interested in the associations among the variables. The simplest and most commonly-used measure of this association is for pairs of variables. The *covariance* between two variables x and y is

$$s_{xy} = \sum_{i=1}^{n} (x_i - \bar{x})(y_i - \bar{y})/(n-1). \qquad (1.54)$$

This quantity is also called the *sample covariance* to emphasize that it is computed from an observed sample, rather than being a related property of a statistical model. We may also denote this as "$\text{Cov}(x, y)$".

The covariance arises in the variance of the sum of two random variables, a fact that we used on page 70 in reference to the sum of two simple returns. Given the two variables x and y, let $w = x + y$. Using equations (1.49) and (1.54), we can see that the variance of w is

$$s_w^2 = s_x^2 + s_y^2 + 2s_{xy}. \qquad (1.55)$$

The covariance depends on the scales of the two variables. A related measure that is independent of the scales is the *correlation* between two variables x and y,

$$r_{xy} = \frac{s_{xy}}{s_x s_y}, \qquad (1.56)$$

if $s_x \neq 0$ and $s_y \neq 0$, otherwise $r_{xy} = 0$. We may also denote the correlation between x and y as "$\text{Cor}(x, y)$". The sample correlation is sometimes called the "Pearson correlation coefficient".

The correlation ranges between -1 for variables related by an exact negative multiplicative constant (if plotted on a graph, they would fall along a straight line with a negative slope) and 1 for variables related by an exact positive multiplicative constant.

In Sections 1.1.1 and 1.1.2, we referred to "autocorrelations" and to the autocorrelation function (ACF) of a time series. They are similar to the correlations of equation (1.54), and we will discuss them more formally in Chapter 5.

In discussing Figure 1.10 showing the relative performance of three indexes from 1987 through 2017, we remarked on the "strong correlations" between the DJIA and the S&P 500. (In that instance, we were using "correlation" in a non-technical sense.) The technical meaning, using the definition in equation (1.54), yields similar substantive conclusions. The actual correlation of the daily closes of the DJIA and the S&P 500 for that period is 0.993.

For more than two variables, we often compute the *sample variance-covariance matrix* by extending the formula (1.54) to all pairs. This is a square, symmetric matrix. We also form the *sample correlation* matrix formula (1.56). An example is given in Table 1.8. Because the matrix is symmetric, we sometimes display only a triangular portion. Also, the diagonal of the matrix does not convey any information because it is always 1.

TABLE 1.8
Correlations of DJIA, S&P 500, and Nasdaq for 1987-2017

	DJIA	S&P 500	Nasdaq
DJIA	1.000	0.993	0.960
S&P 500	0.993	1.000	0.975
Nasdaq	0.960	0.975	1.000

Correlations are bivariate statistics, between two variables, no matter how many variables are involved.

There are other measures of association among (usually between) variables. In Section 1.3.8, we will mention two alternatives to the correlation defined in equation (1.56). Another way of expressing the relationships among variables is by use of copulas, which we will discuss in Section 3.1.7, beginning on page 264.

Correlation Between Prices and Between Returns

Instead of the prices, of course, our interest is generally in the returns.

Although if the prices of two assets are positively correlated, their returns are also often positively correlated, this is by no means necessarily the case.

Sometimes it is useful to examine our statistical measures on a small simple dataset. Consider two assets A1 and A2, and suppose for 14 periods their prices were

$$A1 : \$1, \$3, \$2, \$4, \$3, \$5, \$4, \$6, \$5, \$7, \$6, \$8, \$7, \$9$$

and

$$A2 : \$2, \$1, \$3, \$2, \$4, \$3, \$5, \$4, \$6, \$5, \$7, \$6, \$8, \$7$$

(Such integral prices may not be realistic, but that's not the point.) We see that the prices are generally increasing together, and indeed, their correlation is 0.759. The returns, however, tend to be in different directions from one period to the next; when one goes up, the other goes down. The correlation of the simple returns is -0.731.

Now suppose that we had observed the prices only every other period. We have

$$A1 : \$1, \quad \$2, \quad \$3, \quad \$4, \quad \$5, \quad \$6, \quad \$7$$

and

$$A2 : \$2, \quad \$3, \quad \$4, \quad \$5, \quad \$6, \quad \$7, \quad \$8$$

The correlation of the prices is now 1.000, and the returns are identical. (Because the returns are constant, the correlation is 0; the s_x and s_y in equation (1.54) are 0.) Returns over periods of different lengths may behave differently. Although we will not discuss it in this book, we have referred to the special behavior of returns in very short intervals.

Table 1.9 shows the correlations of the daily log returns of the three indexes shown in Table 1.8. We have also added the daily returns of the stock of Intel.

TABLE 1.9
Correlations of Daily Returns of DJIA, S&P 500, Nasdaq, and INTC for 1987–2017 (Compare Table 1.8)

	DJIA	S&P 500	Nasdaq	INTC
DJIA	1.000	0.966	0.763	0.561
S&P 500	0.966	1.000	0.851	0.606
Nasdaq	0.764	0.851	1.000	0.682
INTC	0.561	0.606	0.682	1.000

The correlations in Table 1.9 are those of the raw log returns, that is, returns that are not adjusted for a risk-free return, as we did on page 61.

We might ask whether or not adjusting the returns by the risk-free return makes any difference in the correlations. From the definition of correlation given in equation (1.56), we see that for three vectors x, y, and r,

$$\text{Cor}(x - r, y - r) = \text{Cor}(x, y) \quad \text{if and only if } \text{Cor}(x, r) = 0 \text{ and } \text{Cor}(y, r) = 0.$$
$$(1.57)$$

If the risk-free return is constant, for example, adjusting by it makes no difference in the correlation. Another thing to note is that the correlation between log returns is the same whether or not they are annualized, because if a is a constant,

$$\text{Cor}(ax, ay) = \text{Cor}(x, y).$$
$$(1.58)$$

(Recall that to annualize daily log returns, we merely multiply by a constant, usually 253 for daily yields on stocks. That does not change the correlations.)

The correlation between the annualized excess returns of Intel and the S&P 500 over the period 2017 shown in Figure 1.17 on page 62 is 0.457. The correlation over this period without this adjustment is 0.455 (which is slightly lower than the correlation of 0.606 over the longer period 1987 to 2017 shown in Table 1.9).

Multivariate Distributions of Returns

No matter how many variables are involved, just as with correlations, it is usually easier to consider bivariate relationships.

Scatterplots, as in Figure 1.17 are useful to see relationships between pairs of variables. When there are several variables, relationships between pairs can be displayed conveniently in a square array.

Figure 1.22 shows a "scatterplot matrix" of the returns on the indexes shown in Table 1.9 along with the returns on Intel and the gold bullion ETF, GLD. Figure 1.22 shows data just for 2017, whereas Table 1.9 is based on data for 1987 through 2017. The shorter period was used just because of a drawback of scatterplots in general; if there are too many points, they become obscured because of overplotting.

Figure 1.22 also shows scatterplots with GLD, the gold bullion SPDF ETF. GLD was not formed until 2004, so it is not included in Table 1.9. The correlations of the daily returns of GLD and DJIA, S&P 500, Nasdaq Composite, and INTC for the period 2004-11-18 through 2017-12-31, respectively, are 0.018, 0.036, 0.017, and 0.0222. Compare these with the much higher correlations in Table 1.9.

The scatterplots indicate bivariate regions of higher and lower frequencies. Just as with univariate data where we represent the relative frequencies with histograms or smooth density curves (Figure 1.21, for example), we can construct a *surface* that represents the bivariate density. (We will discuss the methods in Section 2.3.)

Figure 1.23 shows four different views of the fitted bivariate density surface for Nasdaq and Intel daily returns for 2017. (This is the same data used in the

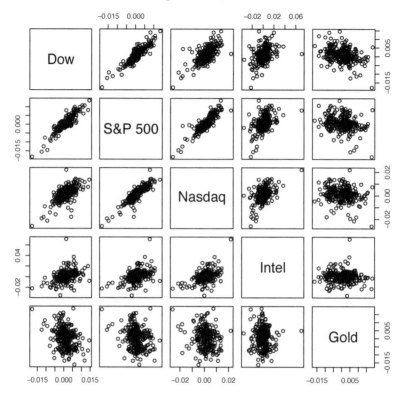

FIGURE 1.22
Scatterplot Matrix of Daily Log Returns (Unadjusted) for 2017; *Source:* Yahoo
Finance

two scatterplots in the lower right side of Figure 1.22.) The two contour plots
in the top row are different only in that the data are plotted in one graph and
not in the other. The contour plot on the bottom right is called an "image
plot".

The heavy tails of these returns cause the outer contours to be rather
ragged. (This is most clearly seen in the upper left-hand graph.)

We note that the outlying point (plotted in the upper right-hand graph)
causes the scale on the displays to obscure some of the information. In this
case, it may be justified to remove that observation in all views, in order to
get a better picture of the frequency density.

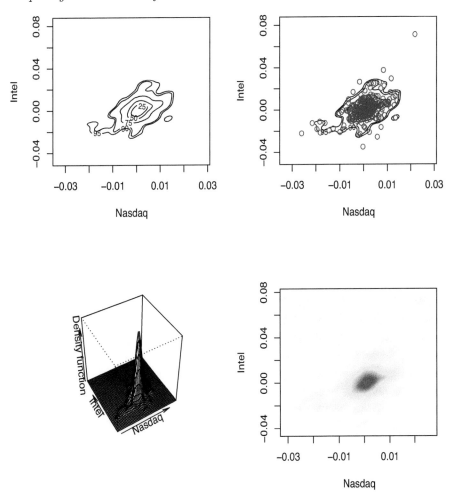

FIGURE 1.23
Four Views of the Bivariate Frequency Density of Nasdaq and Intel Daily Log
Returns (Unadjusted) for 2017; *Source:* Yahoo Finance

We also note that plotting the data points as in the upper right-hand graph
is probably not a good idea because they obscure the contours.

1.3.5 The Normal Distribution

The *normal probability distribution* is a mathematical model that is the basis
for many statistical procedures. There are several reasons for this. Among
other reasons, firstly, the normal distribution is mathematically tractable; it is
easy to derive properties of the distribution and properties of transformations

of normal random variables. Secondly, there are many processes in nature and in economics that behave somewhat like a normal distribution.

The normal distribution is also called the *Gaussian distribution*. I will use "normal", and I will discuss some of its properties in more detail beginning on page 287.

More properly, we refer to the normal or Gaussian *family* of distributions. A member of this family of distributions may have any mean and any variance. Those two parameters fully characterize a specific member of this family. We specify a normal distribution by specifying these two values. We sometimes denote a specific normal distribution by $N(\mu, \sigma^2)$, given values of μ and σ^2. The skewness of any normal distribution is 0, that is, it is symmetric, and the kurtosis is 3. These values are based on mathematical expectation, and for any sample from a normal distribution, the sample skewness and kurtosis, as defined in equations (1.51) and (1.53) would likely be close to this. Because the kurtosis of a normal distribution is 3, we define "excess kurtosis" as the kurtosis in the formula above minus 3.

Because many standard statistical procedures are based on the normal distribution, it is often important to determine whether the frequency distribution of a given sample is consistent with the normal mathematical model. One way of doing this is to compute the skewness and the kurtosis. If the sample is from a normal distribution, both the skewness and the excess kurtosis would be close to 0. ("Close" is not defined here.)

There are formal statistical tests of the hypothesis that the sample came from a normal distribution. These are "goodness-of-fit" tests, and we will discuss some in Section 4.6.1.

Another way to assess whether the distribution of the sample is similar to that of a sample from a normal distribution is to compare a graphical display of the sample distribution, such as Figure 1.20 or Figure 1.21, with a graph of the probability density of the appropriate member of the family of normal distributions. The appropriate normal distribution would be one whose mean is the mean of the sample, and whose variance is the variance of the sample.

The histogram in Figure 1.20 is shown with the superimposed normal probability density in Figure 1.24. (We will consider this same sample of data on page 458, and illustrate a chi-squared goodness-of-fit test for it.)

In Figure 1.24, we see that the empirical distribution of INTC returns for this period has both a sharper peak and more elongated tails than a normal distribution. Because of this, it is said to be *leptokurtic*.

The excess kurtosis of the sample of INTC daily simple returns is 1.533. A distribution with positive excess kurtosis is said to be leptokurtic, and one with negative excess kurtosis is said to be *platykurtic*. A distribution with kurtosis similar to that of a normal distribution is called *mesokurtic*.

Actually, as it turns out, the distribution of the INTC daily returns over that nine-month period is not very dissimilar from a normal distribution.

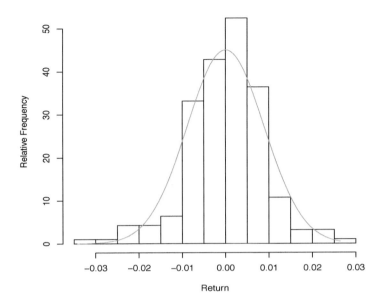

Histogram of INTC Daily Simple Returns

FIGURE 1.24
Frequency of Daily Simple Rates of Return for INTC Compared to an Idealized
Normal Frequency Distribution; *Source:* Yahoo Finance

Let us now consider the daily returns of the S&P 500. We will look at a
longer period, from January 1990 to December 2017. Also, instead of simple
returns, we will use log returns. (This makes little difference; we do it because
log returns are more commonly used.) The approximately 7,000 observations
are shown in a histogram together with a superimposed normal probability
density in Figure 1.25.

The histogram in Figure 1.25 is very leptokurtic. The width of the graph
is due to the fact that there were a few observations in the extreme range
from -0.20 to -0.05 and from 0.05 to 0.10. The overwhelming majority of the
observations were in the range from -0.05 to 0.05, however.

Leptokurtic distributions that have elongated tails are called "heavy-
tailed" distributions. A sample from a heavy-tailed distribution is likely to
contain a few "outliers". We briefly discuss outliers in Section 1.3.7 below.
In Section 3.1.11 beginning on page 273, we discuss heavy-tailed probability
distributions in more detail, and on page 293 we will briefly describe some
specific useful heavy-tailed families of probability distributions.

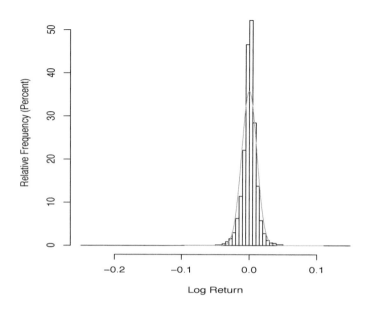

FIGURE 1.25

Frequency of Daily Log Returns for S&P 500 Compared to an Idealized Normal
Frequency Distribution; *Source of S&P Closes:* Yahoo Finance

The histogram in Figure 1.25 of the daily log returns of S&P 500 for the
period 1990 through 2017 is skewed to the left and is very leptokurtic. These
are general characteristics of the distributions of returns.

Instead of using the sample histogram to compare with the normal probability density curve, we could use a kernel density plot as in Figure 1.21. The
problem with using either a histogram or a density plot is that the important
differences in the sample and the normal distribution are in the tails of the
distribution. In Figure 1.25, it is easy to see the big differences in the frequency
and the normal density near the middle of the distribution. We can tell that
the tails are different, but it is difficult to see those differences because the
values are so small. It could be that the differences are actually very large on
a relative difference. This same problem would occur if we used a density plot
of the sample rather than a histogram. For that reason, the better graphic
would be one of the *log* of the density.

Figure 1.26 shows a plot of the log of the density of the sample of returns
with a plot of the log of the density of a normal distribution superimposed.

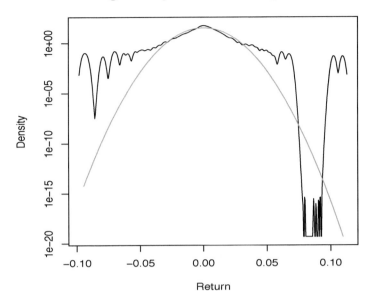

FIGURE 1.26
Log Density of Daily Returns for S&P 500 Compared to an Idealized Normal
Log Density; *Source of S&P Closes:* Yahoo Finance

We can clearly see that the distributions are very different in the tails. The
upper tail of the distribution of the S&P 500 returns is particularly unusual
(and this would not show up in an ordinary density plot of that sample.)
There are outlying observations on both ends of the set of S&P 500 returns.
It so happens that on the upper tail, there are almost no returns between
approximately 0.07 and 0.09, but there are some returns (exactly 2) larger
than 0.10. (This can be seen in the graph on page 99, but it is not easy to
see.) This causes bumps in the kernel density, and the bumps are magnified
in the log density.

In the next section, we will consider another graphical method of comparing a sample distribution with a normal distribution.

1.3.6 Q-Q Plots

A *quantile* is a value in a sample or a value of a random variable that corresponds to a point at which a specified portion of the sample or values of
the random variable lies below. For a number p between 0 and 1, we speak
of the p-quantile; if p is expressed as a percentage, we may call the quantile

a "percentile". (See Section 2.2 beginning on page 211.) A *quantile-quantile plot* or *q-q plot* is a plot of the quantiles of one distribution or sample versus the quantiles of another distribution or sample.

Neither distribution in a q-q plot needs to be normal, but of course our interest is in comparing a given sample with the normal family of distributions.

Figure 1.27 shows a q-q plot of the same data shown in the histogram of Figure 1.25 and the density of Figure 1.26, that is, the daily log return for the S&P 500 from January 1, 1990 through December 31, 2017, compared to a normal distribution. Because there are two degrees of freedom in a curve in two dimensions, a specific normal distribution with mean and variance matching that of the sample does not need to be used in a q-q plot as we did in the histogram and the density.

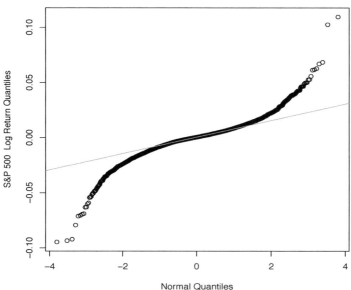

FIGURE 1.27
q-q Plot of Daily Rates of Return for the S&P 500 for the Period January 1, 1990 through December 31, 2017; *Source:* Yahoo Finance

In the q-q plot of Figure 1.27, the horizontal axis represents the quantiles of the normal distribution. The line segment in the plot provides a quick visualization of the differences of the sample of the S&P 500 log returns from what would be expected of a sample from a normal data-generating process.

First of all, we note that the points on the left side of the graph in Figure 1.27 generally fall below the normal line. This means that the smaller values in the sample are smaller than what would be expected if the data were from a normal data-generating process. Likewise, points on the right side of the graph generally lie above the normal line. This means that the larger values in the sample are larger than what would be expected if the data were from a normal data-generating process. The sample has heavier tails than the normal distribution. This is a fact we also observed in the histogram of Figure 1.25.

We will discuss q-q plots further in Chapter 2 and in Figure 2.10 on page 232 we will show q-q plots of the same data in Figure 1.27 except with respect to distributions that are not normal.

See Figure 2.9 on page 230 for illustrations of q-q plots of samples from distributions with various shapes.

Aggregated Returns

In Figure 1.2 we considered closing prices of INTC at various time scales, daily, weekly, and monthly. We observed that price data collected at different sampling frequencies may appear slightly different in plots, but the essential properties of the data were not different. Weekly prices, for example, are not aggregates of daily prices.

Returns computed at lower frequencies, however, do involve an aggregation of returns at higher frequencies, as we see from equation (1.26) on page 31. The weekly return is the sum of the daily returns for all of the trading days in the week. Because of this, we might expect to see some smoothing in weekly or monthly return data, in the same way that means smooth out individual data.

Figure 1.28 shows q-q plots of weekly and monthly rates of return for the S&P 500 for the same period as the daily rates shown in Figure 1.27. The points in these q-q plots are closer to the normal line than those in Figure 1.27.

1.3.7 Outliers

Something that is characteristic of heavy-tailed distributions is the occurrence of very extreme values, or "outliers".

Outliers pose a problem in statistical analyses. A single observation may have a very large effect on any summary statistics. Outliers should not be ignored, of course. Sometimes, however, they can distort an analysis. An extreme method of dealing with outliers is to remove them prior to performing computations. This is particularly appropriate if the outliers are due to possible recording errors. Another way of dealing with outliers is to use statistical methods that are resistant to outliers, such techniques are sometimes called

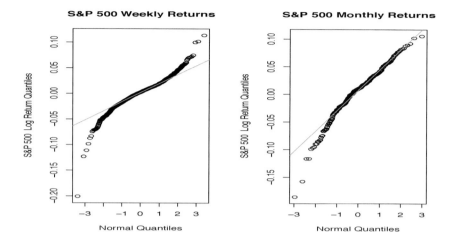

FIGURE 1.28
q-q Plots of Weekly and Monthly Rates of Return for the S&P 500; *Source of S&P Closes:* Yahoo Finance

"robust methods". If the possible presence of outliers causes us to use a special statistical method, we should also use the "standard" method with the outliers included and compare the results.

Outliers should be examined closely because they may have new information, or at least information that is not consistent with our assumptions about the data-generating process.

1.3.8   Other Statistical Measures

The field of *descriptive statistics* is concerned with summary statistical measures that together provide an understanding of the salient characteristics of a dataset. ("Inferential statistics", on the other hand, is concerned with use of a dataset to understand a broader population, which the dataset represents.)

Many of the measures we have mentioned in this section are based on moments, that is, on powers of the data: the mean is based on the first power; the variance and standard deviation, the second power; the skewness, the third power; and the kurtosis, the fourth power. The bivariate measures we mentioned, the covariance and the correlation, are based on a second cross-moment.

The problem with measures based on moments is that a single or a few outlying observations can have a very large effect on the overall measure. A familiar example is the average hourly wage of 9 employees who make $1 and the boss who makes $100; the average is $10, which is not a fair summary.

The median, the amount close to the middle of the ranked data, may be more appropriate. It is often the case, however, that a single number is not adequate.

There are a number of statistics based on ranked data or on the ranks of the data, and they can be particularly useful for data with large fluctuations, such as much of the financial data.

Order Statistics and Quantiles

Ranked data are called *order statistics*. For a set of numeric data, $x_1, \ldots, x_n$, we denote the corresponding set of order statistics as $x_{(1)}, \ldots, x_{(n)}$; that is, $x_{(1)}$ is the smallest, $x_{(2)}$ is the second smallest, and $x_{(n)}$ is the largest among the original set. (Ties can be handled in various ways; there are several possibilities, but the bottom line does not justify our consideration here.)

An alternate notation that I will sometimes use is

$$x_{(i:n)}$$

to denote the ith order statistic in a sample of size n.

Order statistics are related to *sample quantiles* or *empirical quantiles*. A *quantile* associated with a given frequency or probability α is a value such that a proportion of the sample or the population approximately equal to α is less than or equal to that amount. (Because there may be no value that satisfies this relationship exactly, there are various quantities that could be called an α-quantile. An exact definition is not necessary for our purposes here, but we will return to this question on page 215.)

The median is the value at the "middle of a vector". Formally, for a numeric vector x of length n,

$$\text{median}(x) = \begin{cases} (x_{(n/2)} + x_{(n/2+1)})/2 & \text{if } n \text{ is even} \\ x_{((n+1)/2)} & \text{if } n \text{ is odd.} \end{cases}$$

It is a good single alternative to the mean as a measure of location.

Measures of Location Based on Order Statistics

Two related measures of location are the *trimmed mean* and the *Winsorized mean*. Both of these latter measures are based on a chosen percentage of the data in the tails whose effect on the measure must be restrained. For a given dataset, $x_1, \ldots, x_n$, we choose a proportion by α, where $0 \leq \alpha < 0.5$. We determine an integer $k \approx \alpha n$. (How to choose k if αn is not an integer is not important here; just choose a close value.) The α-trimmed mean is

$$\frac{1}{n - 2k} \sum_{i=k+1}^{n-k} x_{(i)}. \tag{1.59}$$

The α-Winsorized mean is

$$\frac{1}{n} \left(k \left(x_{(k+1)} + x_{(n-k)} \right) + \sum_{i=k+1}^{n-k} x_{(i)} \right). \tag{1.60}$$

These measures are more robust to outliers or other effects of heavy tails than the mean.

Measures of Scale Based on Order Statistics

Other useful measures are based on order statistics, often the 25th, the 50th (the median), and the 75th percentiles. These percentiles correspond to quartiles, the first, second, and third quartiles, respectively. In the following, we let q_1 be the first quartile of a univariate dataset, let q_2 be the second (the median), and let q_3 be the third quartile.

For a measure of the *spread* or *scale* of a frequency distribution, instead of the usual variance and standard deviation, the *interquartile range*, also called IQR, may be useful. The IQR is based on quartiles of the data:

$$\text{IQR} = q_3 - q_1. \tag{1.61}$$

Another measure of the variation within a sample is the *median absolute deviation*, also called MAD:

$$\text{MAD} = \text{median}(|x_i - \text{median}(x)|). \tag{1.62}$$

There are variations of this measure in which one or both of the medians are replaced by some other measure of central tendency, such as the mean.

We can also define population or theoretical quantities corresponding to the sample quantities in equations (1.61) and (1.62).

The relationship between the IQR or the MAD and the standard deviation depends on the frequency distribution or the theoretical distribution. For example, in the normal or Gaussian family of distributions, $N(\mu, \sigma^2)$, we have the approximate relationships

$$\sigma \approx 0.741\text{IQR}, \tag{1.63}$$

and

$$\sigma \approx 1.48\text{MAD}. \tag{1.64}$$

Measures of Shape Based on Order Statistics

There are also measures of skewness based on quartiles or order statistics. One quantile-based measure is sometimes called *Bowley's skewness coefficient*. It is

$$\frac{q_1 + q_3 - 2q_2}{q_3 - q_1}. \tag{1.65}$$

Karl Pearson also defined several other measures of skewness, including the standard moment-based one given in equation (1.51). Other Pearson skewness coefficients involved the median and the mode of the distribution.

These measures based on order statistics are more robust to outliers or other effects of heavy tails than the more commonly-used measures based on moments. They are said to be more *robust*; we discuss them further in Section 4.4.6.

Another measure of the skewness is the Gini index, see page 214.

The Rank Transformation

The effects of large deviations in a sample can also be mitigated by replacing the actual data by the rank of the data, which we denote as "rank(x)". For the set of order statistics $x_{(1)}, \ldots, x_{(n)}$, the *rank transformation* is

$$\tilde{x}_j = \text{rank}(x_j) = i \quad \text{where } x_j = x_{(i)}. \tag{1.66}$$

This transformation may seem rather extreme, and it does completely obscure some properties; for example, the mean of the rank-transformed data is always $(n+1)/2$, and the variance is always $n(n+1)/12$. It is, however, useful in measuring relationships among variables.

Measures of Relationships Based on Orders and Ranks

Relationships between two sets of rank-transformed data, however, are similar to relationships between the two sets of raw data; in fact, these relationships are sometimes more informative because they are not obscured by outlying observations.

Let $y_1, \ldots, y_n$ be another set of data, and use the same notation as above. The ordinary correlation ("Pearson's correlation coefficient") defined in equation (1.54) when applied to the ranked data is merely

$$\frac{12}{n(n^2-1)} \sum_{i=1}^{n} (\text{rank}(x_i))(\text{rank}(y_i)) - n(n+1)/4. \tag{1.67}$$

This is called *Spearman's rank correlation* or sometimes "Spearman's rank correlation coefficient" or "Spearman's rho".

Another measure of association focuses on concordant or discordant pairs. The pair (x_i, y_i) and the pair (x_j, y_j), where $i \neq j$ are said to be *concordant* if either $x_i > x_j$ and $y_i > y_j$, or $x_i < x_j$ and $y_i < y_j$. The pair is said to be *discordant* if $x_i > x_j$ and $y_i < y_j$, or $x_i < x_j$ and $y_i > y_j$. If either $x_i = x_j$ and $y_i = y_j$, the pair is neither concordant nor discordant. The difference in the numbers of concordant and discordant pairs normalized by the total number of pairs is called Kendall's tau. That is, Kendall's tau is

$$\frac{2(p-q)}{n(n-1)}, \tag{1.68}$$

where p is the number of concordant pairs and q is the number of discordant pairs. Kendall's tau is also called "Kendall's rank coefficient". It is clear that

Kendall's tau is bounded by -1 and 1. It is also obvious that larger (and positive) values indicates that the two variables move in the same direction.

For more than two variables, we can form a symmetric matrix of Spearman's rhos and of Kendall's taus, just as we formed a correlation matrix in Tables 1.8 and 1.9. Again, because the matrix is symmetric, we sometimes display only a triangular portion. Also, the diagonal of the matrix does not convey any information because it is always 1.

Table 1.10 shows Spearman's rhos and Kendall's taus for the same data on returns for which we gave the correlations in Table 1.9 on page 84.

TABLE 1.10
Spearman's Rank Correlations and Kendall's Tau of Daily Returns of DJIA, S&P 500, Nasdaq, and Intel for 1987-2017

	Spearman's Rho					Kendall's Tau			
	DJIA	S&P	Nasdaq	Intel		DJIA	S&P	Nasdaq	Intel
DJIA	1.000	0.940	0.743	0.538		1.000	0.804	0.567	0.387
S&P	0.940	1.000	0.845	0.599		0.804	1.000	0.671	0.433
Nasdaq	0.743	0.845	1.000	0.679		0.567	0.671	1.000	0.500
Intel	0.538	0.599	0.679	1.000		0.387	0.433	0.500	1.000

Comparing the results in the tables, we see that the interpretations of the results would be very similar. We see stronger associations between the Dow and the S&P than between either of those and the Nasdaq for example. We also see stronger associations between Intel and the Nasdaq than between Intel and either of the less-tech-heavy Dow and S&P. (Of course, we should remember that Intel is a component of each of those indexes.)

Spearman's rho and Kendall's tau, just as the correlation, are bivariate statistics between two variables, no matter how many variables are involved.

1.4 Volatility

Returns represent change; large returns, positive or negative, represent large changes of the basic asset or index. In general terms, "volatility" is the variability of returns. (We will define this term more precisely below.) Volatility is one of the most important characteristics of financial data. Let us first visually examine the variability in the returns in a few datasets.

We can see the variability of returns in the histograms of Figures 1.20, 1.24, and 1.25, and in the q-q plots of Figures 1.27 and 1.28.

1.4.1   The Time Series of Returns

Histograms show a static view of a distribution. Another view of the returns, however, may reveal an interesting property of returns. The time series plot in Figure 1.29 indicates that the extent of the variability in the returns is not constant. (The data in Figure 1.29 are the same as in Figures 1.25 and 1.27 with three additional years. The graph in Figure 1.29 with "Black Monday" is a classic.)

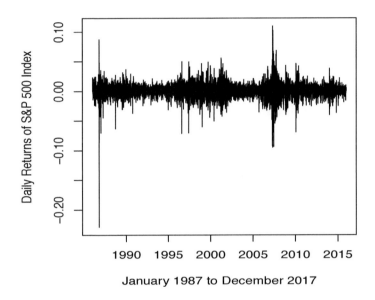

FIGURE 1.29
Daily Returns of S&P 500 Index for the Period January 1987 to December 2017

The large drop on the left side of Figure 1.29 corresponds to "Black Monday", October 19, 1987. On that day, both the S&P 500 and the DJIA fell by larger percentages than on any other day in history. (The year 1987 is notorious in the annals of finance. Both the S&P 500 and the DJIA were actually up for the year, however.)

Heteroscedasticity of Returns

In many academic models of stock returns, the "volatility" is an important component. These models often assume (explicitly or otherwise) that this quantity is constant. The point we want to observe here, however, is that the volatility seems to vary from time to time. There is a clustering of the volatility. Sometimes clusters of large volatility correspond to periods of market decline. The large volatility shown in Figure 1.29 for 1987 seems to be an anomaly, but the periods of relatively large volatility around 2000 and 2008 correspond to periods in which the index declined, as shown in Figure 1.10 on page 51. Over shorter time spans, the volatility of individual stock returns increases prior to a scheduled earnings announcement.

In the financial literature, the property of variable volatility is often called "heteroscedasticity". The volatility that varies is called "stochastic volatility".

Serial Correlations in Returns

The time series plot in Figure 1.29 makes us wonder if there is any serial correlation in the daily returns. Financial common sense would lead us to believe that there are no meaningful autocorrelations, because if there were, traders could take advantage of them. Once traders identify a property that allows them to make consistent profits, that property goes away.

On page 7, we defined the autocorrelation function (ACF) of a time series $\{X_t\}$ as $\text{Cor}(X_t, X_{t-h})$ for the lag h, and we defined a type of stationarity in which the ACF is a function only of the lag h. In that case, we denoted the ACF as $\rho(h)$. We also noted that this type of stationarity implies that the variance $V(X_t)$ is constant for all t. Given data $x_1, \ldots, x_T$, a consistent estimator of $\rho(h)$, with $\bar{x} = \sum_{t=1}^{T} x_t / T$, is

$$\hat{\rho}(h) = \frac{\sum_{t=h+1}^{T} (x_t - \bar{x})(x_{t-h} - \bar{x})}{\sum_{t=1}^{T} (x_t - \bar{x})^2}. \tag{1.69}$$

This is called the *sample autocorrelation function (ACF or SACF)*.

It is rather clear from the time series plot in Figure 1.29 that the variance of the returns is not constant, and hence, the time series is not stationary. The value of $\text{Cor}(X_t, X_{t-h})$ may be different not only for different values of h, but also for different values of t. This would mean that we have no statistical basis for estimating the autocorrelation because we only have one observation for each combination of h and t.

Despite the problems with $\hat{\rho}(h)$ as an estimator of $\rho(h)$ in the absence of stationarity, let us compute the sample ACF for the S&P 500 daily returns. It is shown in Figure 1.30.

Under assumptions of stationarity, and some other assumptions about the underlying time series, the horizontal dotted lines in Figure 1.30 represent asymptotic limits of values of $\hat{\rho}(h)$ corresponding to 95% confidence bands for $\rho(h) = 0$. (We will discuss the bands and other aspects of the analysis in Section 5.2.)

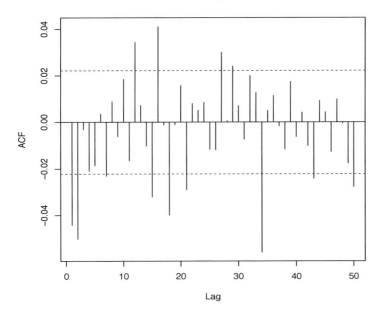

FIGURE 1.30
Autocorrelation Function of Daily Returns of S&P 500 Index for the Period
January 1987 to December 2017

Even with the caveats about the sample ACF, we can take the results
shown in Figure 1.30 as an indication that the autocorrelations of the returns
are not large. There are many empirical studies reported in the literature that
indicate that this is indeed the case. We will discuss these issues further later.

The generally negative autocorrelations at short lags of the daily returns
indicate a tendency of positive returns to be followed by negative returns and
vice versa. The clustering of volatility indicates that there is likely to be a
positive autocorrelation in the absolute returns, however. Figure 1.31 shows
that this is indeed the case, even up to lags of almost two years in length.

We must also be cautious in the interpretation of the ACF in Figure 1.31.
The underlying process is not stationary.

Aggregated Returns

We have observed that the frequency distributions of returns depend on the
time intervals over which the returns are computed. The q-q plots in Fig-

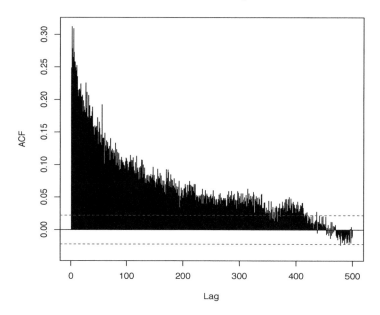

FIGURE 1.31
Autocorrelation Function of Absolute Daily Returns of S&P 500 Index for the
Period January 1987 to December 2017

ures 1.27 and 1.28 show differences in the tails of the distributions of daily
returns, weekly returns, and monthly returns. The data in those plots is the
same as that used in the ACF in Figure 1.30.

The autocorrelations in a time series of monthly returns are also different
from the autocorrelations in a time series of daily returns. Figure 1.32 shows
the autocorrelations of the monthly returns of the S&P 500 Index over the
same period of the daily returns shown in Figure 1.30. The autocorrelations
of the monthly returns are closer to zero than those of the daily returns.

1.4.2   Measuring Volatility: Historical and Implied

Volatility means change, and the simplest measure of the amount of change
within a distribution is the variance or standard deviation. Because the standard deviation, which is the square root of the variance, is in the same units
as the quantity of interest, it is often more appropriate.

The volatility that we are interested in is the volatility of the returns.

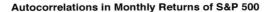

Autocorrelations in Monthly Returns of S&P 500

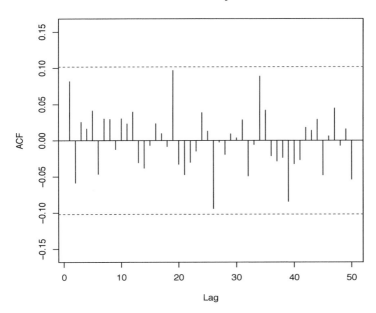

FIGURE 1.32
Autocorrelation Function of Monthly Returns of S&P 500 Index for the Period January 1987 to December 2017 (Compare with Figure 1.30)

While we can "see" the volatility of the S&P 500 in the returns shown in the graph in Figure 1.29, we cannot measure it. We do not observe the volatility directly. We cannot even estimate it unbiasedly because the conditional standard deviation changes at each point and we only have one observation at each point. We now define "volatility".

The *volatility* of an asset is the *conditional standard deviation* of the daily log returns on the asset.

Historical Volatility

One common measure of volatility is the sample standard deviation over some given time period. This is called *historical volatility*, or sometimes "statistical volatility". Although the distribution of this quantity is conditional on the information up to the beginning of the time period, it does provide a measure of the conditional distribution at the end of the period.

Because the volatility may not be constant, however, if we were to use historical volatility, we might want to compute it over moving windows of some relatively short length. At time t we might use the preceding k values to compute the *k-period historical volatility*,

$$\sqrt{\frac{1}{k-1}\sum_{i=1}^{k}(r_{t-i}-\bar{r}_{t(k)})^2},$$

where $\bar{r}_{t(k)}$ is the mean of the k returns prior to time t. (This is what we used in the portfolio example on page 69.)

Using the daily returns shown in Figure 1.29, we compute 60-day historical volatilities and the 22-day historical volatilities at various points in time, and display them in Table 1.11. First is December 1 1987, which was at the end of a brief highly volatile period. The 60-day historical volatility included the big swing on Black Monday, but the 22-day historical volatility did not. Next is December 1 1995, which corresponded to a period of apparent low volatility, as seen in Figure 1.29. Next is December 1 2008, in the protracted 2008 market turmoil, and finally, is the relatively calm market of December 1 2017.

TABLE 1.11
Historical Volatility of S&P 500 Daily Returns

	Dec 1 1987	Dec 1 1995	Dec 1 2008	Dec 1 2017
60-day	0.0387	0.00460	0.0443	0.0040
22-day	0.0218	0.00483	0.0422	0.0039

First, of course, we see the relatively large differences at different points in time. The other thing we note is that, in the case of the short-lived market turmoil of 1987, the 22-day historical volatility is very different from the 60-day historical volatility.

In Figure 1.34 on page 109 we show the 22-day historical volatilities of returns of S&P 500, computed at each day for the period from January 1990 to December 2017. (The volatilities are annualized as described below. The period is chosen for comparison with another quantity to be discussed.)

We might also find some kind of weighted standard deviation to be more useful. Obviously, there are several issues in the use of historical volatility. One is the issue of the effect of a few outliers on the historical volatility. We will discuss some of the other issues further in later chapters.

Effect of Outliers on Historical Volatility

Earlier, we mentioned the effect of outliers on summary statistics. Outliers in returns result in large effects on historical volatility. In the returns shown in Figure 1.29, for example, the large negative return of Black Monday yields a standard deviation of the daily log returns of the S&P 500 for the full year of 1987 as 0.0213. The standard deviation for the full year is 0.0156 if this one day is omitted.

Aggregated Returns

We now compute the weekly and monthly log returns the S&P 500 Index for the period January 1987 to December 2017, and display them in Figure 1.33. (The weekly data are every fourth or fifth value from the data shown in Figure 1.10; recall the discussion in Section 1.1.3 about trading days and weekly closes. The monthly data are roughly every 21st value from the data shown in Figure 1.10. These are the datasets displayed in the q-q plots of Figure 1.28, where we observed that the weekly data are more "normal" than the daily data, and the monthly are more "normal" than the weekly.)

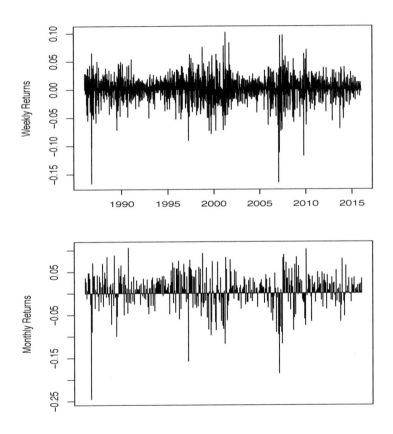

January 1987 to December 2017

FIGURE 1.33
Weekly and Monthly Returns of S&P 500 Index for the Period January 1987 to December 2017

It is instructive to compute means and standard deviations for the weekly and monthly returns over various periods. We would observe some interesting relationships in this empirical study. Rather than exhibit means and standard deviations for the various periods here, I will only show the results for the full period. The means of the daily, weekly, and monthly returns are

$$m_d = 0.000298 \quad m_w = 0.00142 \quad m_m = 0.00601.$$

Note that
$$m_w = 4.76 m_d \quad \text{and} \quad m_m = 20.13 m_d.$$

This is consistent with the relation in equation (1.21) and the fact that with the nine holidays per year, there are about 4.76 days per week on average, and there are about 21 days per month. (There is some variation here due to the beginning and ending times of weeks and months and also for extraordinary market closures such as five days including and following 9/11.)

The standard deviations (volatilities) are more interesting:

$$s_d = 0.0115 \quad s_w = 0.0225 \quad s_m = 0.0430. \tag{1.70}$$

We note
$$s_w = 1.953 s_d \quad \text{and} \quad s_m = 3.73 s_d. \tag{1.71}$$

As we saw on page 31, log returns over short periods are aggregated into a longer period by a simple sum of the individual short-period returns. If we have random variables R_1 and R_2 representing returns over two consecutive periods, then $R_1 + R_2$ would represent the aggregated return over the combined period. The variance of the sum of two random variables is the sum of the individual variances, plus twice the covariance. (We used this fact on page 70 in reference to the sum of two simple returns, and we observed it in the case of sample variance on page 82.) As we observed in Section 1.4.1, returns have very little serial correlation; that is, the covariance of two random variables over consecutive periods is very small, and a good approximation to the variance of $R_1 + R_2$ is $V(R_1 + R_2) = V(R_1) + V(R_2)$.

Now, consider returns over several consecutive periods. Generalizing the discussion above, we see if the variances are the same for the returns over n short periods, the variance of the return over the aggregated period is merely n times the variance of each shorter period.

Volatility is measured as the standard deviation of the returns; that is, as the square root of the variance. Hence, we have the "square-root rule" for aggregating volatility. If the volatility in each of n consecutive periods is σ, then the volatility of the full period is (approximately)

$$\sqrt{n}\sigma. \tag{1.72}$$

There are two aspects to "approximately". The first is the fact that the serial correlations are not really 0. The other is the nature of trading, which determines the volatility. Trading periods are not consecutive. Furthermore, the gaps are not of equal length. Nevertheless, we find the square-root rule to be useful in analysis of financial data. (In the example in equation (1.71) the correspondence is not very good. You are asked to consider it using other historical data in Exercise A1.9.)

Just as with rates (interest rates and rates of return), we can "annualize" volatility. Thus, if σ_d is the daily volatility, $\sqrt{253}\sigma_d$ is the corresponding annualized volatility, because there are approximately 253 trading days in one year.

Implied Volatility

Because the volatility may be changing continuously, the historical volatility may not be a very good estimator of the actual volatility. There are other anomalies in the historical volatility that do not correspond to any reasonable model of the price movements, which we will not discuss here.

For reasons unrelated to the study of volatility, various models have been developed to relate the fair market price of an option to the corresponding strike price, the expiry, and the price of the underlying. (The most famous of these is the Black-Scholes model. It turns out that these models only involve one additional property of the underlying, its volatility (which, of course, is unobservable). If K is the strike price, and at a fixed time, the time to expiry is t and the price of the underlying is S, and the volatility of the underlying is σ, then the fair market price of the option p is given by

$$p = f(K, t, S, \sigma), \tag{1.73}$$

where f is some function. There are several versions of f that have been studied (and used by market participants).

Because there is now a very robust market in stock and index options, p in equation (1.73) is observable. The other quantities except for σ are also observable and f is invertible with respect to σ, we write

$$\sigma = g(K, t, S, p). \tag{1.74}$$

This provides us an indirect way of measuring volatility. A volatility computed in this way is called an *implied volatility*.

For any given underlying, there are various option series at various expiries; hence, at any point in time, there are multiple combinations of K, t, and p. In practice, these different combinations yield different values of σ from equation (1.74). This is not just because volatility is different at different points in time as we observed in Figure 1.29 or Table 1.11. There are some systematic relationships among the values of σ obtained from different combinations of K, t, and p. One common relationship is the fact that for fixed t, p, and S, that is, for σ as a function of K, the curve in equation (1.74) tends to decrease as K increases toward S and then tends to increase as K increases beyond S. The relationship is known as the "volatility smile". Prior to Black Monday in 1987, the curve tended to be symmetric. Since that time, the curve has tended to be shifted beyond the current price S, and skewed so as to indicate larger implied volatilities for $K < S$. We will briefly discuss this skew on page 113 below.

1.4.3   Volatility Indexes: The VIX

The CBOE has developed several indexes of implied volatilities of various segments of the market. A volatility index is a continuously updated measure of the volatility of the market or of some aspect of the market. The most widely followed volatility index is the CBOE Volatility Index, VIX, which is based on the S&P 500 and the prices of options on it. CBOE began developing a volatility index in the 1980s, and by 1990 had a preliminary version. A formal index was introduced in 1993. As indicated above, there are some issues with any version of the formula (1.74) for implied volatility, and the VIX has undergone a few revisions since its introduction. CBOE also has other volatility indexes for various other measures of market performance. Although implied volatility can be computed in various ways, all using the idea of equation (1.74), we often equate the phrase "implied volatility" with the VIX.

The CBOE has described the VIX as a "measure of market expectations of near-term volatility conveyed by S&P 500 Index (SPX) option prices", and states that it is a "barometer of investor sentiment and market volatility".

The VIX itself corresponds to an annualized daily volatility (in percent) of the S&P 500 Index. In Figure 1.34 we compare the VIX with 22-day historical volatilities computed daily over the same period.

There is a close correspondence between the historical volatilities and the implied volatilities in the VIX. In Figure 1.35 below, we will observe this correspondence from a different perspective. Although the maximum 22-day historical volatility over this period was larger than the maximum of the VIX (88% compared to 81%), the historical volatilities tended to be slightly smaller.

Figure 1.35 shows four plots for the period January 1990 to December 2017. The top plot is the S&P 500 itself (which is the same graph as one shown in Figure 1.10), the second plot shows returns of S&P 500 (which is most of the data shown in the graph in Figure 1.29), the third plot is the VIX (which is the same as the bottom plot in Figure 1.34), and the bottom plot is of the SKEW (which we will discuss on page 113 below).

As we observed in the plot in Figure 1.34, from 1990 through December, 2017, the VIX ranged from a low of around 10 to a high of around 80. Here, we want to observe any obvious relationships between the VIX and the S&P 500. From the first and third plots in Figure 1.35, we can see that VIX is

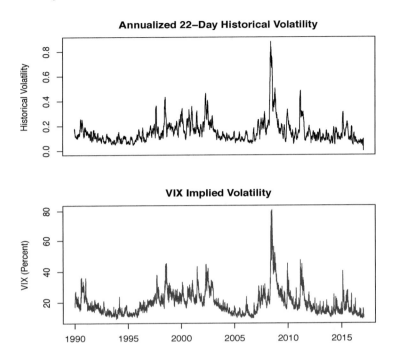

FIGURE 1.34
The 22-Day Historical Volatility of Returns of S&P 500 and the VIX for the
Period January 1990 to December 2017; *Source:* Yahoo Finance

generally smaller during periods when the S&P 500 is increasing. The second
and third plots in Figure 1.35 show that the VIX seems to correspond closely
to the apparent volatility seen in the daily S&P 500 returns (which was the
point of the plots in Figure 1.34).

Comparing the Frequency Distribution of the VIX in Three Years

We have used two different types of graphs to display frequency distributions,
a histogram as in Figure 1.20 and a density plot as in Figure 1.21.

We will now use another type of display, a *boxplot*. Boxplots are particularly
useful for comparing different frequency distributions. (We will discuss this
type of display more fully in Section 2.4.3.)

We have observed in Table 1.11 that the historical volatility of S&P 500
daily returns was very different in the month of December in the years 1995,
2008, and 2017. We also observed in Figures 1.34 and 1.35 that the VIX was
generally smaller and less variable in the years 1995 and 2017 compared to
the year 2008. We can use boxplots to get a quick picture of how the VIX
varied in those three years. The three boxplots in Figure 1.36 show clearly the
frequency distributions of the VIX in those three years.

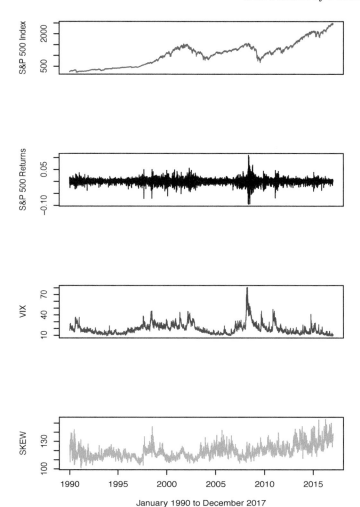

FIGURE 1.35
The S&P 500 (from Figure 1.10), Returns of S&P 500 (from Figure 1.29), the
VIX (from Figure 1.34), and the SKEW; *Sources:* Yahoo Finance and CBOE

We see that the frequency distribution in all three years was skewed positively, and that in all years the distribution was heavy-tailed on the right
(positive) side. We also see that the median value was much greater in 2008
than in 1995 or 2017, and we see that the variation was also much greater in
2008.

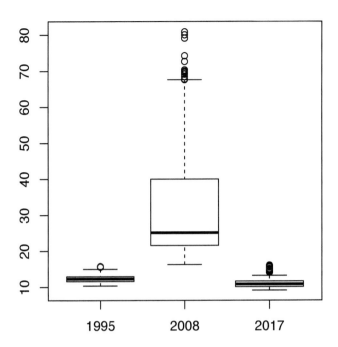

FIGURE 1.36
Boxplots of VIX Daily Closes; *Source:* Yahoo Finance

The Inverse of the VIX

Volatility has become a popular concern for fund managers. Financial services companies have developed ways of monetizing volatility and have developed financial instruments to allow hedging or speculating on volatility, primarily on the VIX, but also on other indexes of volatility.

The SVXY inverse ETF issued by ProShares and traded on Nasdaq seeks to replicate (net of expenses!) the inverse of the daily performance of the S&P 500 VIX Short-Term Futures index. (These are futures on the VIX.)

The XIV ETN issued by Credit Suisse also sought to replicate the inverse of the daily performance of the S&P 500 VIX Short-Term Futures index. (This fund was closed in February, 2018, in the wake of a 200% increase in the VIX within a period of a few days. XIV was similar to an inverse ETF, but as discussed previously, an ETN is not an equity.)

Other Volatility Indexes

There are other indexes of volatility, including ones that emphasize tail behavior, such as SKEW (see below). NationsShares, a Chicago-based financial
research firm, starting in 2013, has introduced a number of volatility indexes.
Some of these indexes are called VolDex® indexes. There are separate indexes
for various market sectors. The most common VolDex focuses on the S&P 500
as does the VIX. This index, called the Large Cap VolDex but is often just
called the VolDex, measures implied volatility by using options on the SPDR
S&P 500 ETF instead of on the S&P itself. The VolDex is thought by many
to be a better measure than the VIX.

1.4.4   The Curve of Implied Volatility

As we have noted, the distribution of returns on stocks is not normal. There
are relatively more very small and very large values than would be expected
in a normal distribution. The tails of the distribution of returns, both simple
and log, are heavier than those of a normal distribution. In addition to the
tails being heavier, the distribution is also skewed to the left, as observed in
Figure 1.25 on page 90.

   An additional property that we have observed is the distribution of the
returns is not constant. We have focused particularly on the volatility, that is,
the standard deviation, of the returns, but other aspects of the distribution,
such as its skewness, also vary in time. At times the skewness is close to
0, that is, the distribution is essentially symmetric, and at other times the
skewness is rather negative. (The actual values depend on the frequency of
the returns, daily, weekly, and so on.) The behavior of traders is what causes
these variations, and furthermore, traders are aware of the properties and of
the fact that they vary.

   These properties together cause the implied volatility from equation (1.74),
$\sigma = g(K, t, S, p)$ for some function g, to behave in interesting ways. We referred
to the "volatility smile" and the "volatility skew" above. In this section we
will briefly discuss the differences in the tails of the distribution, and how to
measure these differences over time. The idea is similar to the way a measure
of the volatility is obtained. We observe the various prices p of put and call
options for various strike prices K at various expires t when the underlying
has a known price S.

   One common relationship is the fact that for fixed t, p, and S, that is, for
σ as a function of K (inverting equation (1.74)),

$$\sigma = g_{t,p,S}(K),$$

tends to decrease as K increases toward S and then tends to increase as K
increases beyond S. The relationship is known as the "volatility smile". Prior
to Black Monday in 1987, the curve tended to be symmetric. Since that time,
the curve has tended to be shifted beyond the current price S, and skewed so
as to indicate larger implied volatilities for $K < S$.

Tail Risk Index; SKEW

There are reasons for this skewness; they arise from the skewness of the returns themselves, which we have observed in multiple contexts. The main reason for the skewed curve, however, is due to the differences in the risks. Risks due to increasing prices are unbounded, whereas risks due to decreasing prices are bounded by the bound of 0 on the price of the underlying. Option traders began using the term "skew" shortly after 1987 to refer to this property. The CBOE began computing a measure of this asymmetry based on observed out-of-the-money option prices on the S&P 500. After some tweaking of the formula, the CBOE began publishing the SKEW index, which they also refer to as a "tail risk index", that is computed from the prices of S&P 500 out-of-the-money options. Since an increase in perceived tail risk increases the relative demand for low strike puts, increases in SKEW generally correspond to an overall steepening of the VIX.

The bottom plot in Figure 1.35 on page 110 shows the CBOE's SKEW Index for the years 1990 through 2017.

The SKEW typically ranges from 100 to 150. It appears from the plot that the SKEW experiences more day-to-day variation than the VIX, and that is true.

According to the CBOE (2019), "a SKEW value of 100 means that the perceived distribution of S&P 500 log-returns is normal, and the probability of outlier returns is therefore negligible. As SKEW rises above 100, the left tail of the S&P 500 distribution acquires more weight, and the probabilities of outlier returns become more significant."

Although some traders read much into the "sentiment" of the market from the SKEW Index, there do not seem to be any strong relationships between the SKEW and the other series shown.

NationsShares, the provider of the VolDex indexes mentioned above, also provides two indexes that focus on the tail behavior called SkewDex® and TailDex®.

1.4.5 Risk Assessment and Management

The value of any set of assets fluctuates over time. Moderate decreases in asset values are rarely a problem. Extreme decreases may have significant, but different, effects on different types of holders, such as an individual, a bank, a pension fund, a mutual fund, or a country. Each such entity seeks to manage the risk of an extreme decrease.

The value of an equity asset fluctuates for various reasons. Sometimes, for whatever reason, the whole market goes down. In this case, the values of most individual assets decrease just because the market went down. The risk of a general market decline is called *systematic risk* or *market risk*.

An individual asset may decrease in value for reasons related specifically to that asset, such as poor execution of a marketing strategy, or an inordinate change in price of a commodity central to the operation of the company. This is called *specific risk*.

An individual stock has a risk component from the market, the systematic risk, and its own risk component, the specific risk. Ordinary diversification, as discussed in Section 1.2.9, can mitigate the specific risk, but not the systematic risk. The specific risk is sometimes called "diversifiable risk". Hedging, as also discussed in Section 1.2.9 and which can be considered as a type of diversification, can reduce both specific and systematic market risk.

There are many other sources of risk. Risk of default by a creditor is a major consideration. For fixed-income assets, this is the most important risk.

Although we generally equate risk with volatility, the risk that is of interest results from variation in one direction only. Variation that increases the value of a portfolio is not a "risk".

An approach to assessing risk of a loss is to model the frequency distribution of the values of the portfolio over some time period, such as the distribution of INTC prices shown in Figure 1.19 on page 77, and then compute the probability of the value at the end of the period being in the lower tail of the distribution. Alternatively, and more commonly, we use the frequency distribution of returns, such as in Figures 1.20 or 1.25, and determine the amount of a negative return corresponding to a specified (small) probability. The returns used in this analysis are generally simple one-period returns.

The obvious problem with this approach is the model of the frequency distribution of the portfolio value or of the returns. As we will emphasize throughout this book, however, a statistical analysis always begins with a *model*. Statistical analysis involves fitting the model, assessing the adequacy, or "correctness", of the model, and using the model to make further inferences about the process. The questions of the validity of the model and of the fitted model remain, however. A model does not need to be correct to be useful. In financial applications, a model that is useful at one point in time, may not be useful at later times.

Value at Risk

In the *value at risk* (*VaR*) approach to risk assessment, we are interested in the performance of the portfolio over some specified time interval. (We use the notation "VaR" because "VAR" is commonly used to mean "Vector AutoRegression"; see page 535.) Commercial banks generally compute a VaR for one day. Pension funds, on the other hand, may be more concerned about the VaR in a one-month period. The values of the portfolio over a given period is a time series and could fluctuate considerably over that interval. The frequency distribution itself is a time series. We focus on the estimated or expected frequency distribution at the end of the period.

Suppose the portfolio consists of a mixture of assets of various types. We use observations from the recent past to develop frequency distributions for each asset, and then we combine them according to their proportion in the portfolio and *according to their correlations with each other* to form a frequency distribution for the total value of the portfolio. (Understanding the correlations and incorporating them into the model is difficult, and failure to do that adequately has been the cause of many financial disasters.)

The model of the frequency distribution may be in the form of a histogram based on historical data as in Figures 1.19, 1.20, or 1.25, or it may be in the form of some standard frequency distribution that corresponds to observed properties of historical data. A third method of developing a model of the frequency distribution at the end of the given period is to model the returns and to simulate future prices starting with the current price.

In any case, the historical data must be taken from the appropriate assets over the relevant time period. We will not consider the details of the development of a model here, but rather, assume that we have a model for the frequency distribution of the value of the portfolio, and use that model to illustrate common simplistic VaR approaches to risk assessment.

In the standard approach, sometimes called PVaR, once the model frequency distribution is decided on, to assess the value at risk, we choose some small probability α, and determine the point at which the probability of a loss is less than or equal to α. The frequency distribution, of course, depends on the time horizon. We sometimes emphasize the dependencies in VaR by use of the notation $\text{VaR}(t, \alpha)$.

The loss can be measured in dollars (or a numeraire) or as a percentage of the current value, that is, as the single period simple return.

Consider, for example, the VaR for one day for a long position in INTC on October 1, 2017. We first need a frequency distribution for the returns. We may use the histogram in Figure 1.20 on page 78 as a model of the frequency distribution of daily simple returns for INTC. Those data were for the period January 1, 2017 through September 30, 2017.

We next choose a probability α, say 5%, and determine what amounts of loss correspond to the worst 5% of days under the given model of the frequency distribution.

Using the relative frequencies of the individual bins, which are stated in percentages, we accumulate the frequencies times the bin widths, which are all equal, until the cumulative total exceeds 5%, and then adjust within that bin so that the cumulative sum of the products is exactly 5%. This is the lower α cutoff point. In this example, that point occurs at -0.01535. Hence, the worst 5% daily returns are -0.01535 or less, or stated informally another way, 5% of the time the INTC holding will lose slightly more than 2% of its value in one day. Figure 1.37 displays these results.

From this analysis, the 5% one-day VaR of a \$100,000 long position in INTC on October 1, 2017 was \$1,535.

The analysis depended on having a good model of the probability distribution of returns for the date in question. October 1 was a Sunday. Is

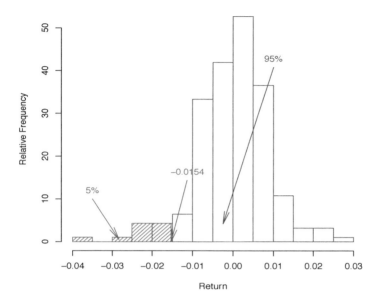

Histogram of INTC Daily Simple Returns

FIGURE 1.37
5% Value at Risk of a Hypothetical Frequency Distribution of Returns, Expressed as a Histogram

the frequency distributions of daily returns of INTC from January through September of that year appropriate? Were there systematic changes in the distribution in the period January through September? Is there a weekend effect? Does this same analysis hold for October 2? These questions do not have simple answers, of course, but they are the kinds of issues that arise in VaR assessments.

The basic idea illustrated in Figure 1.37 starts with a representation of the frequency distribution of the returns over the relevant time interval, and then merely determines the lower α cutoff point. The representation of the frequency distribution is an approximation of the probability distribution. Instead of using a histogram, there are many other nonparametric ways of estimating probability distributions, some of which are discussed in Section 2.3.

Following the same basic methodology, another approach to approximating the probability distribution of the returns is to use some standard probability distribution, such as the normal distribution, and then choose the parameters in that distribution to fit a set of observed data.

To continue with this illustration, we use as a model of the distribution of the returns a normal distribution, $N(0, s^2)$. For the variance s^2 we use an estimate from the same set of data used to form the histogram above. We merely compute the historical volatility; that is, we compute the standard deviation, for which we get $s = 0.008918854$. (The sample mean is 0.000183.) Following the same approach used above on the histogram of daily returns for INTC, we determine the α cutoff point from the normal distribution. Figure 1.38 illustrates this approach, again at the 5% level.

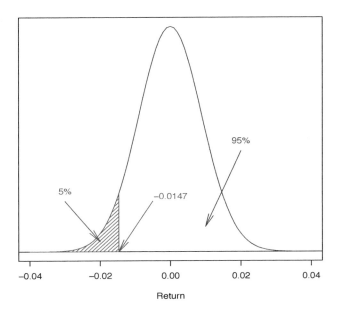

FIGURE 1.38
5% Value at Risk of a Hypothetical Normal Frequency Distribution of Returns, Expressed as a Probability Density Curve

The idealized distribution shown in Figure 1.38 is quite similar to the distribution in the histogram of Figure 1.37. From the analysis using Figure 1.38, the 5% one-day VaR of a $100,000 long position in INTC on October 1 2017 was $1,467, slightly less than the former result.

This approach shown in Figure 1.38 to computing a VaR is sometimes called the "variance-covariance" normal distribution approximation method. In this approach, not only is the adequacy of the model, specifically the normal distribution, in question, but also the relevance of the computed value of s must be considered; recall the various values of historical volatilities shown on page 104.

Because the family of normal distributions has simple linear properties, the α cutoff point for any normal distribution is the α cutoff point for a standard normal distribution multiplied by the standard deviation. For example, the 5% cutoff for a standard normal distribution is -1.645, and the 5% cutoff in the example is $-1.645s$, where s is the standard deviation of the sample of returns; that is, the volatility.

The parametric approach can also be used with a different family of distributions. We have observed in Figures 1.24 and 1.25 that the normal family of distributions does not provide a very good fit to observed distributions of returns. There are several heavy-tailed distributions that may be more appropriate. We will discuss some of them beginning on page 293. The t distribution, for example, often works well as an idealized model.

These same types of analysis can be applied to other assets, including portfolios of holdings of various kinds. The frequency distribution used must be one that is appropriate for the asset being assessed. If the asset is a portfolio of more than one position, it must be analyzed as a whole. Often the distribution of returns for a portfolio is not as extreme as the distribution of a single stock. This is the effect of "diversification".

In Exercise A1.16 you are asked to perform the same analysis of VaR for the SPY ETF, which is effectively a portfolio of stocks that includes INTC.

Variations on Value at Risk and Further Comments

After the financial crises that swept markets across the world in 2008, assessing value at risk became a popular exercise for financial analysts. Some of the regulations put in place after those crises required periodic computations of VaR.

The standard VaR approach is simple: if outcomes whose probability is less than α are excluded, the PVaR is the maximum loss during the given time interval. Thus, this approach does not address the sizes of losses with more extreme probabilities. (Here, I am tempted to say "Duh".) This presents problems with the meaning of the results.

Another problem with VaR in general is that the focus is not on the really large losses. While the loss that occurs with a relatively low frequency may be manageable, the really large losses that also are included in the VaR analyzed may be catastrophic.

Because VaR is based on quantiles, the sum of two VaRs associated with two separate portfolios may be less than the VaR associated with a portfolio consisting of the two separate portfolios. This property would indicate that diversification increases the risk of a portfolio. VaR is not *coherent*. (We will see the reasons for this in Section 3.1.3.) Measures based on expected values or probabilities, such as discussed below, may be coherent.

To view the problem from a slightly different perspective, for a given portfolio and time horizon, let the random variable L denote the loss. The distribution of the loss L is determined by the distribution of the returns. For a given probability α, the VaR, expressed slightly differently, is c_α, given by

$$c_\alpha = \inf\{x \in \mathbb{R} | \Pr(L \geq x) \leq \alpha\} \tag{1.75}$$

More interesting questions about the financial risk arise within those events whose probability is less than α. They are not all of the same magnitude and they do not have equal effects. So for a given probability α and the corresponding critical value c_α, we consider what we can expect in that tail region below c_α. Using the notation above, we form

$$E(L|L \geq c_\alpha). \tag{1.76}$$

This is sometimes called the *conditional value at risk* or CVaR. It is also called the *tail loss* or *expected shortfall*.

Aside from the problems of the model of the probability distribution not corresponding to reality, both the standard VaR and CVaR do not fully describe the random nature of the possible losses. The tail distribution of losses may be of interest. This is sometimes called the "distribution of the value at risk", or DVaR.

$$\Pr(L|L \geq c_\alpha) \tag{1.77}$$

We discuss this probability distribution further on page 275. The point to be made here is that risk assessment is not a one-dimensional problem.

The other problem, of course, with any measure of value at risk is the model of the distribution of the returns, which is the starting point for the assessment of value at risk. Sophisticated statistical techniques cannot make up for unrealistic models.

Financial institutions must often assess the value at risk of their assets. The multivariate distribution of the returns of a portfolio of assets is difficult to model adequately. It is a simple matter to compute the distribution of a linear combination of random variables, if the multivariate distribution of the vector of individual random variables is known. The distribution of a linear combination of random variables cannot be computed from only the distributions of the individual random variables and their pairwise correlations. Even if those entities were sufficient to construct the distribution of the linear combination, in a practical application in which all of those entities must be estimated from historical data, the overall variance of the estimators may render the exercise in combining them useless. One of the major problems in the assessment of value at risk by financial institutions is a lack of understanding of how a portfolio of assets behaves.

The Basel Accords (Basel II, in particular) set forth by the Basel Committee require that banks regularly compute VaR and maintain capital reserves to cover losses determined by the VaR analysis. Various methods and assumptions for the VaR analysis are allowed, however. Basel III built on the structure of Basel II but set even higher standards for capital and liquidity.

Value at risk is a concern for short positions as well as for long positions. The equations above apply unchanged to short positions. The type of analyses illustrated in Figures 1.37 and 1.38 would apply to short positions also, except that the region of interest would be on the right-hand side of the graph instead of on the left-hand side. A major difference in VaR for short positions is that the amount of loss is unbounded.

1.5   Market Dynamics

In Section 1.2.4 we equated prices at which stocks are traded with their fair market values. This equivalence is a tautology if "fair market value" is interpreted as the price in the market and if we assume that there is only "one price". A more difficult question, however, is the relationship of stock prices to some more fundamental "intrinsic value" of the stock. Without being precise in a definition of intrinsic value, we can think of it as what the market value "should be". In the following discussion, we will take *intrinsic value* as an undefined term, but one that carries an intuitive, heuristic meaning. (Recall the difficulties in Section 1.2.4 in attempting to assign a value to stock in terms of book value or the value of future earnings of the stock. We did not conclude with any firm definition of fair or intrinsic value, and we will not develop one here.)

An *efficient market* for securities is one in which the price of every security equals its intrinsic value at all times. This would mean that all who trade stock have access to all information about the universe of tradable stocks that is currently available, and that all who trade stock make optimal decisions in trading. The *Efficient Market Hypothesis (EMH)* states that "the market" is efficient; that is, that current market prices contain all the information that is available to investors. This means that stocks are efficiently priced and prices are always at fair value.

There are three forms of the EMH that specify different levels of the "information" available to investors. The *weak form* of the EMH assumes that all past pricing information is available; the *semi-strong form* assumes that in addition to past pricing information, all "publicly available" information about the companies whose stock is tradable is available; and the *strong form* assumes that in addition to past pricing information and publicly available information, all private or insider information about the companies whose stock is tradable is available.

A related idea about market dynamics is that price changes or returns are essentially random. This is known as the *Random Walk Hypothesis (RWH)*. As with the EMH, there are three forms, relating to the distribution of the random changes. In the weak form of the RHW, the returns are assumed to have zero autocorrelations and to have zero correlation with the price of the stock. In a semi-strong form, the returns are assumed to be serially independent and to be independent of the price of the stock. In the strong form, the returns are assumed to be serially independent and identically distributed, and to be independent of the price of the stock.

Many academic studies of market dynamics implicitly assume the EMH. Furthermore, large financial institutions often set policies and make investment decisions based on the EMH.

Fortunately, most of the discussions about the EMH and the RWH are irrelevant both to understanding the market and to developing a trading strategy. Market participants know that these conditions do not exist, and many traders and analysts have identified patterns in price movements that they believe are indicative of future movements. Because market prices are set by traders, if enough traders believe that the patterns have meaning, the patterns will have meaning. The actual meaning of the pattern, however, is not necessarily the meaning that the traders believe in. In any event, statistical data analysts are interested in identifying patterns in data or functions of the data that relate to interesting properties. In this section we will mention some of the interesting features of stock price data.

Seasonal Effects

In the popular culture, various seasonal effects on stock prices have been identified. The "January effect" has been observed since the mid-twentieth century. It was noticed that stock prices, especially those of companies with smaller capitalization, tended to increase during the month of January. Various explanations have been offered, such as a rebound due the "window dressing" of large mutual funds during December, in which the fund managers tended to take positions in stocks whose prices had increased during the year, and then to move out of those positions. This may cause an increase in prices of smaller or less-noticed companies. The rationale of window dressing is that the presence of these stocks in the portfolio that is published at year-end would give a more positive appearance to the fund. Another explanation is based on supply and demand. Year-end cash bonuses and other infusions of new cash in the markets may result in more stock purchases, causing increases in prices.

During the summer months there is a tendency for smaller volumes of trades, and it is believed by some that the thinner trading results in higher volatility, and that gains or losses realized during that period would be erased by higher volumes of trading in the fall. This seasonal effect resulted in the maxim "sell in May and go away".

Obviously, if these seasonal effects were real and quantifiable, traders would discount them so that they become irrelevant. You are asked to explore these effects using real data in Exercise A1.19.

These two seasonal effects have been incorporated into trading strategies by some. A strategy that involves moving in and out of the market is called *market timing*. Most strategies involving market timing are based on patterns of prices, rather than on seasons of the year.

Mean Reversion

Certain aspects of the market seem to have a "normal" range, and if they deviate too far from the usual range, they tend to return in time to the usual values. This is called "mean reversion". The movements of the VIX is a good example of this. You are asked to explore this mean reversion conceptually in Exercise 1.19 and using real data in Exercise A1.18.

This is not to imply that market behavior does not change over time. New financial products appear from time to time. The ubiquitous ETFs did not exist until the late twentieth century. Market dynamics also change. We have pointed out the change in implied volatility that occurred in 1987. The phrase "this time it's different", however, is heard far more often than any observable change actually occurs.

Market Indicators

The prices of assets and the returns of assets seem to exhibit specific patterns over time. Some people believe that certain patterns are indicative of future prices or returns. Identification and use of patterns of price movements to predict the types of future price changes are called "technical analysis" (as opposed to "fundamental analysis"). Most academic financial analysts do not believe that future prices are significantly related to past price movements, and of course both the EMH and the RWH imply that they are not.

Asset prices tend to be quite noisy, so before looking for patterns, the analyst generally smoothes the data in some way. Moving averages, as in Figure 1.9 on page 40, smooth out the noise. Other ways of smoothing the price fluctuations use straight line segments, tending up or down, over time periods of varying lengths. Often the straight line segments are drawn to connect relative highs or lows of the prices over some time period. Although it may seem highly coincidental for a sequence of relative highs or lows to fall on a straight line, they actually do often, at least within a few pennies. Figure 1.39 shows the daily closing prices of the stock STZ over the year 2017. The prices "bounced off" of that "support line" six times during the year. (See Exercise A1.22 on page 202, however, for what happened next.)

Traders often use "crossovers" of moving averages or of support lines or "resistance lines" (similar to support lines, but connecting a sequence of relative highs) as indicators of future prices. A price crossover is taken as a signal of a potential change in trend.

Crossovers of two moving averages, one longer and one shorter, are also taken as indicative of future directions of price changes. When a shorter-term MA crosses above a longer-term MA it is taken as a buy signal, because it is believed that this crossover indicates that the trend is shifting up. Such a crossover is known as a "golden cross". When the shorter-term MA crosses below the longer-term MA, it is taken as a sell signal because it indicates

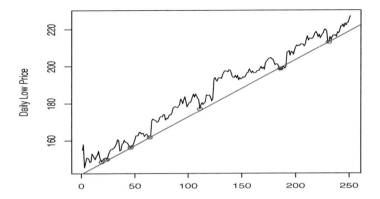

FIGURE 1.39
Programmed Buying at the Support Line

the trend is shifting down. This is known as a "death cross". There were six crossovers of the 20- and 50-day moving averages in the daily INTC closes over the 9-month period shown in Figure 1.9. Whether the moving averages are simple MAs or exponentially weighted moving averages and the lengths of the periods of the two moving averages result in different crosses. (Moving averages are discussed beginning on page 39. In Exercise A1.5 you are asked to compute moving averages for MSFT, and to identify any crossovers.)

Another way of using different moving averages as an indicator of future price movements is called the *moving average convergence-divergence* or MACD, which is usually based on 26-period and 12-period exponentially weighted moving averages. The "MACD line" is just the difference between these two EWMA lines. An exponentially weighted moving average of the MACD line is called the "MACD signal line". A crossing of the MACD line above the MACD signal line is taken as a buy signal and a crossing below is a sell signal.

Other possible indicators of future price movements are based on the relative numbers of recent price advances and declines or the averages of the positive returns and negative returns. Technical analysts have used various measures based on these quantities. One simple way of combining them is called the *relative strength index* or RSI. As with any indicator, this index requires a trading period (often a day) and a number of such periods to be specified. The relative strength index is based on the ratio of the average price increase to the average price decrease over those periods, which is called the *relative strength*. (The decreases are measured as positive numbers.) The RSI is then defined as

$$RSI = 100 - \frac{100}{1 + RS},$$ (1.78)

where the relative strength is given by

$$RS = \frac{\text{ave}(u)}{\text{ave}(d)},$$

and "ave" is an average of the absolute values of the returns. Various types of averages can be used, but the most common type is a variation of an exponentially weighted moving average and the most common number of periods used is 14. The RSI ranges between 0, representing overwhelming average declines, and 100, representing overwhelming average increases. (If the denominator of RS is zero, the RSI is taken as 100.) Small values, say below 30, indicate "oversold" conditions, and may portend future price increases. Large values, say above 70, indicate "overbought" conditions, and may portend future price decreases.

Market indicators such as the relative strength index are called *oscillators* because they are bounded above and below. Where the index is relative to the bounds is believed to indicate future prices. Another oscillator that is widely used is called the *stochastic oscillator* or just "stochastics". (The term "stochastic" in this name has no particular meaning.) The stochastic oscillator is similar to a relative strength index, except that it uses the high and low prices of the trading period, which is usually a day.

There are many patterns in a time series of prices that are believed to portend future price movements of a certain type. One of the most widely-used patterns is the *head and shoulders* and the related *inverted head and shoulders*. The head and shoulders pattern is comprised of a steady rise in prices, followed, in succession, by a decline, a rise above the previous high, a decline, another rise but lower than the immediately previous high, and a decline through the line segment connecting the two previous relative lows. For this general description to become a working definition, we must specify a method of smoothing out the noise of day-to-day fluctuations. There are many ways that this can be done, and different traders or financial institutions have various methods. The key in smoothing is the time frame. A very short time frame could yield a head and shoulders formation in a period of just seven days (day 2, up; day 3, down; day 4 up higher; day 5, down; day 6, up but not as high; and day 7, down), but hardly anyone would believe that such short term patterns have any meaning.

As an example of head and shoulders and inverted head and shoulder, consider again the daily closing prices of INTC for the period January 1, 2017 through September 30, 2017 that we plotted in the upper left panel in Figure 1.2 and have used in other examples. That time series of prices exhibits both a head-and-shoulder pattern and an inverted head-and-shoulder pattern, as shown in Figure 1.40.

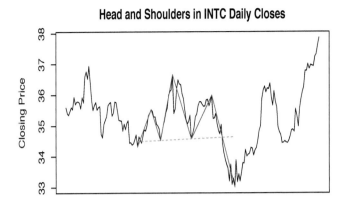

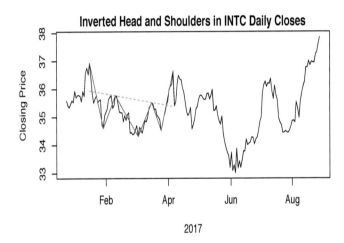

FIGURE 1.40
Patterns in INTC Daily Closes

Each market indicator is based on a specific trading period, usually a day but can be shorter or longer, and on one or more specific numbers of periods. Each also has various definitions, depending on the author or institution using it. Many commercial investment managers have proprietary indexes, some of which are unique only in name.

Institutional Investors

Market dynamics are determined by the trades made. A trader who makes more trades has more effect on market dynamics than a trader who makes fewer. Managers of mutual funds, pension funds, and endowment funds often must buy or sell stock just because of cash flows in or out of the funds. The cash flows may be substantial. Some managers buy or sell because they feel that their portfolio may perform better with a different mix of assets. Institutional investors generally make more and larger trades than individual investors.

The trades made by institutional investors generally follow a discipline, that is, a set of fixed rules for trading. The discipline in the case of certain mutual funds or ETFs is based on the need to maintain a specific balance in the portfolio. When new money comes into an index mutual fund, for example, the buys must be similar to the composition of the index that the fund follows.

In other cases, the institutional investor follows a discipline designed to optimize some aspect of the portfolio, possibly to increase "alpha" (expected return, according to some model) or to maintain "beta" (volatility, again according to some model) within certain bounds.

Programmed Trading

Disciplines used by money managers to optimize performance are called algorithms, or "algos". In some cases, trades are set up to be submitted automatically, resulting in "programmed trading".

The algorithms are based on analyses that use assumptions, models, and data (from the past, obviously). Any of the components may be invalid. The data generally is accurate, but it may be irrelevant. The algorithms are often very complicated, and may include many conditionals.

A simple conditional for buys is a price hitting a "support line". A support line for a stock defines a value below which the price of the stock is not expected to fall, as the line shown in Figure 1.39. An algorithm may include a conditional to buy the stock if the price hits the support line. This conditional would be combined with other conditionals, of course. The other conditionals may also determine the number of shares to be bought.

Support lines and "resistance" lines (bounding from above) are among the simplest components of trading algorithms. They are, however, widely used, both by institutional investors and individual investors. The use of support and resistance lines, and, indeed, any algorithm acts to confirm the validity of the algorithm. Whether or not the indicators can be justified based on financial theory, their use by many large traders gives them meaning.

Often a sequence of highs or lows in a stock price will follow a straight line. One reason for this is the discipline of selling or buying that is based on straight lines.

Correlations and Extreme Events

Markets of similar types of assets tend to move together. We can see this in the graphs of the daily closes of the three major indexes in Figure 1.10 and Figure 1.13 (pages 51 and 55). In a bull market, that is, when prices of stocks are rising, "a rising tide lifts all boats"; but in a bear market, that is, when prices are falling, "there is no place to hide".

Table 1.9 on page 84 shows that the correlations between the major indexes as well as with the closing prices of Intel stock are positive and relatively large.

As we have seen, the distributions of returns are heavy-tailed. This means that there will occasionally be extreme events, that is, outliers or events that are quite different from the ordinary or common occurrences in the market. These events seem to be clustered, as we see in Figure 1.29 on page 99 for the S&P 500 Index. The volatility of the Dow Jones or the Nasdaq would have exhibited similar variability.

We see in the graphs in Figure 1.10 and Figure 1.13 that all three market indexes declined in the period 2008 through 2009, and during that period, the volatility as indicated in Figure 1.29 was very high. What is not quite so clear in the graphs and tables is that when extreme events occur, not only do they occur simultaneously in various parts of the market, but the correlations between asset returns become greater.

Table 1.12 shows the correlations of the daily log returns of the three indexes and of the stock of Intel for the period 2008–2009. These correlations are much higher than the correlations shown in Table 1.9 for the longer period 1987–2017.

TABLE 1.12
Correlations of Daily Returns of DJIA, S&P 500, Nasdaq, and INTC for 2008 and 2009 (Compare Table 1.9)

	DJIA	S&P 500	Nasdaq	INTC
DJIA	1.000	0.988	0.947	0.781
S&P 500	0.988	1.000	0.967	0.791
Nasdaq	0.947	0.967	1.000	0.833
INTC	0.781	0.791	0.833	1.000

This tendency toward larger correlations in the tails of the frequency distribution is called *tail dependence.*

Risk and Uncertainty

In analyzing risk as in Section 1.4.5, we use a model of a probability distribution. We explicitly recognize the randomness inherent in the problem.

Our analyses and conclusions are predicated on the approximate correctness of the model. There is another aspect to the uncertainty of our conclusions that relates to the use of an appropriate model. This is often because some completely unforeseen event has made the model irrelevant. No matter how careful and sophisticated our statistical analyses are, and how good they would be if we used the appropriate model, they cannot compensate for invalid assumptions. The causes of this kind of randomness is often called Knightian uncertainty, after Knight (1921) who wrote extensively on the distinction between the randomness due to a probability distribution and the kind of events that disrupt the probability distribution.

Trading on Insider Information

There are many factors that keep the market from being efficient. Certain practices that are illegal are nonetheless widespread because they are hard to identify. Insider information is knowledge of financial aspects of a corporation that results from communication within a group of employees or corporate officers or governing boards. Any trading either by the insider or by any entity to whom the insider has divulged the insider information is *possibly* illegal. The definitions of legality vary from country to country. They may involve detailed exceptions. Even to the extent that the definitions of legality are clear, identification and prosecution of the crime are difficult.

Biased Financial Reporting

Academic studies and abstract descriptions of market dynamics often include statements about "information". In the real-world market, this term may have different meanings. There is a variety of fora in which information may be dispensed, and some information is not accurate. Many statements about publicly traded companies or about specific securities are not true, yet these statements may appear in various media used by market participants. There are enough traders making decisions on biased reports or inaccurate "information" that they move the market in directions that "efficient" market dynamics would not.

Some misinformation about securities is disseminated by those with short-term positions in the securities. Holders of a long position may promote a security by spreading unfounded positive statements, and then close their position after the security has increased in price ("pump-and-dump"). Shorts may spread unfounded negative statements, and then close their position after the security has decreased in price ("short-and-distort"). Both pump-and-dump and short-and-distort are illegal in the United States. The laws have curtailed this open behavior by large financial institutions, but upon sober reflection about possible behaviors, the hollowness of the laws are seen very quickly. The key issue is the ambiguity in "unfounded". There are two sides to every trade, and both sides often believe that the reasons for taking the opposing side are unfounded.

If a famous money manager or investor makes it known that a position, either long or short, has been taken in a certain security, the price of that security will often move in the direction benefiting the money manager. If, in addition, the money manager presents selective analyses that support the position taken (that is, either pumps or distorts), the price movement is even more beneficial to the money manager.

Front-running

Brokers and dealers have unique insight into trading that is not available at the same time to other market participants. Regulatory agencies forbid use of this kind of information by dealers and brokers to make trades. To forbid, however, is not the same as to prevent.

New bid or ask prices must be put "in the market" before a broker can act on them. "Front-running" occurs when a broker or dealer trades at the new bid or ask and immediately replaces the order with one with a lower bid or higher ask.

The best ask prices and bid prices, are called the *national best bid and offer* (NBBO). Regulation NMS of the SEC requires that brokers guarantee customers this price. The NBBO is updated throughout the day from all regulated exchanges, but prices on dark pools and other alternative trading systems may not appear in these results. Furthermore, very high-frequency trading has made it difficult to detect front-running.

1.6   Stylized Facts about Financial Data

In two papers in 1994, different economists referred to empirical properties of financial data as "stylized facts". Since then, and possibly due to the use of the term in the title of a later widely-cited research article, various sets of properties of the data-generating process for financial data have come to be called "stylized" facts about financial data.

Some of these stylized properties of rates of return that we have observed in examples in this chapter include the following.

- Heavy tails; that is, the frequency distribution of rates of return decrease more slowly than $\exp(-x^2)$, which is the proportionality of the normal distribution (Figures 1.24, 1.25, and 1.27).

- Rates of return are slightly negatively skewed, because traders react more strongly to negative information than to positive information (Figure 1.25).

- Aggregational normality, that is, the distribution of rates of return over longer periods tend to be less leptokurtic (Figure 1.28).

- Autocorrelation of returns are not zero but become closer to zero as the interval of the return increases (Figures 1.30 and 1.32, but recall the caveats regarding use of the sample ACF; see also page 526).

- Large positive autocorrelation of absolute value of returns, resulting in clustering of volatility (Figure 1.31 and Figure 1.29).

- Graphs of stock prices may exhibit geometric patterns (Figure 1.39).

- Correlations between returns become greater in periods of market turmoil (Table 1.12 and see also page 276).

In addition to the facts listed above, there are some others that are often included in canonical lists of "stylized facts", and we will discuss some of them in later chapters.

Notes and Further Reading

There are many books on the markets, finance, and financial data at all levels. A very good and comprehensive text at an introductory level is Sharpe, Alexander, and Bailey (1999).

One of the best ways of learning about finance and financial data in general is through the news media, such as the television channel CNBC and its associated website

> www.cnbc.com/

or various business news websites such as

> money.cnn.com/ (Money.CNN)
> www.msn.com/en-us/money/ (MSN)
> finance.yahoo.com/ (Yahoo)

Another website with general background information on a wide range of financial topics is

> www.investopedia.com/

In addition to reading expositions of financial topics, there is nothing like having a brokerage account with some investments, no matter how small the value of the account is. There are a number of online brokerage houses that allow accounts to be opened with a very small investment. Brokerage firms, in addition to providing the opportunities for real-world experience, also offer educational opportunities, in the form of webinars and research reports.

In the appendix to this chapter, beginning on page 139, we will discuss use of these and other websites for obtaining financial data.

Historical Background

Stock trading in US markets has a long and fascinating history. Some very interesting general accounts are available in Fraser (2005), Fridson (1998), and Gordon (1999). Although these books are rather old now, they provide relevant discussions of market behavior.

Fischer (1996) provides a wide-ranging global perspective on the development of financial markets.

Market crashes have occurred periodically. Galbraith (1955) discusses the

1929 crash and its aftermath. Kindleberger and Aliber (2011) describe several periods of extreme market turmoil, beginning with the Dutch tulip bulb bubble in 1636. Smithers (2009) focuses on the financial crisis of 2008 and blames it, in large part, on the unquestioning acceptance of the Efficient Market Hypothesis and the Random Walk Hypothesis by central bankers.

Financial Instruments and Market Behavior

There are a number of topics we mentioned in this chapter. Some of them we will address further in later sections, but others are outside the scope of this book. One general topic in particular that we will not discuss further is fixed-income instruments. There are many interesting aspects to the determination of value and risk in fixed-income instruments. The term structure of interest rates is not only a factor in the value/risk analysis of the instruments themselves, but also has possible implications for the broader market. Petitt, Pinto, and Pirie (2015) provide an extensive coverage of fixed-income instruments. Choudhry (2004) discusses the yield curve and its relation to other aspects of the economy in detail.

The return/risk approach to portfolio analysis, which is sometimes called "modern portfolio theory" (MPT) was developed primarily by Markowitz (1952). Sharpe (1966) introduced some of the ideas of capital markets that involve the concepts of risk-frees assets and excess return. The use of standard deviation as the measure of risk, and optimizing a portfolio over the curve of risk and return is not without its critics; see Michaud (1989), for example. Jobson and Korkie (1980) addressed some of the effects of the use of statistical estimation in application of modern portfolio theory, and suggested use of the bootstrap to assess the resulting variance. (We discuss the bootstrap in Section 4.4.5.)

Vidyamurthy (2004) describes investment strategies based on pairs trading.

McMillan (2012) describes various types of options positions, including simple spread combinations as we discussed and other more complicated combinations with such fancy names as "butterflies", "iron condors", and so on.

Another topic we discussed, but will only briefly consider further in this book (in Chapter 4), is value at risk. This, of course, is an extremely important area in finance. A standard book on the topic is Jorion (2006). One of the biggest events that underscored the importance of a better understanding of VaR was the 2008 global financial crisis. The events of 2008 were a watershed for VaR. Chen and Lu (2012) review some more recent developments since then, and Hull (2015) includes discussions of some of the effects of the crisis on procedures for risk management.

Although no one who participates actively in the market believes that prices follow a random walk, the Random Walk Hypothesis simplifies many academic studies. Any system put forward as counter to a random walk would fail quickly because all traders would take the profitable side. Market strategies that give a trading advantage suffer from an uncertainty principle; once they are measured, they change. The classic book by Malkiel (1999) argues strongly that market dynamics are essentially a random walk. On the other side, the book by Lo and MacKinlay (1999) shows many instances in which the hypothesis of randomness can be rejected. Rejecting the hypothesis, of course, does not necessarily endorse a specific alternative. Many of the indicators used in technical analysis were developed by Wilder (1978). Achelis (2001) provides an encyclopedic summary of all of the common indicators. Mandelbrot presents an interesting alternative view on market behavior, (Mandelbrot and Hudson, 2004).

The role of human behavior in determining market dynamics and in other economic phenomena is discussed at length by Thaler (2015). The differences between risk and uncertainty were first discussed extensively by Knight (1921). The shape of the frequency distributions of prices and returns causes a separation between "normal" events and "rare" events that affects how differing utility functions direct individual behavior. The term "black swan" has often been used to refer to the rare events; see Taleb (2010).

The general effects of nonsynchronous trading and the weekend effect are discussed in a number of publications. In particular, Lehalle and Laruelle (2013) and other authors in that book, discuss the effects of high-frequency trading, which is the ultimate nonsynchronous trading, on the market microstructure.

Several other topics alluded to in this chapter, such as ways of measuring volatility, and the existence of long-range dependencies, are all worth further study, but we will not discuss them further in this book. Many of these topics are addressed in the articles in Duan, Härdle, and Gentle (2012).

Exercises and Questions for Review

Much of the discussion and many of the examples in this chapter were based on actual data, and involved use of computer software for computations and graphical displays. We have not, however, discussed how to get this data or how to do the computations.

The exercises below generally do not involve actual data, and the computations require only simple calculations.

In the appendix to this chapter beginning on page 139, we describe how to obtain financial data from the internet and bring it into R for analysis. We then give another set of exercises beginning on page 191, in which you are asked to do some simple computations on real data similar to what was done in this chapter.

1.1.**US Treasury yields.**

During the course of a year, the yields of US Treasuries vary. Which do you expect to vary most: Bills, Notes, or Bonds?

Why do you answer as you do?

See also Exercise A1.3a on page 192.

1.2.**Coupon bonds.**

Suppose a coupon bond with a par value of $1,000 pays a semiannual coupon payment of $20. (This is fair-priced at issue at an annual rate of 4%.)

(a) Now, suppose the annual interest rate has risen to 6% (3% semiannually). What is the fair market price of this bond if the remaining time to maturity is 2 years?

What is the fair market price of a zero-coupon bond with everything else being the same as the previous question?

(b) Now, suppose the annual interest rate has dropped to 3%. What is the fair market price of this bond if the remaining time to maturity is 2 years?

What is the fair market price of a zero-coupon bond with everything else being the same as the previous question?

1.3.**Returns.**

Suppose the price of one share of XYZ was $100 on January 1 and was $110 on July 1. Suppose further that at the end of that year, the simple return for the stock was 5%.

(a) As of July 1, what was the simple one-period return?
(b) As of July 1, what was the annualized simple return?
(c) As of July 1, what was the log return?
(d) As of July 1, what was the annualized log return?
(e) What was the year-end price?
(f) What was the simple one-period return for the second half of the year?
(g) What was the log return for the second half of the year?
(h) What was the log return for the full year?

1.4.**Returns.**

Assume that an asset has a simple return in one period of R_1, and in a subsequent period of equal length has a simple return of R_2. What is the total return of the asset over the two periods? (Compare equation (1.26) for log returns.)

1.5.**Returns.**

Assume a portfolio of two assets with weights w_1 and w_2, and having log returns of r_1 and r_2. Show, by an example if you wish, that the log return of the portfolio is not necessarily $w_1 r_1 + w_2 r_2$ as in the case of simple single-period returns in equation (1.39). Can you give some conditions under which the portfolio log return is that linear combination?

1.6.**Moving averages.**

Suppose the stock XYZ has been generally increasing in price over the past few months. In a time series graph of the stock price, its 20-day moving average, and its 50-day moving average, as in Figure 1.9 on page 40, which curve will be highest, the price, the 20-day moving average, or the 50-day moving average? Which curve will be the next highest? Explain why.

See also Exercise A1.5.

1.7.**Price adjustments.**

Suppose a common share of XYZ closed at $40.00 on December 31, and on April 1 of the following year, the stock split two-for-one. The stock closed at $21.00 on December 31 of that following year.

(a) Suppose the company did not pay any dividends, returned no capital, and no other splits were made during the year. What was the adjusted closing price for December 31 of the previous year?

(b) Again, suppose the company did not pay any dividends, re-turned no capital, and no other splits were made during the year. What was the total simple return for that year?

(c) Now, suppose there were no splits except the two-for-one split on April 1, but suppose the company paid quarterly dividends of $0.50 on January 15, April 15, July 15, and October 15. What was the adjusted closing price for December 31 of the previous year?

(d) Again, assume the same conditions with regard to dividends and splits as in the previous question. Give two ways of computing the total simple return for that year.

1.8.**Market indexes.**

On January 3, 2017, the closing price of Goldman Sachs (GS) was 241.57 and the closing value of the DJIA was 19,881.76. The DJIA divisor at that time was 0.14523396877348.

(a) Suppose that GS split two-for-one at the close on that day (and no other components of the DJIA split). What would be the value of the DJIA divisor on January 4?

(GS did not split on that day, or at any other time in 2017.)

(b) Suppose that GS split three-for-one at the close on January 3. What would be the value of the DJIA divisor on January 4?

1.9.**Market indexes.**

Explain the similarities and the divergences in the three indexes shown in Figure 1.10 on page 51.

1.10.**Market indexes and ETFs.**

In Figure 1.14 on page 57, we compared the daily rates of return on the adjusted closing prices of SPY with the daily rates of return on the S&P 500 Total Return. Why is that appropriate (instead of, for example, using the rates of return on the S&P 500 Index)?

1.11.**Portfolio construction.**

The risk of a portfolio consisting of two risky assets with risks σ_1 and σ_2 and correlation ρ in proportions w_1 and w_2 is

$$\sigma_P = \sqrt{w_1^2 \sigma_1^2 + w_2^2 \sigma_2^2 + 2 w_1 w_2 \rho \sigma_1 \sigma_2}.$$

Let us require that $w_1 + w_2 = 1$ and $w_1, w_2 \geq 0$, that is, no short selling.

(a) Show that for $\rho < 1$, the value of w_1 that minimizes the risk is

$$w_1 = \frac{\sigma_2^2 - \rho \sigma_1 \sigma_2}{\sigma_1^2 + \sigma_2^2 - 2\rho \sigma_1 \sigma_2}.$$

What is the minimum if $\rho = 1$?

(b) Suppose there is a risk-free asset with fixed return μ_F.

Determine the tangency portfolio (that is, the weights of the two risky assets).

Determine the expected return μ_T and the risk σ_T of the tangency portfolio.

(c) Consider the case of two risky assets and a risk-free asset as above. Now, assume $\rho = 1$.

What are the efficient frontier and the capital market line? Consider various scenarios with respect to μ_1, μ_2, σ_1, and σ_2.

1.12.**Hedges.**

Explain why each position in the following pairs of positions is a hedge against the other.

(a) long 200 shares of MSFT and long 1 MSFT put (at any strike price and at any future expiry)

(b) short 200 shares of MSFT and short 2 MSFT puts (at any strike price and at any future expiry) (This is not a smart investment strategy!)

(c) short 200 shares of MSFT and long 2 MSFT calls (at any strike price and at any future expiry)

(d) long 200 shares of MSFT and short 2 MSFT calls (at any strike price and at any future expiry)

(e) long a portfolio of gold-mining stocks (ABX, RGLD, FNV, AEM) and long puts at any strike price and at any future expiry in GLD (see page 58)

(f) a short straddle

(g) a long strangle

1.13. **Options.**

Suppose that on 2017-04-04 the closing price of MSFT was 65.00.

On that day, the price of the October 70 call was 1.65.

On that day, the price of the October 60 put was 2.21.

In 2017, the date of expiration of October stock options was October 20.

(Ignore any ambiguities in the "price" of the options. At least one trade during that day occurred at each price quoted here. A common way of stating the price of an asset, especially an option, is to use the mean of the NBBO; that is, the midpoint of the highest bid and lowest ask. Of course, each bid and offer is associated with a quantity. A bid or ask associated with one contract is somewhat different from a bid or ask associated with 100 contracts.)

In each hypothetical case below, state whether the price of each option would be higher or lower than its price on 2017-04-04, and tell why. Assume other market statistics were relatively unchanged.

(a) Suppose that on 2017-05-04 the closing price of MSFT was 65.00.

(b) Suppose that on 2017-05-04 the closing price of MSFT was 71.00.

(c) Suppose that on 2017-05-04 the closing price of MSFT was 59.00.

(d) Suppose that on 2017-10-19 the closing price of MSFT was 65.00.

(e) Suppose that on 2017-10-19 the closing price of MSFT was 71.00.

(f) Suppose that on 2017-10-19 the closing price of MSFT was 61.00.

1.14. **Options.**

(a) Suppose that stock XYZ is trading at $80.00 on February 12, and its price increases to $83.00 on February 14. This $3.00 increase will affect the price of options on the stock. Consider six different call options on the stock: the February 80 call, the February 95 call, the May 80 call, the May 95 call, the August 80 call, and the August 95 call. The prices of all of these call options will increase, but by different amounts. Which of these options will increase the most, which the second most, and so on? List these six options in the order of their price increases, from most increase to least. (Some of these will be indeterminate, but rank them anyway and state why you think the relative increases would be in the order you specify.)

(b) The "Greeks" are quantitative measures of the price movements of an option; in particular, the delta of an option is the relative change in price of an option to a small change in the price of an underlying. The delta of an at-the-money call option is usually of the order of 0.5 and the delta of an at-the-money put option is usually around -0.5. A delta of a deep in-the-money option is usually close to 1 or -1. Furthermore, as the time to expiry decreases, the delta of in-the-money options approach 1 in absolute value.

Suppose that stock XYZ is trading at \$80.00 on February 12. Suppose on February 12 that the price of the May 80 call is \$3.62 (which gives an implied volatility of 0.23) with a delta of 0.533 (according to E*Trade), and suppose that the price of the May 80 put is \$3.50 (which also gives an implied volatility of 0.23) with a delta of -0.475 (again according to E*Trade).

  i. On February 14, the price of the stock increases to \$83.00. What is the implied price of the May 80 call and the price of the May 80 put on February 14? Comment on the delta.

  ii. On February 15, the price of the stock drops to \$79.00. What is the implied price of the May 80 call and the price of the May 80 put on February 14? Comment on the computational method.

1.15. **Options.**

Several widely-cited empirical studies have shown that approximately 75% to 80% of stock options held until expiration expire worthless (out of the money). Based on these studies, many have concluded that long option positions are on the whole unprofitable.

Give two counter-arguments that these data do not show that option buying is a losing strategy.

1.16. **Descriptive statistics.**

Suppose a company has 10 employees who make \$10 an hour, a manager who makes \$100 an hour, and a director who makes \$280 an hour.

(a) What is the mean wage?

(b) What are the 5% trimmed and Winsorized means?

(c) What are the 15% trimmed and Winsorized means?

1.17. **Descriptive statistics.**

Explain why quantile-based measures such as the IQR and Bowley's skewness coefficient are less affected by heavy-tailed distributions than are moment-based measures such as the standard deviation or the ordinary measure of skewness (Pearson's).

1.18. **Autocorrelations of returns.**

Explain how absolute returns can have large autocorrelations, as shown in Figure 1.31, when the returns themselves are uncorrelated. Does this make sense?

1.19. **Volatility Index (VIX).**

Does the VIX seem to exhibit a mean reversion? (See also Exercise A1.18.)

When the VIX is 12 or less, does it appear to be more likely that it will be larger (than 12) within the next 6 months than when it is 25 or greater to be larger (than 25)? We will return to this question in Exercise A1.18.

Appendix A1: Accessing and Analyzing Financial Data in R

In Chapter 1 we discussed several financial datasets, performed statistical computations on them, and displayed them in various graphs. We did not discuss how we obtained the data or how we did any of the computations. That was because we wanted to maintain the focus of the chapter: the nature of the data themselves.

The computer software used in Chapter 1 and throughout the book is R. In this appendix, we will discuss some of the basics of R, and then how R can be used to acquire real financial data, and to do simple computations on it as in Chapter 1. Fragments of the code used will be displayed in this appendix. All of the code used in the figures and examples of Chapter 1 is available at the website for the book.

While much financial data can be processed and analyzed as "little data", in this book, the attitude is one of "big data". This means that we do not want to look up some number or some date, and then manually code it into a program or a webpage. We want to access, process, analyze, and make decisions to the extent possible by computer programs.

The objectives of this appendix are to

- introduce R and some simple R functions useful in analysis of financial data;

- describe some open sources of current real-world financial data, and how to bring the data into R and prepare it for processing.

This appendix only covers some basics, and it is not intended to be a tutorial on R. The best way to learn the system is to look at some simple scripts such as those displayed throughout this appendix, and then run some similar scripts on your computer. Later chapters will require statistical and graphical abilities of R to analyze financial data obtained from internet sources.

To me, it is unacceptable to be surprised by software. For many tasks, the software does exactly what is expected. In those cases, the user needs very little instruction on the software. In other cases, the user must be warned about how the software interprets the tasks. In this appendix, in addition to the basics, I will mention unexpected things, such as how R reads the headers in CSV files (see page 162 if you do not know).

If you know the basics of R, it is likely that you can skip to Section A1.2, page 158, or at least skip to the section on date data beginning on page 156.

A1.1   R Basics

R is a freely-available system currently maintained by a number of volunteers (see R Core Team, 2019). It was originally built by Robert Gentleman and Ross Ihaka based on a design by John Chambers. The system, along with various associated packages, is available at

> www.r-project.org/

Executables for various Apple Mac, Linux, and Microsoft Windows platforms can be downloaded from that site.

R is an object-oriented functional programming system. Although there are "point-and-click" interfaces available, most R users write R statements and then submit them directly to the system for execution. RStudio, which is a free and open-source integrated development environment for R, provides an even more useful interface.

R is portable across the various platforms, except for a few OS-specific functions and except for any graphical user interface of course.

Names of R objects are case-sensitive. R statements are line-oriented. If a statement is syntactically complete at the end of a line, it is considered to be a statement; otherwise, the statement is continued onto the next line. R statements may assign values to objects by use of "<-", control flow of a program or a group of statements, request that a function or group of statements be executed, or merely request that the values of an object be displayed. An R statement may also just be a comment for the reader or the programmer. A comment is any statement that begins with "#". Comment statements may help to understand the other R statements. The readability of a group of R statements may also be improved by indentation. Unlike with some old-fashioned computer languages, indentation or the placement of a statement on a line has no meaning in the execution of R statements.

R has a modular design, and the basic system consists of a number of "libraries" or "packages" that provide specific functionalities. We will discuss a few packages that are especially useful for financial data.

In this appendix, I will display fragments of R code either by itself or together with the output from R. Code will be shown with a shaded background:

```
radius1 <- 3
Area1 <- pi*radius1^2   # pi is a built-in constant
```

Note in this example, that pi is a built-in constant in R. It is shown in a different font to distinguish what comes with R and what the user has defined, like radius1. When any R object is referenced in the text, it will be printed in a typewriter font.

If R output is also shown, I will generally show the statements with the R prompt to represent what would be visible on the console. The R output often contains element numbers, even if the object only has one element, as in "[1]" below:

```
> radius1 <- 3
> Area1 <- pi*radius1^2
> # To print an object, just type its name.
> Area1
[1] 28.27433
```

Names

R provides many useful objects, such as functions and built-in constants. The names of these objects are assigned by R, and I will represent them in a special font, as `pi` in the code above. Names displayed in that style can be used in any R session as it is initially instantiated.

Packages have names and may also provide useful objects with their own names. I will represent names of packages and the objects they provide in a special font, as `xts`. Names displayed in that style may not be available in an R session until some library is loaded into the session.

Names in R are not restricted; if the user assigns a value to any R object, the name of the object will refer to that value assigned. (The only exceptions that I am aware of are the R program control tokens `if`, `else` (but not `elseif`), `for`, `while`, `repeat`, `break`, `next`, and `function`.)

Names of R objects, such as built-in objects or user-assigned objects such as the variables `radius1` and `Area1` above, must begin with a letter or with ".". They cannot contain blanks, arithmetic, logical, or constructor operators ("$", "[", and so on).

Other names may not have these restrictions. For example, names of rows or columns in a dataset, variable names, perhaps, can contain blanks and begin with numerals. The name of the column containing yields on 3-month US Treasuries can be `"3 MO"`, for example. It is almost always a bad idea to put a blank in a name, however. An underscore or a period can often serve whatever purpose would be served by the blank.

Documentation

There are a number of books and tutorials on R, and a vast array of sources of information on the internet.

R has an integrated help system for functions. Documentation for R functions is generally stored in the form of **man** pages and can be accessed through the R `help` function or by the R "?" operator, for example

```
?tapply
```

To find packages and other software that provide functions for a specific task, the Task View webpage at Comprehensive R Archive Network (CRAN) is useful:

cran.r-project.org/web/views/

Areas of application or high-level tasks, such as "Bayesian", "Finance", and so on, are listed at this site, each with a link to a page that lists relevant R packages or other software with links and brief descriptions.

More information and examples for functions and other objects included in packages are often available through the `vignette` function. This function is (or may be) particularly useful for obtaining information on a package or a class of objects used in a package. Note that the argument to `vignette` must be quoted.

```
vignette("dplyr")
vignette("tibble")
vignette("xts")
```

There is considerable variation in the information provided by `vignette`. Invocation of this function may initiate a window with just a few comments, or it may bring into view an extensive PDF document.

Missing Values

One of the most important and useful properties of R is the ways it can handle missing data or values that do not exist as ordinary numbers or characters, such as ∞ or $0/0$. Often when data contain missing values, we wish to ignore those values and process all of the valid data. Many functions in R provide a standard way of requesting that this be done, by means of the logical argument `na.rm`. If `na.rm` is true, then the function ignores the fact that some data may be missing, and performs its operations only on the valid data; see page 185.

Arithmetic Operators

R has the usual scalar arithmetic operators "+", "−", "*", "/", and "^". It also has a sequencing operator, ":" (see Figure A1.42 below), and a mod operator "%%" where x %% y yields $x \mod y$.

Logical Operators

Logical conditions in R that have a value of TRUE or FALSE are defined by logical relational operators. The usual logical relations are ==, !=, <, <=, !<, >, >=, and !>, with obvious meanings. The negation logical operator is !. The conjunctions are & and |. The binary function `xor` performs the exclusive or operation. For examples,

3<5	TRUE
3<2 & 2+2==4	FALSE
3<2 | 2+2==4	TRUE
xor(3<2, 2+2==4)	TRUE
xor(3>2, 2+2==4)	FALSE

A logical condition can be an operand in an arithmetic expression. A TRUE condition has a value of 1 and a FALSE condition has a value of 0; for example, 1+(3<2)*2+(3<5)*3 has a value of 4.

Program Control

R has operators and statements for controlling the execution of other R statements. Statements in R are grouped by "{" and "}"; hence there is no need for an "end" statement.

The statements for conditional execution are if, else, and elseif. Looping in R is controlled by for or while. Subprograms are formed by a function statement. The statements that are included in any of these control structures are grouped by "{" and "}".

Objects

"Object-oriented" design (OOD) is a unified methodology for problem solving that focuses on *objects*, which are entities that are composed of both data and operations (*methods*) that manipulate the data. Within the context of a specific problem, there may be multiple types of objects, and within each type or *object class*, there may be multiple objects. Objects of the same class have similar properties and are subject to the same kind of operations. A class may have subclasses, or *derived classes*. Objects in a derived class *inherit* properties of the *super class*.

Software systems may or may not be built on a strong OOD; that is, the system may or may not provide a well-defined set of object classes with functions that operate on objects within each class in a consistent manner. The operations generally do not require additional information about how the operation is to be performed; the particular operation is performed in the same way on all objects in that class.

Each entity, or object, in R belongs to a specific class that determines the general characteristics of the entity and the methods or computations that can be performed on the object. R incorporates multiple object-oriented designs. In some cases, it may be important to recognize the different OODs, which go by names "S3", "S4", "R6", and so on; but for our purposes we can generally ignore any differences, and just treat all schemes the same. The standard R package methods contains several utility functions for navigating the various R classes and methods.

Some of the more common classes of R objects that we encounter in financial applications are numeric, character, Date, data.frame, and matrix.

All of these are built-in classes, but there are many more. Packages often also define object classes. A class we will use extensively is xts, for financial time series. See page 173 for a description of this kind of object.

An object in R also has a *mode* and a *type*, which determines more fundamental properties of the object. For example, an object in R may be of class matrix, of mode numeric, and of type double. A class and mode and/or type can be the same; that is, the general class may determine the mode and/or type. Two of the most important modes that we encounter in financial applications are numeric and character.

One of the most important built-in classes of objects in R is function. In R, all actions are "functions". R has an extensive set of built-in functions and it also provides the user the ability to define additional functions.

Functions in R

Actions on objects are performed in R by *functions*. There are many "built-in" functions, such as sqrt, exp, and so on; and the user can also write functions.

A function has a name and may accept arguments that are to be operated on. The arguments are enclosed in parentheses, and may be positional or named. Named arguments may have default values.

An R function may operate differently on different types of objects.

Functions in R for Working with Probability Distributions

There are a number of built-in functions for computations involving several common probability distributions. The function names consist of a root name that identifies the family of distributions, such as norm or t and a prefix to determine the type of function, density, CDF, or quantile, or a function to simulate random numbers. Functions whose names begin with "d" compute the density or probability function, with "p" compute the CDF (the probability of a value less than or equal to a specified point), and with "q" compute the quantile. (See Section 3.1 for definitions of these terms.) For example, pnorm(x) computes the probability that a standard normal random value is less than or equal to x.

R also provides functions for simulating random samples from these distributions (see Section 3.3.3). Functions to generate random numbers have names that begin with "r". The root names of common distributions are shown in Table 3.4 in Chapter 3. The probability functions or probability density functions for most of these are given in Tables 3.1 and 3.2 in Section 3.1.

User-Written Functions in R

A simple way a user can extend R is to write a function using the R function constructor called function. An example of an R function called myFun is shown in Figure A1.41. This function determines additional points on a line that is defined by two given points.

```
myFun <- function(x1,y1, x2,y2, x) {
#   Given the points (x1,y1) and (x2,y2) and a set of abscissas x,
#   determine the ordinates at x for the line that goes through all of them.
    slope <- (y2-y1)/(x2-x1)
    intercept <- y1 - slope*x1
    y <- slope*x + intercept
    return(y)
}
```

FIGURE A1.41
A Simple Function in R

Once the function has been entered into the R session, the function can be invoked by assigning values to the arguments x1, y1, x2, y2, and x, called "formal arguments", and then issuing the R statement

```
y <- myFun(x1,y1, x2,y2, x)
```

The example function myFun establishes an *environment* in which the local variable names slope and intercept have meaning. The names x1, y1, x2, y2, and x are *formal arguments* to the function. They are assigned a value when the function is invoked, and then are used within the function. The names of the variables in the environment in which the function is invoked do not have to be the same. Any changes made to them within the function are not passed back to the environment in which the function was invoked. The function, however, inherits variables from the environment in which it was invoked.

An alternative to a user-written R function is an R GUI app, which may be useful in a task in which various values may be easily entered and evaluated; see page 147.

Packages

A group of functions written either in R or in a compiler language can be packaged together in a "library" or "package", and then all of the functions, together with associated documentation, can be loaded into R just by loading the package.

To use a package, first *install* it on the R system being used, and then *load* it into the current R session. There are many R packages stored in the

Comprehensive R Archive Network (CRAN), and an R system program, such as `install.packages`, can install them on a local system. They can be loaded in an R session by the `library` or `require` function. (These two functions are essentially the same. The main difference is how they handle the case of a requested package that is not available.) The `install.packages` function requires that the package name be a character string, but the `library` or `require` function does not require that, although it allows it.

```
install.packages("vars")   #  requires quotes
library(vars)              #  allows, but does not require quotes
```

There are very many R packages available on CRAN. Some packages address a fairly narrow area of application, but some provide a wide range of useful functions. Two notable ones of the latter type are MASS, developed primarily by Bill Venables and Brian Ripley, and Hmisc, developed primarily by Frank Harrell. Another very useful R package, for graphics, is ggplot2, written by Hadley Wickham, who has also written useful packages for data munging and making the data "tidy". The tidyverse suite of R packages includes a number of packages with similar designs that use a common set of objects.

The pracma package, written by Hans Werner Borchers, includes a number of basic mathematical functions for numerical analysis and linear algebra, numerical optimization, differential equations, and time series. The function names, and to the extent reasonable, the arguments, are the same as in the standard MATLAB® functions to perform the operations. This facilitates the conversion of MATLAB code to R. It also provides some useful MATLAB utility functions that are not available in the standard R packages.

For financial applications, one of the most useful packages is quantmod, written by Jeffrey A. Ryan. We will use this package extensively in this book.

Some packages create a graphical user interface (GUI) through which the user interacts with the functions in the package through selecting menu items and entering data in boxes in a computer window. The GUIDE package (for "GUI for derivatives") is an example. The package is a GUI app (application). A GUI provides user convenience.

Care must be exercised in use of packages. A package may redefine a base function, which can cause unexpected results. Two packages may define functions with the same name, so which is used as a default depends on the order the packages are loaded. Which function will be used without further specification can be determined by the `environment` function:

```
> environment(date)
<environment: namespace:base>
> library(lubridate)
```

```
> environment(date)
<environment: namespace:lubridate>
```

A loaded package from which a function is to be used can be specified by use of "::", for example

```
lubridate::date(...)
```

specifies that the date function in the lubridate package is to be used, regardless of the order in which it or any other package is loaded. (The "..." indicates the arguments that are supplied.)

Packages can also change the environment or have other side effects that cause unexpected results. A common example is the functions that invoke the random number generator. If this is done incorrectly (and it is done incorrectly in many R packages), it can destroy the user's random number stream. Another common problem is that the developers or maintainers may make changes that are not backward compatible, or that in other ways yield unexpected results.

R GUI Apps

Using an R GUI app may be very different from using R itself. Once an R GUI app is created, probably by someone conversant in R, the app can be shared with other users who may not even know R, although most R GUI apps allow interaction with the R system. An R GUI app, for example, can have the appearance of an Excel spreadsheet. (Microsoft Excel® is the most widely used program in the world for statistical analyses.)

There are several packages that create GUI apps. The most widely used package is Shiny, developed by RStudio.

Vectors and Other Arrays

Objects in R are organized into individual elements. The basic array structure is a single succession of elements of the same type, called a "vector" or an "atomic vector". A single datum in R is considered to be a vector with one element.

A vector is constructed by the c function, as in

```
x <- c(9,8,7,6,5,4,3,2,1)
```

The length function returns the number of elements in a vector.

Other structures may correspond to arrays with two dimensions, and the dim function returns the numbers of elements in both dimensions. (It returns NULL if the object does not have two dimensions, so care must be exercised when both one- and two-dimensional objects are used in the same program.)

The sizes of arrays are generally limited by the maximum size of integers supported by the computing platform, often $2^{31} - 1$. R has limited support for larger arrays, called generically "long vectors".

Sequences

If the elements of the vector are numeric, and have a simple sequence, the ":" (colon) operator can be used to generate the elements. The colon operator has higher precedence than any of the four standard arithmetic operators ("+", "−", "*", and "/"), but lower than the exponentiation operator ("^"). The seq function provides more options for generating a sequence. Examples are shown in Figure A1.42.

```
> y <- 1:9
> y
[1] 1 2 3 4 5 6 7 8 9
> x <- c(9,8,7,6,5,4,3,2,1)
> x
[1] 9 8 7 6 5 4 3 2 1
> 9:1
[1] 9 8 7 6 5 4 3 2 1
> seq(9,1,-1)
[1] 9 8 7 6 5 4 3 2 1
> -1:2
[1] -1  0  1  2
> -1:2+3
[1] 2 3 4 5
> -1:(2+3)
[1] -1  0  1  2  3  4  5
> 3*-1:2
[1] -3  0  3  6
> (3*-1):2
[1] -3 -2 -1  0  1  2
> -1:2^3
[1] -1  0  1  2  3  4  5  6  7  8
> length((3*-1):2)
[1] 6
```

FIGURE A1.42
Vectors and the Colon Operator

Indexes of Arrays

Indexes of arrays are indicated by "[ ]". Indexing of arrays starts at 1; for example, x[1] refers to the first element of the one-dimensional array x, just as is common in mathematical notation for the first element in the vector x, that is, x_1. (This can result in annoying errors for someone accustomed to using a programming language not designed for numerical computations, in which the first element is the zeroth element.)

Any set of valid indexes can be specified; for example, x[c(3,1,3)] refers to the third, the first, and again the third elements of the one-dimensional array x.

Negative values can be used to indicate removal of specified elements from an array; for example, x[c(-1,-3)] refers to the same one-dimensional array x with the first and third elements removed. The order of negative indexes or the repetition of negative indexes has no effect; for example, x[c(-3,-1,-3)] is the same as x[c(-1,-3)]. Positive and negative values cannot be mixed as indexes.

Lists

A *list* in R is a simple, but very useful structure. It differs from an atomic vector in that the elements of a list may be objects of different types, including other lists. The elements of a list can be named and accessed by their names, or else they can be accessed by an index, similar to an atomic vector, but because of the fact that the elements of a list can be other lists, the elements have a hierarchy, and a simple index in a list refers to a top-level item in the list (see portfolios[1] and portfolios[2] in the example below).

A list is constructed by the list function. The elements of the list may be given names when the list is formed. The names of the elements of the list may contain blanks. The following examples of *bad code* illustrates the usage.

```
portfolios <- list("Fund 1"=c("INTC","MSFT"),
                   "Fund2"=c("IBM","ORCL","CSCO"))
```

The list portfolios contains two elements, each of which is an atomic character vector. An element of a list may be accessed by its name, or by its index.

```
> portfolios$"Fund 1"   # Note the quotes.
[1] "INTC" "MSFT"
> portfolios$Fund2      # Note no quotes.
[1] "IBM"  "ORCL" "CSCO"
> portfolios[1]
$'Fund 1'
```

```
[1] "INTC" "MSFT"
> portfolios[2]
$Fund2
[1] "IBM"  "ORCL" "CSCO"
```

It is almost always a bad idea to put a blank in a name. An underscore can often serve whatever purpose would be served by the blank: Fund_1.

Basic Operations with Matrices and Vectors

A matrix is a two-dimensional, rectangular array. A pair of indexes is used to address an element in a matrix. Indexes of matrices in R are indicated by "[,]". Indexing starts at 1; for example, A[2,1] refers to the second element in the first column of the matrix A, similar to the mathematical notation, $A_{2,1}$.

As with vectors, any set of valid indexes can be specified, and negative values can be used to indicate removal of specified elements from a matrix. Positive and negative values cannot be mixed as indexes.

The dim function returns the number of rows and the number of columns in a matrix. *Note that the argument to* dim *must have two dimensions; it cannot be a vector.*

A matrix is constructed from a vector of numeric values by the matrix function, in which by default the elements of the vector are used to fill the columns of the matrix. Both of the following statements yield the same matrix.

```
A <- matrix(c(1,4,7,2,5,8,3,6,9), nrow=3)
B <- matrix(c(1,2,3,4,5,6,7,8,9), nrow=3, byrow=TRUE)
```

$$A = B = \begin{bmatrix} 1 & 2 & 3 \\ 4 & 5 & 6 \\ 7 & 8 & 9 \end{bmatrix}.$$

A matrix can also be constructed by binding vectors or matrices as the columns of the matrix (the cbind function) or by binding vectors or matrices as the rows of the matrix (the rbind function).

The apply function (or one similar) should be used to avoid looping over the rows or columns. These features are illustrated in Figure A1.43.

Most operators such as "+", "−", "*", and "/" are applied elementwise when the operands are arrays. The symbol "*", for example, indicates the the elementwise product. In the expression

```
A * B
```

```
> r1 <- c(1,2,3); r2 <- c(4,5,6); r3 <- c(7,8,9)
> c1 <- c(1,4,7); c2 <- c(2,5,8); c3 <- c(3,6,9)
> AA <- rbind(r1,r2,r3)
> BB <- cbind(c1,c2,c3)
> AA
   [,1] [,2] [,3]
r1    1    2    3
r2    4    5    6
r3    7    8    9
> dim(AA)
[1] 3   3
> BB
     c1 c2 c3
[1,]  1  2  3
[2,]  4  5  6
[3,]  7  8  9
> AA["r1", ]
[1] 1 2 3
> AA["r2", ]
[1] 4 5 6
> AA["r1",2]
r1
 2
> AA[r1,2]
r1 r2 r3
 2  5  8
> cbind(BB,AA)
   c1 c2 c3
r1  1  2  3 1 2 3
r2  4  5  6 4 5 6
r3  7  8  9 7 8 9
> apply(AA, 1, mean)
r1 r2 r3
 2  5  8
> apply(AA, 2, mean)
[1] 4 5 6
```

FIGURE A1.43
Matrices

the number of rows of A must be the same as the number of rows of B, and the number of columns of A must be the same as the number of columns of B.

Cayley multiplication of matrices, on the other hand, is indicated by the symbol "%*%". The expression

```
A %*% B
```

indicates the Cayley product of the matrices, where the number of columns of A must be the same as the number of rows of B.

Subarrays can be used directly in expressions. For example, the expression

```
A[c(1,2), ] %*% B[ ,c(3,1)]
```

yields the product

$$\begin{bmatrix} 1 & 2 & 3 \\ 4 & 5 & 6 \end{bmatrix} \begin{bmatrix} 3 & 1 \\ 6 & 4 \\ 9 & 7 \end{bmatrix}.$$

The transpose of a vector or matrix is obtained by using the function t:

```
t(A)
```

Data Frames

The matrix class in R can be of mode and type numeric or character, but a matrix object cannot contain elements of both numeric and character modes. The R class data.frame, on the other hand, allows any column to be numeric, character, date, or other mode. Each variable or column can be of any mode.

An example of an R data frame called "Stocks" is shown below.

```
  Symbol    Price Quantity
1   AAPL   155.15      200
2    BAC    31.06      400
3   INTC    43.94      400
4   MSFT    81.92      300
```

There are three variables, "Symbol", which is a character variable, "Price", and "Quantity". This data frame was produced from three vectors, each with a name corresponding to the name of the variable, and containing the corresponding values of the variables, as shown in Figure A1.44 below.

The default action is to convert any character data in a data frame into a variable of class factor. (To prevent this, use stringsAsFactors=FALSE.)

A data frame is a special kind of a list. As a list, its components names (the column names or row names) may contain blanks or may begin with a numeral. It is almost always bad practice to use such names, however.

A variable in a data frame can be accessed by the name of the data frame followed by $ and then by the name of the variable. Note that $ cannot be used to extract columns (or variables) in a matrix.

The rows of data frames can be named. In some cases, it may be more appropriate for a variable to be used as an index to the rows of a data frame than to be a variable on the data frame. This is illustrated in the two data frames, `Stocks` and `Stocks1` in Figure A1.44. A type of financial dataset for which this may be appropriate is one in which the observations (rows) correspond to a sequence of dates, as in a time series. We will illustrate this in Figure A1.50 on page 175.

Figure A1.44 illustrates the creation of a data frame and some simple manipulations with it.

Data frames can be combined or updated using `cbind` and `rbind`, just as with matrices. Data frames can also be combined in other meaningful ways using the `merge` function. We will discuss these operations for special kinds of data frames, beginning on page 178.

A variation of a data frame is a *tibble*, which is an object of class `tibble`. A tibble is similar to a data frame, but some operations such as printing and subsetting work differently. The `dplyr` package provides several functions for manipulating and searching data in a tibble. Tibbles are said to be a "modern" version of data frames.

Functions in R for Sequences

R has a number of functions for operating on time series or other sequences. The `acf` function, for example, computes and optionally plots the ACF, or SACF (sample autocorrelation function), of a series, as in Figure 1.30 on page 101. Analysis of time series is generally performed in the context of a fairly strong model, such as an ARMA or GARCH model. (R has several useful functions for analysis of these models, and there are also a number of packages for parametric time series analysis. We discuss some of these in Chapter 5.)

The `pracma` package also includes a number of utility functions, such as `movavg` to compute various types of moving averages, such as shown in Figure 1.9 on page 40.

Analysis of time series data often involves comparing a value at one time with previous, or lagged, values. Both simple returns, as in equation (1.17), and log returns, as in equation (1.21) involve a present price and the price at a previous time.

Lagged values are easy to obtain in R just by use of the indexes to a vector. For example, if time series data are stored in the R vector `x` of length `n`, then the vector `x[-n]` is the lagged vector of `x[-1]`.

x	x_1	x_2	x_3	$\cdots$	x_{n-2}	x_{n-1}	x_n
x[-n]	x_1	x_2	x_3	$\cdots$	x_{n-2}	x_{n-1}	
x[-1]		x_2	x_3	$\cdots$	x_{n-2}	x_{n-1}	x_n
x[-c(n-1,n)]	x_1	x_2	x_3	$\cdots$	x_{n-2}		
x[-c(1,2)]			x_3	$\cdots$	x_{n-2}	x_{n-1}	x_n

R has a function `diff` that computes the differences between the values in a vector and the previous values. For the R vector `x` of length `n`, `diff(x)` yields the vector of length `n-1` with values `x[2]-x[1]`, ... `x[n]-x[n-1]`.

```
> Symbol    <-  c("AAPL", "BAC","INTC","MSFT")
> Price     <- c(155.15, 31.06, 43.94, 81.92)
> Quantity <- c(  200,    400,   400,   300)
> Stocks    <- data.frame(Symbol, Price, Quantity)
> Stocks
  Symbol  Price Quantity
1   AAPL 155.15      200
2    BAC  31.06      400
3   INTC  43.94      400
4   MSFT  81.92      300
> Stocks$Price
[1] 155.15  31.06  43.94  81.92
> Stocks[1,2]
[1] 155.15
> Stocks[2,3]
[1] 400
> Stocks[1, ]
  Symbol  Price Quantity
1   AAPL 155.15      200
> names(Stocks)
[1] "Symbol"   "Price"    "Quantity"
> colnames(Stocks)
[1] "Symbol"   "Price"    "Quantity"
> Stocks[Price>50,]
  Symbol  Price Quantity
1   AAPL 155.15      200
4   MSFT  81.92      300
> Stocks1 <- Stocks[,-1]
> rownames(Stocks1)<-Stocks$Symbol
> Stocks1
      Price Quantity
AAPL 155.15      200
BAC   31.06      400
INTC  43.94      400
MSFT  81.92      300
> Stocks1[c("BAC","MSFT"),]
     Price Quantity
BAC  31.06      400
MSFT 81.92      300
```

FIGURE A1.44
An R Data Frame

The R function `diffinv` computes the inverse of the `diff` function, starting at `xi=x[1]`. The R function `cumsum` that computes the cumulative sums of a vector can also be used to get the inverse of the `diff` function.

Another useful function in R is `rev`, which reverses the order of the elements in a vector.

Functions in R for Returns

Returns, either simple or log returns, can be computed easily using the `diff` function.

If a beginning price is known, a sequence of prices can be computed from a sequence of returns using the `diffinv` function or the `cumsum` function.

If a sequence of log returns is permuted (as, for example, reversed), the ending price is the same (this follows from equation (1.26)). These functions are illustrated with a simple numeric vector in Figure A1.45.

```
> x <- c(2,5,3,7,1)
> x_difs <- diff(x)
> x_SimpleReturns <- diff(x)/x[-length(x)]
> x_LogReturns <- diff(log(x))
> # the following statements all retrieve the original series
> diffinv(x_difs,xi=x[1])
[1] 2 5 3 7 1
> cumsum(c(x[1],x_difs))
[1] 2 5 3 7 1
> diffinv(x_SimpleReturns*x[-length(x)],xi=x[1])
[1] 2 5 3 7 1
> x[1]*exp(diffinv(x_LogReturns))
[1] 2 5 3 7 1
> x[1]*exp(diffinv(rev(x_LogReturns)))
[1] 2.0000000 0.2857143 0.6666667 0.4000000 1.0000000
```

FIGURE A1.45
Differences and Returns; R functions

We should note that the `diff` function works slightly differently (get it?) on other types of objects, such as `xts` objects. To ensure that `diff` works correctly, we sometimes use the statement

```
x_LogReturns <- diff(log(as.numeric(x)))
```

We must do this for `xts` objects, for example (see below).

Graphics in R

R has several functions for producing graphics. All of the graphics in Chapter 1 were produced by R.

Figures 1.1 through 1.7 were all produced by the `plot` function. These figures have different layouts determined by the `par` function, and different line types, labels, and titles, all determined by arguments to the `plot` function. The argument `type="l"`, for example, yields a plot with line segments from one point to the next, and `type="b"` plots the points as well as connecting line segments.

When there are multiple graphs in the same display, as in Figures 1.5, 1.6, 1.9, and 1.13, it is necessary to distinguish one from another. This can be done by using different line types, different characters for points, and different colors. The argument `lty=1`, for example, yields solid line segments, and `lty=2` yields dashed line segments. The argument `pch` causes a specific character to be used to plot a point. (The possible values of `pch` are small positive integers that represent different characters.) The argument `col` causes a specific color to be used to plot a line or a point. Some colors can be specified by name, for example, `col="red"` specifies some shade of red. Colors can also be specified using an additive RGB scheme. A color in this scheme is specified by six hexadecimal digits preceded by "#", in which the first two digits specify the intensity of red, from 0 (none) to 254 (full saturation), the second two digits specify the intensity of green, from 0 to 254, and the last two digits specify the intensity of blue. For example, `col="#00FF00"` represents green, `col="#FF00FF"` represents magenta, and `col="#780078"` represents a light magenta. The default color for a line and for points is black, `col="#000000"`, that is, no reflection. (White is `col="#FFFFFF"`, full reflection of all colors.)

The histograms in Figures 1.19, 1.20, and 1.24 were produced by the `hist` function. The normal density curve in Figure 1.24 was added using the `curve` function with `col="green4"`.

The q-q plots in Figures 1.27 and 1.28 were produced by the `qqnorm` function, and the straight line segment was added using the `qqline` function with `col="green4"`.

The appearance of a plot produced by the generic `plot` function may depend on the type of object being plotted. It produces slightly different plots for numeric objects, time series objects and `xts` objects, for example. `ggplot` does not plot `xts` objects without transformations. We discuss some of the issues in time series plots in this appendix beginning on page 179.

We will discuss types of graphics and their applications in statistical analysis in more detail in Chapter 2. In Section 2.4.7, we summarize the R functions for various types of graphs.

Dates in R

In analysis of financial data, it is often necessary to determine the length of time between two given dates, to match the dates in two different time series, or to perform other operations on dates. One way to do this would be to convert all dates to numbers representing the time (days, hours, or whatever) from some fixed starting point. This of course is not a very convenient way of specifying dates. We would prefer something that we can immediately relate to a date on the ordinary calendar.

The International Standard ISO 8601 specifies representation of a date in the form "1987-10-19" (but also allows the form "19871019"), and the Portable Operating System Interface (POSIX) standards adopted by the IEEE Computer Society uses dates in this general format. The ISO 8601 format also allows for date plus time in hours, minutes, and seconds. The full format is *yyyy-mm-dd hh:mm:ss*. All fields have a fixed number of integer characters except "ss", which can have an appended decimal fractional part. The fields must be filled from the left, and the last field on the right indicates the full time period specified; for example, "1987-10" refers to the full month of October, 1987. I will generally refer to this form of representation, or most often, just the date portion of this, as the "POSIX format".

The base R package provides a Date class and a function as.Date to work with dates specified in ways that people usually express dates. The as.Date function accepts dates in various formats that the user can define using the symbols in Table A1.13. The formats can also be used in the format function for printing dates in different forms. See Figure A1.46, for examples.

TABLE A1.13
Some Date Formats for as.Date

Code	Meaning
%d	Day of the month (decimal number)
%m	Month (decimal number)
%b	Month (3-letter abbreviation)
%B	Month (full English name)
%y	Year (2 digits; breaks between 68 and 69)
%Y	Year (4 digits)
%j	Day of the year

The date function in the lubridate package provides more flexibility in working with dates, and it is the one I will use most often.

Operations on Date Data

Date ranges can be indicated by "/"; for example, "198710/19871120" represents October 1 through November 20, 1987.

Numeric data can be added to Date data (but not multiplied with it). Objects of class Date cannot be added, but one object can be subtracted from another. When an object of class Date is subtracted from another, an object of class difftime is created. The base R package also provides utility functions, such as weekdays, for working with objects of class Date. The seq function also works with Date data. The by parameter in seq can be "days", "weeks", "months", or "years". The R function strftime is useful for getting dates in different formats. It operates on vectors, and is often used in conjunction with the as.numeric function, as shown in the example. The format argument specifies the format of the output. The functions and operators are aware of leap years and other aspects of the Gregorian calendar.

```
> as.Date("2016-02-28")+1:2
[1] "2016-02-29" "2016-03-01"
> as.Date("2017-03-02")-1:2
[1] "2017-03-01" "2017-02-28"
```

Figure A1.46 illustrates some of these properties and operations, using only the base package. (Figure A1.50 on page 175 illustrates some uses of the date function in the lubridate package.)

The base package of R also includes two functions, as.POSIXct and as.POSIXlt, that handle time as well as dates. The POSIXct and POSIXlt classes carry information about time zones and daylight savings time.

Generating Simulated Data in R

A useful approach for studying methods of analyzing real financial data is to generate pseudorandom data that matches models of the behavior of the real data. We may generate data that are similar to data from a Gaussian distribution or from a distribution with heavy tails, for example. In Sections 3.3.1 and 3.3.2, we mention some ways that "random" data can be generated.

R provides several functions to generate pseudorandom data, and these functions are used in many examples and exercises. The basic methods implemented in these R functions are described in Section 3.3.3. In that section, we describe how a "seed" can be set to control the "random" data and we mention how the basic method of introducing "randomness" can be chosen by use of the RNGkind.

In all of the examples of simulating random data in this book, the **kind** variable in RNGkind is "**Mersenne-Twister**", which is the default.

A1.2   Data Repositories and Inputting Data into R

Accessing data and getting it into an appropriate form for analysis are necessary preliminary steps in any analysis of data. In fact, often these chores constitute the bulk of the computer operations in an analysis. The first consideration in this phase is the organization and format of the data.

Data: Observed or Derived

Some data can be directly observed, such as stock price data, and other data are derived, such as volatility, either historical or implied. This simple dichotomy masks a spectrum of kinds of data. While the last price at which a stock was traded in the open market may be a fixed and observable number,

```
#    "Black Thursday"
> x <- as.Date("Oct 24, 1929",format="%b %d, %Y")
#    "Black Tuesday"
> y <- as.Date("29 October 1929",format="%d %B %Y")
#    "Black Monday"
> z <- as.Date("10/19/87",format="%m/%d/%y")
> class(x)
[1] "Date"
> mode(x)
[1] "numeric"
> format(z, "%b %d, %Y")
[1] "Oct 19, 1987"
> weekdays(c(x,y,z))
[1] "Thursday" "Tuesday"  "Monday"
> x
[1] "1929-10-24"
> x+5
[1] "1929-10-29"
> x-y
Time difference of -5 days
> y
[1] "1929-10-29"
> class(x-y)
[1] "difftime"
> z-x
Time difference of 21179 days
> seq(x, length=6, by="days")
[1] "1929-10-24" "1929-10-25" "1929-10-26" "1929-10-27"
[5] "1929-10-28" "1929-10-29"
> seq(x, length=6, by="weeks")
[1] "1929-10-24" "1929-10-31" "1929-11-07" "1929-11-14"
[5] "1929-11-21" "1929-11-28"
> strftime("2017-12-31", format="%j")
[1] "365"
> as.numeric(strftime(c("2017-01-01","2017-12-31"), format="%j"))
[1]   1 365
```

FIGURE A1.46
Examples of Date Data and Functions to Work with It

the price, at any point in time, may not be observable, but rather may be assumed to be somewhere between two observable numbers, the public bid and ask prices.

Derived data may be the result of simple calculations on observed data, such as historical volatility, or it may be derived by application of rather complex models of observed data, such as the implied volatility as measured by the VIX.

It is important for the data analyst to have at least a subjective feeling for how any data relates to observable data. The PE ratio, for example, is an observable number divided by a reported number in which there is a degree of arbitrariness (see page 41).

Data Structures

Data structure refers to the way data are organized conceptually. (The data structure is not necessarily the way the bits representing the data are organized and stored.) One of the simplest data structures, and one that is best for most statistical analyses, is a two-dimensional structure, similar to a matrix or an R data frame. "Columns" are associated with variables, price, return, and so on, and "rows" are associated with an observational unit or, especially in financial data, the observations corresponding to a particular point in time. This data structure is sometimes called a *flat file*.

Data can often be stored in a better way for presentation purposes. For example, the data on interest rates of Treasuries of various terms at various points in time is probably best displayed as in Table 1.2 on page 23 in which the columns represent different terms, and the rows represent different points in time. As a statistical dataset, however, a better representation identifies the variables, and assigns each to a column, as shown in Table A1.14.

TABLE A1.14
Interest Rates of Treasury Bills, Notes, and Bonds; Same Data as in Table 1.2 on Page 23

Date	Term	InterestRate
January 1982	Mo3	12.92
January 1982	Yr2	14.57
January 1982	Yr5	14.65
January 1982	Yr30	14.22
June 1989	Mo3	8.43
⋮	⋮	⋮

While the display in Table A1.14 may not be as well-structured as the display in Table 1.2, the organization represented in Table A1.14 is more amenable to analysis. An R data frame has this structural arrangement. The standard operators and functions can be used to operate on these data in this form, but not in the structure indicated by Table 1.2.

The dates in the dataset in Table A1.14 could be stored either as character data, or as date data, possibly in a POSIX format, as discussed beginning on page 156. Storing them as date data allows us to use various operations on the dates, such as determining the number of days between two dates.

As we have seen, the observations (rows) in many financial datasets correspond to a sequence of dates. The dates could be stored as a variable in an

R data frame, but the dates in these datasets serve more as indexes to the dataset, rather than as a variable on the dataset.

As an example, consider the closing prices of Intel stock for the period January through September 2017. These data have been used in several examples in Chapter 1. They are stored in two different R data frames shown in Figure A1.47. In the data frame INTC_df_var shown on the left hand side, the dates are stored in a variable called date, and in the data frame INTC_df_index on the right, the dates are just an index to the data. They are the row names of the data frame. Although dates as an index is a more natural storage method, operations on dates as discussed on page 157 cannot be performed on an index. Also, row names of ordinary data frames cannot be used to label axes in plotting functions. An xts data frame, provides the ability to operate on a date index, and the plot.xts function uses the dates as labels on the time axis. We will discuss xts objects beginning on page 173.

```
#   date as a variable         ||     #   date as an index (row name)
> class(INTC_df_var)           ||     > class(INTC_df_index)
[1] "data.frame"              ||     [1] "data.frame"
> head(INTC_df_var)            ||     > head(INTC_df_index)
          Date Close           ||                    Close
1 2017-01-03 36.60            ||     2017-01-03   36.60
2 2017-01-04 36.41            ||     2017-01-04   36.41
3 2017-01-05 36.35            ||     2017-01-05   36.35
4 2017-01-06 36.48            ||     2017-01-06   36.48
5 2017-01-09 36.61            ||     2017-01-09   36.61
6 2017-01-10 36.54            ||     2017-01-10   36.54
```

FIGURE A1.47
Dates Used as a Variable or as an Index

Inputting Data

There are many different ways to input data into R. Data can be read from a file or directly from the computer console. We must tell the system three things: where the data are, how it is stored, and what R object to put it into.

We will consider some of the more common ways of inputting data here, but caution the reader that there are many details we will not go into. In Section A1.3, we will discuss ways of bringing financial data into R directly from the internet.

For data stored in a tabular form in an external file, the R function read.table, which allows various kinds of metadata (headers, field separators, and so on), produces an R data frame from the data in the external file. Because `read.table` produces a data frame, columns in the table can be numeric or character (but not date or complex).

The complementary function `write.table` can be used to create an external file and to store data in an R matrix or data frame in the external file.

An R session has a *working directory*, which can be set by the function `setwd`. (The function `getwd` without an argument obtains the name of the working directory.) Files can easily be read from or written into this directory, and objects created in an R session can be saved in the working directory.

Comma Separated Files and Spreadsheets

A simple but useful structure for storing data is called "comma separated value" (or just "comma separated") or "CSV". The storage is of plain text, and commas are used to separate the fields. The basic CSV structure is the same as that of a spreadsheet, so spreadsheet data are often stored as CSV files because they are portable from one system to another.

Although the R function `read.table` allows the user to specify various things such as field separators, the function `read.csv` is more useful in inputting data from a CSV file. Just as the `read.table` function, `read.csv` produces an R data frame.

The complementary function `write.csv` can be used to create an external CSV file and to store data in an R matrix or data frame in the CSV format in the file.

CSV files provide the simplest means for exchanging data between R and a spreadsheet program, such as Microsoft Excel. The spreadsheet program can input data from a CSV file and can save many types of data in its native format as CSV files. Most spreadsheet programs provide facilities for converting date formats, for example, from "mm/dd/yyyy" to a POSIX format.

Inputting CSV Files

The use of CSV files for inputting or saving financial data is very convenient. While it is even more convenient to input data directly from the web, sometimes it is necessary to go to the website in a browser and manually download data. The `read.csv` function in R can then be used to read the data into an R data frame. Remember the default action of R to convert any character data into a variable of class `factor`.

The first row in many CSV files is a header, that is, a row containing the names of the columns, but there is nothing in the file structure that indicates whether the first row is a header or not. There is a logical argument to the `read.csv` R function, `header`, that controls the action of the function. By default, the argument is `TRUE`, so the `read.csv` R function assumes that the first row is a header, and the values in the first row become column names of the data frame that is produce. If the first row is not to be treated as a header, the `header` argument is set to `FALSE`.

Although variable names, that is, column names, in an R data frame may contain blanks or begin with numerals (neither of which is a good idea!) and of course CSV file headers have names that contain blanks or begin with numerals, the read.csv R function will not convert such names correctly. The read.csv R function will insert a period for one or more blanks in a name in a header, and will prepend any character name that begins with a numeral with an "X.". Any numeric value that read.csv interprets as a header name will be prepended with a "V.". The Quandl dataset referred to in Exercise A1.3 has variable names of the form "3 MO". If the data is brought in using the Quandl R function, the names will be assigned correctly. If, however, the data are stored as a CSV file with a header, read.csv will not assign the column names correctly.

As an example, consider reading in the data (open, high, and so on) for Intel stock for the period January through September, 2017. These data have been used in several examples in Chapter 1. They are stored on the website for the book in the file named INTCd20173Q.csv or you can download them onto your own computer. Figure A1.48 illustrates the use of read.csv to read directly from the website for the book. This produces an R data frame. Note the use of stringsAsFactors. The data frame is converted into an xts object in the manner illustrated in Figure A1.46 on page 159. For date formats and other information on conversions, see Table A1.13.

```
> INTCdat <- read.csv(
+          "http://mason.gmu.edu/~jgentle/books/StatFinBk/Data/INTCd20173Q.csv",
+          header=TRUE, stringsAsFactors=FALSE)
> class(dat)
[1] "data.frame"
> head(INTCdat, n=2)
      Date  Open  High   Low Close Adj.Close    Volume
2017-01-03 36.61 36.93 36.27 36.60  35.35268  20196500
2017-01-04 36.71 36.77 36.34 36.41  35.16915  15915700
```

FIGURE A1.48
Inputting Data from a CSV File

Data Repositories

Many websites provide current financial data for free. Some web portals provide not just current data, but historical data for free, and many of the agencies mentioned in Section 1.2.1 also freely provide financial data. Some websites allow selection of data within a specified time interval, but many do not; the data provided are for the full range of time of the dataset.

In this section, we list some of the most important and useful repositories. In Section A1.3 we discuss R software to access data in the repositories.

Google has built an "online community" called Kaggle that contains a large number of datasets that can be downloaded, mostly as csv files. Google also provides a useful tool to search for datasets, financial and otherwise, at

> `toolbox.google.com/datasetsearch`

Although Google Dataset Search gives priority to Kaggle, it also identifies other sources of data.

Although our emphasis in this book is on dynamic current data, there are static repositories that may be of interest. These occasionally are useful in examples to illustrate statistical techniques. Some of these static datasets are provided in R packages.

Financial and Economic Data

The kinds of data we will be most concerned about in this book are stock prices. Certain stock prices, such as the closing prices, are directly observable. Observable data are available in a variety of repositories. Much other data, such as returns or stock indexes, can be derived simply from the observable data, and some are available in standard repositories.

There are various other kinds of derived data that we will mention from time to time, such as the US Gross Domestic Product (GDP), the Consumer Price Index (CPI), and a Case-Shiller Index of housing prices. These data are *government statistics*. While they purport to measure something that superficially is well-defined, government statistics, from any country, are what the government of the country defines them to be.

Copyrighted Data

Data *per se* is not covered by US copyright law. US copyright law does cover the "organization and presentation" of data, however, and so occasionally there are some murky areas in the use of "copyrighted data". (Information provided by Moody's, for example, includes a statement that it is "protected by copyright and other intellectual property laws". In FRED's graphical presentations of Moody's data, FRED includes a statement that "Data in this graph are copyrighted". It is not clear what that means. The electronic data from Moody's that is supplied by FRED is not copyrighted.)

Yahoo Finance

One of the best sources for historical stock price data is Yahoo Finance, at

> `finance.yahoo.com/`

Most of the data used in the illustrations and examples in Chapter 1 and this chapter were obtained from Yahoo Finance using the R function `getSymbols` in the `quantmod` package written by Jeffrey A. Ryan.

Software for Reading Data from the Internet; quantmod

The getSymbols function allows specific time periods for data from Yahoo Finance to be specified by use of the from and to arguments. The Yahoo Finance data generally go back to January 1962, or to the date of the initial public offering. Unless the from date specified in the getSymbols function is a long time prior to the IPO, getSymbols detects the missing values prior to the IPO, and returns only the valid data. For stock symbols that have been retired, for example BLS (BellSouth) or VCO (Concha y Toro ADR), getSymbols may or may not return some of the valid values. The function in that case may just return an error condition.

The frequency, daily, weekly, or monthly, can be specified in getSymbols using the periodicity keyword. The beginning date can be specified using from and the ending date using to. Daily frequency is the default. The to date is "to", and not "through". See Figure A1.49, for example, and see pages 171 and 186 for more information on using getSymbols to get data from Yahoo Finance.

Figure A1.49 illustrates the use of getSymbols to read stock prices directly from the internet. This produces an xts object with the same name as the stock symbol. Notice that for weekly data, the dates are Sundays. The price

```
> library(quantmod)
> getSymbols("INTC", from="2017-01-01", to="2017-10-1", periodicity="weekly")
> class(INTC)
[1] "xts" "zoo"
> head(INTC, n=2)
           INTC.Open INTC.High INTC.Low INTC.Close INTC.Volume INTC.Adjusted
2017-01-01     36.61     36.93    36.19      36.48    65212200      34.17867
2017-01-08     36.48     37.00    36.32      36.79    93230700      34.46911
```

FIGURE A1.49
Inputting Data from the Internet

data, however, do not exactly match any specific day of the previous week.

In Figure A1.49, the name of the R xts object holding the data is INTC. To produce an xts object with a different name, use the argument env=NULL and assign a name to the object:

```
z <- getSymbols("INTC", env=NULL, from="2017-01-01", to="2017-10-1")
```

The getSymbols function has an argument to specify the source of the data, src, which has as default Yahoo (see page 164).

The Yahoo Finance stock price data can also be obtained by using a web browser, going to the main website, entering the stock symbol in the search box, and then choosing "Historical Data". This brings up a header and a table of stock Open, High, Low, Close prices, and Volume. This is the OHLC data format plus Volume (see page 181). The default time period is the one-year period prior to the current date. The default frequency is daily. Both defaults can be changed on the webpage.

The data can be saved using the "Download Data" button in a web browser. The data are saved in a CSV file, which, of course, can be read into an R data frame using `read.csv`. (It should be noted that the data are organized in reverse chronological, and in most plots and analyses, we want the data to be in chronological order.)

The `to.period` function in the `xts` package can also be used to extract data into a coarser period. If the data are OHLC, the extraction is done in such a way as to obtain valid highs and lows for each period.

The Yahoo Finance data currently (mid-year, 2019) go back to January 1962 for stocks that were trading then. Data on stocks whose public trading began after that date may or may not be available from the initial public offering. Data on delisted stock symbols may or may not be available during their periods of is public trading.

In order to obtain stock or ETF price data, we must know the symbol. These symbols are standard, and are used by brokerage houses and other financial institutions. Yahoo Finance can be used to look up stock symbols. Enter "Intel" and "INTC" is returned, for example.

The stock symbol is also used as the name of a dataset that is accessed by the `getSymbols` function. Many different datasets can be accessed by `getSymbols`, of course. Each must be accessed by the correct name. For stock data in Yahoo Finance, the name is the same as the stock symbol; for other datasets, the name is a character string assigned by the data repository.

Stock indexes are not traded, and hence do not have standard symbols. Data repositories such as Yahoo Finance use their own symbols for indexes. Yahoo Finance prefixes their symbols for indexes with a caret. Table A1.15 shows the symbols used in Yahoo Finance for various indices. Other symbols are also used for some of these indexes both by Yahoo Finance and by other data providers. For example, the symbols .INX and $SPX are also used for the S&P 500 by various data providers.

The character strings shown in Table A1.15 are names of data collections in Yahoo Finance that can be accessed using `getSymbols`.

International Markets

There are indexes for stocks traded in various international markets, such as those shown in Table 1.7. Many of these indexes are available at Yahoo Finance. Table A1.16 shows the symbols for some of these indexes.

These indexes are quoted in the currency of the country of the market, as

TABLE A1.15
Index Symbols (Yahoo Finance)

Dow Jones Industrial Average	^DJI
Dow Jones Transportation Average	^DJT
Dow Jones Utility Average	^DJU
S&P 500	^GSPC
S&P 500 Total Return	^SP500TR
Nasdaq Composite	^IXIC
Russell 2000	^RUT
CBOE Volatility Index	^VIX
Large Cap VolDex®	^VOLI

TABLE A1.16
International Markets Index Symbols (Yahoo Finance)

FTSE-100 (British pound, GPD)	^FTSE
DAX (euro, EUR)	^GDAXI
CAC-40 (euro, EUR)	^FCHI
Nikkei-225 (Japanese yen, JPY)	^N225
Hang Seng (Hong Kong dollar, HKD)	^HSI
SSE Composite (Chinese yuan, CNY)	^SSEC

shown in Table A1.16. Also, in that table, the standard abbreviation of the currency is shown. The currency exchange rates can be obtained from FRED, which uses its own abbreviations, as shown in Table A1.17. For example the daily rate for US dollars in euros is DEXUSEU and the daily rate for Japanese yen in US dollars is DEXJPUS.

How currency exchange rates are stated follows a tradition that began when the United Kingdom was the major economic power. The standard quotation is US dollars per UK pound, for example, but it is Japanese yen per US dollar. The direction of the euro/dollar rate (US dollars per euro), for some reason, was chosen to be the same as the direction of the pound/dollar rate. (This is relevant in the conversions in Exercise A1.25.)

FRED and Other Federal Reserve System Data Sources

The Research Division of the Federal Reserve Bank of St. Louis (the Eighth District of the Federal Reserve System) maintains a number of databases, collectively referred to as "FRED" (Federal Reserve Economic Data), that contain various types of economic data.

The main website is

                    `fred.stlouisfed.org/`

The website is very well organized, and allows browsing by various categories.

A good way to access data from FRED is by use of the `getSymbols` function in the `quantmod` package. The `periodicity` keyword and the `from` and `to` keywords are ignored when `getSymbols` accesses the FRED site; rather each dataset at FRED has a fixed frequency and a fixed beginning and end (usually the business day, week, or month prior to when the site is accessed).

To access data from FRED, one must know the name of the dataset. The various datasets in FRED are given mnemonic names, which can be found manually at the website. For example, the interest rates on the three-month US Treasury Bill in the secondary market shown in Figure 1.4 goes by the name TB3MS.

```
getSymbols("TB3MS", src = "FRED")
```

The datasets provided by FRED cover specific times, usually up to the present. The data in TB3MS, for example, are for the period January 1934 to the present.

Names of some common series available at FRED are shown in Table A1.17. These series begin at different times; some go back as far as the early twentieth century, while others were started as late as the early twenty-first century. Also, the weekly series may be computed on different days of the week; for example, the weekly fed funds rate is as of Wednesday and the weekly Moody's bond rates are as of Friday.

Other data from the Federal Reserve System can be obtained at

                `www.federalreserve.gov/data.htm`

There is a wealth of data at this site. Different types of data may be stored in different formats.

As with data from FRED, to access the data from a computer program, it is necessary to know the dataset name. The name can be obtained by visiting the website and searching for the data of interest. In many cases, for a given search term, say "japanese yen exchange", there will be several matches. In the case of "japanese yen exchange", there are daily, monthly, and annual data files, all going back to 1971.

US Treasury Department

The US Treasury Department provides financial data online. The main website is

                    `home.treasury.gov/`

Yield curve data are probably the most commonly used data from the Treasury, and is available at

TABLE A1.17

Some Economic Time Series in FRED

US Gross Domestic Product (quarterly) annual rates, seasonally adjusted	GDP
US Gross Domestic Product (annually)	GDPA
US Consumer Price Index for All Urban Consumers	CPIAUCSL
Case-Shiller US National Home Price Index	CSUSHPINSA
Case-Shiller US National Home Price Index, seasonally adjusted	CSUSHPISA
15-Year Fixed Rate Mortgage Average	MORTGAGE15US
Effective Fed Funds Rates (monthly)	FEDFUNDS
Effective Fed Funds Rates (weekly)	FF
3-Month LIBOR	USD3MTD156N
3-Month US T-Bill Rates, Secondary (monthly)	TB3MS
3-Month US T-Bill Rates, Secondary (weekly)	WTB3MS
3-Month US T-Bill Rates, Secondary (daily)	DTB3
2-Year Treasury Rates, Constant Maturity (weekly)	WGS2YR
10-Year Treasury Rates, Constant Maturity (weekly)	WGS10YR
30-Year Treasury Rates, Constant Maturity (weekly)	WGS30YR
Moody's Seasoned Aaa Corporate Bonds Rates (weekly)	WAAA
Moody's Seasoned Baa Corporate Bonds Rates (weekly)	WBAA
U.S. / Euro Foreign Exchange Rate (daily)	DEXUSEU
U.S. / U.K. Foreign Exchange Rate (daily)	DEXUSUK
Japan / U.S. Foreign Exchange Rate (daily)	DEXJPUS
Canada / U.S. Foreign Exchange Rate (daily)	DEXCAUS

```
data.treasury.gov/feed.svc/DailyTreasuryYieldCurveRateData
```

The `Quandl` function in the R package `Quandl` accesses data from the Treasury site. See page 26 for an example that uses `Quandl`. Given observed rates and terms of any class of instrument, the R package `YieldCurve` produces yield curves using various methods, such as Nelson-Siegel or Svensson.

Other Sources

Publicly-traded companies and financial institutions are required by law annually to file various reports with the US Securities and Exchange Commission. The reports include financial data on their operations, such as revenues, cost of revenue, earnings per share, and so on, as well as general information about the company. A comprehensive annual report is submitted in a form called *10-K*.

These forms can be accessed through a system called EDGAR, the Electronic Data Gathering, Analysis, and Retrieval system. EDGAR also is used by the reporting companies to perform automated collection, validation, indexing, acceptance, and forwarding of submissions by the companies.

The corporate information filed with EDGAR is publicly and freely available. This is an important resource for fundamental analysis of corporate securities.

The main web site is

> www.sec.gov/edgar/searchedgar/companysearch.html

The R package `finreportr` contains functions to scrape data from the SEC site.

Several other US agencies collect various types of data. The US Census Bureau collects population data; the Bureau of Economic Analysis collects information on economic indicators, national and international trade, accounts, and industry; the Bureau of Labor Statistics measures labor market activity, working conditions, and price changes in the US economy; the Energy Information Administration provides data on US use of coal, natural gas, and other energy sources; and the National Agricultural Statistical Service collects data on food production and supply. The website

> www.usa.gov

is a portal to websites maintained by these and other individual government agencies that make data available publicly.

The World Bank, through its Open Data initiative, provides a wide range of financial and economic data. It is categorized by geographic region as well as by data type. It has a well-designed web site at

> datacatalog.worldbank.org/

The financial services and brokerage company, Oanda, is a good source of data on forex and metal prices. Its main web site is

> www.oanda.com/

FX data can be accessed from Oanda using `getFX` and metal prices can be accessed using `getMetals`, both in the `quantmod` package.

A source of interesting but rather eclectic financial data is StockSplitHistory.com. Its main web site is

> www.stocksplithistory.com/

Another source is Quandl, which is a private company that provides access to a large compilation of financial data from various governmental, quasi-governmental, and corporate sources. Most of the datasets obtained from government agencies and central banks are made freely available. Access to other types of data requires a subscription to Quandl. Access above a certain usage threshold (currently 50 retrievals per day) requires an "API key", which requires a free registration.

Quandl is also the name of the API ("applications program interface") used to access data from Quandl. Quandl supports two data formats, one for time series and the other a tabular format. The structure of the data and options for obtaining the data depend on whether it is time series or "datatable".

Datasets are identified by "Quandl code", of the form *"source/dataset"*; for examples, the Quandl code for the US GDP data from FRED is FRED/GDP, the Quandl code for the US Dollar Euro exchange rate is CURRFX/EURUSD, and the Quandl code for daily rates for US Treasuries is USTREASURY/YIELD. The Quandl function in the R package Quandl accesses data from the web site idendified by the source. The Quandl function returns the data in the form of an R data frame. Dates are in a data frame variable Date, and are stored in standard POSIX format. The Quandl dataset USTREASURY/YIELD begins with data on January 2, 1990, and goes until the last trading day prior to the date of access. It contains many missing data values because some maturities were not offered continuously for that period.

The Center for Research in Security Prices (CRSP) at the University of Chicago sells various types of financial data, including various stock indexes. The CRSP database includes daily prices of all NYSE stocks going back to 1926, as well prices of a number of other US securities, and other financial data.

There are a number of other private firms that provide financial data and software, generally for a fee. Capital IQ, which is a technology and financial services company that acts as the research division of Standard & Poor's, is a widely-used resource for investment managers, investment banks, private equity funds, advisory firms, corporations and universities.

Inputting Data from Internet Data Repositories

There are various sources of financial data on the internet and data can be input to R in a number of ways. The data in many internet repositories are stored in CSV files. These data can often be read directly into R, as illustrated in Figure A1.48 on page 163.

The getSymbols in the quantmod package is user-friendly, and provides access to a variety of financial data at various repositories. Arguments in getSymbols may be specific to the data source and/or to the type of data. For example, the from and to arguments in getSymbols are not implemented when the source is FRED. Whenever specific time periods are not selected in getSymbols data just for those periods can be put into an R object using the subsetting mechanism for xts objects described in Figure A1.53.

There are also other functions in quantmod for retrieval of specific types of financial data, such as getDividends, getFX, getMetals, and getOptionChain.

For a specified stock, the getOptionChain function returns the option chain, which, for each expiry, is a list with two components, one for calls and and one for puts. For each strike price, each component consists of last, bid, and ask prices, volume (of previous trading day), and open interest for the option.

The quantmod functions are also used in other R packages; for example, the tq_get function in the tidyquant package and the get.hist.quote function in the tseries package use the data acquisition functions in quantmod.

Another useful package for processing financial data is TTR by Joshua Ulrich. This package includes several functions for smoothing financial data.

There are also a number of R packages and other software to process HTTP requests and data acquisition from the internet. Some of these are listed and briefly described, with links, under the "WebTechnologies" section of the task views webpage:

cran.r-project.org/web/views/

A1.3   Time Series and Financial Data in R

The R stats package provides extensive capabilities for working with time series, including functions to compute the ACF and to fit various linear time series models (to be discussed in Chapter 5). The stats package also provides a time series class, ts, which causes some R functions to produce results that are more appropriate for a time series (such as the generic plot producing a graph with connecting line segments). There are also special functions in the stats package for working with ts objects, such as the diff function, which produces a time series of differences with a specified lag. The filter function can be used to compute various kinds of filters. For a time series in x, a d-period moving average, for example, can be produced by

```
xdma <- filter(x, filter=rep(1/d,d), sides=1)
```

The first few items in a sequence of moving averages can be computed in various ways. Most R functions, including filter, produce NAs in the first d-1 positions of the output. Another way, as suggested on page 39, is to use the available data from the beginning of the sequence. This is done in the MATLAB function movavg available in the pracma package. Moving averages are easy to produce in R just using the mean function and adjusting the indexes. The rollapply function in the zoo package allows moving or "rolling" applications of general functions over specified window widths. You are asked to compute some moving averages in Exercise A1.5.

There are a number of R packages for working with and analyzing time series. Some of these packages allow more flexibility in date data and some provide special classes for time series objects, and they provide functions for various kinds of analyses of time series. We discuss and illustrate R packages for time series in Chapter 5.

Class ts only deals with numeric time stamps. Various packages implement irregular time series based on POSIXct time stamps. These are especially important in financial applications, because, as we have pointed out in Chapter 1, weekends and holidays make most financial data unequally spaced.

The R package `timeDate` contains functions to provide financial date and time information, including information about weekends and holidays for various stock exchanges.

The package that I will use most often in this book is `xts`, written by Jeffrey A. Ryan, Joshua M. Ulrich, and Ross Bennett. This package is based on the `zoo` ("z- ordered observations") package. It is included in the `quantmod` package.

Time Series Objects

As mentioned above, `ts` is the basic time series class. It provides for regularly spaced time series using numeric time stamps; that is, an integer index. The `xts` class is much more powerful. It allows irregularly spaced time series and uses arbitrary classes for the time stamps, including date classes.

The package `xts` uses objects of class `xts`. Here, we will consider some features of an `xts` R object.

xts Objects

Much of the data used in statistical analyses can be contained in a numeric matrix. The financial data we considered in Chapter 1 consisted primarily of numbers. These data could have been (and generally were) just stored in R atomic vectors or matrices. Data frames allow data of different types and provide more metadata and through it allow for a wider range of operations. For a time series, we may have one column (variable) that contains the date, or we could use dates as names of the rows (observations), but neither of these methods would allow the kinds of operations we might wish to perform on the time series data, such as subsetting by dates or merging two time series by date.

An object of class `xts` has a row index that can be accessed by the `index` function. The row index is usually an object of class `date`, and date operators can be used to manipulate the individual rows.

An object of class `xts` provides not only for date stamping, but it allows for operations on the object via date functions and operators. On the other hand, sometimes computations are easier to perform by first converting `xts` objects to numeric objects using `as.numeric`. Of course, they can also be converted to R data frames, and sometimes this makes processing the data simpler. Conversion of an `xts` object to a data frame is straightforward using `data.frame`. If the index of the `xts` object is a date, the dates become the row names of the data frame. In some cases, for example in using `ggplot` to plot the data with a date scale on the horizontal axis, it may be necessary to create a variable in the data frame to hold the dates. This can be done by

```
df <- data.frame(dataxts, date=as.Date(index(dataxts)))
```

(The ggplot function requires a variable of class Date.)

An object of class xts can be created by the xts function in the xts package. The basic format is

```
dataxts <- xts(x, order.by=ind)
```

where x is a vector, a matrix or a data frame, containing the data, and order.by is an index, which specifies how the data are to be ordered and which becomes part of the xts object. The index can be integers or dates in any of the acceptable R formats. The entries in the xts object (that is, the rows) are ordered so that the index itself is ordered.

The xts object inherits the names of the variables in the data frame or else the names of the matrix columns (if any) or the name of the vector. A variable in an xts object can be accessed by the name of the object followed by $ and then by the name of the variable, just as accessing a variable in a data frame.

One of the most useful features of xts objects is the ability to access elements or subsets by use of an index formed by dates. The date function in the lubridate package is a convenient way of initializing the dates.

Figure A1.50 shows some data from the first three trading days and the last three trading days of the month of January, and the first three trading days of the month of February 2017. The data are not in chronological order.

Figure A1.50 illustrates an advantage of an xts object over a data.frame object.

In the example, a data frame is first built with the dates as row names. The data frame has all of the information, but order is just the order that the data are entered, and it would be difficult to put the data in proper order, except by just manually rearranging the rows.

Next, in Figure A1.50 an xts object is built. It has the data in the order we want.

The xts object in Figure A1.50 also has some useful methods that a data frame does not have. One of the more important methods is the flexible access to elements and subsetting of the elements.

Some standard methods, however, do not work correctly on xts objects. One that we encounter often is in the computation of log returns. As shown before, the simple way of doing this is diff(log(XYZ)), which yields a vector of length one less than the length of XYZ. If XYZ is an xts object, however, diff(log(XYZ)) is an xts object with length the same as the length of XYZ, with an NA in the first position. Figure A1.51 illustrates the problem and two ways to deal with it. I often convert xts objects to numeric objects because I always know exactly what R functions will do to numeric objects. The disadvantage of the numeric object, of course, is that date data are lost. I often remedy this by use of either the as.xts function or the merge function.

```
> library(xts)
> library(lubridate)
> dates17 <- lubridate::date(
+               c("2017-02-01","2017-02-02","2017-02-03",
+                 "2017-01-27","2017-01-30","2017-01-31"))
> INTC <- c(  36.52,    36.68,    36.52,
+             37.98,    37.42,    36.82)
> GSPC <- c(2279.55, 2280.85, 2297.42,
+           2294.69, 2280.90, 2278.87)
> datadf <- data.frame(INTC,GSPC)
> rownames(datadf) <- dates17
> datadf                    #  a data frame
           INTC    GSPC
2017-02-01 36.52 2279.55
2017-02-02 36.68 2280.85
2017-02-03 36.52 2297.42
2017-01-27 37.98 2294.69
2017-01-30 37.42 2280.90
2017-01-31 36.82 2278.87
> dataxts <- xts(datadf,dates17)
> dataxts                   #  an xts object
           INTC    GSPC
2017-01-27 37.98 2294.69
2017-01-30 37.42 2280.90
2017-01-31 36.82 2278.87
2017-02-01 36.52 2279.55
2017-02-02 36.68 2280.85
2017-02-03 36.52 2297.42
```

FIGURE A1.50
xts Objects

Indexing and Subsetting xts Objects

The elements of an xts object can be addressed either by the time index alone or by two indexes, the time and the column. Also, instead of the time index, the row number can be used.

If a time index is used, it must be specified in the ISO 8601 date format order, but the "-" separators may be omitted; "20170202" is the same as "2017-02-02" in the index of an xts object.

The time index of an xts object can be manipulated in the ways described in this appendix, and illustrated in Figure A1.46.

A range of times in an xts object can be specified using the "/" separator in the form "from/to", where both "from" and "to" are optional. If either is missing, the range is interpreted as the beginning or the end of the data

```
> library(xts)
> class(INTCdC)
[1] "xts" "zoo"
> INTCdC[1:2,]
           INTC.Close
2017-01-03      36.60
2017-01-04      36.41
># standard computation of returns
> INTCdCReturns <- diff(log(INTCdC))
># get NA
> INTCdCReturns[1:2,]
               INTC.Close
2017-01-03            NA
2017-01-04 -0.005204724
># problem is diff; not log
> diff(INTCdC)[1:2,]
           INTC.Close
2017-01-03         NA
2017-01-04   -0.189998
># clean NAs
> INTCdCReturnsClean <- na.omit(INTCdCReturns)
> INTCdCReturnsClean[1:2,]
               INTC.Close
2017-01-04 -0.005204724
2017-01-05 -0.001649313
># fix name
> names(INTCdCReturnsClean)<-"INTC.Return"
> INTCdCReturnsClean[1:2,]
               INTC.Return
2017-01-04 -0.005204724
2017-01-05 -0.001649313
># alternatively, get returns in numeric vector
> INTCdCReturnsNum <- diff(log(as.numeric(INTCdC)))
> INTCdCReturnsNum[1:2]
[1] -0.005204724 -0.001649313
```

FIGURE A1.51

Unexpected Behavior in Computing Log Returns in xts Objects

object, as appropriate. Exact starting and ending times need not match the underlying data; the nearest available observation is chosen.

The date index of the ith row of an xts object can be obtained by use of the index function; for example, the date index of the third row in the xts object dataxts is obtained by

```
index(dataxts)[3]
```

The `dim` function can be used to obtain numeric indexes corresponding to date indexes.

Figure A1.52 shows examples of indexing. If a specified time index does not exist in an `xts` object, only the column names are returned. The figure shows a way to determine if a specified date does not exist in an `xts` object.

```
> d1 <- index(dataxts)[1]
> dataxts[d1,]
              INTC      GSPC
2017-01-27 37.98 2294.69
> dataxts[d1+1,]
     INTC GSPC
># Is a date missing (weekend or holiday)?
> missing <- length(dataxts[d1,"INTC"]>0)==0
> missing
[1] FALSE
> missing <- length(dataxts[d1+1,"INTC"]>0)==0
> missing
[1] TRUE
```

FIGURE A1.52
Indexing in `xts` Objects (Using Object Created in Figure A1.50)

Figure A1.53 shows some examples of subsetting `xts` objects.

Changing the Frequency in xts Objects

The `to.period` function in the `xts` package operates on an `xts` object to produce a new `xts` with a coarser frequency; that is, for example, it may take an `xts` object that contains daily data and produce an `xts` object containing monthly data. (This in itself would be difficult to do manually, but if the columns of the `xts` dataset correspond to open, high, low, close prices, and volume all appropriately identified by the names of the columns, then the `to.period` function produces a dataset with a coarser frequency that has the appropriate values of high, low, close prices, and volume for each of the new periods. The `to.period` function assumes that the input is an OHLC dataset (see page 181). If the input is not an OHLC dataset, the `OHLC=FALSE` can be used; otherwise, some of the output of `to.period` may be spurious.)

Figure A1.54 shows some examples of the use of `to.period` on the `xts` object built in Figure A1.50. Notice that the monthly data produced from the weekly data is the same as the monthly data produced from daily data, as it should be.

The monthly and weekly data used in making the graphs in Figure 1.28 on

```
> dataxts[5,2]
              GSPC
2017-02-02 2280.85
> dataxts["20170202", 2]
              GSPC
2017-02-02 2280.85
> dataxts["2017-02-02", "GSPC"]
              GSPC
2017-02-02 2280.85
> dataxts["20170202"]        #  February 2
            INTC    GSPC
2017-02-02 36.68 2280.85
> dataxts["201702"]          #  all observations in February
            INTC    GSPC
2017-02-01 36.52 2279.55
2017-02-02 36.68 2280.85
2017-02-03 36.52 2297.42
> dataxts["20170128/20170201", "INTC"]
            INTC
2017-01-30 37.42
2017-01-31 36.82
2017-02-01 36.52
> dataxts["20170131/"]
            INTC    GSPC
2017-01-31 36.82 2278.87
2017-02-01 36.52 2279.55
2017-02-02 36.68 2280.85
2017-02-03 36.52 2297.42
```

FIGURE A1.53

Subsetting in xts Objects (Using Object Created in Figure A1.50)

page 94 and in Figure 1.33 on page 105 were produced from daily data using to.period.

Merging xts Objects

One of the most useful methods for xts objects is the ability to merge them based on the date index.

Merging of datasets in general is a common activity, and the ability to do so is provided in most database management systems, including those built on SQL.

The merge function in R is quite flexible, and provides for most of the common options. A common application is to match observations in one dataset with those in the other, based on common values of one of the variables. In the case of two time series, generally, we want to match based on the time index,

```
> dataxtsw <- to.period(dataxts, "weeks", OHLC=FALSE)
> dataxtsw
             INTC    GSPC
2017-01-05 36.35 2269.00
2017-01-27 37.98 2294.69
2017-02-03 36.52 2297.42
> dataxtsm1 <- to.period(dataxts, "months", OHLC=FALSE)
> dataxtsm1
             INTC    GSPC
2017-01-31 36.82 2278.87
2017-02-03 36.52 2297.42
> dataxtsm2 <- to.period(dataxtsm1, "months", OHLC=FALSE)
> dataxtsm2
             INTC    GSPC
2017-01-31 36.82 2278.87
2017-02-03 36.52 2297.42
```

FIGURE A1.54
Changing Time Periods in `xts` Objects

and that is what `merge` does for `xts` objects. (Note that the time index is not one of the variables; rather, it is a special type of row name.) The examples in Figure A1.55 are self-explanatory. The merged dataset has missing values in positions for which the value is lacking in one of the datasets. Note the use of the `join` keyword that limits the merged dataset to the observations that have the common values of the time index. The last merge in the examples is processed so that all rows with missing values were omitted. This is often useful in cleaning data (see page 183).

Graphics with xts Objects

In plots of time series, the labels of the tick marks of the time axis are often just the positive integers, corresponding to the indexes of the observations in a numeric vector. The software that chooses these labels usually does so in a "pretty" fashion; the labels are chosen in a reasonable way so that the sequence of printed values follows a regular pattern. In many cases, of course, the time series object does not even contain the actual date information.

For financial data collected over an interval of several days or even of several years, a desirable alternative to using integers is to use labels that actually correspond to the calendar time. One of the major advantages of `xts` objects is that they contain the date information, and the methods for processing these objects can actually use this information; that is, the date information in an `xts` object is more than just names of the rows, as in an ordinary R data frame.

```
> dataxts1
           INTC  MSFT
2017-01-05 36.35 62.58
2017-02-03 36.52 63.68
> dataxts2
              GSPC
2017-01-05      NA
2017-01-27 2294.69
2017-02-03 2297.42
> merge(dataxts1, dataxts2)
           INTC  MSFT    GSPC
2017-01-05 36.35 62.58      NA
2017-01-27    NA    NA 2294.69
2017-02-03 36.52 63.68 2297.42
> merge(dataxts1, dataxts2, join="inner")
           INTC  MSFT    GSPC
2017-01-05 36.35 62.58      NA
2017-02-03 36.52 63.68 2297.42
> na.omit(merge(dataxts1, dataxts2))
           INTC  MSFT    GSPC
2017-02-03 36.52 63.68 2297.42
```

FIGURE A1.55
Merging xts Objects

The plotting functions in the ggplot2 package will not handle xts objects (those functions require a data frame); the ggfortify package, however, extends ggplot2 to handle xts objects, as well as objects created by other packages.

The plot method for an xts object (plot.xts) accesses the dates and uses them to label tick marks on the time axis. It is much more difficult for software to choose the dates to print than it is to choose which integers to print. The plot.xts function does a good job of choosing these dates. The time labels on the graphs in Figure 1.8 on page 34 is an example.

Many xts objects are OHLC datasets (see page 181). The plotting methods for OHLC xts objects provide an option for candlestick graphs (see page 182).

A disadvantage of the plot method for an xts object is that other standard R plot functions such as lines, abline, and legend, cannot be used to add elements to the graph. This is not a problem with the plot method for a zoo object, that is, plot.zoo. Because xts objects inherit from zoo objects, plot.zoo will plot xts objects, and the standard R plotting functions can be used to add graphical elements. The graph produced by plot.zoo does not have the same appearance as one produced by plot.xts, but it does have dates as labels on the time axis.

As mentioned elsewhere, I often convert xts objects to numeric objects because I always know exactly what R functions will do to numeric objects. The disadvantage of the numeric object, of course, is that date data are lost. The indexes of the numeric object are just 1, 2, up to the number of rows in the xts object, which can be obtained from the dim function. It is not straightforward, however, to determine the integer index that corresponds to a given date in the xts object. One way to do this is to use the dim function. Consider, for example, the xts object dataxts shown in Figure A1.50. Suppose we wish to have the row index corresponding to the date 2017-02-02. We can obtain the index using the following statement.

```
> dim(dataxts["/2017-02-02",])[1]
[1] 5
```

That date corresponds to the 5th row of dataxts.

OHLC Data

Datasets containing open, high, low, and close prices are called *OHLC datasets*.

The getSymbols function in quantmod produces an object of class xts, which is also an OHLC dataset. Column names must contain an extension that is either ".Open", ".High", ".Low", or ".Close". For example, the column names in an OHLC dataset for INTC prices are "INTC.Open", and so on. This is the default for objects returned from most calls to getSymbols.

An OHLC dataset may also contain a column for volume for each period and a column for adjusted price. The names of these columns must include the extension ".Volume" or ".Adjusted".

The getSymbols function by default uses Yahoo Finance, which supplies daily open, high, low, and close, as well as volume and adjusted close.

The quantmod package provides functions to obtain the individual values of an OHLC dataset, open and so on. The basic functions are Op, Hi, Lo, and Cl. (The functions operate by accessing the column names, which, as stated above, must conform to a standard format.)

There are also functions that consist of combinations of the Op, Hi, Lo, and Cl functions; for example OpCl returns the simple return within the day, $Cl(\cdot) - Op(\cdot)/Op(\cdot)$, which is not the same as the "daily return".

The quantmod package also provides a number of functions to determine properties of an OHLC dataset, such as seriesLo, seriesHi, seriesAccel, and seriesIncr. These operate on the adjusted price column.

Graphics with OHLC xts Objects

Plots of daily prices often show the opening, high, low, and closing prices. A convenient way of doing this is with a "candlestick" at each day. A candlestick has the appearance of the graphic in Figure A1.56, with the meanings shown.

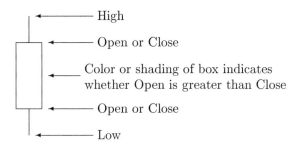

FIGURE A1.56
A Candlestick

An example of a daily candlestick plot, together with a panel showing the trading volume, is in Figure 1.8 on page 34.

Many traders believe that the relationships among the open, high, low, and close, which can be seen very easily in a candlestick, indicate something about the future price moves.

More on getSymbols

Objects created by quantmod functions are in the quantmod environment by default and have the name of the symbol itself. Sometimes this is inconvenient because we may want to process the data in the regular R environment. The R function get can be used to bring the object into a local variable. The quantmod function getSymbols also has an option to form the object in the local environment. Figure A1.57 illustrates these points. Notice the difference in the three local objects z1, z2, and z3.

Dividend Data

Many corporations pay dividends, some on a regular basis, paying out the same amount each quarter, and possibly increasing the amount frequently. Dividends may be an important component of the total return of some stocks. The times and the amounts paid in dividends are recorded by Yahoo Finance. The R function getDividends in the quantmod package obtains the dividend amount for a specified period, as illustrated in Figure A1.58.

```
> library(quantmod)
> z1 <- getSymbols("INTC", from="2017-1-1", to="2017-9-2",
+                 periodicity="daily")
> head(z1)
[1] "INTC"
> z1[1:2,1:2]
Error in z1[1:2, 1:2] : incorrect number of dimensions
> class(z1)
[1] "character"
> class(INTC)
[1] "xts" "zoo"

> z2 <- get("INTC")
> z2[1:2,1:2]
           INTC.Open INTC.High
2017-01-03    37.902    38.233
2017-01-04    38.005    38.067
> class(z2)
[1] "xts" "zoo"

> z3 <- getSymbols("INTC", from="2017-1-1", to="2017-9-2",
+                 env=NULL, periodicity="daily")
> z3[1:2,1:2]
           INTC.Open INTC.High
2017-01-03    37.902    38.233
2017-01-04    38.005    38.067
> class(z3)
[1] "xts" "zoo"
```

FIGURE A1.57
getSymbols and R Environments

A1.4 Data Cleansing

Often the first step in getting data ready for analysis is just to get the data into a proper format. The format required may depend on the model that guides the analysis, or it may depend on requirements of the software. Some R functions, for example, may require that the data be stored in a data frame. This initial step may also involve acquisition and assemblage of data from disparate sources. These data preparation activities are often called "data wrangling" or "data munging".

Individual items in a dataset in a repository such as Yahoo Finance or FRED may be missing or invalid. Also, a program such as getSymbols or Quandl that obtains the data may not work properly. This kind of problem may be due to a change in the structure of the data repository. Such a problem occurred in April 2017 when, due to changes made at Yahoo Finance,

```
> library(quantmod)
> divs <- getDividends("INTC", from="2016-1-1", to="2016-12-31")
> divs
           INTC.div
2016-02-03    0.26
2016-05-04    0.26
2016-08-03    0.26
2016-11-03    0.26
> sum(divs)
[1] 1.04
```

FIGURE A1.58

getDividends

getSymbols and other programs to access data at Yahoo Finance quit working properly. (The initial problems have been fixed, but these programs still sometimes return incorrect data; see the notes to this appendix.)

Unexpected things should always be expected. Consider, for example, the weekly data series from FRED. Suppose we want to regress weekly data from Moody's corporate bond rates on weekly effective fed funds rates (or the weekly differences, as we do in an example in Chapter 4, and as you are asked to do in Exercise 4.27). If we want the weekly data for these two series for the period from 2015 through 2017, an obvious way to get them is to use the following code.

```
getSymbols("WBAA", src = "FRED")
WBAA1 <- WBAA["20150101/20171231"]
getSymbols("FF", src = "FRED")
FF1 <- FF["20150101/20171231"]
```

Surprisingly, however, WBAA1 has 157 rows and FF1 has 156. This is because the WBAA data are for Fridays and the FF data are for Wednesdays, and the number of Fridays in the period from 2015 through 2017 is different from the number of Wednesdays in that period.

Head and Tail

When a dataset is first brought into R, before performing any analyses, it is a good idea to look at the first few and last few values. This can be done using the head and tail functions. The number of rows can be specified by the n argument.

```
head(x, n=2)
tail(x, n=2)
```

Whenever there are known relations that should exist among variables or among observations, it is a good idea to perform some simple consistency checks to ensure that those relations hold in the dataset. This can often be done visually using `head`.

Missing Data

One of the most common problems in data analysis is missing values. Financial data acquired from the internet often has missing values. R has useful ways of indicating data are missing, as we have seen, the most common is assigning the value `NA`.

Many analyses can be carried out with little effect from missing values, but even so it is important to know if there are missing values in the dataset. A very simple way to determine whether there are missing values and if so, how many there are is to use an R statement of the form

```
sum(is.na(x))
```

The function `complete.cases` can be used to determine which cases in a `data.frame` or which rows in a matrix contain no missing values.

The function `na.omit` produces a copy of an R object with the missing values omitted. The logical parameter `na.rm` is available in many R functions to specify that missing values are to be removed before performing any computations, if possible.

```
> x <- c(1, 2, NA, 3)
> mean(x)
[1] NA
> mean(x, na.rm=TRUE)
[1] 2
> mean(na.omit(x))
[1] 2
```

In `var` and `cov`, if `na.rm` is true, then any observation (or "case") with a missing value is omitted from the computations. These functions, however, provide more options through the `use` keyword. If `use="pairwise.complete.obs"` is specified, for example, the computation of the covariance of two variables will use all complete pairs of the two.

The `zoo` package (and hence `quantmod`) has a function `na.approx` that will replace missing values with a linearly interpolated value and a function `na.spline` that will use a cubic spline to approximate the missing value.

Missing Data from `getSymbols` Using Yahoo Finance

The Yahoo Finance data generally go back to January 1962 or to the date of the initial public offering. If the `from` date specified in `getSymbols` is a long time prior to the IPO, some meaningless numerical values may be supplied along with NAs in some fields prior to the date of the IPO. When using `getSymbols` for a period back to the IPO, the `from` date must not be "too long" before the IPO date. Most of the fields before the IPO are NA, but to find where the valid data begin, we cannot just search for the first NA (processing the data in reverse chronological order). As mentioned elsewhere, the data from Yahoo Finance or the data returned by `getSymbols` may occasionally contain NAs, for which I have no explanation.

As emphasized earlier, when processing financial data, we take the attitude of "big data" processing; we do not want to look up the IPO date and hardcode it into the `from` date in `getSymbols`. If a `from` date prior to the IPO is specified, the output can be inspected automatically by the computer and the `from` date can be adjusted (see Exercise A1.24).

Merging Datasets

Problems sometimes arise when merging data sets. Sometimes this is because one of the datasets is messy. Other times it is because some rows are missing in one or the other of the datasets. Figure A1.55 on page 180 illustrates some of these problems. The `join` keyword in `merge` can ensure that only the proper rows are matched with each other. Sometimes, the best cleanup is just to use the `na.omit` function. This was what was done in producing Figure 1.17 on page 62 when the US 3-month T-bills interest rates (from FRED) were matched with the daily S&P 500 and daily INTC returns (from Yahoo Finance). The interest rates are quoted on days that the market is closed.

Another problem in merging datasets by the dates is that the actual dates associated with different comparable time series may never match. For example, the weekly series of FRED data may be computed on different days of the week. The weekly fed funds rate is as of Wednesday and the weekly Moody's bond rates are as of Friday. Hence, if a data frame containing FF (see Table A1.17) is merged with a data frame containing AAA over any given period, the resulting data frame would have twice as many observations as in those for FF and AAA, and each variable would have NAs in every other row.

Data cleansing is an important step in any financial analysis. If the data are garbage, the results of the analysis are garbage. In addition to the issues of misinterpreting the data format, unfortunately, there are many errors in the data repositories.

Notes, Comments, and Further Reading on R

There are many useful publications on R, some introductory and some more advanced, and some oriented toward teaching basic statistics at the same time. Financial analysts who are proficient in the use of spreadsheets may find the book by Taveras (2016) to be a useful introduction. Chambers (2008) and Wickham (2019) cover some of the more advanced topics.

Wickham (2019), Part III, discusses the various OODs incorporated in R and the related topic of object-oriented programming (OOP).

Connecting R with other systems and, in general, extending the capabilities of R are discussed in Chambers (2016), Eddelbuettel (2013), Wickham (2019), and Wiley and Wiley (2016).

Interfacing R with data on internet websites is an important activity in analysis of financial data. (The functions we have used most often for this are the getSymbols function in the quantmod package and the Quandl function in the Quandl package.) Nolan and Temple Lang (2014), especially Chapter 8, and Perlin (2017), especially Chapter 7, discuss sources and methods for obtaining data from the web ("web scraping").

The freely available online journal *Journal of Statistical Software* at

www.jstatsoft.org/

publishes articles on statistical software in general. Many of the articles concern R, and many R packages are first announced and described in this journal.

As you write more R code, the style of writing becomes more important. Wickham (2019), Part I, has some good guidelines.

Cotton (2017) discusses methods of developing and testing larger programs in R.

R Packages

With the thousands of R packages available, there is likely an R function that has been written by someone for almost any task, so we rarely must write any really significant amount of new R code. Nevertheless, the effort in identifying, learning, and, most importantly, gaining confidence in a piece of existing code is sometimes greater than the effort to develop and test new code to perform the task.

Packages can change the environment or have other side effects that cause unexpected results. Another common problem with many packages is that the developers or maintainers may make changes that are not backward compatible, or that in other ways yield unexpected results. For the casual user of that particular software, this can be a major problem. Learning and using software is an investment, and much effort can be wasted when previously working scripts or programs no longer work as expected.

Many packages define and produce useful objects. Well-designed packages often build objects that inherit from other objects, and which are more useful because of the added features. A dataset stored in an xts object is much easier to work with and has more useful metadata than one just stored in a numeric two-dimensional array or in a number of one-dimensional numeric vectors.

Until it doesn't work.

An anecdote:

While working on an example in Chapter 5, I was using some data from FRED. The data were in various xts objects, which was nice because I had to do things such as matching dates and converting monthly data to quarterly data. Then after doing a linear regression, I needed to fit the residuals to an AR model, and then use generalized least squares to refit the regression model with AR errors.

The errors from gls in nlme baffled me and wasted 30 minutes of my life, until I realized the errors were occurring only because the data were in an xts object. I was able to move on only after converting data to a lower common denominator object.

I often choose *lowest common denominator* objects if they make me more efficient.

For this book, most of the use of R involved getting financial data from the internet and doing statistical analysis, primarily graphical displays, of that data. The functions in the quantmod package are well-suited to this purpose. The xts object class defined in quantmod is very useful in dealing with financial time series. The package extends the generic R plot function to produce good displays of these objects. The plots produced by the quantmod::plot function may "look better", but quantmod::plot lacks the flexibility of making labels look even better according to the user's criteria of what looks good. It also does not allow use of standard R convenience functions such as abline.

The plotting functions in the ggplot2 package will not handle xts objects directly; those functions require a data frame. There are two options. One is the ggfortify package, which extends ggplot2 to handle xts objects, as well as objects created by other packages. Converting an xts object to a data.frame is another possibility. The dates, which are indexes in the xts object, become row names in an ordinary data.frame object, and the date functions cannot be used on the row names. The dates must be stored in a data frame variable. Using date data to index a dataset can of course be done if the dates are stored in a data frame variable, but it is much more complicated.

There is no free lunch. The plots produced by functions in ggplot2 may look better, but the logical hierarchy of this package imposes a rigid structure on the graphics elements; hence, simple tasks may not be as simple as they could be. Also standard R convenience functions such as abline may not work as expected.

Side effects are a major issue in software. Saving plots produced by either quantmod::plot or functions from ggplot2 results in unexpected sizing problems for both the plots produced by those packages and plots produced by other functions in the same R session. These problems do not occur in the graphics windows in the console; they appear when, for example, the plots are incorporated into a LATEX 2$_\varepsilon$ document. (These problems could be solved; all *expected* problems can be solved. These problems were *unexpected* side effects of the functions in the packages. The effects then show up when other software is used. The text-generating packages like knitr do not solve these problems.)

Popular, widely-used packages are "safer" than ones less commonly used. Both quantmod and ggplot2 are safe investments, but they do not play well together.

Issues with Accessing Data from the Internet; Reality Checks

Several of the references that I have mentioned describe R functions to get financial data from Yahoo Finance, FRED, and so on. In this appendix and elsewhere, I have blithely mentioned ways of getting financial data from the web. Functions like getSymbols do not always yield the expected results. Data cleansing is a necessary preliminary step for any analysis.

I had to change various pieces of code during the work on this book because of changes made at the data repositories. Yahoo Finance made major structural changes sometime in 2017. The US Treasury site made changes sometime in 2018. The R package ustyc, which contains functions such as getYieldCurve to access data from the Treasury site, no longer works reliably. The data for the graph on page 26 was originally obtained using getYieldCurve. In early 2019, when I tried to reproduce the graph, getYieldCurve was not working. I switched to Quandl, which, however, has a limit on how many times I can use it without making a payment.

Some pieces of code that are shown in the references that I cite do not work *today*. The problem is not with the internet connection. The problem is that the internet, by its very nature, is fluid, and the owners of the sites may decide to make changes in the URL or in the structure of the data.

Yahoo Finance has made various changes over the years. The adjusted close column once included dividend adjustments; it no longer appears to do so. The open, high, and low columns are now adjusted for splits; formerly they were not. The raw data seems to contain more missing values and more errors, than formerly — but who knows?

Missing values often cause problems. For several months during 2017, the stock price data at Yahoo Finance contained missing values in seemingly random places. Data retrieved from FRED often contains missing values, especially on US holidays. I recommend checking data acquired from the internet for missing values. (Use, for example, sum(is.na(xxx)).)

When producing the graph in Figure 1.8 on page 34, I used getSymbols to access data at Yahoo Finance. After looking at the graph, I realized that the closing prices were less than the low prices. I checked the data (on January 31, 2018):

```
> library(quantmod)
> z <- getSymbols("INTC", env=NULL, from="2017-1-1",
+          to="2017-3-31", periodicity="daily")
> z[1, ]
           INTC.Open INTC.High INTC.Low INTC.Close
2017-01-03    37.646    37.975   37.297       36.60
           INTC.Volume INTC.Adjusted
                20196500      35.59252
```

The stated closing price is lower than the stated low price.

I then used tq_get in the tidyquant package and get.hist.quote in the tseries package:

```
> library(tidyquant)
> zz <- tq_get("INTC", get="stock.prices", from="2017-1-1",
+                    to="2017-3-31")
> zz[1, ]
#  A tibble: 1 x 7
  date         open  high   low close   volume adjusted
  <date>      <dbl> <dbl> <dbl> <dbl>    <dbl>    <dbl>
1 2017-01-03  37.9  38.2  37.6  36.6 20196500     35.6
> library(tseries)
> zzz <- get.hist.quote(instrument="INTC", start="2017-1-1",
+                    end="2017-3-31", compression="d")
time series starts 2017-01-03
time series ends    2017-03-30
> zzz[1, ]
            Open    High   Low Close
2017-01-03 37.902 38.233 37.55  36.6
```

Those are the same erroneous results; tq_get and get.hist.quote use getSymbols.

So did Yahoo Finance have incorrect data? No, accessing Yahoo Finance directly, I got

Jan 03, 2017 36.61 36.93 36.27 36.60 35.59 20,196,500

The rule we can draw from these experiences is always check your data. It is not possible, or at least not easy, to know whether the numbers 36.61, 36.93, 36.27, 36.60 are correct, but we can know that either 37.3 (the low) or 36.6 (the close) from getSymbols, tq_get, and get.hist.quote, is incorrect.

This is just one of the (often overlooked) hazards of data science.

Exercises in R

These exercises cover some of the basic uses of R in accessing and manipulating financial data. Exercises in statistical analysis in later chapters assume the ability to perform these basic operations.

Many of the exercises in this appendix require you to use R (or another software system if you prefer) to acquire and explore real financial data, performing similar computations and producing similar graphics as those in Chapter 1. For exercises that require stock or ETF price data or stock index values, it is recommended that you use getSymbols in the quantmod package and get the data from Yahoo Finance.

To perform an action in R, the first thing of course, is to decide what R functions to use. For most of these exercises, the R functions have been discussed in the text in this appendix. Many were used in the Chapter 1 examples, although the R code itself was not shown.

After deciding on the functions, the next thing is to decide how the arguments of the function are to be specified. This text obviously is not a "user's manual" for R, so it is expected that you will need to make frequent use of R's online help system. You may know, for example, that you need to use the function rt, but you may not know the arguments. In that case, type ?rt.

The vignette function may also be useful, although relatively few packages or general topics have vignette documents.

You should also be willing to experiment with various R commands.

Some of these exercises require you to produce simple graphical displays. For these you may use the R functions plot, hist, and qqnorm. We will discuss and illustrate use of these functions in more detail in Chapter 2.

A1.1.**Market indexes.**

Obtain the daily closes of the Dow Jones Industrial Average, the Dow Jones Transportation Average, and the Dow Jones Utility Average for the period January 1, 1986 through December 31, 2018, and make a scaled plot of all three indexes on the same set of axes.

Use different line types or different colors to distinguish the graphs.

(For some reason, in mid-2019 the Yahoo Finance data for the DJIA was not available prior to January 29, 1985. If you can get the data as far back as 1929, you could make a more interesting graph. In particular, the relative performance of the Transportation Average would be of interest over the longer period.)

A1.2.**Market indexes.**

Use R to solve Exercise 1.8a.

A1.3.**Yield curves.**

The yields of various US treasuries can be obtained from the US Treasury Department website. The R function getYieldCurve in the ustyc package is designed to return a data frame df containing yield data, but due to formatting changes at the site, it may not work correctly.

Alternatively, the Quandl R package can be used.

The USTREASURY/YIELD dataset in Quandl contains all of the available Treasury yield data from 1990 to the present (the date of access). The R function Quandl returns an R data frame.

The variables in the R data frame are Date and then 1 MO, 2 MO, 3 MO, 6 MO, 1 YR, 2 YR, 3 YR, 5 YR, 7 YR, 10 YR, 20 YR, and 30 YR corresponding to the terms. Note the blank space in the variable names. Because Treasuries of certain terms were not sold at all times since 1990, many variables, especially 1 MO, 2 MO, and 20 YR, have missing values.

In Exercise A1.3b, you are just to compute the standard deviations of the yield for three different series.

In Exercises A1.3c and A1.3d, you are to plot yield curves. The horizontal axis in a yield curve is the term of the Treasury and the vertical axis is the yield.

(a) Obtain the daily yield data from 1990 for the 1-month, 2-month, 3-month, 6-month, 1-year, 2-year, 3-year, 5-year, 7-year, 10-year, 20-year, and 30-year Treasuries for each day they are available.

Display the first few observations in your dataset.

(b) During the course of a year, the yields of US treasuries vary. For the year 2008, compute the standard deviations of the yields of the 3-month bills, of the 2-year notes, and of the 30-year bonds.

Now do the same for the year 2017.

Comment on your results.

(c) i. Write an R function to plot a yield curve for any specified day using Treasury yield data in a data frame in the form that is returned by Quandl (see the description of the names and structure of the dataset above).

The horizontal axis must correspond to the relative lengths of the terms of the Treasury series. The plot should include all Treasuries that were available on the day specified. (On some days, there were twelve different US Treasuries: 1-month, 2-month, 3-month, 6-month, 1-year, 2-year, 3-year, 5-year, 7-year, 10-year, 20-year, and 30-year. On other days some were not offered. The data frame has missing values in the corresponding fields.)

Your function must account for the missing values. (This is the whole point of this part of the exercise.)

ii. Use your function to plot a yield curve for all available Treasury series for June 5, 2018, using the data frame obtained in Exercise A1.3a.

Comment on the yield curve.

Were any Treasury series missing on that date?

iii. Use your function to plot a yield cure for all available Treasury series for June 5, 2019, using the data frame obtained in Exercise A1.3a.

Comment on the yield curve.

Were any Treasury series missing on that date?

(d) Determine at least two days during the year 2000 in which the yield curve was flat or declining, as defined by the 3-month and 30-year rates. (Here, we are just choosing two maturities arbitrarily; we could use other criteria to identify flat or declining yield curves.) See Figure 1.5 on page 24.

On a single set of axes, plot the yield curves for two days that had flat or declining yield curves, along with a legend to identify the different days. If there were more than two such days, choose days during which the curve was most "interesting". (How do you define "interesting"?) See Figure 1.6.

Comment on the yield curves in the context of the general economic conditions of 2000. What occurred soon after that?

A1.4. **Reading and writing CSV files.**

In a web browser, visit the Yahoo Finance web site. Get daily historical data for Microsoft Corporation stock (MSFT) for the full years 2010 through 2017.

Save the data as a CSV file.

(a) Read the data from the CSV file into an R data frame and convert it to an xts object. Inspect the xts object.

(b) Plot a time series just for the year 2010 of the (unadjusted) closing prices and the daily highs and lows on the same graph, using different colors or line types.

Print a legend on the graph.

(c) Make a bar plot of the volume just for January, 2010. Shade the bars on down days.

(d) Now, compute the daily log returns (using the unadjusted closing prices). In Chapter 1, we illustrated graphical methods for assessing how similar (or dissimilar) a sample frequency distribution is to a normal distribution.

For the MSFT returns for the period 2010 through 2017, make a histogram with a normal density graph superimposed and a q-q plot with a normal reference distribution.

A1.5. Moving averages.

(a) Obtain the MSFT adjusted daily closing prices for the year 2017.

For the adjusted closing prices, compute the 20-day and the 50-day moving averages, and the exponentially weighted moving average using a weight (α) of 0.1. For first days of the moving averages, compute the means of the available data. (The first two days are just the price on the first day; the third and subsequent days are averages of available previous data.) For the exponentially weighted moving average, just start with the second value.

Plot the prices and the smoothed averages on the same set of axes.

The prices of MSFT were generally increasing over this period. Compare the moving averages in this exercise with Exercise 1.6.

(b) There were no crossovers in the 20- and 50-day moving averages of the MSFT adjusted closing prices during 2017, because the price was generally increasing through the year.

Obtain the MSFT adjusted daily closing prices for the year 2015, and the adjusted closes for 50 days prior to the start of the year.

Now, compute the 20- and 50-day moving averages for the full year, using the necessary data from 2014.

Identify the dates of any crossovers of the 20- and 50-day moving averages.

Which were golden crosses and which death crosses?

What was the price of the stock 3 days, 10 days, and 50 days after any golden crosses?

Were these prices higher or lower than on the day of the crossover?

What was the price of the stock 3 days, 10 days, and 50 days after any death crosses?

Were these prices higher or lower than on the day of the crossover?

A1.6. Returns.

Obtain the MSFT adjusted daily closing prices for the year 2017.

(a) Compute the simple daily returns for this period and make a histogram of them.

(b) Using the simple returns computed in Exercise A1.6a and the closing price of the first trading day (2017-01-03), reconstruct the adjusted daily closing prices for the year 2017. (Use `diffinv`.)

Plot these prices as a time series. They should be the same as the original prices, of course.

(c) Compute the daily log returns for this period and make a histogram of them.

Are the log returns essentially the same as the simple returns?

(d) Using the log returns computed in Exercise A1.6c and the closing price of the first trading day, reconstruct the adjusted daily closing prices for the year 2017.

Plot these prices as a time series. (They should be the same as those obtained in Exercise A1.6b, of course.)

A1.7.**Excess returns.**

Using the 3-month T-Bill rates, compute the excess daily return of SPY for the period January 1, 2008 through December 31, 2010. Use adjusted closes. Check for missing values. Make a time series plot of the SPY returns and of the SPY excess returns.

A1.8.**Adjusted prices.**

Obtain unadjusted and adjusted prices for MSFT for the year 2018, and obtain dividend data for the year.

Use the dividend data and unadjusted prices for that year to compute adjusted prices. (These are the adjusted prices as of December 31, 2018.)

Compare with the adjusted prices reported now (the present date) in Yahoo Finance.

A1.9.**Returns over different time intervals;** changing frequency in xts objects.

Obtain the daily closes of MSFT for the period 1988 through 2017.

(a) Compute the daily log returns.

Compute the mean, standard deviation, skewness, and excess kurtosis of the daily returns.

Produce a q-q plot of the daily returns (with respect to the normal distribution).

(b) Convert the daily closes data to weekly closes.

Compute the weekly log returns.

Compute the mean, standard deviation, skewness, and excess kurtosis of the weekly returns, and comment on the differences from the daily data. (How well do the historical volatilities correspond to the $\sqrt{n}\sigma$ relation given on page 106?)

Produce a q-q plot of the weekly returns (with respect to the normal distribution).

(c) Convert the daily closes data to monthly closes.

Compute the monthly log returns.

Compute the mean, standard deviation, skewness, and excess kurtosis of the monthly returns. Comment on the differences from the weekly and daily data. (How well do the historical volatilities correspond to the $\sqrt{n}\sigma$ relation given on page 106?)

Produce a q-q plot of the monthly returns (with respect to the normal distribution).

(d) Summarize the results; specifically describe the observed differences in daily, weekly, and monthly returns.

A1.10. **Weekend effects.**

Obtain the INTC adjusted daily closing prices for the year 2017 and compute the daily returns. Now form two groups of returns, those between consecutive days (for example, Monday to Tuesday, or Thursday to Friday) and those with weekends or holidays intervening (for example, Friday to Monday or Monday, July 3 to Wednesday, July 5).

Analyze these two samples separately (compute summary statistics, form exploratory graphs, and so on).

Comment on any differences you observe between the returns.

A1.11. **Returns of ultra and short volatility ETFs.**

Obtain the daily closes for SVXY, TVIX, and VIX for the week of July 10, 2017, and compute the daily returns (for four days). Plot the returns of SVXY, TVIX, and VIX on the same graph, with a legend to identify them. Each of these graphs is similar to one of the plots shown in Figure 1.15 or Figure 1.16, except all three are on the same graph.

Comment on your plots.

A1.12. **Option chains.**

In this exercise, you are to get current (meaning on the day you work the exercise) price data for two option chains for two different expiries.

Define the price of an option to be the average of the bid and ask prices.

(a) Obtain price data for MSFT for the option chain at the next expiry (from the time you work the exercise). You may have to search on the internet to determine this date. It will be the third Friday of a month or else the last trading day before the third Friday.

Make a chain to consist of 5 strikes at or below the current price, and 5 strikes at or above the current price. (The chain will consist of either 9 or 10 strikes.) For each strike determine the intrinsic value, using the closing price of the underlying on the previous trading day.

Now make a plot with the strike price as the horizontal axis and the option price/value as the vertical axis. Plot the prices/values for both the calls and puts on the same graph, using different line types or colors, and a legend to identify the lines.

The preliminary code below shows the structure of the list that getOptionChain returns.

```
> str(getOptionChain("MSFT"))
List of 2
 $ calls:'data.frame':  48 obs. of  7 variables:
  ..$ Strike: num [1:48] 45 47.5 50 55 60 65 70 ...
  ..$ Last  : num [1:48] 79.9 61 83.2 71.5 63.2 ...
  ..$ Chg   : num [1:48] 0 0 1.05 0 0 ...
  ..$ Bid   : num [1:48] 85.5 84.4 82.4 75.5 ...
  ..$ Ask   : num [1:48] 89 86.3 82.5 79.9 74 ...
  ..$ Vol   : int [1:48] 7 0 15 33 1 7 2 38 5 8 ...
  ..$ OI    : int [1:48] 136 0 214 43 13 43 ...
 $ puts :'data.frame':  46 obs. of  7 variables:
  ..$ Strike: num [1:46] 45 47.5 50 55 60 65 ...
  ..$ Last  : num [1:46] 0.01 0.01 0.02 0.02 ...
  ..$ Chg   : num [1:46] 0 0 0 0 0 0 0 0 0 0 ...
  ..$ Bid   : num [1:46] 0 0 0 0 0 0 0 0 0 0 ...
  ..$ Ask   : num [1:46] 0.01 0.02 0.01 0.01 ...
  ..$ Vol   : int [1:46] 35 5 20 4 4 17 40 ...
  ..$ OI    : int [1:46] 4226 42 697 178 356 ...
```

(b) Now repeat the exercise above but for an expiry date at least 6 months hence. (Search the internet for a date.)

A1.13. Data and beta.

Financial analysts using the common formulas such as the "market model", equation (1.35) on page 61, or the simple formula for beta, equation (1.36), must make decisions about how to assign values to the terms in the formulas.

To use historical data (is there any other way?) to assign values to the terms in these formulas requires decisions: (1) which market, M; (2) frequency of returns; (3) time period of the data; and (4) returns of adjusted or unadjusted prices. For some quantities, there may also be a choice of formulas. Simple returns are more commonly used in this kind of analysis, but log returns may also be used.

In this exercise, you are asked to explore the effects of these choices on calculation of beta for INTC, just using the simple correlation formula (1.36), $\beta_i = \text{Cov}(R_i, R_M)/\text{V}(R_M)$.

On January 11, 2019, TD Ameritrade, on its website linked from users' accounts, quoted a beta for INTC of 0.8. On the same day, E*Trade, on its website linked from users' accounts, quoted a beta for INTC of 1.3.

The difference (assuming each used a "correct" formula correctly) would be due to the formula used, which market is used, the frequency of returns, and the time period.

We have compared the daily returns of INTC with those of the S&P 500 on page 62 in Figure 1.17, but a more relevant market may be the Nasdaq composite or a more focused sector index.

Compute betas for INTC based on the 18 combinations of the following

- 3 benchmarks: S&P 500, Nasdaq composite, and VGT (an information technology ETF owned by Vanguard)
- 2 frequencies: daily returns and weekly returns;
- 3 time periods: 2018-07-01 through 2018-12-31, 2018-01-01 through 2018-12-31, and 2017-01-01 through 2018-12-31.

Organize your program to be efficient in terms of coding (not necessarily in terms of execution).

Summarize your results.

A1.14. **Portfolio expected returns versus risk.**

Consider a portfolio consisting of two risky assets A_1 and A_2 in the relative proportions w_1 and w_2. Assume that the mean returns are $\mu_1 = 0.000184$ and $\mu_2 = 0.000904$, and the risks are $\sigma_1 = 0.00892$ and $\sigma_2 = 0.00827$. We now consider the return/risk profiles of portfolios consisting of these two assets under different assumptions of their correlation. This is similar to the example in Figure 1.18 on page 71. In that case the correlation was $\rho = 0.438$, and the minimum-risk portfolio had $w_1 = 0.434$.

Consider four cases:
$\rho = 0.800$; $\rho = 0.000$; $\rho = -0.438$; and $\rho = -0.800$.

What value of w_1 yields the minimum-risk portfolio in each case? (See Exercise 1.11a.)

Produce a two-by-two display of the graphs of the return/risk profiles under the different values of the correlation.

A1.15. **VaR with a normal distribution.**

In the VaR example shown on page 117, a normal distribution with

0 mean and standard deviation of 0.00892 used to model the distribution of returns and a 5% critical value was computed. Use R to compute the 1% and 10% critical values using that normal distribution.

A1.16.**VaR.**

   (a) Consider a long position of $100,000 in SPY on October 1, 2017.
      i. Using the SPY daily returns for that year to date, form a histogram of the frequency distribution of the returns and determine the one-day 5% value at risk in this single position. (This is similar to the procedure illustrated in Figure 1.37.)
      ii. Using a normal distribution to model the SPY daily returns, determine the one-day 5% value at risk in this single position. (This is similar to the procedure illustrated in Figure 1.38.)
   (b) Now consider a short position of −$100,000 in SPY on October 1, 2017.
      i. Using the SPY daily returns as above, use the histogram frequency distribution to determine the one-day 5% value at risk in this single short position. (This is the amount that the value of SPY might *increase*.)
      ii. Using a normal distribution to model the SPY daily returns, determine the one-day 5% value at risk in this short position.

A1.17.**Correlation of the VIX and the volatility of the S&P 500.**

Compute the correlation of the VIX daily closes with absolute S&P 500 returns for the period 2000 through 2017.

Comment on the results.

A1.18.**Properties of the VIX.**

Does the VIX seem to exhibit a mean reversion? (See Figure 1.34 on page 109 and Exercise 1.19 on page 138.)

Obtain the closing VIX prices for the period January 1, 2000 to December 31, 2017.

   (a) Over this period, for each day when the VIX is 12 or less, compute the proportion of time that the VIX was larger (than 12) 120 days (approximately 6 months) later.

      Now do the same for a VIX of 11; now do the same for a VIX of 10.

      Likewise over this period, for each day when the VIX is 25 or greater, compute the proportion of time that it was less (than 25) 120 days later.

      Now do the same for a VIX of 30; now do the same for a VIX of 35.

      Do these proportions confirm your visual inspection in Exercise 1.19?

(b) Compute the mean value and the median value of the VIX for the entire period. How do they compare? What does this tell you about the distribution of the VIX prices?

For each day, determine whether the value is above or below its median.

For all days when the value is above the median, compute the proportion of times the VIX 5 days later is smaller than on that day. This can be interpreted as an estimate of the probability that the VIX will be lower 5 days after any day when it is higher than its median value.

Now do the same except for 15 days out; now do the same except for 25 days out.

Likewise, for all days when the value is below the median, compute the proportion of times the VIX 5 days later is larger than on that day.

Now do the same except for 15 days out; now do the same except for 25 days out.

How do the proportions compare?

A1.19.**Seasonal effects.**

On page 121, the "January Effect" and the maxim "sell in May and go away" were mentioned.

(a) For the years 1998 through 2017 determine the returns of the large-cap S&P 500 Index and the small-cap Russell 2000 Index during the month of January and also for each of the full years. Discuss the results. That is, comment on the differences in the performance of each index in January compared to the whole year, and on the differences between the S&P 500 and the Russell 2000,

(b) For the years 1998 through 2017, consider a strategy based on "sell in May and go away".

Specifically, assume a beginning investment of $100,000, assume that the SPY ETF can be bought in any dollar amount (that, is fractional shares, can be purchased), assume no trading costs, and assume cash returns 3% per annum. (While the assumptions cannot satisfied or would not have been satisfied over that period, they do not distort the results meaningfully.) Consider full investment in SPY on the first trading day of 1998; full liquidation (into cash) on the last trading day of May, 1998; full reinvestment on the first trading day of October, 1998; full liquidation (into cash) on the last trading day of May, 1999 (that is, no additional January investment); and repetition of these trades for each of the subsequent years through 2017.

Compare this strategy with full investment in SPY on the first trading day of 1998 with no subsequent trades. Compare the returns for each year separately and for the full period.

A1.20.**Pairs trading.**

Because of a possible trade war between the US and China in April, 2018, many traders believed that American companies with significant exposure to China may suffer. One such company was Yum China Holdings, which had been been spun off from Yum! Brands (YUM), the fast food chain. The parent company, which owned KFC, Pizza Hut, Taco Bell, and other fast food companies, was generally viewed positively by financial analysts. A possible pairs trade, would be to short YUMC and buy YUM. Obtain the prices of these stocks in the months of April and May, 2018, and evaluate the strategy by computing the gains or losses accruing from exercising such a strategy at various times during the months. (During this period, YUMC was a hard-to-borrow security.)

A1.21.**Pairs trading.**

Consider a simple intuitive strategy of pairs trading. First, identify a pair to trade as two stocks in the same sector, say, General Motors (GM) and Ford (F). The strategy we use is based on the stocks' divergence from their 50-day moving averages.

Obtain unadjusted daily closing prices for GM and F for the years 2016 and 2017.

Compute the correlation coefficient for that period.

Now determine the curves of the 50-day moving averages, and compute the relative price differences from their moving averages; that is, if P_t is the price at time t and M_t is the moving average price at time t, the relative price different at time t is $r_t = (P_t - M_t)/M_t$. (For the first days of the moving averages, compute the means of the available data.)

The trading strategy is to buy a fixed dollar amount of the stock that has a small relative price difference and to sell the same fixed amount of the stock with a large relative price difference. The price of the stock bought will often be below its moving average and the price of the stock sold will often be above its moving average.

This strategy means buying the "cheaper" stock (the one with the smaller relative residual) and shorting the "more expensive" stock. It is counter to a strategy based on "momentum", which gives preference to stocks with prices above their moving averages.

Assume the commission for a trade (buying or selling) is a fixed \$5.

Begin with a balance $B = \$100,000$. Determine the first point in time when the difference between the relative price differences is at least 0.05. Establish a long position of B, less the commission, in the stock with the smaller relative difference and a short position of B in the stock with the larger relative difference. (The receipt from the short sale is B minus the commission, but the position is $-B$.) In both of these transactions, we assume that a fixed dollar amount of stock can be traded; that is, we assume that fractional shares can be traded.

Now, identify the next point in time at which the difference in the relative differences in 0.01 or less, and close out both positions on that day. Here, as in establishing the positions, we assume the trades are made on the day on which the closing prices satisfied the conditions for making the trades. A commission is paid for each closed position.

How many trades are made during those two years?

What is the running balance and the running total of the gains/losses over that time?

Notice also that while in this exercise, all data were available from the beginning, the strategy can be implemented as a continuing process as data become available.

A1.22. **Working with** xts **objects;** indexing and graphing.

In Figure 1.39 on page 123 we showed the daily closing prices of STZ over the year 2017 when the prices bounced off of a support line six times. Following that period, in early 2018, the price "broke through the support". Technical analysts generally view breaking through "support" as a bearish signal and breaking through "resistance" as a bullish signal. The support line shown in the figure is $p = 142 + 0.308t$. (This was determined using a smoothing technique not discussed here.)

(a) Obtain the daily closes of STZ for the two full years 2017 and 2018. Plot these prices using the generic plot function for an xts object, and try to plot the support line. The plot should be similar in appearance to Figure 1.39, except that the horizontal axis will be indexed by actual dates, and the abline R function does not work.

(b) Now form a numeric object containing the closing prices. Determine the date at which support was broken. Make a time series plot using different line types before and after the support was broken. Using abline, add the support line for the period of rising prices in 2017. (The line is $p = 142 + 0.308t$.)

A1.23. **Working with** xts **objects;** changing frequency, indexing, and graphing.

Obtain the Consumer Price Index for All Urban Consumers: All Items for the period January 1, 2000 to December 31, 2018. These monthly data, reported as of the first of the month, are available as CPIAUCSL from FRED. Compute the simple monthly returns for this period. (The first return is as of February 1, 2000.)

Now obtain the 15-Year Fixed Rate Mortgage Average in the United States for each month in the same period, January 1, 2000 to December 31, 2018. Weekly data, reported on Thursdays, are available as MORTGAGE15US from FRED. This series contains four values. The "close" is the fourth variable (column). To convert these data to correspond approximately to the same monthly dates as those of the CPI data, we take the data for last Thursday in each month as the data for the first day of the following month; hence, the first "monthly" data is for the last Thursday in December 1999.

Compute the simple monthly differences for this period. (The first difference corresponds to February 1, 2000.) Also, note that these differences are percentages, so to make them comparable to the returns of the CPI, we divide by 100.

Now plot these two series of monthly returns and differences (from February 1, 2000 to December 1, 2018) as time series on the same set of axes. Plotting the series as zoo objects allows both series to be plotted together, and allows a legend to be added.

A1.24. **Accessing Yahoo Finance data.**

Stock price data at Yahoo Finance generally go back to 1962. For stocks that did not begin publicly trading until after that year, the data generally go back to the IPO date. If stock trading data are requested for a long time before the IPO date, Yahoo Finance may return spurious numeric data in some fields. In this case, the getSymbols function may not be able to identify the IPO date.

Consider, for example, TWLO, which IPOed on June 23, 2016. If we specify a date a few years before the IPO date, getSymbols returns spurious results.

```
> library(quantmod)
> z1 <- getSymbols("TWLO", env=NULL, from="2010-01-01", to="2018-1-1")
Warning message:
TWLO contains missing values. Some functions will not work if
objects contain missing values in the middle of the series.
Consider using na.omit(), na.approx(), na.fill(), etc to remove
or replace them.
> head(z1)
           TWLO.Open TWLO.High TWLO.Low TWLO.Close TWLO.Volume TWLO.Adjusted
2010-01-04 1e-03     1e-03     8e-04    8e-04      125800      8e-04
2010-01-05    NA        NA       NA       NA          NA          NA
2010-01-06 1e-03     1e-03     1e-03    1e-03        1000      1e-03
2010-01-07    NA        NA       NA       NA          NA          NA
2010-01-08 8e-04     8e-04     8e-04    8e-04       35000      8e-04
2010-01-11 8e-04     8e-04     8e-04    8e-04       40000      8e-04
```

On the other hand, if the `from` date is only a few years prior to the IPO, Yahoo Finance supplies only null values prior to the IPO, and `getSymbols` correctly identifies the IPO date and supplies only valid data.

```
> library(quantmod)
> z2 <- getSymbols("TWLO", env=NULL, from="2015-01-01", to="2018-1-1")
> head(z2, n=2)
           TWLO.Open TWLO.High TWLO.Low TWLO.Close TWLO.Volume TWLO.Adjusted
2016-06-23     23.99    29.610    23.66      28.79    21203300         28.79
2016-06-24     27.54    28.739    26.05      26.30     4661600         26.30
```

As mentioned earlier, our attitude is that we are dealing with "big data"; we do not want to look up some date in order to hard-code it into a program. If the IPO date is necessary in our program, we would prefer to scrape the web to obtain it.

Assume that the stock trading data obtained from Yahoo Finance using `getSymbols` has very few NAs after the IPO, but immediately prior to the IPO, all data are NAs.

Write an R function to determine the date of the first record of valid trading within an `xts` object. The `xts` object possibly contains spurious data or missing data at the beginning. If all data appear to be valid, for instance if the IPO occurred before the first date on the file, then the function should return that first date.

Test your function using the TWLO data.

A1.25. **Accessing Yahoo Finance data and foreign exchange data.**

Obtain the daily closing values of the S&P 500, the FTSE-100, the DAX, and the Nikkei-225 for the period January 1, 2000 through December 31, 2018. Use the exchange rates as quoted on FRED to normalize all indexes to US dollars. (See Table A1.17, and note the direction of the conversion.) Then normalize the data again so that all indexes start at the same value on January 1, 2000.

Plot all indexes on the same set of axes.

Comment on the relative performances of portfolios indexed to these four markets. Both the returns and their volatility (annualized) are relevant. International investment inevitably involves consideration of currency exchange rates. Comment also on the performance of portfolios indexed to these four markets in US dollars.

2

Exploratory Financial Data Analysis

Statistical data analysis may be conducted for a variety of reasons, and the methods may be different depending on the purpose. For example, if the purpose is litigation or if it is to gain approval from a medical regulatory agency such as the FDA, the methods may include formal calculation of a "p-value", which may be interpreted in a rather strict sense.

In "exploratory data analysis", or "EDA", on the other hand, the purpose is to gain a general understanding of observable variables and their relationships with each other. The purpose is not to test statistical hypotheses or to make other formal statements of inference. This usually involves performing some simple computations and looking at various graphical displays, as we did in Chapter 1 with financial data. We form *descriptive statistics* for the data.

A basic technique in exploratory data analysis is to view the same dataset from various perspectives or transformations. In Chapter 1, for example, we displayed the same data on S&P 500 log returns in time series plots and in various plots representing their frequency distribution. We also plotted their absolute values.

Another basic technique in exploratory data analysis of financial data is to view data from the same data-generating process at different frequencies. A daily perspective of a process may reveal different features from a weekly or monthly perspective of the same process. S&P 500 weekly or monthly returns, as shown in Figure 1.33 on page 105, may not appear very different from daily returns, as shown in Figure 1.29 on page 99, except that there are more of them. The q-q plots of weekly and monthly returns in Figure 1.28 (page 94), however, reveal important differences from the daily returns shown in a q-q plot in Figure 1.27 (page 92).

A general objective in exploratory data analysis often is to determine whether a set of data appears to have come from a normal population. Are the data skewed? Are the tails heavier than normal? In exploratory analysis, we assess these questions visually or by simple statistics, rather than by performing formal tests of hypotheses of normality. (We will discuss statistical tests for normality in Section 4.6.1, beginning on page 456.)

Much of Chapter 1 is an exercise in exploratory data analysis of financial data. The stylized facts in Section 1.6 are summaries of the results of the exploratory analysis. In this chapter, we will consider those exploratory methods in more detail.

Models

A statistical analysis of data begins with a *model.* A model can be a simple description of the frequency distribution of an observable random variable, or it can be a very detailed description of the relationships among multiple observable variables.

The statistical model in exploratory data analysis has a general form, with few details or parameters specified. It is often preliminary to a more formal analysis with a more definite model. The model may just be a simple statement that the observable data are positive numbers observed in a time sequence, such as closing stock price data may be.

A fundamental distinction in data is whether the order in which the data were collected is important. Data on the daily closing prices of a stock obviously have an order that is important, and any analysis must take that order into account. The daily log returns on that stock, however, seem to have a weaker dependence on the order. As we have seen in Chapter 1, they do not seem to depend strongly on each other from day to day, although we did observe that they go through periods of high volatility and other periods of low volatility. We also saw that a single return that is very large in absolute value is usually followed by another return, often of opposite sign, that is very large in absolute value.

A statistical model, even a very simple one, must include a time component if the data are dependent on time.

Another fundamental distinction in data is whether it is "discrete" or "continuous". Discrete data can take on only a countable, often finite, number of distinct values. When tossing a coin n times, for example, the number of heads is an integer between 0 and n. The number of market selloffs ("corrections") in a given period is a nonnegative integer with no obvious upper bound. The number of heads in the coin tosses and the number of market corrections in a given period of time are examples of discrete data.

Continuous data can take on any value within an interval; for example, a person's height is a continuous datum.

Although much financial data are discrete, such as the number of stock shares traded or the prices of those shares, we often treat them as if they were continuous.

Most data-generating processes that are interesting enough to merit study have a random component (or at least, a component that appears to be random). Sometimes the nature of the random component itself is the topic of interest; that is, we may wish to describe the frequency distribution of the random component. In other cases, we are more interested in the relationship between different observable variables, and the random component is an inconvenience that complicates our understanding of the relationships. Our approach in those cases is to partition the observed variation in one variable into variation due to its relationship to the other observable variable and residual, "unexplained" variation.

In more formal statistical analyses, the model includes a family of probability distributions. We will discuss models for probability distributions in Chapter 3.

2.1   Data Reduction

Data analysis generally results in a reduction of the amount of data. A dataset containing thousands of observations may be summarized by just a few values.

2.1.1   Simple Summary Statistics

For numerical data, most summary statistics are based either on moments or on quantiles. A few moments and a few quantiles may be sufficient for a general understanding of the full dataset. In Sections 1.3.1 through 1.3.3, we used the first four moments to describe the general shape of the frequency distribution of returns; and in Section 1.3.8, we defined various other quantities based on sample quantiles to characterize the shape of frequency distributions. Because the moment-based measures are more sensitive to tail values, comparison of moment-based measures with quantile-based measures can yield further information about the shape of the frequency distribution; for example, if the mean is larger than the median, the distribution is likely to be positively skewed, although that is not necessarily the case.

The quantile-quantile plots used in Section 1.3.6 were useful to identify differences in the distribution of a given sample from a normal distribution. In Section 1.4.5, we also used quantiles to characterize value at risk.

We use two different types of terms in describing statistical distributions; some terms are general, possibly without a specific technical definition, and others are quite precise. These types of terms are often confused in the nontechnical literature. Some general terms that we often use without precise definitions are *location, scale, spread, dispersion, precision,* and *shape*. Some terms that we use with precise meanings are *mean, variance, standard deviation,* and so on. These technical terms are often defined in terms of mathematical expectation or probability. A general term may be assigned a specific meaning in a given context, often as a parameter in a mathematical formulation. For example, given an expression for the probability density function (PDF) of a Pareto distribution, one of the quantities may be called "the dispersion". In other cases, a general term may be given a technical meaning within the context of a specific statistical methodology. For example, in Bayesian analyses, "precision" may be used to mean the reciprocal of the variance. (In other, less technical, contexts, it often means the reciprocal of the standard deviation.)

In addition to the variance or standard deviation which are measures of the spread or scale of a frequency distribution, we may use measures based on quantiles, such as the median absolute deviation (MAD) which is based on the median, or the *interquartile range* (IQR), which is based on the 25th and the 75th percentiles (the 1st quartile and the 3rd quartile). The relationship between the MAD, the IQR, and the standard deviation depends on the frequency distribution. In the case of a normal distribution, the MAD in the population (that is, using the median of the population model, rather than the sample quantiles) is about 0.674 times the standard deviation, and the IQR in the population (that is, using the quartiles of the population model, rather than the sample quantiles) is about 1.35 times the standard deviation.

We may reduce either a large sample or a small sample to the same small number of summary statistics, but those statistics from the larger dataset usually provide more precise information about the underlying population.

A simple fact is that the variance of summary statistics generally decreases as the sample size increases. For example, if the variance of an underlying population is σ^2, then the variance of the mean of a simple random sample of size n is σ^2/n; that is, the variance of the statistic decreases proportionately to the increase in the sample size. The standard deviation, which is often a more appropriate measure of the deviation from the true population value, decreases more slowly as the inverse of the square root of the sample size, $\sqrt{n}$.

In exploratory data analysis, it is generally desirable to compute several summary statistics, not just the mean and variance. The skewness and kurtosis help us get an idea of the shape of the distribution of observable data. Order statistics, such as the median and the extreme values are also informative.

2.1.2   Centering and Standardizing Data

It is often useful to subtract the sample mean of each variable in a dataset from all individual values of that variable. This is called "centering" the data. Centering the data is especially useful when analyzing relationships among different variables in a dataset. We can then analyze the relationships between the deviations of the individual variables from their means.

In addition, because differences in the variation of different variables can obscure how the variability of two variables move together, it is often a good idea to divide the centered variables by their sample standard deviations. This is called "standardizing" the variables. When we refer to a variable as standardized, we mean that the variable was centered prior to the scaling:

$$\tilde{x} = \frac{(x - \bar{x})}{s_x}. \tag{2.1}$$

2.1.3   Simple Summary Statistics for Multivariate Data

In Section 1.3.4, we considered several measures of the strength of the relationships among the different variables in multivariate data. Simple multivariate analysis focuses on relationships between two variables at a time. The most commonly used measures are the covariance and correlation, which are moment-based measures. In an initial exploration of data relating to multiple assets, the correlations among pairs should be examined.

The correlations between two assets or indexes may vary over time, as we observed in Tables 1.9 and 1.12. This led to the important summary statement in Section 1.6: correlations between returns of financial assets or indexes become greater in periods of market turmoil. This relationship between correlations and extreme conditions is a "tail dependency", which we will discuss in Chapter 4.

Also in Chapter 4, we will discuss some alternatives to correlation, Spearman's rank correlation and Kendall's tau, which we introduced in Chapter 1.

2.1.4 Transformations

Transformations often help in understanding data because they give a different perspective on the data or because they allow us to make more meaningful comparisons among different sets of data. There are many types of transformations.

Standardization of a variable or of data is an example of a *linear transformation*. A linear transformation of a variable x is one of the form

$$z = a + bx, \tag{2.2}$$

where a and b are constants and $b \neq 0$. Linear transformations, while very simple, are useful in many statistical methods.

For data that range from negative to positive, such as asset returns, the absolute value transformation is useful in studying relationships. For example, the serial autocorrelations of daily returns may vary only slightly from zero, both positively and negatively (see Figure 1.30, page 101), but the serial autocorrelations of the daily absolute returns are strongly positive (see Figure 1.31, which helps explain the volatility clustering observed in Figure 1.29).

For data representing change over time, a log transformation is often useful. Figure 1.10 on page 51 shows the growth of the three major stock indexes for a period of about thirty years. It appears that the rate of growth is increasing. (The curve is "parabolic", as financial analysts might say.) Figure 1.13 on page 55, which shows the log-transformed data reveals that the rate of growth, while it has varied considerably over that time period, is essentially no different in the earlier part of the period than in the latter part. The curve is exponential instead of parabolic; most growth curves are.

Linear Transformations of Multivariate Data

Data for use in statistical analyses is often of the form of a two-dimensional array, or matrix, where the columns correspond to individual variables, such as closing price and volume of transactions, and the rows correspond to different observations, such as on different days. This structure is very common for financial data, and even when the data may naturally have a different structure, it can usually be put in the form of a matrix, as we illustrated on page 160.

A common notation for data in this structure is to denote the variables with single subscripts, such as $x_1, \ldots, x_m$. The observations on the variables are then represented by another subscript; for example, observations on the variable x_1 are denoted as $x_{11}, \ldots, x_{n1}$. A natural representation for a dataset in this structure is as a matrix, which we denote as X:

$$X = \begin{bmatrix} x_{11} & \cdots & x_{1m} \\ \vdots & \vdots & \vdots \\ x_{n1} & \cdots & x_{nm} \end{bmatrix} \tag{2.3}$$

Linear transformations on data in this representation are matrix additions and matrix multiplications. The dataset consisting of the centered observations, for example, is

$$X_{\mathrm{c}} = X - \overline{X}, \tag{2.4}$$

where $\overline{X}$ is an $n \times m$ matrix whose first column consists of the constants $\bar{x}_1$, the second column consists of the constants $\bar{x}_2$, and so on. The dataset consisting of the centered and standardized observations is

$$X_{\mathrm{cs}} = X_{\mathrm{c}} D^{-1}, \tag{2.5}$$

where D is an $m \times m$ diagonal matrix whose nonzero entries on the diagonal are the sample standard deviations $s_{x_1}, \ldots, s_{x_m}$. These representations of X_{c} and X_{cs} are useful conceptually, but of course we would never actually form the $\overline{X}$ and S matrices.

For a dataset X as above, the covariances referred to in Section 2.1.3 are the elements of the matrix

$$S = X_{\mathrm{c}}^{\mathrm{T}} X_{\mathrm{c}} / (n - 1), \tag{2.6}$$

and the correlations are the elements of the matrix

$$R = X_{\mathrm{cs}}^{\mathrm{T}} X_{\mathrm{cs}} / (n - 1). \tag{2.7}$$

Identification of these transformations as matrix operations and use of this notation facilitate our discussions of linear statistical models.

2.1.5 Identifying Outlying Observations

An important objective in exploratory data analysis is determining whether there are some observations that stand out from the others. Such outliers can cause simple summary statistics to be misleading. Outliers are sometimes merely erroneous data points, but often outliers point to interesting occurrences in the data-generating process.

Graphical methods, in particular q-q plots and half-normal plots (see Figure 2.11), are useful in identifying outliers.

Outliers occur naturally in heavy-tailed distributions. As we have seen, returns tend to follow heavy-tailed distributions, and so statistical methods based on a normal probability distribution are often inappropriate.

Exploratory data analysis may suggest the need for robust statistical methods, which we will discuss in Section 4.4.6.

2.2 The Empirical Cumulative Distribution Function

For a random variable, the *cumulative distribution function* or *CDF* is the probability that the random variable takes on a value less than or equal to a specified amount. If X is a random variable, and x is some real number, the CDF of X evaluated at x is

$$\Pr(X \leq x),$$

where "$\Pr(\cdot)$" is the probability of the event. We often denote the CDF using an upper-case letter, $F(x)$, for example.

The CDF is one of the most useful functions in probability and statistics, and we will discuss it further in Section 3.1.

For a given sample of numerical data, we can form a function that is very similar to the CDF, but rather, one that is based on the relative frequencies of the observed data. For the numerical data, $x_1, \ldots, x_n$, we define the *empirical cumulative distribution function* or *ECDF* as

$$F_n(x) = \frac{1}{n} \#\{x_i \leq x\}, \tag{2.8}$$

where "$\#$" is the number of items specified. Thus, the ECDF is the number of order statistics up to and including the point at which it is evaluated. The ECDF evaluated at an order statistic has a very simple value:

$$F_n(x_{(i:n)}) = \frac{i}{n}. \tag{2.9}$$

(We defined the notation for order statistics on page 95.)

The ECDF is constant between two successive order statistics of the sample; thus, it is a "step function". Figure 2.1 shows an ECDF of the daily returns of INTC for the first three quarters of 2017. (These are the INTC data used in Figure 1.24 but here we chose a smaller set for clarity of the graph. Even for a sample of this size, the steps in the ECDF are difficult to make out. We will consider this same sample of data on page 458, and illustrate a Kolmogorov-Smirnov goodness-of-fit test for it.)

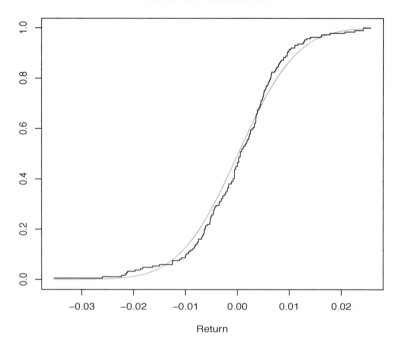

FIGURE 2.1
ECDF of Daily Simple Rates of Return for INTC with Superimposed Normal
CDF

Superimposed on the ECDF plot is a plot of a normal CDF whose mean
and standard deviation are the same as the sample mean and standard deviation of the INTC returns. Figure 1.24 showed a histogram of these data and
a normal PDF. If the objective is to explore differences in the sample and a
normal distribution, the histogram and PDF in Figure 1.24 are better than
the ECDF and CDF in Figure 2.1.

There is a close connection between the ECDF and the CDF. The ECDF
is one of the most important statistical summaries for any sample of numeric
data. It is an estimator of the CDF. If we denote the CDF as $F(x)$, we might
also denote the ECDF as $\widehat{F}(x)$. (Technically, there is a difference between
"estimator" and "estimate", and I generally respect this difference. If the
reader does not know or understand the difference, however, it does not matter
much to the overall understanding of this material.)

The ECDF gives a quick view of the general shape of a frequency distribution. Figure 2.2 shows ECDFs of samples from four types of distributions.

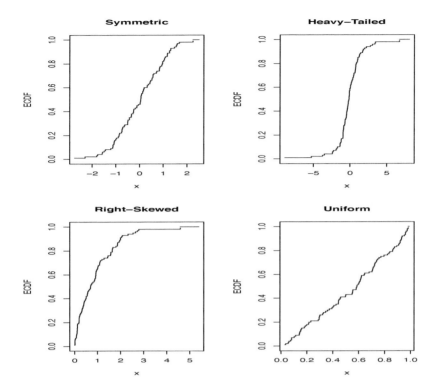

FIGURE 2.2
ECDFs of Various Samples

The ECDF is useful in graphical comparisons of two samples or with a sample and a model distribution, but as Figure 2.1 shows, it is difficult to see the differences in the ECDF and the CDF of a reference distribution.

The ECDF defines a discrete probability distribution. We discuss such probability distributions in general in Section 3.1.1. The distribution defined by the ECDF is the discrete uniform distribution, which we discuss briefly on page 285. This distribution forms the theoretical basis of the bootstrap, which we will discuss in Section 4.4.5.

The R function ecdf produces an object of class ecdf, which can be plotted with the generic plot function.

Related Plots and Other Applications

The q-q plot, which we used several times in Chapter 1 and will discuss in Section 2.4.6, is a plot of the inverse of an ECDF against the inverse of the CDF of a reference distribution or against another ECDF at the sample points.

It is much easier to see differences in the inverse ECDF and inverse CDF, as in Figure 1.24.

Because in financial applications, extreme events are particularly important, focusing just on the tail behavior of a distribution is useful. We form a conditional ECDF, given that the observations are above or below a chosen threshold. The shape of the ECDF (or the q-q plot) in the tails may indicate important features of the data. The conditional distribution above or below a threshold is called the exceedance distribution (see page 275).

A variation on an ECDF plot is a *folded ECDF* or *mountain plot*, which focuses on the two tails separately. In this variation, the ECDF is flipped above the median; that is, it is subtracted from 1. The folded ECDF is useful in visually assessing the symmetry of a distribution. The idea could be generalized so that the flip occurs at point other than the median, but in that case, the curve would have a gap.

Another way to view the ECDF is in terms of its relative increase. This is particularly informative in right-skewed distributions. It shows the proportionate amount of a total corresponding to proportionate amount of the total populations. This graph, called a *Lorenz curve*, is often used in social economics to show the relative amount of income or wealth corresponding to some specific proportion of the population. A Lorenz curve corresponding to a left-skewed distribution is concave, but such distributions are not often interesting. A Lorenz curve corresponding to a uniform distribution is obviously a straight line from $(0,0)$ to $(1,1)$, hence a straight line is often plotted with a Lorenz curve better to illustrate "inequality". Figure 2.3 shows two Lorenz curves, the less convex of which corresponds to the skewed distribution in Figure 2.2. (We will refer to this graph again on page 293.)

The greater the deviation of the Lorenz curve from the straight line, the less uniform is the distribution. A measure of this deviation is the area between the Lorenz curve and the straight line. The maximum of this area is 0.5, so a normalized measure is twice the area. This measure is called the *Gini index*. The Gini index of the less convex Lorenz curve in Figure 2.3 is 0.209 and the Gini index of the more convex one is 0.639. For right-skewed distributions, the Gini index varies from 0 to 1. For a frequency distribution of wealth among households in a given country, a highly convex Lorenz curve, and hence a relatively large value of the Gini index, indicates large inequalities in household wealth.

The ECDF is the basis for several goodness-of-fit tests. The basic Kolmogorov test and the Kolmogorov-Smirnov test are based on the ECDF. For testing for normality specifically, the Lilliefors test is based directly on the ECDF, and the Cramér-von Mises and Anderson-Darling tests are based on the ECDF with a quadratic transformation. The Shapiro-Wilk test is also ultimately based on the ECDF. See Section 4.6.1 beginning on page 456 for discussion of these tests.

Lorenz Curves

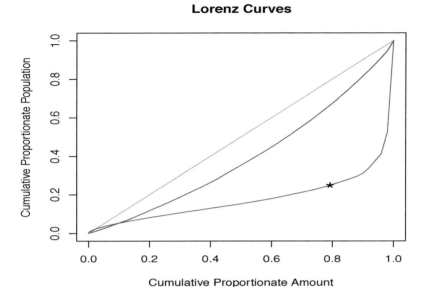

FIGURE 2.3
Lorenz Curves Corresponding to ECDFs of Skewed Data

Sample Quantiles

In a general sense, a quantile is the inverse of a CDF that corresponds to a specific probability. In a standard normal probability distribution, for example, the 0.25 quantile is -0.6744898. Conversely, in the standard normal probability distribution, the probability associated with -0.6744898 is 0.2500000.

The problem in a finite population is that there is no value that corresponds to just any given probability. Hence, while we frequently use the term "quantile" and related terms such as "quartile", "median", "percentile", and so forth, there are some problems with defining the terms.

We may first consider the question from the other side; that is, given a specific value, what "probability" should be associated with it? This question is of particular interest in the case of a sample. Given a specific item in the sample, what quantile is it?

One simple answer is that the ith order statistic in a sample of size n is the i/n-quantile. That may seem reasonable for small values of i, but for $i = n$, the n/n-quantile does not make sense. A simple adjustment serves well, however. We associate the ith order statistic with the $i/(n + 1)$-quantile.

The other issue is, for some given probability α, what is the α sample quantile?

Most useful definitions of sample quantiles are weighted averages of some of the order statistics. Some definitions use all of the order statistics; others use only the two order statistics around the exact quantile.

We can form a flexible definition of a sample quantile corresponding to a given probability α based on two order statistics $x_{(j:n)}$ and $x_{((j+1):n)}$, where j is chosen so

$$\frac{j - m}{n} \leq \alpha < \frac{j - m + 1}{n}, \tag{2.10}$$

where m is a constant such that $-1 \leq m \leq 1$, and which depends on the particular definition of "sample quantile".

For given j, the corresponding α-quantile is

$$q_\alpha = (1 - \gamma)x_{(j:n)} + \gamma x_{((j+1):n)}, \tag{2.11}$$

where $0 \leq \gamma \leq 1$ depends on the particular definition. For some definitions, γ is either 0 or 1, meaning that the quantiles are discontinuous; for other definitions, γ ranges between 0 and 1, yielding a continuous range of values for quantiles.

Consider, for example, the small set of numbers

$$11, 14, 16, 17, 18, 19, 21, 24$$

For the 0.25 quantile, the various definitions yield values between 14 and 15.5.

In some cases, we may prefer a definition of quantile that requires the quantile to be a member of the sample. An alternative simple definition is just to take the α sample quantile in a sample of size n to be $x_{(j:n)}$, where j is an integer such that j/n is as close to α as any other value of the form i/n.

The R function `quantile` provides nine different sample quantiles following the definitions of expressions (2.10) and (2.11). You are asked to compute the sample quantiles under the different definitions in Exercise 2.5.

Sample Quartiles and Percentiles

The same ambiguity in "quantile" generally, however, attaches to other terms such as "quartile" and "percentile". Quartiles are important in some summary statistics, such as the interquartile range (IQR), and a type of quartile is used in boxplots. One simple definition of quartile is that the first quartile is the median of the half of the sample that is less than the median, and the third quartile is defined similarly. This "median of the halves" definition is the one generally used in forming boxplots.

The quartiles used in the interquartile range, are generally the 0.25 and 0.75 quantiles, however those are defined.

The R function `IQR` that computes the interquartile range allows the same nine different definitions allowed in the `quantile` function.

The issue of a precise definition also applies to the *sample median*; that is, the 0.5 sample quantile. In this case, however, all reasonable considerations lead to the same definition. If the sample size n is odd, the median is the

$((n+1)/2)^{\text{th}}$ order statistic; if the sample size n is even, the median is the mean of the $(n/2)^{\text{th}}$ and $(n/2+1)^{\text{th}}$ order statistics.

The R function `median` computes the sample median this way.

2.3 Nonparametric Probability Density Estimation

Along with the CDF, the *probability function* or *probability density function* (*PDF*) is an important component in models used in probability and statistics. These functions describe the idealized frequency distribution of observable random variables. We discussed these in a general way for the frequency distribution of returns in Section 1.3, and we will discuss the CDF and PDF further in Section 3.1.

While the ECDF is an estimator of the CDF, the histogram based on binned data and the density curve based on a kernel function are estimators of the PDF. We gave examples of histograms and density curves in Chapter 1. In this section, we will discuss these two estimators further.

In many cases, we assume that a sample of data follows some simple probability distribution, such as, say, a normal distribution. (We speak of the "frequency distribution" of observed data, and of the "probability distribution" of a mathematical model. In either case, if the data can assume any value within an interval, or if the mathematical model is continuous, we refer to the "density".) In other cases, however, we do not attempt to identify the distribution of the data with any given mathematical form of a probability distribution. The study of the frequency distribution of a set of data from this approach is called "nonparametric probability density estimation".

2.3.1 Binned Data

A simple way of getting an overview of a frequency distribution is to divide the observed sample into bins, and to count the number of observations in the sample that fall into each of the bins. The results can be displayed in histograms, of which we displayed several examples in Chapter 1.

Although generally "histogram" refers to a graphical display, we will also use the term to refer to the set of bins and counts.

The R function `hist` produces a graph by default, but it also produces an object of class `histogram`, which contains atomic vectors of endpoints of the bins, counts within the bins, and so on.

In binning data, a choice of bin sizes must be made. For discrete data that takes on only a small number of distinct values, each individual value may constitute a bin.

For continuous data, the range is divided into intervals, maybe, but not necessarily, of equal length. The size of the bins determines the "smoothness" of the histogram; the larger the bin size, the smoother the histogram.

Smoothing is part of any type of nonparametric probability density estimation. With each estimation technique is associated a *smoothing parameter*, sometimes called a *tuning parameter*. The smoothing parameter may be a vector, as it would be in the case of a histogram in which the bin sizes may be different.

For discrete data, the relative counts in the bins can be interpreted as an estimate of the probability of the occurrence of any of the values represented by the bin. For continuous data, the relative counts in the bins can be interpreted as a scaled estimate of the probability density within the intervals represented by the bins. Specifically, if the breakpoints of the bins are

$$b_0, b_1, \ldots, b_k,$$

and x is some point in the jth bin, that is, $b_{j-1} < x \le b_j$, then the estimate of the probability density at the point x is

$$\widehat{f}_H(x) = \frac{\#\{x_i \text{ s.t. } x_i \in (b_{j-1}, \, b_j]\}}{n(b_j - b_{j-1})}. \tag{2.12}$$

(In the first bin, we use the closed interval $[b_0, b_1]$.) If all of the bins have a constant width, say $h = b_j - b_{j-1}$, this estimate is just

$$\widehat{f}_H(x) = \frac{1}{nh} \#\{x_i \text{ s.t. } x_i \in (b_{j-1}, \, b_j]\}. \tag{2.13}$$

Different values of the smoothing parameter h yield different estimates, as we can see in Figure 2.5 on page 224. The width of the bins, or equivalently the number of bins, is a *tuning parameter*. There are various rules of thumb for choosing the bin width or number of bins. A common one is called "Sturges's rule", which takes the number of bins as $k = 1 + \log(n)$, rounded to the nearest integer. Sturges's rule tends to yield an oversmoothed histogram.

In tuned statistical methods, the tuning parameter often provides a bias-variance tradeoff. In the case of histograms, as the bin width increases, the variance decreases, but the bias increases, because larger regions are required to have the same density but there is likely to be variation within the region.

There are several possibilities for smoothing the data within the intervals. In a simple type of smoothing, the probability density is estimated by linear interpolators of the counts between two bins. (These variations are not important for our discussion here.)

For multivariate data, bins can be formed as a grid or a lattice. The bins are rectangles (or hyperrectangles). A problem with binning of multivariate data is that the number of bins necessary to achieve the same resolution of bin sizes must grow exponentially in the number of dimensions; 10 bins in one dimension become 100 bins in two dimensions, and 1,000 in three.

2.3.2 Kernel Density Estimator

Another approach to nonparametric probability density estimation for continuous data is to estimate the probability density at one point at a time. The estimate at a given point is similar to the probability density estimate based on a histogram if the point happens to be in the center of a bin. Instead of the estimate just being the relative frequency of the number of observations in the bin, as it would be in a histogram, the individual points in the bin may be weighted based on their distances to the point where the probability density is to be estimated.

Another difference in this approach and a histogram estimator is that a bin can have infinite width because of the weighting that is applied to the observations in the bin.

The bin is defined by a function called a "kernel", and we call this approach *kernel density estimation*. We illustrated this in Figure 1.21, but did not describe how it was formed.

Kernel methods are probably best understood by developing them as a special type of histogram. The difference is that the bins in kernel estimators are centered at the points at which the estimator is to be computed. The problem of the choice of location of the bins in histogram estimators does not arise, but we still have a smoothing parameter, which is either the width of the bin or the way that the observations are weighted.

Suppose that we want to compute the estimate of the density at a specific point. We could use something like an ordinary histogram, but if we are free to move the bin anywhere, it might make sense to center a bin over the point at which the density is to be estimated. So now, given a sample $x_1, \ldots, x_n$, the estimator at the point x is

$$\widehat{f}_K(x) = \frac{1}{nh} \#\{x_i \text{ s.t. } x_i \in (x - h/2, \quad x + h/2]\,\}. \tag{2.14}$$

This is very similar to the ordinary histogram estimator in equation (2.13).

Notice that this estimator can also be written in terms of the ECDF:

$$\widehat{f}_K(x) = \frac{F_n(x + h/2) - F_n(x - h/2)}{h}, \tag{2.15}$$

where F_n denotes the ECDF, as in equation (2.8).

Another way to write this estimator over bins all of the same length h is

$$\widehat{f}_K(x) = \frac{1}{nh} \sum_{i=1}^{n} K\left(\frac{x - x_i}{h}\right), \tag{2.16}$$

where $K(t)$ is the uniform or "boxcar" kernel:

$$K_u(t)) = \begin{cases} 1 & \text{if } |t| \leq 1, \\ 0 & \text{otherwise.} \end{cases} \tag{2.17}$$

The width h, often called the window width, is a *smoothing parameter* or a *tuning parameter*, just as in the case of the histogram. Different values of the smoothing parameter h yield different estimates, as we can see in Figure 2.7 (although the kernel in those plots is the same Gaussian kernel; see below). As mentioned in the discussion of histograms, the tuning parameter often provides a bias-variance tradeoff. In the case of kernel density estimation, just as with histograms, as the tuning parameter increases, the variance decreases, but the bias increases because variation in the true density is smoothed out.

Other kernel functions could be used, and equation (2.16) is the general form of the univariate kernel probability density estimator.

One possibility is the standard normal PDF,

$$K_{\mathrm{G}}(t) = \frac{1}{\sqrt{2\pi}} e^{-t^2/2}, \tag{2.18}$$

which, in this context, is often called the "Gaussian kernel".

The window width smoothing parameter makes more difference than the form of the kernel. There are various rules of thumb for choice of the window width. The optimal window width depends on the underlying distribution, which of course is not known. Some rules for choosing the window width require fairly extensive computations to measure certain aspects of the distribution. (These methods use cross-validation, a general method described in Section 4.6.2.) A simpler rule, which is optimal in a certain sense if a normal kernel is used and the underlying distribution is normal, is

$$h = 1.06 s n^{-1/5}, \tag{2.19}$$

where s is the sample standard deviation and n is the sample size.

Although the kernel may be from a parametric family of distributions, in kernel density estimation, we do not estimate those parameters; hence, the kernel method is a nonparametric method.

The R function `density` computes univariate kernel density estimates over a grid of points. The Gaussian kernel is the default kernel in `density`, but that function also allows other kernels. It also allows the user to choose the window width, as well as other options.

2.3.3   Multivariate Kernel Density Estimator

The kernel density estimator extends easily to the multivariate case. In the multivariate kernel estimator, we usually use a more general scaling of $x - x_i$,

$$H^{-1}(x - x_i),$$

for some positive-definite matrix H. The determinant of H^{-1} scales the estimator to account for the scaling within the kernel function. The kernel estimator is given by

$$\widehat{f}_K(x) = \frac{1}{n|H|} \sum_{i=1}^{n} K\left(H^{-1}(x - x_i)\right), \qquad (2.20)$$

where the function K is the *multivariate kernel*, and H is the *smoothing matrix*. The determinant of the smoothing matrix is exactly analogous to the bin volume in a histogram estimator.

The kernel for multivariate density estimation is often chosen as a product of the univariate kernels. The product Gaussian kernel, for example, is

$$K_{G_k}(t) = \frac{1}{(2\pi)^{k/2}} e^{-t^\mathrm{T} t/2},$$

where t is a k-vector.

In practice, H is usually taken to be constant for a given sample size, but, of course, there is no reason for this to be the case, and indeed it may be better to vary H depending on the number of observations near the point x. The dependency of the smoothing matrix on the sample size n and on x is often indicated by the notation $H_n(x)$.

The R packages ks, sm, and KernSmooth provide functions for multivariate kernel density estimation. The sm and KernSmooth packages require diagonal smoothing matrices, but ks allows general (positive definite) smoothing matrices. We will illustrate ks in Figure 2.8.

2.4   Graphical Methods in Exploratory Analysis

Graphical displays and visualization play major roles in exploratory data analysis. There are many basic types of plots, and each has variations.

Graphs are often used to compare a sample with a reference distribution, such as we have done with the normal family of distributions, using line plots, histograms, and q-q plots.

We used several different types of plots in Chapter 1 to illustrate various aspects of financial data. Basic types of plots are listed in Section 2.4.7 below, and for each type, at least one R function to produce a plot of that type is listed, and a reference to an example of the type illustrated in Chapter 1 is given.

In addition to variations in the forms of the basic graphical elements, such as colors and types of lines, there are tuning choices such as positioning of the data and relative scales of components. These tuning factors can affect the overall display and its effectiveness in conveying a message about the data being displayed. The size of the dataset being displayed is an important consideration. A good graphical layout for a small dataset may not be at all appropriate for a large dataset.

In this section, we discuss some of the basic types of graphs and their variations.

2.4.1   Time Series Plots

As mentioned in Chapter 1, many financial datasets are time series, which may mean that the data are not iid. The data are not collected at the same instant in time, and the population from which the data are collected possibly changes with time. Many simple statistical procedures, however, assume that the data are a "random sample" from a stable population. Although the time dependency may not invalidate the standard procedure, the analyst should be aware of it.

Often a dataset may not have a date or time associated with each observation. Even if an explicit time is not present in the dataset, however, it is possible that the sequential index is a surrogate for the time when the observation was made.

In an exploratory phase of the analysis, a time series plot should be made of each of the variables being analyzed. If no explicit time is present in the dataset, the time series plot should use the index of the vector, matrix, or data frame containing the data. Surprising patterns, which indicate lack of independence of the observations will often emerge. The time series plots of the closing prices of INTC shown in Figure 1.2 on page 9 show patterns that clearly indicate time dependencies. (Of course, we knew that to be the case because of what the data represents.)

Another simple plot that might indicate time dependencies even if the time or date is not an explicit variable is one that plots a variable against its lagged value. In R, for example, this has the form

```
plot(x[1:(n-1)], x[2:n])
```

Figure 2.4 shows the daily closing prices of INTC in the first three quarters of 2017 (the upper-left plot in Figure 1.2) plotted against the price of the previous day. This pattern is indicative of a time dependency generally of the form of a random walk. (Here and elsewhere, we have used the term "random walk" informally; we will define it technically on page 490.)

2.4.2   Histograms

A histogram is a pictorial representation of binned data. We used histograms extensively in Chapter 1 to illustrate various characteristics of financial data.

As mentioned above, the number and sizes of the bins determine the "smoothness" of the display. The bins are usually chosen to be of equal sizes,

Intel Prices on Successive Days

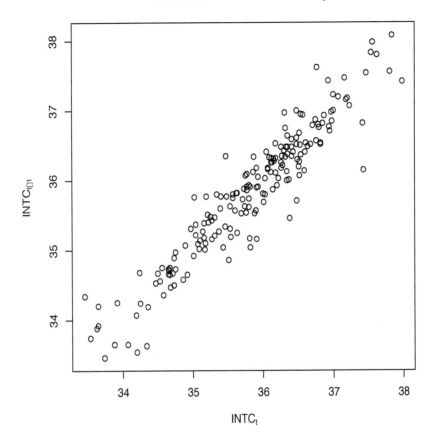

FIGURE 2.4
Daily Closing Prices and Lagged Closing Prices of INTC

but that is not necessary. Figure 2.5 shows some examples of histograms of the same dataset using different bins. The bins can be specified in the R function `hist` using the `breaks` keyword.

Histograms represent either frequencies (counts) or relative frequencies (densities) of bins of a given set of data. Histograms showing densities, as those in the one in Figure 2.5, are based on relative frequencies that account for both the width and the length of the bins in such a way that the sum of the products of the densities and the areas of the bins is 1. Whether frequencies

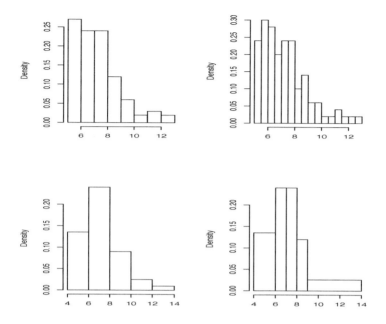

FIGURE 2.5
Examples of Histograms of the Same Dataset, a Random Sample from a
Gamma Distribution

or densities are to be produced by the R function `hist` is specified using the
logical `freq` keyword.

2.4.3 Boxplots

Another type of display is called a *boxplot* or a *box-and-whisker plot*. Boxplots
are particularly useful for comparing different frequency distributions, as in
Figure 1.36 on page 111 for the distribution of the VIX in different time
intervals.

A boxplot in its simplest form, consists of four components: a rectangular
box that encloses the range of the data from the first to the third sample
quartile (that is, it encloses half of the data), a line segment through the box
at the sample median, "whiskers" (line segments) that extend from the top
and bottom of the box to points representing "most" of the remaining data,
and individual points representing outlying data. What constitutes "most" of
the data, can be chosen differently. In some boxplots, "most" is all; in others
it may mean a substantial percentage, say 80% to 99%. In a sample from

a heavy-tailed distribution there will likely be more points that are plotted individually.

The length of the central box is the interquartile range, or the IQR.

The boxplot can also be drawn with a notch at the median. Notches can be used to determine the "significance" of the differences between medians of the different groups being compared with boxplots. If the notches overlap, the differences are probably not "significant". (We will sometimes use the terms *significant* and *significance* in a technical sense in statistical inference (see page 238 and Section 4.2); here, the terms are just descriptive.

Boxplots are usually drawn vertically, as in Figure 1.36, but they can also be drawn horizontally, as in Figure 2.6.

There are also many other variations in types of boxplots.

Figure 2.6 shows examples of boxplots of the same dataset shown in the histograms of Figure 2.5. The boxplots are drawn horizontally, for better comparison of the frequency distributions with the graphs of Figure 2.5. (Note that the boxplots in Figure 2.6 are all of the same dataset; they are not being used to compare different distributions, as were the boxplots of Figure 1.36.)

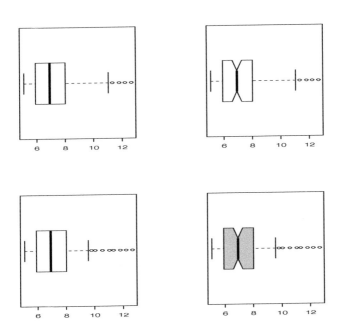

FIGURE 2.6
Examples of Boxplots of the Same Dataset, a Random Sample from a Gamma Distribution: Notched or Not, and Different Length Whiskers

The general shapes of the boxplots in Figure 2.6 indicate that the data are skewed positively, as was also clear in the histograms of Figure 2.5. We see that all of the data are positive. The vertical lines in the boxes in Figure 2.6 represent the median point of the data. The whiskers in the upper two boxplots extend to the most extreme data point that is beyond the box by no more than 1.5 times the length of the box (its width when it is drawn horizontally as in the figure); and the whiskers in the lower two boxplots extend to the most extreme data point that is beyond the box by no more than 0.8 times the length of the box. We see that there are four points in these data that are outside the whisker range in the upper plots and there are eight points outside the whisker range in the lower plots. This distribution is rather heavy-tailed on the right (positive) side.

Boxplots are generally more useful for relatively small datasets (fewer than 100 observations or so).

The R function `boxplot` produces boxplots. The notches and lengths of whiskers can be specified using the `notch` and `range` keywords. The logical `horizontal` keyword determines whether the R function `boxplot` produces horizontal or vertical boxplots.

2.4.4   Density Plots

The kernel probability density estimator in equation (2.16) could be evaluated at many points along the horizontal axis, and then a curve could be drawn through those values. This results in a density plot, such as the one illustrated in Chapter 1 on page 79.

The `density` function in R produces an object of class `density` and the `plot` function produces a density plot from that object. The density on page 79, for example, was produced by

```
plot(density(INTCd20173QSimpleReturns))
```

As mentioned above, the smoothing parameter has a major effect on the appearance of the density plot. Figure 2.7 shows density plots of the same dataset as used in Figures 2.5 and 2.6. The only difference in the plots in Figure 2.7 are the different values of the smoothing parameter. (The Gaussian kernel was used in each.)

Sometimes salient features in a sample can be seen more clearly following a transformation of the data, or even a transformation of a function of the data. In the tails of a distribution, the values of the PDF are likely to be very small. Relative differences in values of the PDF in different regions can sometimes be seen more readily by plotting the log of the density. We used a log density

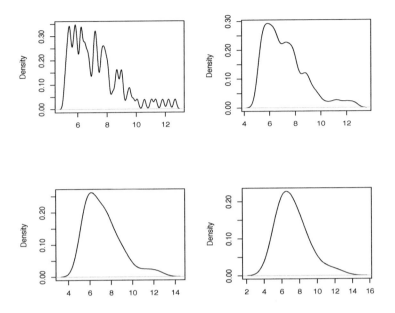

FIGURE 2.7
Examples of Kernel Density Plots of the Same Dataset

plot in Figure 1.26 on page 91 to focus on the differences in a density plot of a sample and the normal PDF when both densities are very small.

2.4.5 Bivariate Data

Figure 1.22 on page 86 shows a scatterplot matrix of the returns on the major indexes along with the returns on Intel common stock. Each scatterplot gives an indication of the general shape of the bivariate frequency distribution.

Other ways of displaying bivariate data are by an extension of the boxplot to two-dimensions and by an extension of the kernel density estimate. One way of the boxplot is to construct two or three nested polygons each of the minimal area that encloses a given proportion of the data. Such a plot is called a bagplot or a starburst plot. The kernel density estimate can also be extended to a to yield a surface representing the density. That surface can then be shown in a contour plot.

Figure 2.8 shows contour plots of bivariate kernel densities for two of the scatterplots in Figure 1.22. The estimates were computed by the R function kde in the ks package. The kde function by default chooses the smoothing matrix H equation (2.20) using a "plug-in" method, described, for example, in Scott (2015). The function also allows the user to select the smoothing matrix. Just as with the univariate densities in Figure 2.7, different pictures of these bivariate densities could be obtained by selecting different values of the smoothing matrix (see Exercise 2.11). The labels on the contours represent the percentage of the data included within the contour.

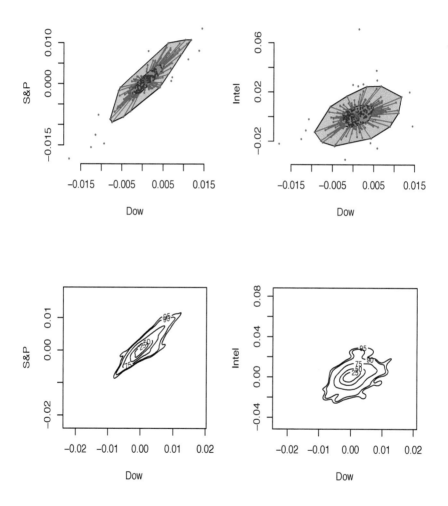

FIGURE 2.8
Bagplots of S&P versus th DOW and of Intel versus the Dow, and Contours of the Estimated Probability Bivariate Densities

When the correlation between two variables is very large, as between the Dow returns and the S&P returns, the bivariate density forms a narrow ridge, as in the plot on the left in Figure 2.8. The bagplots show the outlying points. As we mentioned in Chapter 1, a few outlying points can distort a contour plot so that it does not convey as much information as it would if we removed a few points. Both of the lower plots in Figure 2.8 could represent the central distributions better if we removed some points (see Exercise 2.11).

2.4.6   Q-Q Plots

A *quantile-quantile plot* or *q-q plot* is a plot of the quantiles of one probability distribution or sample versus the quantiles of another distribution or sample. It is most often a plot of sample quantiles versus the quantiles of a probability distribution, which is called the *reference distribution.*

A q-q plot is a parameterized plot in which the parameter is a probability ranging from 0 to 1. For a continuous random variable X, the quantile corresponding to the probability p is the value x_p such that $\Pr(X \leq x_p) = p$, where $\Pr(\cdot)$ means the probability of the event. For discrete random variables and for samples, the definition of quantile is similar, but different authors define it slightly differently.

A linear transformation of either of the distributions or samples leaves the general shape of the plot unchanged.

If the q-q plot forms a straight line, then the two distributions or samples are the same within linear transformations. (They are members of the same *location-scale family.*) A straight line segment determined by two quantiles of both distributions provides visual clues of the deviations of the other distribution or sample from the one used in forming the line.

One axis in a q-q plot represents quantiles of a reference probability distribution. Any distribution could be used as a reference distribution in a q-q plot, but a normal or Gaussian distribution is ordinarily most useful. A q-q plot can be formed with either axis representing the quantiles of the reference distribution.

Shapes of Q-Q Plots

A q-q plot with a normal reference distribution gives us a quick picture of the symmetry and kurtosis of a sample. The general shapes that we can observe in the examples of q-q plots are shown in Figure 2.9 along with the corresponding histograms. The top three pairs of plots use the same data as in the first three plots of the ECDFs shown in Figure 2.2 on page 213.

Figure 3.11 on page 307 shows examples of q-q plots of samples from mixtures of distributions.

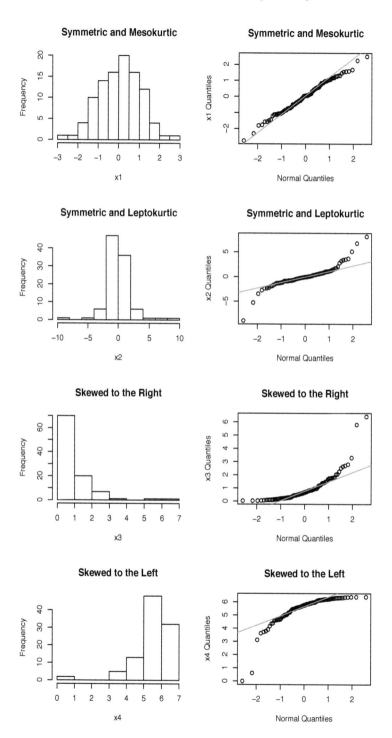

FIGURE 2.9
Some Examples of Q-Q Plots

Figure 1.27 on page 92 shows a q-q plot of the daily log returns for the S&P 500 from January 1, 1990, through December 31, 2017, compared to a normal distribution. That q-q plot indicated that the returns had heavier tails than the normal distribution. Because, as we have emphasized, the distribution of returns is not stationary, we may question the relevance of q-q plots or any other analysis that uses data over a period of 28 years.

At the least, the returns over a long period of time can be considered to have a mixture of distributions. Q-q plots may have different shapes over different time periods. In general, the shapes over longer periods of time have less variation than those over shorter periods. A period of relatively constant high volatility (large standard deviation) will generally have slightly heavier tails than one over a period of relatively constant low volatility; however, a q-q plot is invariant to the standard deviation. In Exercise 2.12, you are asked to explore some of these issues using real data.

Other Reference Distributions

The q-q plots that we have exhibited and discussed here and in Chapter 1 all use the normal distribution as the reference distribution; hence the skewness and kurtosis are both with respect to the normal distribution. Any distribution can be used as the reference distribution; however, and the indications of skewness and kurtosis in a q-q plot are with respect to whatever reference distribution is used in constructing the plot.

In Figure 2.10 we show q-q plots of the same data in Figure 1.27 with respect to t distributions with varying degrees of freedom. (As we will discuss on page 296, the kurtosis of the t distributions increases as the number of degrees of freedom decreases. Compare the q-q plots of the same data with respect to t distributions in Figure 2.10 with the q-q plots of the samples from the same t distributions with respect to the normal distribution in Figure 3.7 on page 297.)

As we remarked in Chapter 1, the distribution of returns is heavy-tailed, so a q-q plot with respect to a normal reference distribution will exhibit a particular shape, as in Figure 1.27. A t distribution with 100 degrees of freedom is very similar to a normal distribution, so the shape of the q-q plot in the upper left panel in Figure 2.10 is similar to the shape with respect to a normal reference distribution. (In the limit as the degrees of freedom of a t distribution approaches infinity, the t distribution is a normal distribution. We will discuss properties of the family of t distributions on page 296.)

All of the q-q plots in Figure 2.10 are of the same sample of returns. The shape becomes closer to a straight line for a reference distribution that is more leptokurtic, as we see in the other plots in Figure 2.10. The bottom two plots seem to indicate that this sample of returns is not as leptokurtic as a t distribution with less than 5 degrees of freedom. We also see, especially in

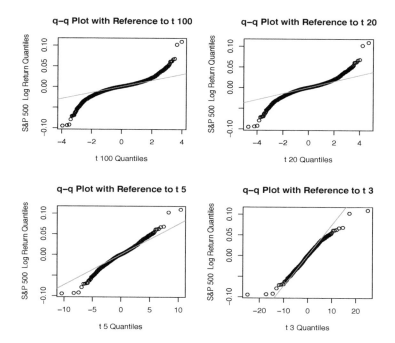

FIGURE 2.10
Q-Q Plots of Daily Rates of Return for the S&P 500 with Respect to t Distributions with Various Degrees of Freedom (Compare with Figure 1.27)

the bottom two plots of Figure 2.10, that the distribution of this sample of returns is slightly skewed. (We had observed this also in the histograms in Chapter 1.)

Most of the examples of q-q plots that we have shown have used the normal distribution as reference. They were produced by the R function qqnorm. The more general R function qqplot allows use of a different reference distribution.

Q-Q Plots of Exceedance Distributions

In financial applications, we are often interested in the tail behavior. To study tail behavior, we choose a tail threshold or exceedance threshold, perhaps based on a probability, say the most extreme 5%. This is the standard approach in VaR analysis, as discussed in Section 1.4.5.

We call the conditional distribution above (or below) this threshold the exceedance distribution (see also Section 3.1.11).

Q-q plots of the data that meets the threshold limits against a reference distribution (which must be conditioned on the same threshold) can be more informative of tail behavior. The qplot function in the evir package is useful for producing q-q plots for data above or below the threshold. The qplot function uses a standard generalized Pareto distribution as the reference distribution (see page 301). The ξ parameter controls the heaviness of the tail of the reference distribution, with larger values yielding heavier tails.

It is important to note that qplot plots the quantiles of the reference distribution on the vertical axis, while qqnorm and qqplot plot the quantiles of the reference distribution on the horizontal axis. In the q-q plots produced by qplot, positive departures from a straight line on the left side of the graph and negative departures on the right side are a sign of heavy-tailed samples. (See Exercise 2.17c.) This is just the opposite of the interpretation of the q-q plots produced by qqnorm and qqplot.

Q-Q Plots to Compare Two Samples

Q-q plots are usually used to compare a sample with a theoretical distribution. They can also be used in the same way to compare two samples. Instead of quantiles from a reference distribution, the quantiles of each of the samples are used.

The interpretation of the shapes that we have discussed above are the same, except that the comparison is between the two samples, or the two data-generating processes that yielded the samples.

Half-Normal Plots

The *half-normal plot* is a variation of a normal q-q plot. It is a plot of the sorted absolute values of the data against the quantiles of a folded normal distribution; that is, the quantiles of the positive half of a normal distribution. A half-normal plot, unlike the normal q-q plot, does not provide evidence of whether the sample comes from a distribution similar to a normal distribution, but, rather, it is effective in identifying outliers of either extreme.

Figure 2.11 shows a half-normal plot of the same data shown in the q-q plot of Figure 1.27, the histogram of Figure 1.25, and the density of Figure 1.26, that is, the daily log return for the S&P 500 from January 1, 1990 through December 31, 2017. The comparisons in all cases are with respect to a normal distribution.

The R halfnorm function in the faraway package produces a half-normal plot.

The R halfnorm function displays the case numbers of the two most extreme observations, in this example cases 4735 and 4746, corresponding to days in the extreme market turmoil of October 2008. Case 4735 corresponds to October 13, a Monday, and case 4746 corresponds to October 28. The returns on both of these dates were actually positive, although this period was generally a bear market.

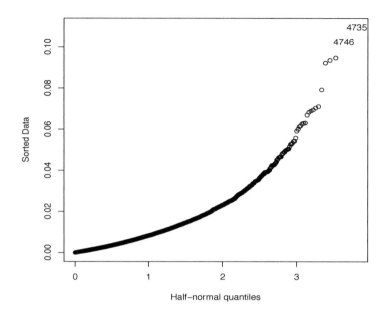

FIGURE 2.11
Half-Normal Plot of Daily Rates of Return for the S&P 500 for the Period
January 1, 1990 through December 31, 2017

2.4.7 Graphics in R

As we have seen, R provides several functions to produce graphical displays of
various types. Most of these functions have a very simple interface with default
settings, but they also allow specification of several graphical characteristics,
such as line color and thickness, labeling of the axes, size of characters, and
so on. All of the graphics in this book were produced using R.

A graphical display has many elements. A well-designed graphics software
system (R is such a system!) allows the user to control these elements, but does
not impose an onerous burden of requiring specification of every little detail.
Most of the graphics functions in R have simple interfaces with intelligent
defaults for the various graphical components. The R function par allows the
user to set many graphical parameters, such as margins, colors, line types,
and so on. (Just type ?par to see all of these options.)

I highly recommend use of the ggplot2 package for graphics in R (see
Wickham 2016). Some plotting functions in ggplot2 are similar to those in
the basic graphics package of R, but they make more visually appealing
choices in the displays.

Graphical displays in ggplot2 are built by adding successive *layers* to an initial plot. Options can be set in ggplot2 by use of the opt function. General graphics elements are determined by "themes", and there are a number of functions with names of the form theme_xxx to control the themes.

Common Types of Statistical Graphics

Two-dimensional graphs display the relationship between two variables or else they display the frequency distribution of one or more variables. For displays of the relationship between two variables, there are different kinds of graphs depending on the interpretation of one of the variables.

A display surface is essentially two-dimensional, whether it is a sheet of paper or a monitor. When there are more than two variables, we generally display them two at a time, possibly in a square array as in Figure 1.22.

There are some situations, however, in which we wish to display a third dimension. A common instance of this occurs when we have just two variables, but we wish to represent another special variable, such as a frequency density. (In the density plot of Figure 1.21, there was just one variable, but associated with that variable was another measure, its frequency density.) Likewise, with two variables, we may have a third variable, measuring their joint frequency density, as in Figure 1.23. Those graphs were produced by the standard generic plot applied to a kde object produced by the kde function in the ks package using the keyword argument display to specify "persp" or "image".

When the object to be viewed is a surface, as in this case, a *contour plot* is useful. The R function contour produces contour plots. The R function image also produces special, colored contour plots. The function contourplot in the lattice package provides more functionality for contour plotting. Another R function useful for viewing a three-dimensional surface is persp, which shows a surface from an angle that the user specifies. As mentioned above, the standard generic plot function has been instrumented to produce contour and three-dimensional surface plots of objects of certain classes.

There are also some R packages that provide functions for producing three-dimensional scatterplots, for example, scatterplot3d in the scatterplot3d package.

General Types of Plots, R Functions, and Examples

- **scatterplot:** points in two dimensions that represent two variables; example: Figure 1.1
  R: plot, ggplot{ggplot2}, points

  - **matrix of scatterplots** (or other bivariate plots)
    example: Figure 1.22
    R: pairs

  - **superimposed scatterplots of the columns of two matrices**
    R: matplot

- **line plot:** continuous functions in two dimensions

example: Figure 1.7
R: `plot`, `ggplot{ggplot2}`, `lines`, `curve`

- **parametric line plot** (parametized continuous functions in two dimensions)
  example: Figure 1.18
  R: `lines`

- **bar plot:** bars whose heights represent values of a single variable
  example: lower panel in Figure 1.8
  R: `barplot`

- **grouped data:** one variable is a factor (an indicator of group membership)
  example: Figure 1.36 or Figure 1.10
  R: `hist`; `boxplot`; or use graphical parameters to distinguish groups

- **time series:** one variable represents time over an interval
  example: Figure 1.2
  R: `plot`, `plot.ts`

- **frequency distribution:** one variable represents a frequency count or density

  - **histogram**; example: Figure 1.19
    R: `hist`

  - **ECDF plot**; example: Figure 2.1
    R: `ecdf` together with `plot`

    * **folded ECDF plot**
      R: `mountainplot{mountainplot}`, `ecdf` together with `plot`

  - **q-q plot**; example: Figure 1.27; also see below
    R: `qqplot`, `qqnorm`, `qplot{evir}`

    * **half-normal plot**; example: Figure 2.11
      R: `halfnormal{faraway}`

  - **boxplot**; example: Figure 1.36
    R: `boxplot`

  - **bivariate boxplot (bagplot, starburst plot)**; example: Figure 2.8
    R: `bagplot{aplpack}`

  - **density plot**; example: Figure 1.21
    R: `density`, `plot.density`

  - **bivariate density plot**; example: Figure 1.23
    R: `kde{ke}`, `plot` with kde methods

- **special graphics for financial data:** candlesticks, bars, bands, and so on.
  example: Figure 1.8
  R: plot, plot.xts, chartSeries{quantmod}

We used most of these types of graphics in Chapter 1 to illustrate various aspects of financial data. The exact R code used to produce these graphs is available at the website for the book.

Plots of different types of R objects may have slightly different appearances. For example, a scatterplot of an R time series will have line segments connecting the points, whereas a scatterplot of a vector will not have those lines, unless the user expressly requests them.

Q-q Plots in R

The R function qqnorm plots quantiles of a sample versus a normal reference distribution. The more general R function qqplot plots quantiles of a sample versus another sample or a reference distribution, specified by its quantile function. For example, the plots in Figure 2.10 were produced by the R code below, with n and nu initialized appropriately. (They are respectively the length of GSPCdReturns1990 and the degrees of freedom of the t distribution. GSPCdReturns1990 as identified in Chapter 1.)

```
qqplot(qt(1:n/n-1/(2*n),df=nu), GSPCdReturns1990,
    main=paste("q-q Plot with Reference to t", as.character(nu)),
    xlab=paste("t", as.character(nu), "Quantiles"),
    ylab="S&P 500  Log Return Quantiles")
```

The reference line is drawn by qqline in R. By default it goes through the 0.25 and 0.75 quantiles of the reference distribution, although the user can specify other quantiles. The qqline function is not designed to work with two samples.

Building a Graph in R

A graph in R may be completed with one function call. For example, the histogram in Figure 1.19 was produced by one simple call to hist.

Many graphs are composed of multiple plots or multiple graphic components such as titles or legends. These graphs are built up from calls to various R functions. Some R functions, such as plot, barplot, qqplot, and hist, are designed to initiate a plot. Other R functions, such as lines, text, title, and legend, are designed to add components to an existing plot. Some R functions, such as curve, can initiate a plot or add to a plot. (The logical keyword add determines this.) Each separate component is a layer on the graph.

One R function is called to produce the basic plot and then other functions are called to add other plots or components. For example, Figure 1.24 was produced by a call to `hist` to make the basic histogram followed by a call to `curve` to add the normal probability density function.

To produce the graph in Figure 1.13, first of all, the data were scaled so that each series would begin with the same value. Next, `plot` was called to make the basic line graph using `type="l"`. In this call, the `ylim` parameter was set so that it would accommodate the largest value in any of the series. The labeling of the axes and the tick marks for the horizontal axis were set, either in the call to `plot` or immediately afterwards. Next, two calls to `lines` were made to plot the logs of the other two series. In this case, the lines had the same type, specified by `lty=1`, which is the default. The colors were set by `col="#FF0000"` (red) and `col="#0000FF"` (blue). The default color for a line and for points is black (`col="#000000"`). Finally `legend` was called to add the legend.

Layouts of Graphs

Many displays, such as Figures 1.1 and 1.2, consist of rectangular arrays of multiple plots. These can be set up in R by use of the multiple figure parameter `mfrow` or `mfcol` in `par`. (The difference is the order in which the individual plots are put into the rectangular array; `mfrow` indicates that they should be produced row by row.) The side-by-side plots in Figure 1.1 were produced using `par(mfrow=c(1,2))`, and the four plots in Figure 1.2 were produced using `par(mfrow=c(2,2))`. The `par` function requires multiple figures in the array to have the same size. The R function `layout` can also be used to control a rectangular array of multiple figures. The figures can have different sizes and not all positions in the rectangular array need be filled. The settings made by `par` or `layout` remain in effect until they are reset.

The `mar` parameter in `par` can be used to control the margins around a plot. This is particularly useful in multiple figure graphs.

More general layouts can be made by use of the `grid` package. In the methods of this package, the various plots are built and stored as R objects of a class called `grob`. They are displayed by use of a separate function, such as `grid.draw`, which may accept a single `grob` object or an array of `grob` objects.

Notes and Further Reading

Exploratory data analysis is often not appreciated. Academic journals look for "significant p-values", and so many scientists and economists who do data analysis perform chi-squared tests or other formal statistical procedures in search of "significance". Statisticians in general are concerned with the role that "statistical significance" has come to play in science; see, for example, Wasserstein, Schirm, and Lazar (2019). In exploratory data analysis, significantly unusual values of computed values serve only to suggest further analysis.

John Tukey in the early 1960s (Tukey, 1962) emphasized that statistics is much more than just formal inferential procedures, and that significance has less relevance in an age of massive data. ("Massive" was much smaller then.) Tukey developed many methods of exploratory data analysis, many suitable for hand calculations.

Scott (2015) discusses nonparametric density estimation and provides a good coverage of both histograms and kernel estimators for univariate and multivariate data. Scott (2015) also discusses graphical displays of density estimates.

Wickham (2016) discusses the design principles that ggplot2 is based on, and describes functions and objects in the package. Keen (2018) discusses various issues for making effective statistical graphical displays using R. Murrell (2018) describes the base graphics system in R and the grid package and its relation to the higher-level lattice and ggplot2 packages.

Exercises

Many of the exercises in this chapter require you to use R (or another software system if you prefer) to acquire and explore real financial data. The same stock price datasets will be used in multiple problems. If you get tired of looking at the same datasets, just choose data on a different stock and use it.

Often the exercises do not state precisely *how* you are to do something or *what type* of analysis you are to do. That is part of the exercise for you to decide.

In all cases, you are invited to choose other real-world datasets for the same exercises and for additional explorations. Instead of a specified date, say, 2018, change it to, say, 2019.

2.1. **Simple summary statistics.**

Obtain closing data for NFLX for the first calendar quarter 2018 (January 2 through March 30). Compute the simple returns and the log returns (unadjusted).

Compute the mean, median, standard deviation, skewness, and kurtosis for both types of returns.

Based on those statistics, what can you say about the distribution of the NFLX returns?

2.2. **Simple summary statistics.**

(a) For a few different values of the standard deviation in normal distributions, compute the MAD. What is the ratio of the standard deviation to MAD? (Perform these computations for the theoretical normal distribution; not for samples.)

Comment on your findings.

(b) For a few different values of the standard deviation in normal distributions, compute the IQR. What is the ratio of the standard deviation to IQR? (Perform these computations for the theoretical normal distribution; not for samples.)

Comment on your findings.

2.3. **Simple multivariate summary statistics.**

Obtain the SPY, UPRO, and SDS daily closing prices (unadjusted) for the periods from their inception through 2018. SPY tracks the S&P 500; UPRO daily returns track three times the S&P 500 daily returns; and SDS daily returns track twice the negative of the S&P 500 one day returns. (SPY is a SPDR ETF begun in 1993, UPRO is a ProShares ETF begun in 2009, SDS is also a ProShares ETF begun in 2006.) Obtain all daily closes for their longest periods.

Also obtain the ˆGSPC prices for the same length of time as that of the oldest of the ETFs (SPY)

Note that the data series are of different lengths.

Now compute the daily returns for these four series and produce a correlation matrix of the daily returns.

For each correlation, use all of the relevant data. State the numbers of observations used for each correlation.

Are the correlations as would be expected?

2.4. **The ECDF.**

Plot the ECDF of the daily simple returns for NFLX for the first calendar quarter 2018. (This is the same dataset used in Exercise 2.1.)

Superimpose on the same plot a plot of the normal CDF with mean the same as the sample mean of the NFLX daily simple returns and standard deviation the same as the standard deviation of the NFLX daily simple returns.

2.5. **Quantiles and quartiles.**

Consider the small set of numbers

$$11, 14, 16, 17, 18, 19, 21, 24$$

(a) Compute the 0.25 quantile of this sample using all nine different interpretations of equation (2.11) that are provided in the R function `quantile`. (Use the `type` argument.)

(b) What are the first and third quartiles using the "median of the halves" definition, and what is the IQR by that definition?

(c) Compute the IQR using the `quantile` R function (with only the default type).

Compute the IQR using the `IQR` R function.

(d) Compute the IQR and the standard deviation for the daily returns for NFLX for the first calendar quarter 2018.

What is the ratio of the standard deviation to IQR?

How does this compare to the ratio in a normal distribution?

(e) Compute the MAD for the daily returns for NFLX for the first calendar quarter 2018.

What is the ratio of the standard deviation to MAD?

How does this compare to the ratio in a normal distribution?

2.6. **The time component.**

On page 222, we emphasized the importance of recognizing whether time is a fundamental component of data, and we suggested a simple plot, illustrated in Figure 2.3 to use to decide how important time may be.

Generally, we know that price data on a stock are highly dependent on time, but the returns on a stock are not so dependent on time, although their frequency distribution may vary over time.

(a) For the NFLX daily closes in the first calendar quarter of 2018, produce a scatterplot of the lagged daily prices versus the prices. (This is the same dataset used in Exercise 2.1.)

Are these data similar to a random walk?

(b) Produce a scatterplot of the lagged daily returns versus the returns of NFLX for the first calendar quarter 2018. Are these data similar to a random walk? Do there appear to be other time dependencies?

2.7. **Simple time series plots.**

Obtain the INTC, MSFT, and ^IXIC adjusted closing prices for the period 1988 through 2017.

Plot these prices as time series data on the same set of axes. Use different colors or line types for the different series, and print a legend on your graph.

2.8. **Histograms.**

Produce three or four histograms of the daily returns for NFLX for the first calendar quarter 2018, using different bin widths of your choice.

Which bin width seems best?

2.9. **Histograms.**

Using the bin width you selected in Exercise 2.8 (or just any old bin width; it is not important for the purpose of this exercise), produce two histograms of the daily log returns for NFLX for the full year 2018. For one, make a frequency histogram and for the other, a density histogram.

What are the differences? When would one kind of histogram be more appropriate than the other?

2.10. **Univariate kernel density estimators.**

Using the daily returns for NFLX for the first calendar quarter 2018, fit a kernel density using four different smoothing parameter values and two different kernels. (Choose the values and the kernels "intelligently".)

Comment on the differences.

Describe the distribution of the NFLX returns.

2.11. **Bivariate density displays.**

Obtain closing data and compute daily returns for NFLX and for GLD for the first calendar quarter 2018.

(a) **Bivariate boxplot; bagplot.**

Produce a bagplot of the returns of NFLX and GLD for that period.

(b) **Bivariate kernel density estimators.**

Choose four different smoothing matrices for a bivariate kernel density estimator. For the four different matrices, choose three that are diagonal (with equal values, then the first one larger, then the second one larger) and choose one proportional to the inverse of the sample variance-covariance matrix.

Fit a bivariate kernel density using four different values for the smoothing matrix. (This is the H argument in the R function kde.) Make contour plots of the fitted densities.

(c) **Bivariate densities.**

Now, compute some relevant summary statistics, and in light of those statistics, as well of the density plots, describe the bivariate distribution of the NFLX and GLD returns.

2.12. **Q-q plots and S&P returns.**

Obtain the daily closes of the S&P 500 from January 1, 1990, through December 31, 2017, and compute the daily log returns. These are the data used in Figure 1.27 on page 92 and a subset of those shown in Figure 1.29 on page 99. Q-q plots of the data are shown in Figure 1.27 and Figure 2.10.

(a) Compute the historical volatility of the daily returns.

Produce a new set of "returns" by multiplying each of the observed returns by 10. Compute the historical volatility of these "returns".

Now produce two q-q plots of these two datasets with respect to a normal distribution, side by side.

Do you see any differences? Comment.

(b) Divide the S&P 500 daily log returns into four periods,
January 1, 1992, through December 31, 1995;
January 1, 1997, through December 31, 2002;
January 1, 2003, through December 31, 2005;
January 1, 2006, through December 31, 2009.

Compute the historical volatility for each period. (See the plot in Figure 1.29 on page 99.)

Now, produce a two-by-two display of q-q plots with respect to a normal distribution for the returns in these four periods.

Comment on your results.

2.13. **Q-q plots and half-normal plots.**

Obtain closing data and compute daily returns for the full year 2017 for the S&P 500 and for First Solar (FSLR), which was a component of the index during that period. FSLR has historically been one of the most volatile stocks in the index.

(a) Produce a q-q plot with respect to a normal distribution and a half-normal plot of daily returns of FSLR.

Comment on the differences in these two kinds of displays. Which of these is more useful in getting a picture of the data?

(b) Produce a q-q plot of the FSLR returns versus the S&P 500 returns. *This is a two-sample plot*; it is not a plot against a reference distribution, although by putting the quantiles of the S&P 500 returns on the horizontal axis, the S&P 500 returns can be considered a "reference sample".

Interpret the plot. Which distribution has heavier tails?

2.14. **Frequency distribution of Nasdaq returns.**

Obtain the daily closes of the Nasdaq Composite (^IXIC) for the period 1988 through 2017. Compute the daily log returns.

(a) Compute the mean, standard deviation, skewness, and kurtosis of the returns.

(b) Produce a histogram of the returns, properly labeled.

(c) Produce a density plot of the returns, properly labeled.

(d) Produce a boxplot of the returns, properly labeled.

(e) Produce a q-q plot of the returns (with respect to the normal).

2.15. **Simple plots of the VIX.**

   (a) Obtain daily closes of the VIX for the period 2000 through 2017. Produce a histogram of the VIX closes for this period.

   (b) Obtain the weekly closes of the VIX for the four years 2014, 2015, 2016, and 2017. Produce boxplots of the weekly closes for each of these years on the same set of axes. (Consider a week to be in the year in which it begins.)

2.16. **Q-q plots with different reference distributions.**

Obtain closing data and compute daily returns for the full year 2017 for FSLR (Exercise 2.13).

Produce four q-q plots with respect to four different reference distributions; a t with 100 degrees of freedom, a t with 20 degrees of freedom, a t with 5 degrees of freedom, and a t with 3 degrees of freedom, as in Figure 2.10.

Comment on the q-q plots. Is there anything unusual? If so, explore the data further.

2.17. **Q-q plots of exceedance distributions.**

Obtain closing data and compute daily returns for the full year 2017 for FSLR (Exercise 2.13).

   (a) Produce four q-q plots of the sample of FSLR larger than the 0.9 quantile with respect to four different reference distributions as in Exercise 2.16; a t with 100 degrees of freedom, a t with 20 degrees of freedom, a t with 5 degrees of freedom, and a t with 3 degrees of freedom.

   (b) Produce four q-q plots of the sample of FSLR smaller than the 0.1 quantile with respect to the four different reference distributions in Exercise 2.17a.

   (c) Use the `qplot` function in the `evir` package to produce q-q plots of the FSLR data with standard location-scale generalized Pareto distributions as the reference distribution (see equation (3.100) on page 301).

Make two q-q plots with the (lower) threshold at the 0.9 sample quantile. For one plot, use the exponential as the reference distribution ($\xi = 0$), and for the other, use $\xi = 0.5$ in the generalized Pareto distribution.

Now make two q-q plots using the same reference distributions and an upper threshold at the 0.1 quantile; that is, lower 10% of the data. Note that the data are from the left side of a distribution more-or-less symmetric about zero.

See also Exercise 3.8 for more work with threshold exceedances.

3

Probability Distributions in Models of Observable Events

The models in exploratory data analysis are generally vague descriptions of the data-generating process. The results of the analysis are also rather vague. We may conclude, for example, that the data are skewed and that they are heavy-tailed, and from this we may infer that the data-generating process yields skewed and heavy-tailed data, but we may not decide that the data follow some specific distribution, such as a t or a Pareto, for example.

In formal statistical inference, we not only provide general descriptions of the data-generating process, we also provide quantitative measures of the process. An important component of statistical inference is quantitative assessment of our confidence of the measures. Statements about confidence in statistical measures are based on assumed probability models. We will discuss how to form those statements of confidence in the next chapter. In this chapter we develop the underlying probability models for confidence statements.

The objectives of this chapter are to

- describe the basics of random variables and probability theory;

- define some specific probability distributions useful for modeling financial data;

- describe methods of simulating data from various probability distributions;

- describe and illustrate R software for working with probability distributions.

These are important because in order to study financial variables, we need to use appropriate models of their probability distributions.

In statistical analysis of financial data, we model the variables of interest such as returns, losses, interest rates, and so on, as random variables. We then proceed to study and analyze observed data in the context of probability distributions of the random variables.

We may assume, for example, that the log returns of a particular index follow a normal distribution. If R represents those returns, we may state

$$R \sim \mathrm{N}(\mu, \sigma^2).$$

Given this model and some data, we may proceed to estimate μ and σ^2.

Also given the model, we may state that we are "95% confident" that μ is between a and b, where we have computed the values a and b based on known properties of the model. We may test the hypothesis that $\mu = 0$ at the 5% "significance level".

We can also perform tests at the 5% significance level that an observed set of data follow that $N(\mu, \sigma^2)$ model, which may lead us to decide that a different probability distribution model is more appropriate. (This is a topic for Section 4.6.1.)

We are often concerned with the extreme cases; that is, the tails of the probability distributions, or the exceedance distributions. In some cases, it is appropriate to use a different probability model in the tails of the distribution.

In some situations, we may need more complicated models that have a probability distribution model as a subcomponent. For example we may take

$$y = g(x, \theta) + \epsilon \tag{3.1}$$

as a model of a data-generating process for observations on the variables y and x. In this model, g is some function that relates the variables y and x through a parameter θ, and ϵ is an adjustment or an "error". A key property of ϵ in this case is that it is not observable.

We may assume that ϵ in equation (3.1) is a random variable. We may use data to infer properties of g and θ, but for formal statistical inference on g and θ, we need to have a model for the probability distribution of ϵ. Once we assume a probability distribution for ϵ, we can make inferences on g and θ with quantified statements of confidence or significance. We will defer these considerations to Section 4.5 in Chapter 4, and in this chapter focus just on the probability distributions.

Probability models also allow us to give precise meaning to concepts such as "independence", which we have already mentioned informally.

In this chapter we will consider some of the fundamental concepts of probability. We will avoid any advanced mathematics, but at the same time we will maintain rigorous accuracy.

We will take "probability" itself as a primitive concept; that is, we will not formally define it. A probability is a number between 0 and 1 inclusive, and relates in an unspecified way to our heuristic understanding of how likely an event is. We denote the probability of an event A by $\Pr(A)$. The probability of event A and event B occurring simultaneously is denoted $\Pr(A \cap B)$, and the probability of event B occurring given the event A occurs, called the *conditional probability* of event B given A is denoted $\Pr(B|A)$. If $\Pr(A) > 0$, we have

$$\Pr(B|A) = \Pr(A \cap B)/\Pr(A). \tag{3.2}$$

In this chapter, we will state several facts about probability, random variables, and probability distributions without proof.

3.1 Random Variables and Probability Distributions

A *random variable* is a mathematical abstraction that can serve as a model for observable quantities. A *probability distribution* expresses the probability that a random variable takes values in a given range. We treat a dataset as a set of observations on a random variable, and we may also speak of the *frequency distribution* of the dataset. We treat the frequency distribution of a dataset in a similar way that we treat a probability distribution for a random variable.

We will call a probability distribution *nondegenerate* if it is not concentrated at a single point.

A probability distribution for a random variable X can be defined in terms of the probability

$$\Pr(X \leq x) \tag{3.3}$$

for any real x, which is similar to the frequency in a numerical dataset, that is, the empirical cumulative distribution function (ECDF) defined in equation (2.8).

If two random variables X and Y are such that

$$\Pr(X \leq x) = \Pr(Y \leq x)$$

for any real x, then we say the random variables are the same. (For technical reasons, however, we may prefer just to say that they have the same distribution.)

In Section 1.3, in the context of returns, we discussed properties of the *frequency distribution of a sample*. We discussed various measures such as mean, variance, skewness, and so on to describe a sample of observations. We also compared various frequency distributions with specific families of probability distributions, often normal distributions. In this section, we will discuss *probability distributions for random variables*, and we will define similar measures for random variables as the measures for observations.

Families of Probability Distributions

It is convenient to identify families of probability distributions that have similar properties. Within a family of probability distributions, *parameters* may determine specific properties. For example, the family of normal distributions, denoted as $N(\mu, \sigma^2)$ above, all have a bell-shaped curve of probability densities. The parameter μ determines the center of the density curve, and the parameter σ^2 determines how spread out the density is.

The *parameter space* for a family of distributions is the set of all possible values of the parameters. For example, the parameter space for the $N(\mu, \sigma^2)$ family is the two-dimensional set of real numbers

$$\{(\mu, \sigma^2) : -\infty < \mu < \infty, 0 \leq \sigma^2\}.$$

The parameter space is often defined either as all real numbers or as a closed subset of real numbers. This may mean that members of a family of probability distributions may be degenerate, as in the case here with $\sigma^2 = 0$. Often, we restrict the parameter space to an open set that ensures that the distributions are not degenerate. (In Tables 3.1 through 3.3, the parameter spaces are restricted so that the distributions are nondegenerate.)

Discrete and Continuous Random Variables

We will consider two types of random variables, "discrete" and "continuous". Most random variables used in applications are either discrete or continuous. For random variables that have both discrete and continuous components, we separate them into two parts corresponding to two sets of random variables; in one set all are discrete, and in the other, all are continuous.

3.1.1　Discrete Random Variables

A discrete random variable is one that takes on values only in a countable set.

A simple example is the number of market corrections in a given year. This number could be $0, 1, 2, \ldots$, most likely less than 2, of course. Let X represent the number. We consider this number to be "random", and X is a random variable. A description of the probability distribution of X would specify the probability that $X = 0$, $X = 1$, and so on. We will use the notation of the form $\Pr(X = 0)$, and so on. Conversely, a specification of $\Pr(X = x)$ for $x = 0, 1, 2, \ldots$ would be a complete description of the probability distribution of X.

The set of all real numbers for which the probability function is positive is called the *support* of the discrete distribution.

We call a function that specifies the probability that a random variable takes on any specific value a *probability function*. In general, the probability function for the random variable X, which we may denote as $f(x)$, is

$$f(x) = \Pr(X = x). \tag{3.4}$$

A probability function $f(x)$ is defined for all real numbers, and $f(x) \geq 0$. A probability function must sum to 1 over the full range of real numbers (that is the "total probability"); and so obviously, the probability function is 0 at most real numbers.

An example of a probability function is

$$f(x; \lambda) = \begin{cases} \lambda^x e^{-\lambda}/x! & x = 0, 1, 2, \ldots \\ 0 & \text{elsewhere.} \end{cases} \tag{3.5}$$

In this equation, λ is a parameter, which must be a positive real number.

We will also call a probability function a *probability density function* or *PDF*. (We use this term for convenience; technically, there is another type of function that is a PDF.)

Equation (3.5) is the probability function for a *family* of probability distributions, each member of the family being determined by the specific value of λ.

It is convenient to define probability functions over all real numbers, as in equation (3.5). It is not convenient, however, always to include the statement about 0, so we will adopt the convention that the specification of a probability function implicitly includes the "0 elsewhere".

In discussing random variables, we will find it convenient to give the random variables names, often upper case letters, X, Y, and so on, and then to use the names of the random variables to distinguish the different functions associated with the different random variables. Furthermore, we may include parameters in the function names usually separated by a semicolon, as in equation (3.5). For example, we may use "$f_Y(y; \beta)$" to denote the probability function for the random variable Y, where β a parameter characterizing the distribution.

Although the probability function itself provides a complete description of the distribution of a random variable, there may be some specific things that can be determined from the probability function that merit evaluation.

A very important element of a description of a probability distribution is the *cumulative distribution function*, or *CDF*. For a random variable X, the CDF, perhaps denoted by $F_X(x)$, is defined as

$$F_X(x) = \Pr(X \leq x). \tag{3.6}$$

For a given probability function $f_X(x)$, the corresponding CDF is obviously

$$F_X(x) = \sum_{t \leq x} f_X(t). \tag{3.7}$$

The CDF corresponding to the probability function in equation (3.5), for example, is

$$F_X(x) = \sum_{t=0}^{x} \lambda^t e^{-\lambda}/t!$$

A useful quantity for any random variable is the *quantile* for a given probability. Let X be a random variable with CDF $F_X(x)$, and let p be a number such that $0 < p < 1$. Then the p-quantile of X is

$$x_p = \inf\{x, \text{ s.t. } F_X(x) \geq p\}. \tag{3.8}$$

Note that the quantile function is continuous from the left. The CDF is continuous from the right.

The quantile function in equation (3.8) is similar to the inverse of F_X; in fact, if F_X is an absolutely continuous function, it would be the ordinary inverse of F_X. We therefore define the quantile function as the *generalized inverse* of F_X, and denote it as $F_X^{\leftarrow}$; hence, we have

$$x_p = F_X^{\leftarrow}(p). \tag{3.9}$$

In general, for any increasing function g, we define the *left-continuous generalized inverse* $g^{\leftarrow}$ of g as

$$g^{\leftarrow}(y) = \inf\{x, \text{ s.t. } g(x) \geq y\}. \tag{3.10}$$

Another thing that may be of interest is the average of the values that a
random variable may take on, weighted by the probability. This value, which
we call the *mean* or the *expected value*, is

$$\sum_x x f_X(x).$$

(The summation is just taken over all values of x for which the probability
function is positive.)

Expectations

More generally, if g is any (measurable) function of the random variable X
with probability function $f(x; \lambda)$, we define the *expectation* or *expected value*
of $g(X)$, which we denote by $E(g(X))$, as

$$E(g(X)) = \sum_x g(x) f_X(x), \tag{3.11}$$

where again the summation is taken over all values of x for which the probability function is positive.

We see that the mean of X is $E(X)$. We often denote the mean of X as
μ_X. As an example, consider the random variable with the probability function
given in equation (3.5). The mean is

$$
\begin{aligned}
E(X) &= \sum_{x=0}^{\infty} x \lambda^x e^{-\lambda}/x! \\
&= \sum_{x=1}^{\infty} \lambda \lambda^{x-1} e^{-\lambda}/(x-1)! \\
&= \lambda \sum_{y=0}^{\infty} \lambda^y e^{-\lambda}/y! \quad \text{letting } y = x - 1 \\
&= \lambda. \tag{3.12}
\end{aligned}
$$

The distribution in this case is called a Poisson distribution. We will encounter
it again later. In this example, we rearranged terms so that we had an infinite
sum that we recognized as being 1, because the terms were the values of a
probability function. This illustrates a technique often used in computations.

Another expectation of a random variable that is of particular interest is
the *variance*. The variance of the random variable X with mean μ_X is

$$E\left((X - \mu_X)^2\right) = \sum_x (x - \mu_X)^2 f_X(x). \tag{3.13}$$

We often denote the variance of X as σ_X^2. The square root of the variance is called the *standard deviation*, and of course these measures of a random variable correspond to measures with the same names for observed data that we defined on page 80. In Exercise 3.3a, you are asked to work out the variance of a Poisson random variable, as above.

The skewness and kurtosis of a random variable or a probability distribution are defined in terms of expectations that are similar to the sums used to define the sample skewness and kurtosis (see page 81). The *skewness* is

$$\mathrm{E}\left(\left(\frac{X - \mu_X}{\sigma_X}\right)^3\right),\tag{3.14}$$

and the *kurtosis* is

$$\mathrm{E}\left(\left(\frac{X - \mu_X}{\sigma_X}\right)^4\right).\tag{3.15}$$

The *excess kurtosis* is the kurtosis minus 3 (which is the kurtosis of a normal distribution).

Transformations of Discrete Random Variables

Random variables are used as models of variables, whether observable or not. Those variables, for various reasons, may be transformed; they may be multiplied by some quantity, some quantity may be added to them, they may be squared, and so on. The question is what happens to the distribution of a random variable if the random variable is transformed in some way. For a discrete random variable, the answer is fairly simple.

Given a random variable X with probability function $f_X(x)$, let

$$Y = g(X),\tag{3.16}$$

where g is some function that defines the transformation. Then the probability function for Y is

$$f_Y(y) = \sum_{x \ni g(x) = y} \Pr(X = x).\tag{3.17}$$

This expression is very simple if the transformation is one-to-one; there is only one term in the sum.

Consider the simple random variable X with the probability function

$$f_X(x) = \begin{cases} 1/4 & \text{for } x = -1 \\ 1/2 & \text{for } x = 0 \\ 1/4 & \text{for } x = 1 \\ 0 & \text{otherwise.} \end{cases}\tag{3.18}$$

Now consider the transformation $Y = g(X) = 2X + 1$. Then

$$
f_Y(y) = \begin{cases} 1/4 & \text{for } y = -1 \\ 1/2 & \text{for } y = 1 \\ 1/4 & \text{for } y = 3 \\ 0 & \text{otherwise.} \end{cases} \tag{3.19}
$$

Consider another transformation; let $Z = h(X) = X^2 + 1$. Then

$$
f_Z(z) = \begin{cases} 1/2 & \text{for } z = 1 \\ 1/2 & \text{for } z = 2 \\ 0 & \text{otherwise.} \end{cases} \tag{3.20}
$$

3.1.2 Continuous Random Variables

A continuous random variable is one that takes on values in an uncountable interval, such as the interval $(0, 1)$. To simplify the discussion, we will use the phrase "continuous random variable" to mean that there is an interval (a, b), where a may be $-\infty$ and b may be ∞, such that for any nonempty subinterval (c, d), the probability that the random variable is in the subinterval is positive. (This excludes random variables with "gaps" in their distribution. Such random variables are not very useful in applications, but they can be handled by defining multiple continuous random variables.)

We first describe a very basic continuous distribution. The *uniform distribution* over the interval $(0, 1)$ is the distribution of a random variable that takes on a value in the interval (c, d), where $0 \le c < d \le 1$, with probability $d - c$. We denote this uniform distribution as $U(0, 1)$, and a similar uniform distribution over the finite interval (a, b) for any $a < b$ as $U(a, b)$.

While the mathematical model allows an uncountable number of different values within a continuous interval, in reality, our observations are always confined to a countable set. For example, consider scores on a test of mental abilities. While in theory, a score could be any positive real number, in fact, these scores are probably constrained to be integers, and probably have a lower bound greater than 0, and a finite upper bound. Nevertheless, a continuous random variable is often a useful concept.

Because the number of different values a continuous random variable can take on is uncountable, the probability that it takes on a specific value must be 0. The probability function, as we defined for discrete random variables, would not make much sense, but the CDF for a continuous random variable, defined the same way as for a discrete random variable in equation (3.6),

$$
F_X(x) = \Pr(X \le x),
$$

would be meaningful. The CDF for a continuous random variable could not be expressed as a sum as in equation (3.7), however.

The approach is to define a *probability density function* or *PDF* in terms of the CDF. Given a CDF $F_X(x)$, the function analogous to $f_X(t)$ in equation (3.7) is the derivative of $F_X(x)$; hence, we define

$$f_X(x) = \frac{d}{dx} F_X(x) \tag{3.21}$$

as the PDF of X. We also see that the analogue to the sum in equation (3.7) holds:

$$F_X(x) = \int_{-\infty}^{x} f_X(t)\, dt. \tag{3.22}$$

In the simple example of a random variable X with the $U(0,1)$ distribution, the PDF is 1 over the interval $(0,1)$ and 0 otherwise, and the CDF is 0 for $x < 0$, x for $0 \le x < 1$, and 1 for $1 \le x$.

It is convenient to define PDFs over all real numbers, and a PDF must integrate to 1 over the full range of real numbers.

The set of all real numbers for with the PDF is positive is called the *support* of the continuous distribution.

Sometimes for a complicated PDF, it may be of interest to focus on how the function depends on the value of the random variable, so we may express a PDF as

$$f_X(x; \theta) = g(\theta)k(x).$$

In this decomposition, $k(x)$ is called the *kernel* of the PDF. In some cases, the kernel also contains some function of θ that cannot be separated out.

The specification of a PDF may state a range over which the formula applies. Whether stated explicitly or not, we will assume that the PDF is "0 elsewhere". An example of a PDF is

$$f(x; \mu, \sigma) = \begin{cases} \dfrac{1}{\sqrt{2\pi}\sigma} x^{-1} e^{-(\log(x) - \mu)^2 / 2\sigma^2} & x > 0 \\ 0 & \text{elsewhere.} \end{cases} \tag{3.23}$$

In this equation, μ and σ are parameters, with $-\infty < \mu < \infty$ and $\sigma > 0$.

The CDF and Quantiles of a Continuous Random Variable

The CDF of a continuous random variable is invertible (because of our assumption of no "gaps"), and so the quantile defined in equation (3.8) for given probability p has the simpler form

$$x_p = F_X^{-1}(p). \tag{3.24}$$

In the case of continuous random variables, $F_X^{\leftarrow} = F_X^{-1}$, so we can write the quantile in the same form as equation (3.9):

$$x_p = F_X^{\leftarrow}(p).$$

If $F_X(x)$ is the CDF of the continuous random variable X, then $F_X(X)$ itself is a random variable with distribution $U(0,1)$. This is easy to see:

$$\Pr\left(F_X(X) \leq x\right) = \Pr\left(X \leq F_X^{-1}(x)\right) = F_X\left(F_X^{-1}(x)\right) = x; \qquad (3.25)$$

hence, the CDF of $F_X(X)$ is the same as the CDF of a $U(0,1)$ random variable, and therefore $F_X(X)$ is distributed as $U(0,1)$. This is one of the most useful facts about the CDF of a continuous random variable.

Expectations

With the PDF taking the role of the probability function and with summation being replaced by integration, the terms such as mean and variance as defined for discrete random variables are similarly defined. In general, if g is any (measurable) function of the random variable X with PDF $f_X(x)$, the expectation of $g(X)$ is

$$\mathrm{E}(g(X)) = \int_{\mathcal{D}} g(x) f_X(x) \, \mathrm{d}x, \qquad (3.26)$$

where the integration is taken over the domain of x for which the PDF is positive.

This definition of expectation is consistent with the definition of expectation for functions of discrete random variables given on page 250, and the same properties, such as linearity, hold.

Also, as with discrete random variables in the discussion above, two special functions yield important measures of the distribution: $g(X) = X$ corresponds to the mean, which we often denote as μ_X, and $g(X) = (X - \mu_X)^2$ corresponds to the variance, which we often denote as σ_X^2. We also denote the mean as $\mathrm{E}(X)$ and the variance as $\mathrm{V}(X)$. The expected value of a random variable or of a power of the random variable is a *population moment*; the mean is the first moment, the variance is the second (centered) moment, and so on.

The skewness and kurtosis of random variables, whether continuous or discrete, are also defined in the same way. Those definitions are given in equations (3.14) and (3.15). The *excess kurtosis* is the kurtosis minus 3.

Because the skewness and kurtosis of any normal distribution are constant (and known), the sample skewness and kurtosis (page 81) form the basis for goodness-of-fit tests for normality. Both the D'Agostino-Pearson test and the Jarque-Bera test are based on the sample skewness and kurtosis. (See Section 4.6.1 beginning on page 456 for discussion of these tests.)

Transformations of Continuous Random Variables

Transformations of continuous random variables are similar to those of discrete random variables, but they also require some adjustments due to the fact that the probability density applies to intervals, not just to discrete points. The effect of a transformation affects the density at all points in the interval.

Consider a continuous random variable X with PDF $f_X(x)$. Assume $f_X(x)$ is positive over the range $\mathcal{D}$ and zero elsewhere. Now, consider a one-to-one transformation, $Y = g(X)$ that is differentiable on $\mathcal{D}$. Since the transformation is one-to-one, we have $X = g^{-1}(Y)$. We want to work out the distribution of Y. We have, for any $a < b$,

$$\Pr(a < Y < b) = \Pr\left(g^{-1}(a) < X < g^{-1}(b)\right) = \int_{g^{-1}(a)}^{g^{-1}(b)} f_X(x)\mathrm{d}x.$$

A change of variables within the integral must involve a rescaling of $\mathrm{d}x$ consistent with the changed scale of $g(x)$. Since $\mathrm{d}x/\mathrm{d}y = \mathrm{d}g^{-1}(y)/\mathrm{d}y$, the corresponding scale is $|\mathrm{d}g^{-1}(y)|$, where we take the absolute value because the scale in an integral is positive.

Hence, when the transformation satisfies the conditions, we have

$$\Pr(a < Y < b) = \int_{g^{-1}(a)}^{g^{-1}(b)} f_X(g^{-1}(y))|\mathrm{d}g^{-1}(y)|;$$

that is, the PDF of Y is

$$f_Y(y) = f_X(g^{-1}(y))|\mathrm{d}g^{-1}(y)/\mathrm{d}y|. \tag{3.27}$$

This is positive over the range $g(\mathcal{D})$ and zero elsewhere. (A careful development of these ideas requires some basic facts about how definite integrals are affected by transformations on the limits of integration; see Gentle, 2019, Section 0.1.6, for example.)

Equation (3.27) provides a direct method, called the *change of variables* technique for forming the PDF of the transformed variable, assuming the conditions are satisfied.

For example, let X be a random variable with PDF

$$f_X(x) = \begin{cases} 1 - |x| & \text{for } -1 < x < 1 \\ 0 & \text{otherwise.} \end{cases} \tag{3.28}$$

(This is the "triangular distribution".)

Now let $Y = 8X^3$. This transformation is one-to-one and differentiable over $-1 < x < 1$ (except at the single point $x = 0$), $X = Y^{1/3}/2$, and $\mathrm{d}x/\mathrm{d}y = 1/(6y^{2/3})$. Using (3.27), we have

$$f_Y(y) = \begin{cases} 1/|6y^{2/3}| - 1/|12y^{1/3}| & \text{for } -8 < y < 0 \text{ or } 0 < y < 8 \\ 0 & \text{otherwise.} \end{cases} \tag{3.29}$$

Note that there is a singularity at $Y = 0$. The probability of this point (or any other single point), however, is 0.

If the transformation is not one-to-one or if it is not differentiable over the full range of X, then the transformation must be handled in separate parts where those conditions exist. In most transformations of interest, there are only one or two regions that must be dealt with.

3.1.3 Linear Combinations of Random Variables; Expectations and Quantiles

Linear combinations of random variables arise often in financial modeling. Analysis of portfolios, as in Section 1.2.9, for example, is based on linear combinations of random variables that represent either asset prices or returns.

Linear combinations of any finite number of variables can be analyzed following the same methods as for just two variables. For two given constants w_1 and w_2, let us consider a general linear combination of two random variables, X_1 and X_2,

$$Y = w_1 X_1 + w_2 X_2. \tag{3.30}$$

Assuming $E(X_1)$ and $E(X_2)$ are finite, without any further restrictions, from the definitions (3.11) or (3.26), it immediately follows that $\mu_c = w_1\mu_1 + w_2\mu_2$, where $\mu_1 = E(X_1)$, $\mu_2 = E(X_2)$, and μ_c is the mean of the combination.

The fact that the combined mean is the same linear combination of individual means that the variable is of the individual variables is because the expectation operator is a *linear operator*. More generally, if we let $g(X) = ah(X) + k(X)$ (where, technically, h and k are measurable functions and a is a constant real number), we have

$$E(ah(X) + k(X)) = aE(h(X)) + E(k(X)). \tag{3.31}$$

(A *linear operator* is defined as a real-valued function ψ such that if u and v are in the domain of ψ and a is a real number such that $au + v$ is in the domain of ψ, then $\psi(au + v) = a\psi(u) + \psi(v)$.)

Let us now restrict w_1 and w_2 in equation (3.30) so that $w_1 + w_2 = 1$. This is the case in portfolio analysis. Let us also restrict w_1 and w_2 so that $w_1, w_2 > 0$. (For a portfolio, this would be equivalent to not allowing short selling.) Under these restrictions, we see that μ_c is between μ_1 and μ_2; this is, if $\mu_1 < \mu_2$ then $\mu_1 < \mu_c < \mu_2$, as shown in Figure 3.1. Also, as we have seen, $\sigma_c^2 = w_1^2\sigma_1^2 + w_2^2\sigma_2^2 + w_1 w_2 \rho \sigma_1 \sigma_2$, where σ_c^2, σ_1^2, and σ_2^2 are the variances and ρ is the correlation between X_1 and X_2; hence, the variance of the combination is generally smaller than either individual. The density of the combination generally is more compact, and is "between" that of the individual components, again, as seen in Figure 3.1. This speaks to the benefits of diversification.

Not all statistical functions behave like expectations. A quantile of a linear combination, for example, is not a linear combination of the quantiles of the individual components. In Figure 3.1, the 5% quantiles of all three random variables are shown as q_1, q_2, and q_c, and we see that $q_1 < q_2 < q_c$. A result of this is that if the random variables X_1 and X_2 modeled the distribution of the prices of two assets, the VaR of a holding in either asset alone would be smaller than a more diversified holding consisting of smaller holdings in

Probability Densities of Variables and Linear Combination

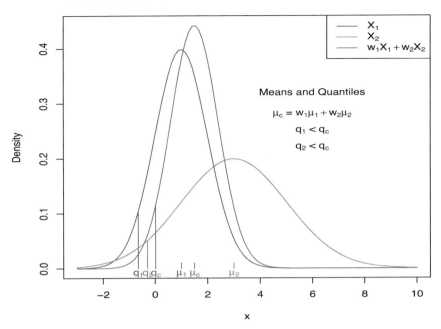

FIGURE 3.1
Means and Quantiles of Linear Combinations"

each asset. That is why we said in Section 1.4.5 that this measure of the value at risk is not *coherent*. (Recall the VaR depends on some parameters, one of which is the probability that determines the quantile. In this example, we are using a 5% quantile.)

3.1.4 Survival and Hazard Functions

It is often of interest to model the time interval between events, such as the time between jumps of the price of an individual asset of jumps of the market, or the duration of a condition, such as the length of a bear market.

Given a random variable X, the *survival function* for X is

$$S_X(x) = \Pr(X > x). \tag{3.32}$$

If the CDF of X is $F_X(x)$, we have the relationship

$$S_X(x) = 1 - F_X(x). \tag{3.33}$$

Use of the survival function is almost always restricted to random variables with positive support; that is, $x > 0$, is often part of the definition. The

Weibull and gamma distributions, which have positive support, (see Section 3.2.2), often serve as useful models of "interarrival" times or of durations of events.

The survival function is also called the complementary cumulative distribution function (CCDF).

For a random variable X with PDF $f_X(x)$, a related function is the *hazard function*, defined as

$$h_X(x) = \frac{f_X(x)}{S_X(x)}, \quad \text{for } S_X(x) > 0. \tag{3.34}$$

The hazard function can be interpreted as a model of the "intensity" of X. In the sense of death and survival, it is the probability density of death around some point, given survival to that point.

3.1.5   Multivariate Distributions

In many applications in which random variables are used as models of observable events, the probability distributions of two random variables are related. For example, the rates of return of INTC and of the S&P 500 considered on page 61 appear to be related. The scatterplot in Figure 1.17 indicates that at any time t, if the S&P return $R_{M,t}$ was relatively large, the Intel return $R_{i,t}$ was more likely also to be relatively large.

For two or more random variables, we are interested in their *joint distribution*, or the *multivariate distribution*. The PDF and CDF discussed above for single random variables extend readily to the joint PDF and joint CDF for two or more random variables. Using similar notation as above, we can define a joint CDF for two random variables X and Y, denoted by $F_{XY}(x, y)$, as

$$F_{XY}(x, y) = \Pr(X \le x \text{ and } Y \le y). \tag{3.35}$$

For continuous random variables X and Y, other functions of the joint distribution are similar to those defined for univariate distributions. The joint PDF is the derivative of the joint CDF:

$$f_{XY}(x, y) = \frac{d}{dx}\frac{d}{dy}F_{XY}(x, y) \tag{3.36}$$

We also have

$$F_{XY}(x, y) = \int_{-\infty}^{y} \int_{-\infty}^{x} f_{XY}(t_1, t_2)\, dt_1 dt_2. \tag{3.37}$$

The expectation of a function of X and Y, $g(X, Y)$, is

$$E(g(X, Y)) = \int_D g(x, y) f_{XY}(x, y)\, dx\, dy, \tag{3.38}$$

where the integration is taken over the domain of x and y for which the PDF is positive. For discrete random variables, summation takes the place of integration.

Marginal Distributions

We speak of the "joint distribution" of X and Y. We call the distribution of either X or Y separately, without any involvement of the other one, the *marginal distribution*. We also use the term "marginal" to refer to any aspect of the marginal distribution, such as the "marginal CDF of X", for example. For continuous multivariate random variables, the marginal PDF of one component is obtained by integrating out all the other components, for example

$$f_X(x) = \int f_{XY}(x, y)\, dy,\tag{3.39}$$

where the integration is taken over the full range of Y. For discrete random variables, summation takes the place of integration.

Means and variances of the two random variables, as in the case of univariate random variables, are defined as special cases of the expectation in equation (3.26) and they are often denoted in similar ways, for example,

$$\begin{aligned}\mu_X &= \int x f_{XY}(x, y)\, dx\, dy\\ &= \int x f_X(x)\, dx,\end{aligned}\tag{3.40}$$

and

$$\begin{aligned}\sigma_X^2 &= \int (x - \mu_X)^2 f_{XY}(x, y)\, dx\, dy\\ &= \int (x - \mu_X)^2 f_X(x)\, dx.\end{aligned}\tag{3.41}$$

Here, the integrations are taken over the full ranges, and we assume that the integrations are interchangeable.

The quantiles of multivariate random variables have similar definitions as in equations (3.8) and (3.24), but they are not as useful in practice. The marginal quantiles, however, are the same as in the univariate case, and just as useful.

If the random variables X and Y are independent, the value taken on by one of the random variables in no way affects the probability distribution of the other one. In that case, the integration above does not involve x, and conversely integrating out x does not involve y; hence, if (and only if) X and Y are *independent*, then

$$f_{XY}(x, y) = f_X(x) f_Y(y).\tag{3.42}$$

This relation also holds for discrete random variables, where f_{XY}, f_X, and f_Y are the respective probability functions. We have referred to "independence" or "independent variables" several times previously without giving a definition. We can take the relation (3.42) above as a definition.

The concepts and terms above that relate to two random variables can be extended in a natural way to more than two random variables. In general, we consider a *random vector*, which is a vector whose elements are random variables. If $X = (X_1, \ldots, X_d)$ is a random vector with joint CDF $F_X(x)$, analogous to $F_{XY}(x, y)$ in equation (3.35) in which X and Y are scalars, the PDF $f_X(x)$ is analogous to $f_{XY}(x, y)$ in equation (3.36). The mean of X is the vector given in equation (3.38) with $g(X) = X$,

$$E(X) = \int_D x f_X(x)\, dx, \tag{3.43}$$

where x is a d-vector and D is the range of the random vector X. The variances and covariances are given in the $d \times d$ variance-covariance matrix

$$V(X) = \int_D (x - E(X))(x - E(X))^{\mathrm{T}} f_X(x)\, dx, \tag{3.44}$$

which is equation (3.38) with $g(X) = (X - E(X))(X - E(X))^{\mathrm{T}}$.

In statistical applications, we often assume that we have a set of random variables that are independent and also identically distributed. In that case, if the individual PDFs are all $f(x)$, then the joint distribution is of the form

$$f(x_1)f(x_2)\cdots f(x_n). \tag{3.45}$$

We say that the random variables are iid, for "independently and identically distributed".

Conditional Expectations and Distributions

Another distribution formed from a joint distribution is a conditional distribution. This is the distribution of one random variable at any given point in the range of the other random variable. (As above, the discussion will involve just two random variables, but each reference to "one" random variable could equally be a reference to multiple random variables.)

Let X and Y be random variables with joint PDF f_{XY}, and let f_X be the marginal PDF of X. Then the *conditional PDF of Y given X*, denoted $f_{Y|X}$ is

$$f_{Y|X}(y|x) = \frac{f_{XY}(x, y)}{f_X(x)}, \quad \text{if } f_X(x) > 0. \tag{3.46}$$

This can be rewritten as

$$f_{XY}(x, y) = f_{Y|X}(y|x)f_X(x). \tag{3.47}$$

A careful development of the concept and theory of conditional distributions, proceeds differently from the development here (see a reference on probability theory, such as, for example, Gentle, 2019, Chapter 1), but the equations above are correct given the assumptions stated.

For discrete random variables, the PDFs in the equations above are replaced by the corresponding probability functions.

The *conditional expectation* of a function of Y and, possibly of X also, given X can be defined in the usual way using the conditional PDF in equation (3.46):

$$E(g(Y, X)|X) = \int_D g(y, X) f_{Y|X}(y|X) \, dy,$$

where the integration is taken over the domain of y for which the PDF is positive.

We note that *conditional expectation* of a function of Y is a function of X, and hence, it is a random variable itself. A careful development of conditional distributions usually begins with this function of the conditioning random variable; see Gentle (2019), Section 1.5, for example.

3.1.6 Measures of Association in Multivariate Distributions

For multivariate distributions, the univariate measures are usually not as interesting as measures that involve both or all of the random variables. If the random variables are independent, these univariate measures can be treated separately. If the variables are not independent, our interest is in assessing the association of one random variable with others.

Most measures of association are *bivariate*; that is, they involve only two random variables at a time, even if we have a joint distribution with more than two random variables.

Covariances and correlations are direct measures of linear association among two variables. We will discuss them first. Next we will discuss copulas, which provide an alternative method of fitting multivariate models of association.

Covariance and Correlation

The most common and most useful measure of association between two random variables is the *covariance* or its related measure, the *correlation*. In Section 1.3.4 (page 82), for paired observations on two variables x and y, we defined the sample covariance, which we denoted as s_{xy} and the sample correlation, which we denoted as r_{xy}.

We now define corresponding measures of association between random variables, which we often denote as σ_{XY} and ρ_{XY}. (Notice the convention of naming properties of random variables with Greek letters, and properties of a sample with Latin letters.) These definitions of association between random variables are expectations, in which the "g" in equation (3.26) is in a form similar to the sample quantities:

$$\sigma_{XY} = E((X - \mu_X)(Y - \mu_Y)). \tag{3.48}$$

We may also denote this as "$\text{Cov}(X, Y)$".

The correlation is

$$\rho_{XY} = \frac{\sigma_{XY}}{\sigma_X \sigma_Y}, \tag{3.49}$$

if $\sigma_X \neq 0$ and $\sigma_Y \neq 0$, otherwise $\rho_{XY} = 0$. We may also denote the correlation between X and Y as "Cor(X, Y)", so

$$\text{Cor}(X, Y) = \frac{\text{Cov}(X, Y)}{\sqrt{\text{V}(X)\text{V}(Y))}}, \tag{3.50}$$

In Section 1.3.8 we defined some sample statistics that could serve as alternatives to the sample correlation for measuring the association between two variables. These statistics, Spearman's rho and Kendall's tau, were based on a given multivariate sample (using order statistics or numbers of pairs of observations) and depended on the size of the sample. Although we could define corresponding measures for probability models, they would not be as useful. Hence, we will not define a Spearman's rho or a Kendall's tau for random variables.

For multivariate distributions with more than two random variables, it is often useful to form a matrix of covariances or correlations, just as we did with the sample correlations of returns in Table 1.8 on page 83. We call these the "variance-covariance matrix" (or just the "covariance matrix") and the "correlation" matrix, and often denote them by the upper-case Greek letters Σ and P (Rho). They are both symmetric, and they are nonnegative definite. (A *nonnegative definite* matrix is a symmetric matrix A such that for any conformable vector x, $x^{\text{T}}Ax \geq 0$. If for any conformable vector $x \neq 0$, $x^{\text{T}}Ax > 0$, A is said to be *positive definite*. A variance-covariance matrix and a correlation matrix are positive definite unless the distribution is degenerate; that is, unless a subset of the elements can be expressed as a linear combination of the others.)

Covariance and Correlation as Linear Measures of Association

It is important to recognize the fundamental properties of the covariance and correlation. A covariance or correlation large in absolute value represents a strong relationship. The converse of this is not true; a strong relationship does not necessarily indicate a large value of the covariance or correlation.

Consider two random variables X and Y with a very strong relationship. For example, suppose

$$Y = X^2.$$

Suppose further that X has mean of 0 and is symmetric about 0. Then from equation (3.48), we see that $\sigma_{XY} = 0$, despite the very strong relationship between X and Y.

The covariance and correlation are measures of *linear* association.

Covariance, Correlation, and Independence

If two random variables are independent, their covariance and, hence, correlation are 0. This is easily seen from the definition (3.48) and the definition of independence (3.42) (using the form for continuous random variables):

$$
\begin{aligned}
\sigma_{XY} &= \mathrm{E}((X - \mu_X)(Y - \mu_Y)) \\
&= \int (x - \mu_X)(y - \mu_Y) f_{XY}(x, y) \mathrm{d}x \mathrm{d}y \\
&= \int (x - \mu_X)(y - \mu_Y) f_X(x) f_Y(y) \mathrm{d}x \mathrm{d}y \\
&= \int (x - \mu_X) f_X(x) \mathrm{d}x \int (y - \mu_Y) f_Y(y) \mathrm{d}y \\
&= 0.
\end{aligned}
$$

The example above with $Y = X^2$ shows that the converse is not the case. A covariance of 0 does not imply that the variables are independent.

There is an important case in which 0 covariance does imply independence, however. That is when the variables have a joint normal distribution. If two random variables with a joint normal distribution have a covariance (and, hence, a correlation) of 0, then they are independent. (This is easily seen using the normal PDF, given in Table 3.2.)

Elliptical Distributions

A multivariate distribution whose joint PDF has ellipsoidal level contours is called an *elliptical distribution*, or an elliptically contoured distribution. The kernel of the PDF of an elliptical distribution is of the form

$$
g\left((x - \mu)^{\mathrm{T}} \Sigma^{-1}(x - \mu)\right),
$$

where Σ is a positive definite matrix. If the variance-covariance of the elliptical distribution exists and is finite, it is proportional to Σ. The multivariate normal distribution is an elliptical distribution, and statistical procedures based on the normal distribution often work fairly well in nonnormal elliptical distributions.

Elliptical distributions are important in various financial applications. In portfolio theory, if the returns on all assets have joint elliptical distributions, then the location and scale of any portfolio completely determines the distribution of returns on the portfolio.

Notation

We have generally used "X" and "Y" for bivariate random variables. Obviously, this notation does not generalize well, and in the following, we will often use the notation, $X_1, \ldots, X_d$ to refer to the elements of a d-variate random variable. Also, as we have already been doing, we will use a single integral notation $\int$ to indicate integration over any number of dimensions. The context makes clear the dimension. Also we generally mean integration over a fixed domain, even when we do not specify it; thus the integrals are definite.

We do not use any special notation to denote a vector or matrix; hence, a single symbol such as X may represent a vector. When a letter represents a vector, we often use subscripts to denote the elements of the vector, for example, $X = (X_1, \ldots, X_d)$.

The first moment (mean) of a multivariate random variable is also a vector:

$$\mu_X = \int x f_X(x) \mathrm{d}x, \qquad (3.51)$$

where all symbols except f_X represent vectors, and the integral is a multidimensional definite integral over the range of all elements of the multivariate random variable.

The variance of a multivariate random variable is a symmetric matrix formed as a definite integral of an outer product:

$$\Sigma_X = \int (x - \mu_X)(x - \mu_X)^{\mathrm{T}} f_X(x) \mathrm{d}x. \qquad (3.52)$$

We call this the variance-covariance matrix (or sometimes just the variance matrix or the covariance matrix). The diagonal elements of Σ_X are the ordinary univariate variances of the elements of X, and the off-diagonals are the covariances of the elements.

Higher order moments of multivariate random variables can be defined, but they are less useful.

It is worth reiterating the common convention that we follow of using upper-case letters to represent random variables and corresponding lower-case letters to represent observations of the random variable. Thus, $X = (X_1, \ldots, X_d)$ may be a vector random variable, and $x = (x_1, \ldots, x_d)$ may be a realization of the random vector. (I must, however, also reiterate the caveat that I'm not entirely consistent in following this notational convention.)

3.1.7   Copulas

We now consider a special multivariate transformation that has wide-ranging applications in statistical modeling. We begin with a d-variate random variable $X = (X_1, \ldots, X_d)$, which in applications may represent various types of economic quantities, prices of stocks or other assets, interest rates, and so on.

In statistical modeling and analysis, we must assume some kind of multivariate probability distribution. We want a distribution family that fits the observed quantities; that is, the frequency distributions of the observations match the probability distributions. (Recall the comparisons of univariate histograms of frequency distribution of returns with idealized univariate normal probability densities in Figures 1.24 and 1.25.) In the multivariate case, we also want a distribution family that expresses the relationship among the individual variables in a meaningful way. (Covariances and correlations often do this, but they do not adequately model relationships in the tails of the distribution.) Finally, we want a distribution family that is mathematically tractable. Unfortunately, there are not many multivariate probability distribution families that satisfy these requirements. The multivariate normal distribution, which we will use often, is easy to work with, and it does serve as a useful model in a wide range of settings. (Covariances and correlations are very effective for relationships among normally distributed random variables.) In financial applications, however, the normal distribution often does not fit the extremes of the distribution, as we have seen. Those extremes are often of most interest. There are other multivariate distribution families, and we will discuss some of them in later sections.

In this section, we develop a very general multivariate family using transformations involving marginal CDFs. There are several subfamilies of this family of distributions, and members of different families are useful as models of distributions of different types of variables.

Let $X = (X_1, \ldots, X_d)$ be a d-variate continuous random variable. (Each component is continuous.) Let $F_X(x)$ be the CDF of X, and let $F_{X_i}(x_i)$ be the marginal CDF of X_i. (Note that it is not the case that $F_X(x) = (F_{X_1}(x_1), \ldots, F_{X_d}(x_d))$; a CDF is a scalar-valued function.) Because of the assumption of continuity, the quantile function is the ordinary inverse $F_{X_i}^{-1}(x_i)$. (*In the following, pay attention to the notation. X_i denotes a random variable; x_i denotes a realization, that is, x_i is just an ordinary real number.*)

Now, for the random variables $X_1, \ldots, X_d$, consider the joint CDF of $F_{X_1}(X_1), \ldots, F_{X_d}(X_d))$. Denote it as $C_X(u_1, \ldots, u_d)$ (where the subscript does not follow our convention for notation of the form "$F_X(x)$"). The arguments of C_X are in the d-dimensional unit hypercube. It is defined as

$$
\begin{aligned}
C_X(u_1, \ldots, u_d) &= \Pr(F_{X_1}(X_1) \leq u_1, \ldots, F_{X_d}(X_d) \leq u_d) \\
&= \Pr\left(X_1 \leq F_{X_1}^{-1}(u_1), \ldots, X_d \leq F_{X_d}^{-1}(u_d)\right) \\
&= F_X\left(F_{X_1}^{-1}(u_1), \ldots, F_{X_d}^{-1}(u_d)\right).
\end{aligned} \tag{3.53}
$$

We call the CDF $C_X(u_1, \ldots, u_d)$ a *copula*. (We also call the distribution with that CDF a copula.)

A copula can also be defined in terms of discrete random variables by using the generalized inverse in equation (3.10), but most modeling applications in finance are based on continuous random variables.

A copula is a CDF, and so C_X is increasing in the u_is, and

$$
C_X(0, \ldots, 0) = 0,
$$

and

$$
C_X(1, \ldots, 1) = 1.
$$

It also follows that

$$
0 \leq C_X(u_1, \ldots, u_d) \leq 1.
$$

Because each marginal CDF of the random d-vector X evaluated at the corresponding random variable, $F_{X_i}(X_i)$, is a random variable with distribution $U(0, 1)$ as we showed in equation (3.25), each marginal distribution of the copula distribution is $U(0, 1)$.

Because of these simple relations, copulas are useful in fitting multivariate models. Copulas provide an alternative to other measures of association such as covariances and correlations. Copulas are useful in simulation modeling, for example, see page 320.

There are many classes of the copula family of multivariate distributions depending on $F_X(x)$ and the $F_{X_i}(x_i)$.

Normal or Gaussian Copulas

If $F_X(x)$ is the CDF of a multivariate normal distribution with mean vector 0 and whose variance-covariance matrix Σ has 1s on the diagonal (that is, it is a correlation matrix) so that the $F_{X_i}(x_i)$ are CDFs of the univariate standard normal distribution, which we often denote as Φ, then the copula

$$C_{\text{Normal}}(u_1, \ldots, u_d | \Sigma) = \Phi_\Sigma \left(\Phi^{-1}(u_1), \ldots, \Phi^{-1}(u_d) \right) \qquad (3.54)$$

is called a *normal copula* or *Gaussian copula*.

Archimedean Copulas

Another class of copulas that is particularly easy to work with are the *Archimedean copulas*. (The name is given to this class because of a certain property of these copulas that is reminiscent of the Archimedean continuity principle that states that for any two real numbers, some integer multiple of either is larger than the other.)

An Archimedean copula is based on a single continuous, strictly decreasing, and convex univariate function φ, called the *strict generator function* that maps the closed unit interval $[0, 1]$ onto the extended nonnegative reals $[0, \infty]$, and the copula is defined as

$$C(u_1, \ldots, u_d) = \varphi^{-1}(\varphi(u_1) + \cdots + \varphi(u_d)). \qquad (3.55)$$

Note that if the mapping φ is onto $[0, \infty]$, then it must be the case that $\varphi(1) = \infty$. (If the mapping is onto $[0, a]$ for $a \neq \infty$, the copula is also called an Archimedean copula, but the generator is not strict and an adjustment to φ^{-1} must be made.)

Various forms of the generator lead to different subclasses. Many subclasses are indexed by a single parameter, θ, and many have been given names (although there are some variations in the names). Some common Archimedean copulas are

- Clayton: $\varphi(u) = \frac{1}{\theta} \left(u^{-\theta} - 1 \right)$ for $\theta > 0$
- Gumbel: $\varphi(u) = (-\log(u))^\theta$ for $\theta \geq 1$

- Frank: $\varphi(u) = \log \left(\dfrac{e^{-\theta u} - 1}{e^{-u} - 1} \right)$ for $\theta \neq 0$

- Joe: $\varphi(u) = \log \left(1 - (1 - u)^{\theta} \right)$ for $\theta \geq 1$

These are all strict generators. In the Clayton copula, however, if $\theta \in (-1, 0)$, the generator is not strict, but the copula is still useful. (For a generator φ that is not strict, φ^{-1} in equation (3.55) must be modified to be a "pseudo-inverse", in which φ^{-1} is taken as 0 wherever $\varphi < 0$.)

The R copula package provides functions for fitting copula models to observed data and for random number generation with various classes of copulas including Gaussian, various Archimedean copulas, and other copula families.

3.1.8 Transformations of Multivariate Random Variables

How we work with transformations of multivariate random variables depends on whether the random variables are discrete or continuous and whether the transformations are one-to-one and differentiable. Most of the variables we will encounter in financial applications are continuous, so we will consider only continuous random variables. Also, as on page 255, we will assume that the transformations are one-to-one and differentiable over the full range of the multivariate random variable. As in the univariate case, if the transformation is not one-to-one or if it is not differentiable, then the transformation must be handled in separate parts in which those conditions exist. In most transformations of interest, there are only one or two regions that must be dealt with.

Using the compact notation for the d-vector X, we consider a one-to-one transformation, $Y = g(X)$ that is differentiable on the full range of X:

$$
\begin{aligned}
Y_1 &= g_1(X) \\
&\vdots \\
Y_c &= g_c(X).
\end{aligned}
\tag{3.56}
$$

If $c = d$ and the transformation is one-to-one, we have $X = g^{-1}(Y) = h(Y)$, where $h(\cdot) = g^{-1}(\cdot)$. We have

$$
\begin{aligned}
X_1 &= h_1(Y) \\
&\vdots \\
X_d &= h_d(Y).
\end{aligned}
\tag{3.57}
$$

It is not the case, however, that the h_i correspond to inverses of any g_i.

We can work out the distribution of Y following the same methods we used on page 255. The additional complexity in this case is the generalization of the derivative transform, $|dg^{-1}(y)/dy|$. The extension to the multivariate case is the absolute value of the determinant of the *Jacobian* of the inverse transformation:

$$|\det(J(h))| = \left|\det\left(\begin{bmatrix} \frac{\partial h_1}{\partial y_1} & \cdots & \frac{\partial h_1}{\partial y_d} \\ \cdots & \ddots & \cdots \\ \frac{\partial h_d}{\partial y_1} & \cdots & \frac{\partial h_d}{\partial y_d} \end{bmatrix}\right)\right|. \tag{3.58}$$

We omit the details here, and only state the result. We refer the interested reader to a text on probability theory or on mathematical statistics for a careful development (see Gentle, 2019, Section 1.1.10, for example).

If X is a d-variate continuous random variable with PDF $f_X(x)$ and $Y = g(X)$ is a differentiable one-to-one transformation with inverse transformation h, then the PDF of Y is

$$f_Y(y) = f_X(h(y))|\det(J(h))|. \tag{3.59}$$

This is positive over the range where $f_X(x)$ is positive and zero elsewhere. We will give a simple example in the next section.

Often the transformations of interest are linear transformations. If X is a d-variate random variable and A is a constant matrix with d columns, the linear transformation

$$Y = AX \tag{3.60}$$

is often useful. This linear transformation yields a multivariate random variable Y with as many elements as there are rows of A. If the transformation is one-to-one, A is a square matrix of full rank. In that case, the Jacobian of the transformation is

$$\frac{\partial^2}{\partial y(\partial y)^{\mathrm{T}}} A^{-1}.$$

If the mean of the d-variate random variable X is μ_X and its variance-covariance matrix is Σ_X, and $Y = AX$, then the mean and variance of Y are given by

$$\mu_Y = A\mu_X \tag{3.61}$$

and

$$\Sigma_Y = A\Sigma_X A^{\mathrm{T}}. \tag{3.62}$$

We can see this very easily by use of the definitions in equations (3.51) and (3.52) on page 264. (These facts are worked out in detail in Gentle, 2017, Section 4.5.3.)

In a common application of equation (3.61), we begin with d independent variates in $X_1, \ldots, X_d$ and form $X = (X_1, \ldots, X_d)$. In that case, Σ_X in equation (3.62) is the identity, so to form a vector random variable Y with variance-covariance matrix Σ_Y, we determine A such that

$$\Sigma_Y = AA^{\mathrm{T}}. \tag{3.63}$$

Because the variance-covariance matrix Σ_X is nonnegative definite, there exists such a matrix A as above. (See Gentle, 2017, Section 5.9.) One such matrix is lower triangular. It is called the Cholesky factor, and the R function `chol` computes it.

Given a vector random variable X, we are often interested in the distribution of a linear combination of its elements, $Z = a_1 X_1 + \cdots + a_d X_d$. An example of this is a principal component, which we will discuss on page 409. The variable Z is a scalar, and this is just a special case of the situation above. Let a be the vector with elements $(a_1, \ldots, a_d)$. Then

$$Z = a^\mathrm{T} X, \tag{3.64}$$

and if Σ_X is the variance-covariance matrix of X, then the variance of Z is the quadratic form

$$\mathrm{V}(Z) = a^\mathrm{T} \Sigma_X a. \tag{3.65}$$

This also applies to the sample variance. If instead of the variance-covariance matrix Σ_X of the distribution of the random variable X, we have the sample variance-covariance matrix S of a set of data, as in equation (1.54), and if a univariate variable is formed as a linear combination of all the variables in the dataset, $z = a^\mathrm{T} x$, the *sample variance* of z is $a^\mathrm{T} S a$. This is the same as that computed directly by the formula for the sample variance, equation (1.49). We use these facts in principal components analysis, discussed in Chapter 4.

3.1.9 Distributions of Order Statistics

On page 95, we described order statistics and some sample descriptive statistics based on them.

The distributions of order statistics of a random variable are obviously related to the distribution of the underlying random variable. It is easy to see some simple relationships of measures of the distribution of certain order statistics to the corresponding measures of the underlying random variable. For example, if the moments of the underlying random variable are finite, the expected value of the minimum order statistic should be less than the expected value of the underlying random variable, and the expected value of the maximum order statistic should be greater. The variance of the median, that is, the central order statistic, should be less than the variance of the underlying random variable. If the range is infinite, the variance of either of the extreme order statistics should be greater than the variance of the underlying random variable.

Given the PDF $f_X(x)$ (or the PF if the random variable is discrete) and the CDF $F_X(x)$, it is simple to work out the PDF of an order statistic in a sample of that random variable. We will consider the ith order statistic in a sample of size n, $X_{(i:n)}$, where $i \neq 1$ and $i \neq n$ (that is, $n > 2$). (After working this out, we will see that the results apply without restrictions on i and n.) Given a random sample of order statistics, $X_{(1:n)}, \ldots, X_{(i:n)}, \ldots, X_{(n:n)}$, the idea is first to divide them into three sets, those smaller than $X_{(i:n)}$, $X_{(i:n)}$ itself, and those larger than $X_{(i:n)}$. This can be done in $\binom{n}{i}$ ways; the probability that there are i sample elements less than or equal to a value of $x_{(i:n)}$ is $\left(F_X(x_{(i:n)})\right)^{i-1}$; and the probability that there are $n - i$ sample elements greater than $x_{(i:n)}$ is $\left(1 - F_X(x_{(i:n)})\right)^{n-i}$. The density of $X_{(i:n)}$ itself, conditional on those probabilities, is $f_X(x_{(i:n)})$. Hence, the density of $X_{(i:n)}$ is

$$ f_{X_{(i:n)}}(x_{(i:n)}) = \binom{n}{i} \left(F_X(x_{(i:n)})\right)^{i-1} \left(1 - F_X(x_{(i:n)})\right)^{n-i} f_X(x_{(i:n)}). \quad (3.66) $$

By the same argument, we see that this holds for the first or the nth order statistic also.

3.1.10 Asymptotic Distributions; The Central Limit Theorem

In statistical data analysis, generally, the more data, the better. One reason, of course, is that more data is more information.

There is another reason, however. As the sample size increases, the probability distributions of statistics such as the mean or variance computed from the sample become closer to known probability distributions.

We can illustrate this very simply. Consider a population with a probability distribution as shown in Figure 3.2.

This is a gamma distribution with parameters 3 and 1. (This curve was produced using the R function dgamma(x,3,1).)

Now, suppose we collect a sample of observations from this population, and we compute the mean from the sample. (We produce the sample using the R function rgamma(size,3,1).)

Suppose we do this many times. What is the frequency distribution of the means of the samples?

That depends on the sizes of the samples.

If the samples are all of size 1 (not very big!), the distribution of the means is the same as the distribution of the data itself. A histogram of the means of samples of size 1 should look similar to the probability distribution shown in Figure 3.2. We did this 1,000 times and so got a set of 1,000 means. The histogram in the upper left side of Figure 3.3 shows the frequency distribution of the means. It is essentially the same as the probability distribution of the original data.

We next collected many samples of different sizes. We used sizes of 3, 5, and

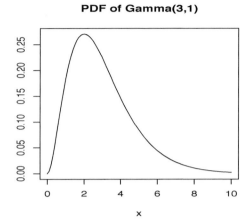

PDF of Gamma(3,1)

FIGURE 3.2
Assumed Probability Distribution of the Population

100. We computed the means for 1,000 repetitions of each size and produced
a histogram of the means for each size.

The histograms are shown in Figure 3.3.

As expected, the distribution of the means of samples of size 1 is very
similar to the distribution of the underlying population itself. It is positively
skewed, and some values are as large as 10 or 11. As the sample sizes increase,
however, the means have fewer extreme values, and their distributions become more symmetric. For samples of size 100, the means have a distribution
similar to a normal distribution. We have superimposed a curve of a normal
probability density to emphasize this.

The normal probability density curve in the figure has a mean of 3 and a
variance of 3/100. (The mean of the gamma(3,1) is 3 and its variance is also 3.
The actual sample mean for the simulated data in the lower right histogram
is 3.006443 and the sample variance is 0.03031104.)

Asymptotic Distributions

The idea that the distribution of the sample mean changes in a systematic
way leads us to a more general concept of an *asymptotic distribution*.

The distributions of a sequence of random variables will converge under
certain conditions, and when they do, the asymptotic distributions are often
useful in statistical inference.

We are interested in the distribution of a sample statistic, such as the

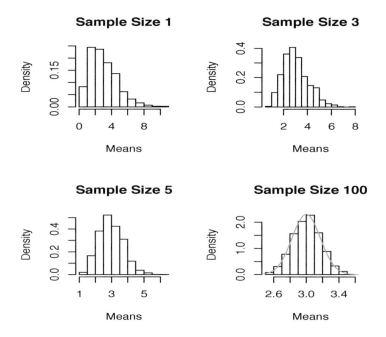

FIGURE 3.3
Means of Samples from a Gamma(3,1) Distribution

mean, as the sample size increases without bound, that is, as $n \to \infty$. In some cases that distribution will *converge* to some fixed distribution.

The asymptotic distribution of many sample statistics based on moments or on central order statistics, properly normalized, often is the normal distribution, as stated in the central limit theorem below. The asymptotic distribution of extreme order statistics is discussed on page 308.

Central Limit Theorem

Consider a random sample of size n from a population whose mean is μ and whose standard deviation is σ. We can represent the sample as a set of iid random variables, $X_1, \ldots, X_n$.

Let

$$\overline{X}_n = \sum_{i=1}^{n} X_i / n.$$

Now form

$$T_n = \frac{\overline{X}_n - \mu}{\sigma}. \qquad (3.67)$$

The *central limit theorem* states that as n increases without bound, in the limit, the asymptotic distribution of T_n is $N(0, 1)$.

This is equivalent to stating that the asymptotic distribution of $\overline{X}_n$ is $N(\mu, \sigma^2)$.

That is what we observe in the sample of means of samples of size $n = 100$ from the gamma(3,1) distribution in the lower right histogram in Figure 3.3.

There are various other forms of the central limit theorem (see Gentle, 2019, Section 1.4 for different sets of conditions).

3.1.11 The Tails of Probability Distributions

In Chapter 1 we saw in the histograms of Figures 1.24 and 1.25 and in the q-q plots of Figures 1.27 and 1.28 that the distributions of rates of returns for stocks tend to have heavier tails than that of the normal distribution. We observed this phenomenon for INTC and the S&P 500 index (which is an average of several stock prices) over specific periods.

The nature of the tail probability distribution is important in many areas of finance, but in particular, in value at risk assessments. Spectacular inaccuracies of those assessments in recent history underscore the need for better understanding of tail probability distributions.

In this section we will discuss various measures for describing the tail behavior of distributions with infinite support. With these measures we could formally define *heavy-tailed distributions*, or *fat-tailed distributions*. (Some authors define the latter term as a subclass of the former; I will, however, use both terms informally to mean generally that the distribution has tails heavier than the normal.)

Because numbers that are large in absolute value have relatively larger densities of probability in heavy-tailed distributions, the higher order moments of heavy-tailed distributions are very large, often infinite. (Technically, depending on the definition of the integral that defines the moments, the moments themselves may not be defined. We will ignore these mathematical subtleties in this book, and just refer to all such moments as "infinite". The reader interested in the differences in nonexistent and infinite is referred to Section 1.1.3 of Gentle, 2019.)

Beginning on page 293, we will briefly describe some useful heavy-tailed families of probability distributions, such as t, Cauchy, and stable. Another approach is just to model the tail of a distribution separately from the central region of the distribution. The generalized Pareto distributions (see page 301) are useful for this purpose. First, we will describe a way of measuring how "heavy" the tail is.

Tail Weight

Because a PDF must integrate to 1 over the full range of real numbers, the PDF of a distribution whose range is infinite must approach 0 sufficiently fast in the tails, that is, as the argument of the PDF approaches $-\infty$ or ∞.

There are several standard distributions, such as the normal and the gamma, with infinite ranges. Let us just consider the positive side of a distribution, even if the range is infinite on both sides. We can identify three general forms of the part of the PDF that determines the tail behavior (that is, the kernel, the part of the PDF that contains the argument):

$$\mathrm{e}^{-|x|^{\lambda}} \quad \text{with } \lambda > 0, \tag{3.68}$$

as in the normal, exponential, and double exponential;

$$|x|^{-\gamma} \quad \text{with } \gamma > 1, \tag{3.69}$$

as in the Pareto (also the related "power law distributions") and the Cauchy (with some additional terms); and

$$|x|^{\lambda}\mathrm{e}^{-|x|^{\beta}}, \tag{3.70}$$

as in the gamma and Weibull. (The PDFs of these distributions are shown in Table 3.2 on page 286.)

What happens as $x \to \infty$ determines whether the moments of the distribution are finite.

The form (3.70) is a combination of forms (3.68) and (3.69), so we will consider only forms (3.68) and (3.69).

Exponential Tails

Consider first the form (3.68). We will call this an *exponential tail*.

Notice that for any $\lambda > 0$ and any finite k

$$E(|X|^{k}) \propto \int_{0}^{\infty} |x|^{k}\mathrm{e}^{-|x|^{\lambda}}\,\mathrm{d}x \;<\; \infty; \tag{3.71}$$

that is, all moments of a distribution with an exponential tail are finite.

In expression (3.68), λ determines how rapidly the probability density decreases in the tails of the distribution. Some distributions with exponential tails therefore have heavier tails than others. For example, the double exponential distribution has heavier tails than a normal distribution because

$$\mathrm{e}^{-|x|^{2}} \to 0$$

faster than

$$\mathrm{e}^{-|x|} \to 0.$$

Polynomial Tails

Now consider the form (3.69). We will call this a *polynomial tail* or a *Pareto tail* because of the form of the Pareto PDF.

For the moments we have

$$E\left(|X|^k\right) \propto \int_0^\infty |x|^k |x|^{-\gamma} \mathrm{d}x, \tag{3.72}$$

and so the moments $E(|X|^k)$ are finite only for $k < \gamma$. In the Pareto distribution, for example, the mean is finite only for $\gamma > 1$ and the variance is finite only for $\gamma > 2$.

In expression (3.69), γ determines how rapidly the probability density decreases in the tails of the distribution.

We see in general that a polynomial tail is "heavier" than an exponential tail. We therefore emphasize it more in measures of tail behavior, such as the tail index.

Tail Index

Because γ in expression (3.69) is greater than 1, we will rewrite the expression as

$$|x|^{-(\alpha+1)} \quad \text{with } \alpha > 0, \tag{3.73}$$

and define the *tail index* of a distribution with a polynomial tail as α in (3.73). (We should note that some authors define the tail index as $\alpha + 1$, that is, as γ in expression (3.69). Furthermore, some authors define the tail index as the reciprocal $1/\alpha$ or $1/\gamma$. Any of these quantities determine the other three, so the differences are not meaningful; the reader, however, needs to be clear on which is being used.)

The PDF $f(x)$ of a distribution with tail index α is of order $|x|^{-(\alpha+1)}$; that is, for large $|x|$,

$$f(x) \propto |x|^{-(\alpha+1)}.$$

For asymmetric distributions, we speak of the left tail index (for $x \to -\infty$) and/or the right tail index (for $x \to \infty$).

The larger the tail index, the more rapidly the PDF will approach 0. (That is why the reciprocal may be a more logical definition.)

Threshold Exceedance

To refer to the "tail behavior", we need to establish thresholds that define the tails. A *tail threshold* or *exceedance threshold* is a value that delineates a conditional *threshold exceedance* distribution.

This threshold could be specified either as a fixed value of the observable random variable or as a proportion. In VaR, for example, the threshold is generally specified as a proportion, often 5%, as in the examples in Chapter 1. The simple approach of PVaR is essentially the conversion of a threshold stated as a proportion to a threshold stated as a value of the random variable. The more detailed analysis of the distribution of the value at risk, or DVaR, focuses on the distribution in the tail, as on page 119.

In the following, let us focus on the right tail. Given a distribution with CDF F and a threshold x_u such that $0 < F(x_u) < 1$, the probability of exceeding this threshold is $1 - F(x_u)$. The *exceedance distribution* corresponding to F over the threshold x_u is the distribution with CDF

$$F_{x_u}(x) = \frac{F(x) - F(x_u)}{1 - F(x_u)}, \quad \text{for } x > x_u. \tag{3.74}$$

The exceedance distribution is the conditional distribution given that the random variable is greater than the threshold. The exceedance distribution is useful in modeling extreme losses.

The mean of the exceedance distribution is the conditional mean, as a function of the threshold, x_u,

$$e(x_u) = \mathrm{E}(X - x_u | X > x_u). \tag{3.75}$$

The empirical cumulative distribution function (ECDF) formed from the order statistics above (or below) a threshold, either a specific value or a proportion of the data, corresponds to the CDF of the exceedance distribution. The exploratory techniques using the ECDF discussed in Section 2.2, or those using the inverse of the ECDF are all relevant when the ECDF is computed using only the largest (or smallest) order statistics. For q-q plots computed using only the largest (or smallest) order statistics, on the other hand, the reference distribution must be conditioned on the threshold, whether it is a quantile or a probability (see Exercise 2.17).

Tail Dependence

The relationship between two random variables in the tails of their respective distributions is of interest. One obvious measure of their relationship is the correlation conditional on each random variable being greater than specified values or specified quantiles. This measure is symmetric in the variables. As we saw on page 127, the correlations of returns are often greater in the tails of the distributions, which may correspond to periods of market turmoil.

Another bivariate measure of the relationship of two random variables in the tails of their distributions is based on conditional distributions and, hence, is asymmetric in the variables.

For two random variables X and Y with CDFs F_X and F_Y, and for a given quantile level p, we define the *upper tail dependence function* of Y on X as

$$\lambda_u(p) = \Pr(Y > F_Y^{\leftarrow}(p) \mid X > F_X^{\leftarrow}(p)), \tag{3.76}$$

where, as usual, $F_X^{\leftarrow}$ and $F_Y^{\leftarrow}$ denote the respective quantile functions, that is, the generalized inverses of the CDFs. Likewise, we define the *lower tail dependence function* of Y on X as

$$\lambda_l(p) = \Pr(Y \le F_Y^{\leftarrow}(p) \mid X \le F_X^{\leftarrow}(p)). \tag{3.77}$$

These functions can also be expressed in terms of the copula $(U, V) = F_{XY}(F_X(X), F_Y(Y))$ as

$$\lambda_u(p) = \Pr(V > p \mid U > p).$$

and

$$\lambda_l(p) = \Pr(V \le p \mid U \le p).$$

It is clear from the definitions (3.76) and (3.77), that $0 \le \lambda_u(p) \le 1$ and $0 \le \lambda_l(p) \le 1$ and furthermore, if X and Y are independent, $\lambda_u(p) = 1 - p$ and $\lambda_l(p) = p$.

In applications, these quantities may be of interest for a given level p, but for the upper tail, we are interested in the limit as $p \to 1$, and for the lower tail, we are interested in the limit as $p \to 0$. We define the *coefficient of upper tail dependence* as

$$\lambda_u = \lim_{p \to 1^-} \lambda_u(p), \tag{3.78}$$

We define the *coefficient of lower tail dependence* as

$$\lambda_l = \lim_{p \to 0^+} \lambda_l(p). \tag{3.79}$$

We see that if X and Y are independent, then $\lambda_u = \lambda_l = 0$, and if $X = Y$ with probability 1, then $\lambda_u = \lambda_l = 1$.

Heavy-Tailed Distributions and Outliers

A graph of the PDF of a heavy-tailed distribution often appears very much like that of a distribution with normal or light tails. The relatively large probability in the tails, however, means that as we collect samples from a heavy-tailed distribution, occasionally a few observations will be very extreme; that is, they will be outliers, as we discussed and saw in examples in Section 1.3.7 in Chapter 1. We say that a heavy-tailed distribution is an outlier-generating distribution.

We can see an example of this by collecting random samples of size 100 from a Cauchy distribution. To do this, will use the `rcauchy` function in R. Figure 3.4 shows four different random samples. The first two samples each had one or two very large values, but no very small value. The third sample had a very small value, but no particularly large value.

This illustrates the difficulty presented to the data analyst by an outlier-generating distribution. A summarization based on either one of the samples, or even by all four, may not correctly describe the underlying population.

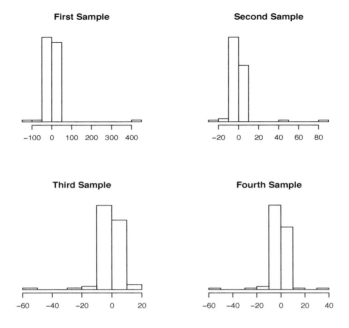

FIGURE 3.4
Random Samples of Size 100 from a Cauchy Distribution

3.1.12   Sequences of Random Variables; Stochastic Processes

Often in financial applications, the variables of interest are in a sequence in which one observation depends on the observations that preceded it. An obvious example is a time series of prices of a stock or of a portfolio.

We often use the notation $\{X_t\}$ to denote the random variables in a stochastic sequence.

One of the first considerations in analyzing a stochastic process is whether or not it is *stationary*, that is, whether there is some constancy in the distributional properties of different subsequences. There are different degrees of stationarity. If each element in the process has the same distribution and each is independent of all others, there is no sequential aspect to be considered, and that of course would be the strongest degree of stationarity. The degree of the stationarity can be made weaker in various ways, either in terms of the sequential associations or the constancy of various aspects of the distribution. Although there are only two types of stationarity that are commonly assumed in models of time series, there are various degrees of the property.

An important characteristic of a stochastic process is how the random variable at point t depends on the values that went before it. Various types of stochastic processes are characterized by the nature of these dependencies.

For example, a *Markov chain* is a stochastic process in which X_t given X_{t-1} is conditionally independent of $X_{t-2}, X_{t-3}, \ldots$.

A *martingale* is a stochastic process in which the conditional expected value of X_t, given $X_{t-1}, X_{t-2}, \ldots$, is X_{t-1}. The assumption of a martingale process underlies much of the pricing theory in finance.

A commonly-used measure of sequential relationships is *autocorrelation*, that is, $\mathrm{Cor}(X_t, X_{t-h})$, which we discussed briefly in Chapter 1 and will return to in Chapter 5.

3.1.13   Diffusion of Stock Prices and Pricing of Options

A stochastic process that is continuous in time is a *diffusion process*. The movement of particles described as *Brownian motion*, for example, is a diffusion process in continuous time in which the differential is taken to have a normal distribution. A stochastic diffusion process can be modeled by a stochastic differential equation or SDE.

The trajectory of the price of a stock or an index can be modeled as a diffusion process, although because prices are not continuous in time, a diffusion process model could only be used as an approximation to the discrete-time process. Letting S represent the price of a stock, and dS represent its infinitesimal change in time dt, we can express the change as a function of S and dt. If we assume some random noise, we can add a term for the noise, which is scaled by the volatility (standard deviation) of the stock. Hence, we have a differential equation of the form

$$dS = \mu S dt + \sigma S dB, \tag{3.80}$$

where μ represents a general drift or rate of growth, σ represents the volatility of the random noise, and dB represents that random noise, which we often assume is Brownian motion.

The intent here is not to derive this equation. Equation (3.80) is shown only to present the general flavor of this approach to modeling stock prices and to suggest its plausibility based on observed behavior of stock prices. We will show, without derivation, two well-known consequences of equation (3.80) that result from a basic theorem in stochastic differential equations known as Ito's lemma, which is a kind of chain rule for stochastic differentiation. In the first equation, we remove S from the right-hand side by dividing, giving

$$d\log(S) = \left(\mu - \frac{\sigma^2}{2}\right) dt + \sigma dB. \tag{3.81}$$

Whereas equation (3.80) represents a proportional Brownian motion with drift, equation (3.81) represents *geometric Brownian motion* with drift.

An important application of this diffusion model of stock prices is to model the price, f, of a derivative asset contingent on S and also depending on time t. In pricing a forward contract or option, we must consider any yields from the asset, such as dividends, and then we must discount its future price back to the present. A common way of discounting prices is to use an assumed risk-free rate of return, say r_F, and dividend yields can be used as an adjustment on the risk-free rate. *In the following, we will assume no dividends or other yields.* Putting these considerations together, another application of Ito's lemma and equation (3.80) yield

$$\frac{\partial f}{\partial t} + r_F S \frac{\partial f}{\partial S} + \frac{1}{2}\sigma^2 S^2 \frac{\partial^2 f}{\partial S^2} = r_F f \qquad (3.82)$$

for the price of the derivative f.

Equation (3.82) is known as the *Black-Scholes-Merton* differential equation. An interesting fact to note is that the mean drift of the prices does not appear in the equation; only the volatility appears. The equation is the basis for the Black-Scholes and related methods of pricing options. The fixed boundary conditions of European options can be used to obtain solutions to the differential equation for either call or put options. That solution is an approximate solution for American options, which because of lack of fixed boundary conditions, do not have a closed form solution. Two immediately obvious drawbacks to models following this approach are the assumptions of constant volatility and of the normal distribution of the steps in Brownian motion dB.

We show the development up to this point only to indicate the general approach. We will not pursue it further in this book.

Approximating the Diffusion Process of Stock Prices

A simple model to approximate the diffusion of a stock price begins with an initial price; adjusts the price for a return in one period, perhaps one day; adjusts that price for the return in the next period; and so on until some future point in time. This is a generalization of a random walk (which we define and discuss on page 490 in Chapter 5). The returns for the periods may follow some kind of smooth distribution, such as the returns of INTC shown in the histogram in Figure 1.20 on page 78.

The probability distribution of the ending prices of these stock price trajectories is the probability distribution of the predicted price at that future point in time. (In Exercise 3.19b, you are asked to simulate such a process. The result is similar to the trajectory of closing prices of INTC shown in the plots of Figure 1.2 on page 9, for example.)

A discrete-time diffusion process with random returns that follow some kind of smooth distribution can be modified to allow for jumps in the price. (See Exercise 3.21b, for example.)

The probability distribution of the ending prices can be used to determine a fair price of an option on the stock. The option value must be discounted to the present, and the nature of the option must be considered. An American option that can be exercised at any time may have a slightly different value from a European option, for example.

Another common way of approximating the diffusion process of a trajectory of stock prices is to form a rooted tree of prices. Various types of trees could be used, but the simplest is a binary tree; that is, two branches from each node. The root node is the beginning price. The next two nodes, at one time period in the future, represent two possible prices, one up, one down. The next four nodes, at two time periods in the future, represent two possible prices, one up, one down, from the two possible prices after the first step. This process continues until a future point in time, perhaps at the expiry of an option on the stock in question.

The general binary tree model has several parameters: the lengths of the time intervals, the probabilities of up movements at each node, and the amounts of a up and down movements.

In an application to option pricing, we make simple choices for the parameters in the binary tree model. First, we choose all time intervals as equal. Then we assume constant probabilities across all nodes. Next, we assume that all price movements up or down and across all nodes are equal in magnitude. This reduces the number of nodes at step k from 2^k to just $k+1$. These two simple choices also make the binary tree a *binomial tree*, as shown in Figure 3.5.

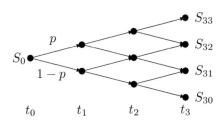

FIGURE 3.5
A Binary Tree Model for Stock Price Trajectories

At time t_3, the price of the stock is S_3, which is one of the four values S_{30}, S_{30}, S_{32}, or S_{33}. The intrinsic value of a call option with strike price K at time t_3 is $(S_3 - K)_+$.

The assumption of equal magnitude price movements up or down is unrealistic. It might be better to assume that all price movements up or down and across all nodes are equal proportionate amounts; that is, at any node with price S_k, the up price at the following node is uS_k, where u is a positive constant greater than 1, and the down price at that following node is S_k/u. Even this is a big assumption. Notice that under this model, the vertical scale in the Figure 3.5 is not constant.

If T is the time of expiry of the option, there is no time value, so the value of a call option is just its intrinsic value, $(S_T - K)_+$, and the value of a put option is just $(K - S_T)_+$. Hence, a probability distribution on the ending prices, that is, S_3 in the case that $T = t_3$, provides a probability distribution for the value of an option at the time of its expiry. This is true for either an American option or a European option.

This model allows us to use the binary tree to determine the expected value of the option at expiry, and together with a risk-neutral assumption, we can determine the fair value of a European option at time t_0 by discounting the expected value back to that time, taking into account any dividends, which would serve to reduce the risk-free discount rate. This model for pricing options is called the Cox-Ross-Rubinstein (CRR) model, after Cox, Ross, and Rubinstein (1979), who first proposed this approach.

To use the model, we need only choose the probability up or down and the proportionate amount of the move. Cox, Ross, and Rubinstein (1979) assumed a proportionate move up of

$$u = e^{\sigma\sqrt{\Delta t}},$$

where σ is the (annualized) volatility of the stock and Δt is the length of the time interval between nodes (measured in years). This is one standard deviation of the distribution of log returns. Assuming no dividends or other returns from the stock, we choose the probability of an up move as

$$p = \frac{e^{r_F \Delta t} - 1/u}{u - 1/u},$$

where r_F is the risk-free rate of return. (If there is a dividend yield of, say, q, the discount term is $e^{(r_F - q)\Delta t}$.) This choice of probability places a restriction on how large the intervals can be. In order for p to be between 0 and 1, we must have

$$\Delta t < \sigma^2/r_F^2.$$

With these choices and a given starting price S_0, we can now construct the tree, or simply compute the price at any node. We can also compute the probability at any node, using the binomial distribution (see page 284).

The price at any node can be discounted k steps back by multiplying the price by $e^{-k r_F \Delta t}$.

In Exercise 3.6, you are asked to use this approach to determine the fair prices of a call option.

3.2 Some Useful Probability Distributions

There are several specific forms of probability distributions that serve well as models of observable events, such as rates of return or prices of stocks. We give names to these useful probability distributions and study their properties. Tables 3.1, 3.2, and 3.3 list some of these standard distributions. In this section, in addition to the standard distributions, we will discuss the problem of choosing a flexible family of distributions or a mixture of distributions that matches a set of observed data.

The R functions for the common distributions are shown in Table 3.4 on page 313. In some cases, there are alternative parameterizations for the distributions. The parameterizations in Tables 3.1 through 3.3 generally match those of the corresponding R functions. There are also R packages for some of the nonstandard distributions. In many cases, a distribution with desirable properties for use as models can be formed from other standard distributions.

Notation

For a few common distributions, we will adopt simplified notation. We have already introduced the notation $U(0, 1)$ for the standard uniform distribution and $U(a, b)$ for a uniform distribution over (a, b). For the standard normal distribution, we introduced the notation $N(0, 1)$, and for the normal distribution with mean μ and variance σ^2, the notation $N(\mu, \sigma^2)$. (Note the second parameter is the *variance*; not the standard deviation.)

To indicate the distribution of a random variable, we will use the notation "$\sim$"; for example,

$$X \sim N(\mu, \sigma^2)$$

means that X is a random variable with the distribution $N(\mu, \sigma^2)$.

3.2.1 Discrete Distributions

Table 3.1 shows some common discrete distributions. All of these distributions are univariate except for the multinomial.

The *binomial distribution* is the distribution for the number of successes on n independent trials, in which each trial has a probability of success of π. A binomial distribution with $n = 1$ is called a *Bernoulli distribution*; that is, a Bernoulli process is a distribution in which the random variable takes a value of 1 with probability π and otherwise it takes a value of 0.

A discrete-time diffusion process for modeling prices, such as referred to on page 280, can be approximated by a sequence of binomial (actually, Bernoulli) events, called a *binomial tree*. A binomial tree is sometimes used in determining fair prices for options.

If the Bernoulli process consists of independent steps with constant probability π, after one step, there is one 1 with probability π; after two steps, there are two 1's with probability π^2, there is one 1 with probability $\pi(1 - \pi)$, and there are zero 1's with probability $(1 - \pi)^2$. This is the same as a binomial distribution with $n = 2$ and π as in the Bernoulli process.

Bernoulli events with varying probabilities can also be chained. In the illustration below, at the first step, the probability of "up" is π_1; at the second step, if the first step was "up", the probability of an "up" is π_{21}; whereas if the first step was "down", the probability of an "up" is π_{22}. The probability of outcome A, B, or C can be expressed in terms of conditional probabilities.

TABLE 3.1

Probability Functions of Common Discrete Distributions

binomial	$\binom{n}{x}\pi^x(1-\pi)^{n-x}, \quad x = 0, 1, \ldots, n$
	$n = 1, 2, \ldots; \quad 0 < \pi < 1$
negative binomial	$\binom{x+n-1}{n-1}\pi^n(1-\pi)^x, \quad x = 0, 1, 2, \ldots$
	$n = 1, 2, \ldots; \quad 0 < \pi < 1$
multinomial	$\dfrac{n!}{\prod x_i!}\displaystyle\prod_{i=1}^{k}\pi_i^{x_i}, \quad x_i = 0, 1, 2, \ldots, n \quad \displaystyle\sum_{i=1}^{k}x_i = n$
	$k = 1, 2, \ldots; \quad n = 1, 2, \ldots; \quad 0 < \pi, \ldots, \pi_k < 1$
Poisson	$\lambda^x e^{-\lambda}/x!, \quad x = 0, 1, 2, \ldots$
	$\lambda > 0$
geometric	$\pi(1-\pi)^y, \quad x = 0, 1, 2, \ldots$
	$(0 < \pi < 1$
hypergeometric	$\dfrac{\binom{M}{x}\binom{N-M}{n-x}}{\binom{N}{n}}, \quad x = \max(0, n - N + M), \ldots, \min(n, M)$
	$N = 2, 3, \ldots; \quad M = 1, \ldots, N; \quad n = 1, \ldots, N$

For this reason (that is, the conditioning), this model is sometimes called a "Bayesian network". The probability of event A, for example, is the probability of an "up" at step 2, given an "up" at step 1, or $\pi_{21}\pi_1$. The probability of event B is the sum of two conditional probabilities, $(1 - \pi_{21})\pi_1 + \pi_{22}(1 - \pi_1)$.

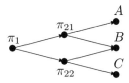

If the probabilities are all equal, say π, and the events are independent, then the distribution of the final states is binomial. With

$$A \Leftrightarrow X = 2 \quad B \Leftrightarrow X = 1 \quad C \Leftrightarrow X = 0,$$

in the diagram above, the PDF of X is that of the binomial $(2, \pi)$,

$$\binom{n}{x} \pi^x (1 - \pi)^{n-x}, \quad x = 0, 1, 2.$$

The *Poisson family* of distributions serves well as a model for the frequency distributions of many things, such as the number of fatal accidents in a given place in a given time period.

The Poisson distribution is sometimes also a good model for occurrences of extraordinary events such as sudden unexplained selloffs in the stock market. (There will never be an "unexplained" event in the market, because there will always be "explanations".) The effect on the price of a stock or an index is referred to as a *jump*. A discrete-time diffusion process for modeling prices, such as referred to on page 280, may include some continuous distribution of "ordinary" returns plus a Poisson distribution for extraordinary returns. The extraordinary returns are sometimes referred to as Poisson jumps.

The mean of the Poisson distribution is λ, and the variance is also λ. (We showed that the mean is λ on page 250, and you are asked to derive the variance in Exercise 3.3a.)

The *discrete uniform distribution* is the distribution of a random variable that takes on any value in a set of k different values with equal probability. The CDF of a discrete uniform distribution is the same as an ECDF of a sample with distinct elements. The discrete uniform distribution forms the theoretical basis of the bootstrap, which we will discuss in Section 4.4.5.

The discrete uniform is a fairly simple distribution, and there are no R functions for the probability function (it is the constant $1/k$) or for the CDF or the quantile function.

3.2.2   Continuous Distributions

Table 3.2 shows some common univariate continuous distributions and Table 3.3 shows two multivariate continuous distributions.

Four additional distributions not shown in Table 3.2 are the chi-squared, t, F, and the stable distributions. The chi-squared, t, and F are derived from normal distributions. Their PDFs are rather complicated, and so we will just define them in terms of the normal distribution. These three families of distributions are widely used in tests of statistical hypotheses, and we will mention them briefly in Section 4.4.2 and again in various applications. There are R functions for each, which are listed in Table 3.4.

The stable distributions are heavy-tailed distributions. The t distributions, although identified primarily with tests of statistical hypotheses, are also heavy-tailed, and so are useful in financial modeling applications. I will discuss these distributions briefly beginning on page 296.

TABLE 3.2
Probability Density Functions of Common Univariate Continuous Distributions

normal; $N(\mu, \sigma^2)$	$\dfrac{1}{\sqrt{2\pi\sigma^2}}e^{-(x-\mu)^2/2\sigma^2}$
(Gaussian)	$-\infty < \mu < \infty;\ \sigma^2 > 0$
lognormal	$\dfrac{1}{\sqrt{2\pi}\sigma}x^{-1}e^{-(\log(x)-\mu)^2/2\sigma^2}\quad$ for $x > 0$
	$-\infty < \mu < \infty;\ \sigma > 0$
beta	$\dfrac{\Gamma(\alpha+\beta)}{\Gamma(\alpha)\Gamma(\beta)}x^{\alpha-1}(1-x)^{\beta-1}\quad$ for $0 < x < 1$
	$\alpha, \beta > 0$
Cauchy	$\dfrac{1}{\pi\beta\left(1+\left(\frac{x-\gamma}{\beta}\right)^2\right)}$
(Lorentz)	$-\infty < \gamma < \infty;\ \beta > 0$
gamma	$\dfrac{1}{\Gamma(\alpha)\beta^\alpha}x^{\alpha-1}e^{-x/\beta}\quad$ for $x > 0$
	$\alpha, \beta > 0$
exponential	$\lambda e^{-\lambda x}\quad$ for $x > 0$
	$\lambda > 0$
double exponential	$\frac{1}{2\beta}e^{-|x-\mu|/\beta}$
(Laplace)	$-\infty < \mu < \infty;\ \beta > 0$
logistic	$\dfrac{e^{-(x-\mu)/\beta}}{\beta(1+e^{-(x-\mu)/\beta})^2}$
	$-\infty < \mu < \infty;\ \beta > 0$
Pareto	$\alpha\gamma^\alpha x^{-\alpha-1}\quad$ for $x > \gamma$
	$\alpha, \gamma > 0$
Weibull	$\dfrac{\alpha}{\beta}x^{\alpha-1}e^{-x^\alpha/\beta}\quad$ for $x > 0$
	$\alpha, \beta > 0$

Uniform Distribution

One of the simplest continuous random variables is one with a *uniform distribution* over the interval $[a, b]$. We denote this distribution as $U(a, b)$. The most common uniform distribution and the "standard one", is $U(0, 1)$. It plays a major role in simulation of random data (see Section 3.3).

TABLE 3.3
Probability Density Functions of Multivariate Continuous Distributions

multivariate normal	$\dfrac{1}{(2\pi)^{d/2}|\Sigma|^{1/2}} e^{-(x-\mu)^{\mathrm{T}}\Sigma^{-1}(x-\mu)/2}$
$N_d(\mu, \Sigma)$	$-\infty < \mu < \infty$; Σ positive definite, $d \times d$
Dirichlet	$\dfrac{\Gamma(\sum_{i=1}^{d+1}\alpha_i)}{\prod_{i=1}^{d+1}\Gamma(\alpha_i)} \prod_{i=1}^{d} x_i^{\alpha_i-1} \left(1 - \sum_{i=1}^{d} x_i\right)^{\alpha_{d+1}-1}$ for $0 < x_i < 1$
	$\alpha > 0$

The CDF of the uniform distribution is

$$F(x) = \begin{cases} 0 & x < a \\ (x-a)/(b-a) & a \le x < b \\ 1 & b \le x \end{cases}$$

The mean of the $U(a, b)$ distribution is $(a + b)/2$ and the variance is $(b^2 - 2ab + a^2)/12$.

Although the "d", "p", and "q" functions in R for the uniform distribution are rather straightforward and rarely used, the "r" function, runif, is used quite often.

The Normal or Gaussian Family of Distributions

The most commonly used probability distribution is the *normal* or *Gaussian* distribution. (There is no difference in "normal" and "Gaussian" in this case.) I will denote the normal distribution with mean μ and variance σ^2 as $N(\mu, \sigma^2)$.

One of the important properties of the family of normal distributions is that all members of the family are stochastically the same as any other member by a linear transformation. If $X \sim N(\mu, \sigma^2)$, and a and b are any numbers such that $b \neq 0$, and

$$Y = a + bX,$$

then $Y \sim N(a + b\mu, b^2\sigma^2)$.

The skewness of any normal distribution is 0 and the kurtosis is 3.

One of the reasons the normal distribution is so important is the central limit theorem (see Section 3.1.10).

The graph of the PDF is the familiar "bell-shaped curve". The graph of the PDF of a normal distribution with mean equal to the sample mean and variance equal to the sample variance is shown as the reference distribution in Figures 1.24 and 1.25 in Chapter 1. The quantiles of a normal distribution are used as references in the q-q plots of Figures 1.27 and 1.28.

We often denote the PDF of the standard normal distribution as $\phi(x)$ and the CDF as $\Phi(x)$.

One thing to note about the R functions is that they specify the normal distribution in terms of the standard deviation, instead of the variance. Hence, for example to evaluate the PDF of the $N(0, 9)$ distribution at the point x, we would use `dnorm(x, 0, 3)`.

The Lognormal Family of Distributions

The distribution of a positive random variable whose logarithm has a normal distribution, is called the *lognormal distribution*; that is, if (and only if) X has a lognormal distribution, then $Y = \log(X)$ has a normal distribution. The lognormal arises in financial applications whenever the normal distribution is used to model log returns. If the log returns have a normal distribution, then the prices have a lognormal distribution. As we have emphasized, the distribution of the log returns generally has a heavier tail than a normal distribution; nevertheless, the normal is often used as a quick approximation.

The two normal parameters fully characterize the lognormal distribution, and so those are the parameters usually specified in order to identify a particular member of the lognormal family of distributions.

If Y has a normal distribution with parameters μ and σ^2, then $X = e^Y$ has a lognormal distribution with parameters μ and σ. (Obviously we could just as well choose σ^2 as the parameter.) The transformation is one-to-one and differentiable, so it is a trivial matter, using the methods described on page 254, to work out the PDF of the lognormal, which is shown in Table 3.2. It is clear from the PDF or from the definition itself that the median is e^{μ} (because the median of the normals is μ), and that the distribution is highly skewed.

The mean of the lognormal can be evaluated as $E(e^Y)$, where $Y \sim N(\mu, \sigma^2)$. Taking this expectation, we get the mean of the lognormal(μ, σ) to be $\exp(\mu + \sigma^2/2)$. We get the variance by first evaluating $E(X^2) = E(e^{2Y})$, and then subtracting the square of the mean. The variance of the lognormal(μ, σ) is $\exp(2\mu + \sigma^2)(\exp(\sigma^2) - 1)$. To perform the necessary computations, complete the square in the exponent in the normal PDF so as to get an integral that is known to be 1. For example, for the mean, completing the square, we have

$$- (y - \mu)^2/2\sigma^2 + y = -(y - (\mu + \sigma^2))^2/2\sigma^2 + \mu + \sigma^2/2, \qquad (3.83)$$

so then the integral in the expectation evaluates to $\exp(\mu + \sigma^2/2)$. These simple computations yield a general expression for the raw moments of a lognormal random variable: $E(X^k) = \exp(k\mu + k^2\sigma^2/2)$. The method of rearranging terms so as to get an integral or sum that we know to be 1 is the same method we used on page 250 to work out the mean of a Poisson random variable.

The Chi-Squared, t, and F Distributions

We first give the three simple relationships to normal random variables that define these distributions.

If $X_1, \ldots, X_n$ are independent and each distributed as $N(0, \sigma^2)$, then

$$V = \frac{X_1^2 + \cdots + X_n^2}{\sigma^2} \tag{3.84}$$

has the *chi-squared distribution with n degrees of freedom* (df). The degrees of freedom is often denoted by ν. The degrees of freedom appear as a parameter in the PDF and CDF of the chi-squared distribution and there it does not need to be an integer. The support of the chi-squared distribution is the nonnegative real numbers, and the distribution is positively skewed.

The important result of these facts (along with a couple more facts we will not get into here) is that if S^2 is the sample variance (equation (1.49), page 80) from sample of size n from an underlying $N(\mu, \sigma^2)$ distribution, then

$$(n-1)S^2/\sigma^2$$

has a chi-squared distribution with $n - 1$ degrees of freedom.

If $X \sim N(0,1)$ and $V \sim$ chi-squared(ν) and X and V are independent, then

$$T = \frac{X}{\sqrt{V/\nu}} \tag{3.85}$$

has the *t distribution with ν degrees of freedom* (df). This is also called "Student's t distribution". (The name "Student" comes from a pseudonym used by the author of the first scientific paper that described the distribution; we will just refer to it as the t distribution.)

The t distribution is symmetric about 0. (We will later consider generalizations of the t distribution to allow for different means, scales, and skews.)

The t distribution has one parameter, the degrees of freedom, ν. The name of the parameter, "degrees of freedom", is suggested intuitively and heuristically by the transformations forming the t variables, but we will not motivate the choice of the name here. We will often call the parameter "df". The chi-squared and F distributions, also derived from the normal distribution, have similar parameters, which are also called df.

In statistical analysis of data that has a normal distribution with unknown mean and variance, we often estimate the variance using a statistic that has a scaled chi-squared distribution. The formation of a t variable by dividing a normally distributed statistic by the square root of a chi-squared distributed statistic and its degrees of freedom is called "studentization".

The important result of these facts (along with a couple more facts we will not get into here) is that if $\overline{X}$ is the sample mean and S^2 is the sample variance from sample of size n from an underlying $N(\mu, \sigma^2)$ distribution, then

$$\frac{\overline{X} - \mu}{S/\sqrt{n}}$$

has a t distribution with $n - 1$ degrees of freedom. (Notice that this does not involve σ^2; the numerator is divided by σ to result in a variance of 1, and the denominator is also divided by σ to form a chi-squared variable.)

A very important property of the normal distribution, which we will not prove here, is that for a sample $X_1, \ldots, X_n$ from a $N(\mu, \sigma^2)$ distribution, the sample mean and the sample variance are independent.

A t random variable formed by the ratio of a standard normal random variable and the square root of a chi-squared random variable divided by its degrees of freedom is a *central* t distribution. If the normal random variable has a nonzero mean, the variable formed by the ratio has a *noncentral* t distribution. The mean of the normal random variable is the *noncentrality* parameter of the noncentral t distribution. The noncentral t distribution is often used in statistical applications to assess the power of a statistical hypothesis test.

If $V_1 \sim$ chi-squared(ν_1) and $V_2 \sim$ chi-squared(ν_2) and V_1 and V_2 are independent, then

$$F = \frac{V_1/\nu_1}{V_2/\nu_2} \tag{3.86}$$

has the *F distribution with ν_1 and ν_2 degrees of freedom* (df). Thus, the F distribution has two parameters, the two degrees of freedom, ν_1 and ν_2. The support of the F distribution is the nonnegative real numbers, and the distribution is positively skewed.

There are many simple relationships among these distributions; for example, from the definition of the chi-squared distribution, we can see that if $V_1 \sim$ chi-squared(ν_1) and $V_2 \sim$ chi-squared(ν_2) and V_1 and V_2 are independent, then $V_1 + V_2$ has the chi-squared distribution with $\nu_1 + \nu_2$ df.

We also see that if T has the t distribution with ν df, then T^2 has the F distribution with 1 and ν df.

These distributions based on an underlying normal distribution with a mean of 0 are "central" distributions. There are noncentral versions that are formed in a similar manner, but rather beginning with $N(\mu, 1)$, with $\mu \neq 0$.

If $X \sim N(\mu, 1)$ and $V \sim$ chi-squared(ν), that is, V has a central chi-squared distribution, and X and V are independent, then

$$\frac{X}{\sqrt{V/\nu}} \tag{3.87}$$

has the *noncentral t distribution with ν degrees of freedom and noncentrality parameter μ*. The noncentral t distribution is skewed to the right if $\mu < 0$, and to the left if $\mu > 0$. Although we sometimes need a skewed version of a t distribution for modeling purposes, the noncentral t does not work well for these purposes; rather, the skewed t distributions formed by the methods discussed on page 302 are more useful.

Beta Distribution

The beta distribution is a very flexible distribution over a finite range, usually $[0, 1]$, but it can be generalized to any finite range. The beta has two parameters that allow it to take on a variety of shapes. The density can be symmetric or skewed in either direction, and it can be concave or convex. If both parameters are 0, the beta distribution is the uniform distribution.

The PDF of the beta distribution is

$$f(x) = \frac{\Gamma(\alpha + \beta)}{\Gamma(\alpha)\Gamma(\beta)} x^{\alpha-1}(1 - x)^{\beta-1}, \quad \text{for } 0 \le x \le 1, \tag{3.88}$$

The parameters of the beta are shape parameters, as seen in Figure 3.6 below. The mean of the beta is $\alpha/(\alpha + \beta)$ and the variance is $\alpha\beta/((\alpha + \beta)^2(\alpha + \beta + 1))$.

Figure 3.6 shows the versatility of the beta distribution with different parameters to provide a variety of shapes. All distributions shown in the graphs have range $(0, 1)$, but they could be scaled to have any finite range.

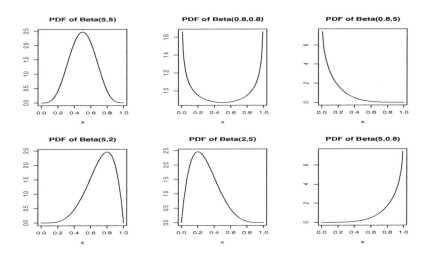

FIGURE 3.6
Beta Probability Distributions

Gamma, Exponential, Weibull, and Double Exponential Distributions

The gamma family of distributions is a very flexible skewed distribution over the positive numbers. There are important special cases and also useful generalizations.

The PDF of the basic gamma distribution is

$$f(x) = \frac{1}{\Gamma(\alpha)\beta^{\alpha}} x^{\alpha-1} e^{-x/\beta}, \quad \text{for } 0 \leq x, \tag{3.89}$$

where Γ is the complete gamma function,

$$\Gamma(\alpha) = \int_{0}^{\infty} t^{\alpha-1} e^{-t} \, dt.$$

The mean of the gamma is $\alpha\beta$ and the variance is $\alpha\beta^{2}$.

There are other common parameterizations of the gamma. The version given in equation (3.89) corresponds to that used in R.

Special cases of the gamma distribution are the exponential distribution, for the case when $\alpha = 1$, and the chi-squared distribution, when $\beta = 2$. In the chi-squared distribution, $\nu = 2\alpha$ is called the "degrees of freedom". As mentioned above, the chi-squared distribution is the distribution of the sum of ν squared independent $N(0,1)$ random variables.

A related distribution is the Weibull distribution, with PDF

$$f(x) = \frac{\alpha}{\beta^{\alpha}} x^{\alpha-1} e^{-(x/\beta)^{\alpha}}, \quad \text{for } 0 \leq x. \tag{3.90}$$

Another related distribution is the double exponential distribution, also called the Laplace distribution. This is a mirrored exponential distribution. Its PDF is

$$f(x) = \frac{1}{2\beta} e^{-|x-\mu|/\beta}. \tag{3.91}$$

It is a symmetric distribution about μ (which in the standard case is 0, the lower bound of the exponential distribution).

The tails of the double exponential distribution are exponential, so all moments are finite. The double exponential distribution, however, is a heavy-tailed distribution relative to the normal distribution.

Pareto Distribution

The Pareto distribution is widely used in economics to model such things as the distribution of incomes. With proper choice of the parameters, it models the "80/20 rule". In a given country, for example, 80% of the wealth may be owned by 20% of the population.

One formulation of the PDF of the Pareto distribution is

$$f(x) = \alpha\gamma^{\alpha} x^{-\alpha-1} \quad \text{for } x \geq \gamma, \tag{3.92}$$

where the location parameter $\alpha > 0$ and likewise the shape $\gamma > 0$. The mean of the Pareto is $\alpha\gamma/(\alpha-1)$ for $\alpha > 1$, and the variance is $\alpha\gamma^{2}/((\alpha-1)^{2}(\alpha-2))$ for $\alpha > 2$.

The Pareto distribution is often used as a model for the distribution of household wealth or income because it is highly skewed to the right. In these applications it is often associated with a Lorenz curve and the Gini index. The more convex Lorenz curve in Figure 2.3 (page 215) is for a sample of size 100 from a Pareto distribution with $\gamma = 1$ and $\alpha = \log_4(5) \approx 1.16$. This value of α corresponds to the 80/20 rule, and the point at which 80% of the total corresponds to 20% of the population is shown on the graph in Figure 2.3.

A simple instance of the Pareto distribution, called the "unit Pareto", has $\alpha = 2$ and $\gamma = 1$. This is the distribution of $X = 1/U$, where $U \sim U(0, 1)$.

The EnvStats R package contains functions for the Pareto distribution that correspond to the parametrization in equation (3.92). There are other formulations of the Pareto PDF, however. For example, the Pareto functions in the rmutil package use a completely different formulation, with a "dispersion" parameter.

The Pareto distribution is a *power law* distribution, which under various parametrizations, is characterized by a kernel of the form

$$x^\rho$$

The tails of these distributions are polynomial or Pareto. (Obviously, for distributions with infinite support, ρ must be negative.)

There are also other variations of the Pareto distribution that do not relate to the formulation above (see page 301). The "generalized Pareto" (see page 301) is not a generalization of this formulation.

Cauchy Probability Distributions

The PDF of the standard Cauchy distribution is

$$f(x) = \frac{1}{\pi(1 + x^2)}. \tag{3.93}$$

A t distribution with 1 degree of freedom is a Cauchy distribution; that is, a Cauchy random variable is stochastically equivalent to the ratio of two independent $N(0, 1)$ random variables.

The mean and variance of the Cauchy distribution are infinite. (Alternatively, we may say that they do not exist.)

Heavy-Tailed Distributions

In Section 3.1.11 we discussed heavy-tailed distributions and various ways of assessing the behavior of random variables in the tail of their distributions. Since many financial variables seem to have heavy-tailed distributions, it is of interest to identify some specific heavy-tailed distributions and study their properties.

Some of the distributions discussed above have heavy tails. A family of distributions whose tail weight varies in a systematic way is very useful in modeling empirical financial data. There are two such families that we will discuss, the t family and the stable family. The tail weight of the members of the t family increases as the degrees of freedom decreases, until for one degree of freedom, the t distribution is the same as the Cauchy distribution. In the limit as the degrees of freedom increases without bound, the t distribution is the same as the standard normal distribution. The location-scale t distributions, as discussed on page 297, provide more flexibility to allow for differing first and second moments.

The members of the stable family also range systematically from a normal distribution to a Cauchy distribution.

In Section 3.2.4 we discuss these and other flexible families of distributions that can be used to model observations with various shapes of frequency distributions.

The standard forms of many families of distributions are symmetric, but they can be modified to be skewed. We discuss two methods for forming skewed distributions beginning on page 302.

3.2.3 Multivariate Distributions

Any univariate distribution can be extended to a multivariate distribution of independent variates just by forming a product of the probability functions or probability density functions. Our interest in multivariate distributions, however, generally focuses on the associations among the individual random variables. There are several ways that the standard distributions can be extended to multivariate distributions, but most of these serve as useful models only in limited settings. There are four standard multivariate distributions, however, that are useful in a range of applications.

Multinomial Probability Distribution

The multinomial distribution is a generalization of the binomial distribution. Whereas the binomial models how n events are parceled into two groups ("successes" and "failures") with probabilities π and $1 - \pi$, the multinomial models the distribution of n events into k groups, with probabilities $\pi_1, \ldots, \pi_k$. The groups are mutually exclusive, and so $\sum \pi_i = 1$. The numbers going into the groups, x_i, sums to n.

The ith and j^{th} (with $i \neq j$) elements of a k-variate multinomial random variable have a negative covariance, $-n\pi_i\pi_j$.

The multinomial can easily be formed recursively from $k - 1$ independent and identical binomial distributions.

Dirichlet Probability Distribution

The Dirichlet is a multivariate extension of the beta distribution. We have seen how the parameters of the beta allow flexibility in the shape of the distribution (Figure 3.6, on page 291). The Dirichlet likewise has various shapes, as determined by the vector parameter α. The d-variate Dirichlet has $d+1$ parameters, just as the univariate beta has 2 parameters.

Given a $(d+1)$-vector of positive numbers α, the PDF of the d-variate Dirichlet is

$$f(x;\alpha) = \frac{\Gamma(\sum_{i=1}^{d+1}\alpha_i)}{\prod_{i=1}^{d+1}\Gamma(\alpha_i)} \prod_{i=1}^{d} x_i^{\alpha_i-1} \left(1 - \sum_{i=1}^{d} x_i\right)^{\alpha_{d+1}-1} \quad \text{for } 0 < x_i < 1. \quad (3.94)$$

In this parametrization, the α_{d+1} parameter is special; it is similar to the β parameter in the beta distribution with PDF given in equation (3.88).

The ith and j^{th} ($i \neq j$) elements of a k-variate Dirichlet random variable have a negative covariance, $-\alpha_i\alpha_j/(a^2(a+1))$, where $a = \sum_{k=1}^{d}\alpha_k$.

Multivariate Normal Probability Distribution

The family of multivariate normal distributions, which we denote by $N_d(\mu, \Sigma)$, are determined by a d-vector μ, which is the mean, and a positive definite $d \times d$ matrix Σ, which is the variance-covariance.

The PDF is

$$f(x;\mu,\Sigma) = \frac{1}{(2\pi)^{d/2}|\Sigma|^{1/2}} e^{-(x-\mu)^{\mathrm{T}}\Sigma^{-1}(x-\mu)/2}. \quad (3.95)$$

As mentioned previously, in the case of the normal distribution, if two variables X_1 and X_2 have 0 covariance, they are independent.

Multivariate t Probability Distribution

A multivariate t random variable can be formed from a multivariate normal random and an independent univariate chi-squared random variable in a similar way as a univariate t is formed. Let $X \sim N_d(0, I)$, let $V \sim$ chi-squared(ν), and let X and V be independent. Then the distribution of

$$Y = X/\sqrt{V/\nu}$$

is *multivariate t distribution* with ν degrees of freedom. If $\nu > 1$, the mean of Y is the 0 vector. If $\nu > 2$, the variance-covariance matrix of Y is $\nu I/(\nu - 2)$.

The multivariate t can also be generalized to a location-scale multivariate t in the way described for a univariate t distribution on page 297.

This is the central multivariate t distribution. A noncentral multivariate t distribution is defined as above, except $X \sim N_d(\delta, \Sigma)$, where $\delta \neq 0$.

Both the multivariate normal and the central multivariate t are elliptical distributions.

3.2.4 General Families of Distributions Useful in Modeling

A major problem in statistical data analysis is selecting a probability model to use in the analysis. We want a distribution that is simple and easy to work with, yet one that corresponds well with frequency distributions in observed samples. In the univariate case, using histograms and other graphical displays together with simple summary statistics, we can usually select a suitable distribution among those listed in Tables 3.1 through 3.3.

An easy way of specifying a probability distribution is in terms of observed quantiles. Knowing just a few quantiles, we can seek a distribution that matches them. This is similar to seeking a reference distribution in a q-q plot. For example, we have seen that a t distribution with 4 degrees of freedom fits the distribution of some returns better than a normal distribution does. (This was the point of Figure 2.10.) There may not be a "standard" distribution that fits very well, however.

t Probability Distributions

As we mentioned above, a t random variable is formed as the ratio of a standard normal random variable and the square root of a chi-squared random variable divided by its degrees of freedom.

The t family of distributions is a symmetric family with a mean of 0, and a kurtosis that varies with a single parameter called the degrees of freedom. It is one of the two families of distributions we have mentioned that range from a standard normal distribution to a Cauchy distribution, depending on its parameter.

The mathematical properties of the t family of distributions are rather complicated, and we will only summarize some important properties. The PDF of a t distribution has a polynomial or Pareto tail, with tail index equal to the degrees of freedom.

A t random variable with ν degrees of freedom is stochastically equivalent to a random variable formed as the ratio of a standard normal random variable and the square root of an independently distributed chi-squared random variable with ν degrees of freedom divided by ν. If $\nu \leq 2$, the variance of the t random variable is infinite (or undefined); otherwise, the variance is $\nu/(\nu-2)$. Notice that as ν increases without bound, the variance approaches 1, which is the variance of the standard normal distribution.

If $\nu \leq 4$, its kurtosis is infinite (or undefined); otherwise, the kurtosis is $3 + 6/(\nu - 4)$ and the excess kurtosis is $6/(\nu - 4)$. Notice that as ν increases without bound, the kurtosis approaches 3, which is the kurtosis of the normal distribution.

In Figure 3.7 below, we show q-q plots of samples from the t distribution with varying degrees of freedom with respect to a normal reference distribution. We see that the tails become progressively heavier as the number of degrees of freedom decreases.

Figure 2.10 on page 232 shows similar properties, but from a different perspective.

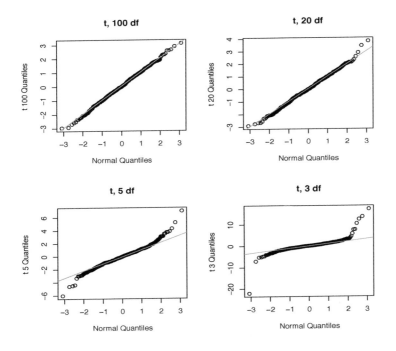

FIGURE 3.7
Q-Q Plots of Samples from t Distributions with a Normal Reference Distribution

The random samples in Figure 3.7 were generated in R. If different samples were generated using the same program, the q-q plots would have the same general shapes, but they may look very different from those in Figure 3.7. This is because the variance of extreme order statistics in samples from heavy-tailed distributions is large, so there is a lot of variation from one sample to another, as we observed in Figure 3.4, for example.

The Location-Scale t Distributions

In financial applications, the t distribution is often used as a model of the distribution of returns because it has a heavy tail.

In modeling applications generally, it is useful to begin with a distribution that has a standard deviation of 1. The standard deviation of a t random variable is $\sqrt{\nu/(\nu - 2)}$ if $\nu > 2$. (If $\nu \le 2$, the standard deviation is not finite.) Dividing a t random variable by $\sqrt{\nu/(\nu - 2)}$ yields a random variable that has a standard deviation of 1. This standardizes the random variable, in the same sense as standardizing a set of observations, as discussed on page 208.

If V has a t distribution with ν df and $\nu > 2$, then

$$W = \sqrt{(\nu - 2)/\nu}V. \tag{3.96}$$

is a standardized random variable whose distribution has the same shape as the t distribution. The random variable W is easy to transform to another random variable with any specified mean and variance.

The location-scale family of random variables,

$$Y = \mu + \lambda W, \tag{3.97}$$

is useful in financial applications, because it has the heavy tails of the t distribution, yet has a flexible mean and variance. The distribution of Y is the *location-scale t distribution*.

The location-scale t distribution has three parameters, μ the mean, λ the scale (standard deviation), and ν the degrees of freedom which determines the shape, that is, the kurtosis.

The location-scale t distribution, while also possibly not centered at 0, is not the same as a noncentral t. The location-scale t distribution is symmetric. A skewed version of the location-scale t distribution can be formed using either of the methods discussed later, shown in equations (3.103) and (3.104). (In financial applications, the location-scale t distribution or one similar to it, depending on the author, is called a *"standardized" t distribution*. I use quotation marks in referring to this distribution because it is not standardized in the ordinary terminology of probabilists or statisticians.)

Stable Probability Distributions

The stable family of probability distributions is a versatile family of heavy-tailed distributions. Unfortunately, the PDFs of most members of this family of distributions cannot be written in closed form.

The stable family is indexed by a parameter usually called α (and that is the name in the R function mentioned below) that ranges from 0 (exclusive) to 2 (inclusive). A value of 1 is a Cauchy distribution, and a value of 2 is a normal distribution. These are the only two members of the family of distributions whose PDFs can be written in closed form.

Although as α ranges from 1 to 2, the distributions have progressively lighter tails, the only stable distribution with finite mean and variance is the normal, with $\alpha = 2$.

Despite the fact that the mean and variance are infinite, the stable family is often used in financial modeling applications.

Members of the family may also be skewed, positively or negatively. This is determined by a parameter in the defining equations, usually called β (and that is the name in the R function mentioned below). In financial applications, a symmetric version is used most commonly, and it corresponds to $\beta = 0$.

Generalized Lambda Distribution, GLD

A general approach to modeling observational data is to define a distribution whose quantiles can be made to fit the observation by proper choice of the parameters of that distribution. The quantiles are determined by the quantile function, that is, the inverse CDF, and so an effective way is to define the inverse CDF of the distribution.

John Tukey suggested this approach with what came to be called Tukey's lambda distribution. A flexible extension is the *generalized lambda distribution* (GLD). The inverse CDF of the GLD is

$$F^{-1}(p) = \lambda_1 + \frac{p^{\lambda_3} - (1-p)^{\lambda_4}}{\lambda_2}. \tag{3.98}$$

The four parameters, λ_1, λ_2, λ_3, and λ_4 allow very flexible shapes of the distribution. There are other parameterizations of the PDF.

The order statistics in a dataset provide quantiles for that set of data. The specification of the inverse of the CDF in equation (3.98) makes the GLD particularly simple to use in simulation studies using the inverse CDF method described on page 317.

The gld R package contains functions for the GLD that also allow an extended form with a fifth parameter. There are also functions in the gld package for fitting the GLD using observed data.

Generalized Error Distribution, GED

Another distribution that allows a range of shapes is the generalized error distribution (GED). The kernel of the GED PDF is

$$e^{-c|x|^\nu}.$$

From this we see that all members of the GED family have exponential tails, and the exponent ν determines the tail weight. The smaller ν is, the heavier the tail is.

A specific formulation of the PDF of the GED family is

$$f(x) = \frac{\nu}{\lambda_\nu 2^{1+1/\nu}\Gamma(1/\nu)}e^{-\frac{1}{2}\left|\frac{x}{\lambda_\nu}\right|^\nu} \tag{3.99}$$

where

$$\lambda_\nu = \left(\frac{2^{-2/\nu}\Gamma(1/\nu)}{\Gamma(3/\nu)}\right)^{1/2}.$$

There are other equivalent formulations. In the form of equation (3.99), the distribution is standardized with a mean of 0 and a variance of 1.

Unless $\nu = 1$, the CDF of the GEDs cannot be written in closed form; it must be evaluated numerically.

The normal distributions are GEDs with $\nu = 2$, and the double exponential (Laplace) distributions are GEDs with $\nu = 1$.

Figure 3.8 shows four different GEDs, including the normal ($\nu = 2$) and the double exponential ($\nu = 1$).

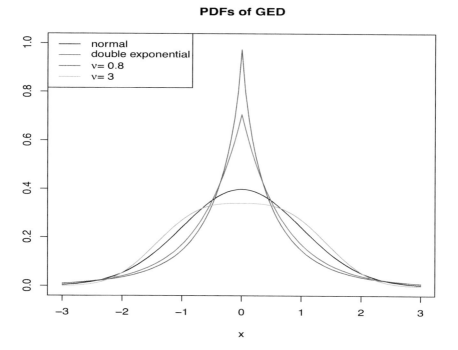

PDFs of GED

FIGURE 3.8
Standard Generalized Error Probability Distributions

There are R functions in the `fGarch` package for the GED.

A skewed version of the GED can easily be formed using either of the methods shown in equation (3.103) or (3.104) by replacing the normal PDF and CDF by those of the GED.

For $\nu = 1$, that is, for the double exponential or Laplace distributions, another type of generalization is useful. There are two ways the distribution is extended. Firstly, the PDF of the GED is multiplied by a shifted gamma PDF. This is effectively a mixing operation. Secondly, the kernel, which for $\nu = 1$ is a special case of a modified Bessel function of the second kind, is generalized to other Bessels of the second kind. The resulting distribution is called a *variance gamma distribution*. (That name comes from viewing the distribution as a continuous mixture of the variance of a GED. Another name for this distribution is *generalized Laplace distribution*.) The R package `VarianceGamma` contains several functions for working with this distribution.

Variations and Generalizations of the Pareto Distributions

The Pareto distribution described on page 292 is highly skewed and hence is often used as a model in social economics because the frequency distributions of some economic variables such as income and wealth within a given population are characterized by a strong skewness. Social economists often refer to the governing rule of the characteristic as a "power law", which states that the frequency of a given value is proportional to a power of the value.

There are various other power law distributions, and for the Pareto itself there are four common forms extending the PDF given in equation (3.92), see Arnold (2008).

Some of these distributions have been developed as better models of income or wealth distributions. A family of such distributions are the Dagum distributions (see Kleiber, 2008).

Another widely-used family of distributions are the so-called generalized Pareto distributions.

Generalized Pareto Distribution, GPD

The single parameter γ in the Pareto PDF (3.92) determines both the scale and the shape. A variation on the specification of the distribution is to begin with a location-scale variable, $x = (y - \mu)/\sigma$ and to form a power-law distribution based on that variable. This is the way a family called the *generalized Pareto distributions* (GPD) is formed.

The PDF of the standard location-scale GPD is

$$f(x) = \begin{cases} (\xi x + 1)^{-(\xi+1)/\xi} & \text{for } \xi \neq 0 \text{ and } x \geq 0 \\ e^{-z} & \text{for } \xi = 0 \text{ and } x \geq 0 \end{cases} . \qquad (3.100)$$

The PDF is 0 for $x < 0$, and there is a further restriction if $\xi \neq 0$; in that case, the PDF is also 0 if $x > -1/\xi$. For $\xi = 0$, we note that the standard location-scale GPD is the standard exponential distribution.

The standard location-scale GPD density function has a negative exponential shape over the positive quadrant. As positive values of ξ increase, the tail becomes heavier, and the shape of the density approaches a right angle.

The location-scale GPD is often used to model just the tail of a frequency distribution.

The R package `evir` contains *pregpd* functions for the probability density function, the CDF, the quantiles, and to generate random values.

The R package `evir` also contains a function, gpd, to determine parameters of the GPD that fits a given set of data of threshold exceedances; that is, that fits the upper tail of a given set of data. The user specifies in gpd the threshold in the data that defines the upper tail. Methods for fitting the GPD to the upper tail are described by McNeil, Frey, and Embrechts (2015).

Skewed Distributions

Financial data such as returns, although more-or-less symmetric about 0, are often slightly skewed. (See, for example, Figure 1.20.)

Most of the common skewed distributions, such as the gamma, the log normal, and the Weibull, have semi-infinite range. The common distributions that have range $(-\infty, \infty)$, such as the normal and the t, are symmetric.

There are several ways to form a skewed distribution from a symmetric one. Two simple ways are

- *CDF-skewing:* take a random variable as the maximum (or minimum) of two independent and identically symmetrically distributed random variables, or

- *differential scaling:* for some constant $\xi \neq 0$, scale a symmetric random variable by ξ if it is less than its mean and by $1/\xi$ if it is greater than its mean.

In each case, it may be desirable to shift and scale the skewed random variable so that it has a mean of 0 and a variance of 1. (We can then easily shift and scale the random variable so as to have any desired mean and variance.)

CDF-Skewing

If a random variable has PDF $f(x)$ and CDF $F(x)$, from equation (3.66), the PDF of the maximum of two independent random variables with that distribution has the PDF

$$2F(x)f(x). \tag{3.101}$$

Intuitively, we see that the maximum of two symmetrically distributed random variables has a skewed distribution.

We can generalize this form by scaling the argument in the CDF,

$$2F(\alpha x)f(x). \tag{3.102}$$

A negative scaling, that is, $\alpha < 0$ yields a negative skewness. (In the case of the normal distribution, the value $\alpha = -1$ is equivalent to the minimum of two independent normal random variables.) Values of α larger in absolute value yield greater degrees of skewness; see Exercise 3.9. The scaling also changes the kurtosis, with larger absolute values of α yielding a larger kurtosis.

Specifically, for the standard normal distribution, we form the PDF of a skewed normal as

$$f_{\mathrm{SN}_1}(x; \alpha) = 2\Phi(\alpha x)\phi(x), \tag{3.103}$$

$\phi(x)$ denotes the PDF of the standard normal distribution and $\Phi(x)$ denotes the CDF.

Obviously, CDF-skewing can be applied to other distributions, such as the t or the generalized error distribution, and of course including normals as in equation (3.103) with other means and variances.

The R functions in the sn package compute the various properties, as well as estimate parameters from a given sample for a skewed normal distribution with PDF (3.103).

Differential Scaling

Another way of forming a skewed distribution is by scaling the random variable differently on different sides of the mean or some other central point. Specifically, for the normal distribution, we can form the PDF of a skewed normal as

$$f_{SN_2}(x;\xi) = \frac{2}{\xi + \frac{1}{\xi}} \begin{cases} \phi(x/\xi) & \text{for } x < 0 \\ \phi(\xi x) & \text{for } x \geq 0, \end{cases} \quad (3.104)$$

where, as before, $\phi(x)$ denotes the PDF of the standard normal distribution. Values of ξ less than one produce a positive skew, and values greater than one produce a negative skew. In either case, the excess kurtosis is positive.

Obviously, differential scaling can be applied to other distributions, such as the t or the generalized error distribution, and of course including normals as in equation (3.104) with other means and variances, where the two scales are used on different sides of μ.

The R fGarch package contains functions to compute the various properties, as well as to estimate parameters from a given sample for a skewed normal distribution with PDF (3.104). The R functions in the fGarch package for this skewed normal distribution are of the form

$$presnorm(arg, \text{mean, sd, xi})$$

where xi is the ξ in equation (3.104).

Figure 3.9 shows the standard normal distribution and two skewed normal distributions, skewed by differential scaling.

Mixtures of Distributions

Another way of forming a probability distribution to fit observed data is by using mixtures of given distributions.

If $f_1(x;\alpha)$ and $f_2(x;\beta)$ are PDFs and p_1 and p_2 are positive numbers such that $p_1 + p_2 = 1$, then

$$f(x;\alpha,\beta) = p_1 f_1(x;\alpha) + p_2 f_2(x;\beta) \quad (3.105)$$

is a PDF, and the distribution that it defines shares characteristics of the other two distributions. We say it is a mixture distribution with two *components*, which are the distributions combined to form it.

Any number of distributions can be mixed in this way. By combining distributions, a new distribution with various specified features, such as the mean, variance, skewness, and kurtosis can be formed.

PDFs of Normal and Skewed Normal

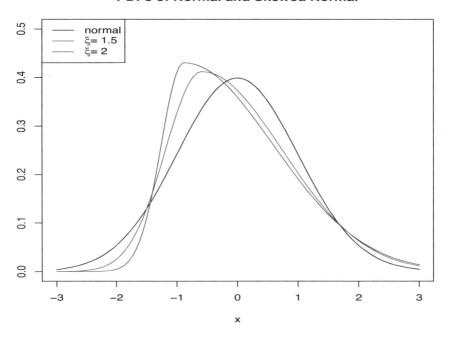

FIGURE 3.9
Skewed Normal Probability Distributions

We must emphasize the difference in a mixture of distributions and a linear combination of random variables. The mixture is defined by a linear combination of the PDFs, as in equation (3.105), but this is not to be confused with a random variable X that is a linear combination of the random variables in the two component distributions.

Although a linear combination of two normal random variables is a normal random variable, the mixture of two normal distributions is not a normal distribution as we can see in Figure 3.10. A linear combination of normal PDFs does not form a normal PDF.

Figure 3.10 shows a standard normal distribution, $N(0, 1)$, together with various mixtures of two normal distributions. Let the notation

$$pN(\mu_1, \sigma_1^2) + (1 - p)N(\mu_2, \sigma_2^2) \tag{3.106}$$

represent a mixture of two normal distributions, with the obvious meanings of the symbols p, μ_i, and σ_i^2. The mixtures in Figure 3.10 are two *scale mixtures*

$$0.6N(0, 1) + 0.4N(0, 9)$$

PDFs of Mixtures of Normals

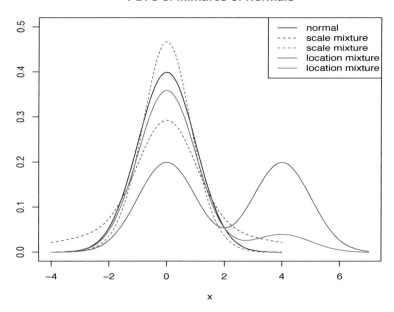

FIGURE 3.10
Mixtures of Normal Probability Distributions

and
$$0.6\text{N}(0,1) + 0.4\text{N}(0,0.49),$$

and two *location mixtures*

$$0.5\text{N}(0,1) + 0.5\text{N}(4,1)$$

and

$$0.9\text{N}(0,1) + 0.1\text{N}(4,1).$$

Notice that although it is not stated explicitly, the formulation (3.106) implies the existence of two *independent* random variables. It would be possible to form mixture distributions of nonindependent random variables, but we will not pursue that topic here.

The moments of the components determine the moments of the mixture. For example, for two components, if μ_1 is the mean of the first distribution and μ_2 is the mean of the second distribution, then $p\mu_1 + (1-p)\mu_2$ is the mean of the mixture. It is clear that for a scale mixture of normal distributions, all odd moments are unchanged. (Note the unchanged mean and skewness of the scale mixtures in Figure 3.10.) This is not the case for scale mixtures of other distributions.

Even moments for both scale and location mixtures are more complicated to work out. For example, the variance of both a scale mixture and a location mixture of normal distributions is different from the variance of either of the components. By the definition, the variance is

$$
\begin{aligned}
\mathrm{E}\left((X - \mathrm{E}(X))^2\right) &= \mathrm{E}(X^2) - (\mathrm{E}(X))^2 \\
&= \int x^2 \left(p f_1(x; \alpha) + (1-p) f_2(x; \beta)\right) \mathrm{d}x \\
&\quad - (p\mu_1 + (1-p)\mu_2)^2 \\
&= p \int x^2 f_1(x; \alpha)\mathrm{d}x + (1-p) \int x^2 f_2(x; \beta)\mathrm{d}x \\
&\quad - p^2\mu_1^2 - 2p(1-p)\mu_1\mu_2 - (1-p)^2\mu_2^2 \\
&= p(\sigma_1^2 + \mu_1^2) + (1-p)(\sigma_2^2 + \mu_2^2) \\
&\quad - p^2\mu_1^2 - 2p(1-p)\mu_1\mu_2 - (1-p)^2\mu_2^2 \\
&= p\sigma_1^2 + (1-p)\sigma_2^2 \\
&\quad + p(1-p)\mu_1^2 - 2p(1-p)\mu_1\mu_2 + (1-p)\mu_2^2 \\
&= p\sigma_1^2 + (1-p)\sigma_2^2 + p(1-p)(\mu_1 - \mu_2)^2,
\end{aligned}
$$

where σ_1^2 is the variance of the first distribution and σ_2^2 is the variance of the second distribution. (For those familiar with the statistical method of partitioning variation as in analysis of variance, we note that the variation of the mixture is composed of two parts, the *variation within* the two distributions, $p_1\sigma_1^2 + p_2\sigma_2^2$, and the *variation between* the two distributions, arising from the difference in their means, $(\mu_1 - \mu_2)^2$. We will discuss partitioning variation in Section 4.1.2 among other places in this book.)

We see that for a simple scale mixture, that is, a mixture with equal means, the variance of the mixture is the linear combination of the variances, $p_1\sigma_1^2 + p_2\sigma_2^2$, just as the mean is a linear combination, no matter whether the variances are equal or not.

The third and fourth moments of a sample are shown in a q-q plot. (Recall that q-q plots do not show location or scale of the sample.) Figure 3.11 shows q-q plots with respect to a normal distribution of random samples from the mixtures of normals in Figure 3.10 above. We notice that scale mixtures result in heavier tails. A location mixture may yield a skewed distribution and one with either lighter or heavier tails. A location mixture essentially yields two different sections in a q-q plot.

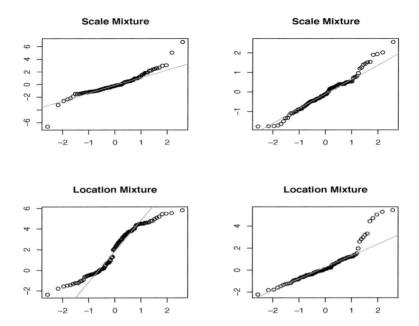

FIGURE 3.11
Q-Q Plots of Samples from Mixtures of Normal Probability Distributions

Compound Distributions

A special case of a mixture distribution is a *compound distribution*, which is a distribution in which a parameter is itself a random variable. Bayesian models, to be discussed in Section 4.4.4, can be regarded as compound models.

The normal location mixture model for the random variable X, $p\mathrm{N}(\mu_1, \sigma^2) + (1-p)\mathrm{N}(\mu_2, \sigma^2)$, for example, could be written as the compound distribution,

$$X \sim \mathrm{N}(\mu_2 + B(\mu_1 - \mu_2), \sigma^2), \quad \text{where } B \sim \text{Bernoulli}(p). \tag{3.107}$$

This is the joint distribution of X and B. The conditional distribution of X given B is either $\mathrm{N}(\mu_1, \sigma^2)$ or $\mathrm{N}(\mu_2, \sigma^2)$.

This is the same distribution as in (3.106) with p being a Bernoulli random variable instead of a constant.

Continuous Mixtures of Distributions

The compound distribution with a Bernoulli random variable that determines which of two normal distributions applies is a discrete mixture of normals. Suppose, instead of the 0-1 Bernoulli, we use a $\mathrm{U}(0, 1)$ random variable. We have a *continuous mixture*:

$$X \sim \mathrm{N}(\mu_2 + U(\mu_1 - \mu_2), \; \sigma^2), \quad \text{where } U \sim \mathrm{U}(0,1). \tag{3.108}$$

The conditional distribution of X given $U = u$ is $\mathrm{N}(u\mu_1 + (1-u)\mu_2, \; \sigma^2)$. The conditional mean varies continuously.

Continuous mixtures arise often in financial applications. Our exploratory analyses in Chapter 1 suggested that many things of interest do not follow a distribution with fixed parameters. The plot of returns in Figure 1.29 on page 99 certainly suggests that the variance of the returns is not constant. One way of modeling this is with a distribution that has a variance that is scaled by a positive random variable S:

$$S\sigma^2.$$

The conditional variance, given S, varies; that is, the process is *conditionally heteroscedastic*. Autoregressive time series models that include a continuous mixture of variances are called ARCH or GARCH models, and we will discuss them in Section 5.4.

Extreme Value Distributions

In Section 3.1.10 we discussed asymptotic distributions of random variables formed from random samples, as the sample size increases without bound. The common example, of course, is the mean of the sample, and the central limit theorem states that under certain conditions, the asymptotic distribution of the mean is a normal distribution. It is clear that certain random variables could not have a nondegenerate limiting distribution. A sum, for example, would not have a limiting distribution. What about order statistics from the sample? We may reason that central order statistics, such as the median, may have a limiting distribution under certain conditions, and indeed that is true. What about extreme order statistics? It is clear that the extreme order statistics from samples of distributions with infinite support cannot have a limiting distribution. If, however, the extreme order statistics are normalized in some way by the size of the sample, similar to the normalization of a sum that results in a mean, it may be possible that the normalized extreme order statistics do have a limiting distribution.

For a sample of size n, we consider a normalizing linear transformation of the maximum order statistic,

$$Z_n = c_n X_{(n:n)} + d_n,$$

where $\{c_n\}$ and $\{d_n\}$ are sequences of constants.

Consider a random variable with an exponential distribution; that is, with CDF

$$F_X(x) = (1 - e^{-\lambda x}), \quad 0 \le x,$$

where $\lambda > 0$. Suppose we have a random sample of size n of exponential variates, and we consider the maximum order statistic, $X_{(n:n)}$. Let $\{c_n\}$ be the constant sequence $\{\lambda\}$ and $\{d_n\}$ be the sequence $\{-\log(n)\}$, and form

$$Z_n = \lambda X_{(n:n)} - \log(n).$$

Substituting $X_{(n:n)} = (Z_n + \log(n))/\lambda$, we have the CDF of Z_n to be

$$F_{Z_n}(z_n) = \left(1 - \frac{1}{n}e^{-z_n}\right)^n.$$

Taking the limit of this, we have the CDF of $Z = \lim_{n\to\infty} Z_n$,

$$F_Z(z) = e^{-e^{-z}}, \qquad \text{for } 0 \le z. \tag{3.109}$$

(The limit in this is the same one used in deriving the formula for continuous compounding, equation (1.15).)

This distribution is called the (standard) Gumbel distribution, or type I extreme value distribution. It is a location-scale family of distributions with linear transformations of the argument.

It turns out that there are only a few different forms that the limiting distribution of a normalized extreme value can have. They can all be expressed as a *generalized extreme value distribution*. The form of the CDF of a generalized extreme value distribution is

$$F_G(x) = \begin{cases} e^{-(1+\xi x)^{-1/\xi}}, & \text{for } \xi \neq 0 \\ e^{-e^{-x}}. \end{cases} \tag{3.110}$$

The second form, of course, is the Gumbel distribution.

Extreme value theory, along with threshold exceedances discussed on page 276, is important in the study of value at risk as well as other applications in finance. The R package evd contains a number of functions for extreme value distributions.

3.2.5 Constructing Multivariate Distributions

As suggested in Section 3.2.4, in statistical data analysis, we must select a probability model that is simple and easy to work with, yet one that corresponds well with frequency distributions in observed samples. For multivariate data, a suitable multivariate distribution must correspond not only to the univariate marginal frequencies, but it must also exhibit the same kinds of relationships among the random variables as observed in data.

As we mentioned in Section 3.2.3, there are only a few "standard" families of multivariate distributions. We can construct others, however.

The simplest way of building a multivariate distribution is to begin with a collection of univariate distributions that model the individual variables.

Independent Variates; Product Distributions

If the individual variables are independent of each other, their multivariate distribution is called a *product distribution*, because the joint PDF is just the product of the individual marginal PDFs, as in equation (3.42).

Dependencies

Most variables in financial applications are dependent, however. Quantities that are seemingly unrelated are often dependent because they are related to a common quantity. Most stock prices, for example, are stochastically dependent because they are related to the overall market.

One approach is to start with independent random variables and then induce relationships by transformations, either a transformation of the variables themselves or indirectly through a copula.

Transformations

In a set of random variables, we generally consider just the bivariate relationships. One of the simplest measures of bivariate association is a linear one, the covariance or correlation.

Given a set of independent variables in a random vector X, we can form a new random vector Y with any given variance-covariance matrix Σ_Y by computing the Cholesky decomposition $\Sigma_Y = AA^T$ and applying the transformation $Y = AX$ (see page 268).

Copulas

Another method of expressing the relationships among a set of random variables is by means of a copula, which is a multivariate distribution of the CDFs of a set of random variables.

Individual random variables that are independent can be combined in a copula to yield a multivariate random variable in which the individual variates are related by the copula distribution.

Conditional Distributions

A simple way to build a PDF of a multivariate distribution is as a product of conditional and marginal PDFs, as an extension of equation (3.47). The challenge is to combine marginal PDFs in such a way as to build a joint PDF that reflects not only the distributions of the individual variables, but also their associations with each other.

3.2.6 Modeling of Data-Generating Processes

Much of statistical data analysis is exploratory, as in Chapters 1 and 2. This exploratory analysis involves computation of simple summary statistics and various types of graphical displays.

We can go much further in understanding a data-generating process and in predicting future observations if we develop mathematical probability models of the data-generating process. This involves use of probability functions and probability density functions that match the frequency distribution of observed data. In statistical data analysis, we *estimate* parameters in these functions or we *test statistical hypotheses* regarding those parameters.

Modeling observable data as random variables allows us to use properties of random variables in the analysis of the data. For example, considering returns to be random variables, we easily arrive at the square-root rule for aggregating volatility, stated on page 106.

Equation (3.5), for example, is the probability function for a *family* of distributions called the Poisson family of distributions. Each member of the Poisson family is determined by the specific value of the parameter λ. If we assume the Poisson family as a model for some data-generating process, we use observed data to estimate λ. We may also use the data to test a statistical hypothesis about λ. For example, the hypothesis to be tested may state that λ is greater than or equal to 5. There is a wealth of statistical theory to guide us in finding good estimators from an observed sample or in developing a good test procedure.

Equation (3.23), as another example, is the PDF for a family of probability distributions called the lognormal family. Each member of the lognormal family is determined by specific values of μ and σ. Again, the existing statistical theory can guide us in finding good estimators of the parameters μ and σ using observed data, or in developing good test procedures for statistical hypotheses concerning μ and σ.

To use data to estimate the parameters in a family of distributions is to *fit* the distributional model. For example, we may assume that the process that generated a random sample $x_1, \ldots, x_n$ is $N(\mu, \sigma)$. Using the given data, we may estimate the parameters μ and σ. Call these estimates $\hat{\mu}$ and $\hat{\sigma}$. Then the fitted distribution is $N(\hat{\mu}, \hat{\sigma})$.

We will often use the "hat notation" to indicate an estimate (a specific value) or an estimator (a random variable). In Chapter 4, we will discuss various estimators and various testing procedures.

3.2.7 R Functions for Probability Distributions

R provides a collection of built-in functions for computations involving several common probability distributions. These computations include the values of the density or probability function at specified points, the values of the CDF at specified points, and the quantiles for given probabilities. These functions have names that begin with "**d**" for the density or probability function, with "**p**" for the CDF (the probability of a value less than or equal to a specified point), and with "**q**" for the quantile. (See Section 3.1 for definitions of these terms.) R also provides functions for simulating random samples from these distributions (see Section 3.3.3). Functions to generate random numbers have names that begin with "**r**". The remainder of the name is a mnemonic root name relating to the name of the distribution. The root names of common distributions are shown in Table 3.4. The probability functions or probability density functions for most of these are given in Tables 3.1 through 3.3 in Section 3.1.

The arguments to the R function are the point at which the result is to be computed, together with any parameters to specify the particular distribution within the family of distributions. For example, the root name of the R functions for the Poisson distribution is `pois`. The Poisson family of distributions is characterized by one parameter, often denoted as "λ" (see Table 3.1 on page 284).

Hence, if `lambda` (which may be a numeric array) is initialized appropriately,

$$\text{dpois}(\text{x}, \text{lambda}) \quad = \quad \lambda^x e^{-\lambda}/x!$$

$$\text{ppois}(\text{q}, \text{lambda}) \quad = \quad \sum_{x=0}^{q} \lambda^x e^{-\lambda}/x! \tag{3.111}$$

$$\text{qpois}(\text{p}, \text{lambda}) \quad = \quad q, \quad \text{where } \sum_{x=0}^{q} \lambda^x e^{-\lambda}/x! = p,$$

where all elements are interpreted as numeric arrays of the same shape. The values of the mean and standard deviation are taken from the *positional relations* of the arguments.

Various R functions for computations involving a Poisson random variable with parameter $\lambda = 5$ are shown for example in Figure 3.12.

```
> dpois(3, lambda=5)
[1] 0.1403739      #  probability that a random variable equals 3
> ppois(3, lambda=5)
[1] 0.2650259      #  probability less than or equal to 3
> qpois(0.2650259, lambda=5)
[1] 3              #  quantile corresponding to 0.2650259
```

FIGURE 3.12
Values in a Poisson Distribution with $\lambda = 5$

For the univariate normal distribution, there are two parameters, the mean and the variance (or standard deviation); see Table 3.2 on page 286. The

TABLE 3.4

Root Names and Parameters for R Functions for Distributions

Continuous Univariate Distributions		Discrete Distributions	
unif	uniform	binom	binomial
	min=0, max=1		size, prob
norm	normal	nbinom	negative binomial
	mean=0, sd=1		size, prob
lnorm	lognormal	multinom	multinomial
	meanlog=0, sdlog=1		size, prob
chisq	chi-squared		*only the* r *and* d *versions*
	df, ncp=0	pois	Poisson
t	*t*		lambda
	df, ncp=0	geom	geometric
f	*F*		prob
	df1, df2, ncp=0	hyper	hypergeometric
beta	beta		n, m, k
	shape1, shape2		
cauchy	Cauchy		
	location=0, scale=1		
exp	exponential		
	rate=1		
gamma	gamma		
	shape, scale=1		
gumbel	Gumbel		
{evd}	loc=0, scale=1		
laplace	double exponential		
{rmutil}	m=0, s=1		
logis	logistic		
	location=0, scale=1		
pareto	Pareto		
{EnvStats}	location, shape=1		
stable	stable		
{stabledist}	alpha, beta,		
	gamma=1, delta=0		
weibull	Weibull		
	shape, scale=1		
Generalized Distributions		Continuous Multivariate Distributions	
gl	generalized lambda	mvnorm	multivariate normal
{gld}	lambda1=0, lambda2=NULL,	{mvtnorm}	mean=0, sigma=*I*
	lambda3=NULL, lambda4=NULL,		*only the* r *and* d *versions*
	param="fkml", lambda5=NULL)	dirichlet	Dirichlet
ged	generalized error	{MCMCpack}	alpha
{fGarch}	mean=0, sd=1		*only the* r *and* d *versions*
gpd	generalized Pareto	mvt	multivariate t
{evir}	xi, mu=0, beta=1	{mvtnorm}	sigma, df, delta=NULL
snorm	skewed normal		*only the* r *and* d *versions*
{fGarch}	mean=0, sd=1, xi=1.5		

root name of the R functions is norm, and the names of the parameters are mean and sd (for standard deviation *not* the variance). Hence, if the variables mean and sd (which may be numeric arrays of the same shape) are initialized to m and s,

$$\text{dnorm}(\mathtt{x}, \mathtt{mean}, \mathtt{sd}) \;=\; \frac{1}{\sqrt{2\pi}\,s}\mathrm{e}^{-(x-m)^2/2s^2}$$

$$\text{pnorm}(\mathtt{q}, \mathtt{mean}, \mathtt{sd}) \;=\; \int_{-\infty}^{q} \frac{1}{\sqrt{2\pi}\,s}\mathrm{e}^{-(x-m)^2/2s^2}\,\mathrm{d}x$$

$$\text{qnorm}(\mathtt{p}, \mathtt{mean}, \mathtt{sd}) \;=\; q, \quad \text{where } \int_{-\infty}^{q} \frac{1}{\sqrt{2\pi}\,s}\mathrm{e}^{-(x-m)^2/2s^2}\,\mathrm{d}x = p,$$

$$(3.112)$$

where all elements are interpreted as numeric arrays of the same shape. (Because this is a continuous distribution, the value `dnorm(x, mean, sd)` is not a probability, as `dpois(x, lambda)` is.)

The values of the mean and standard deviation are taken from the *positional relations* of the arguments, but the keyword arguments allow different forms of function calls. If the mean is stored in the variable `m` and the standard deviation is stored in the variable `s`, we could use the function reference

```
dnorm(x, sd=s, mean=m)
```

The parameters often have defaults corresponding to "standard" values. For the univariate normal distribution, for example, the default values for the mean and standard deviation respectively are 0 and 1. This is the "standard normal distribution". If either parameter is omitted, it takes its default value. The function reference

```
dnorm(x, sd=s)
```

is for a normal distribution with a mean of 0 and a standard deviation of `s`.

3.3 Simulating Observations of a Random Variable

An effective way to study a probability model for a data-generating process is to obtain samples of data from the process and study their frequency distributions. Rather than obtaining actual data from the process, it may be possible to simulate the data on the computer.

If the PDF or probability function of the random variable of interest is f_X, the objective is to generate a sample $x_1, \ldots, x_n$, so that the x_i are independent and each has PDF or probability function f_X.

A method that uses simulated data to study a process is often called a *Monte Carlo* method. Beginning on page 355 in Chapter 4, we discuss Monte

Carlo estimation, and in Chapter 5 we discuss Monte Carlo simulation of time series.

In this section we will briefly describe some of the mathematical properties of probability distributions that underlie the algorithms for random number generation. In Section 3.3.3 below, we list R functions to simulate data from various distributions.

Our objective is to generate on the computer a sequence of n numbers that appear to be a simple random sample of size n from a specified probability distribution. By "appear to be" we mean that most formal statistical procedures would not indicate that the numbers are not a simple random sample from that probability distribution. There are various statistical procedures used to test that a set of numbers are a random sample from some specified distribution. We will discuss general "goodness-of-fit" tests in Section 4.6.1 beginning on page 456, and discuss tests for normality specifically beginning on page 456.

With this objective, we avoid the question of what constitutes "randomness". We seek to generate "pseudorandom numbers". Having made this distinction, I will henceforth, except occasionally for emphasis, refer to the pseudorandom numbers as "random numbers".

Using techniques for transforming one random variable into another one, if we have a random variable with a known distribution, we can form a random variable with almost any other specified distribution.

The first task is to generate numbers on the computer that appear to be random from a known distribution. The second task is to apply a transformation of random variables from that known distribution to the distribution that we are really interested in.

3.3.1   Uniform Random Numbers

The uniform distribution is particularly easy to make a transformation to some other specified distribution.

Random samples from the uniform distribution are also particularly easy to simulate. There are several simple functions that operate on a small set of numbers and yield a result that seems to have no connection with the input number; the result just appears to be random. Not only that, the result may appear to be equally spread over some interval. That is a uniform distribution.

For example, suppose we start with some integers, choose m to be some big integer, and let j and k be integers $1 \leq j < k < m$. Start with k integers $x_1, \ldots, x_k$, "randomly chosen". Now, for $i = j+1, j+2, \ldots$, form

$$x_i = (x_{i-j} + x_{i-k}) \mod m, \qquad (3.113)$$

where "mod m" means to divide by m and take the remainder. (Recall mod in R is performed by %%.) Finally, divide each x_i by m. (See Exercise 3.16.)

For the results to have statistical properties similar to a random sample from a U$(0, 1)$ distribution, they must not only appear to be equally spread over the interval $(0, 1)$, they must also appear to be independent; that is, there must not be any apparent ordering or any patterns in the subsequences.

The generator in equation (3.113) is called a lagged Fibonacci generator, and for certain choices of m, j, and k, the sequences generated do appear to be random samples from a $U(0, 1)$ distribution.

The "random" number generator follows a deterministic algorithm. Notice two characteristics. If the generator is started with the same set of integers (the *seed*) the exact same sequence of "random" numbers will be generated. Secondly, because the are a finite number of numbers on the computer, eventually any given subsequence that is generated once will be generated again; hence, the generator has a finite *period*. (It is not necessarily the case that an arbitrary deterministic generator will eventually yield a given seed, but most will.)

There are several other types of deterministic generators and there have been many studies of what kinds of functions yield sequences that look random and that pass statistical tests for randomness. The apparent randomness can sometimes be improved by combining the output of two or more basic generators or by scrambling the output.

It is usually not a good idea to write a program to generate uniform numbers; rather, it is best to use a well-tested generator. (The exercise above is just to illustrate the idea.)

The R function `runif` implements several good choices. The other random number generators in R generally use `runif` to get uniform variates and then transform them so that they have other distributions. We will mention two types of transformations below.

Quasirandom Numbers

In drawing statistical samples, often instead of simple random sampling, we constrain the process so as to obtain a more representative sample of the population. This same idea can be applied in random number generation. Random numbers generated by restrictions that make them more evenly spread out are called "quasirandom numbers". Methods for generating such numbers are based on deterministic sequences. Two of the more common sequences are the Halton sequence and the Sobol' sequence. (See Gentle, 2003, Chapter 3, for a discussion of these sequences and a general discussion of quasirandom numbers.)

The R package `randtoolbox` has three functions that generate quasirandom numbers: `halton`, `sobol`, and `torus`. Each function has various arguments that control how the underlying sequences are used. By default, the functions yield the same output on each invocation (for the same length requested).

The `randtoolbox` package also contains a number of utility functions and functions for performing statistical tests on the randomness of the generated samples.

3.3.2 Generating Nonuniform Random Numbers

The first problem in simulating random numbers on the computer is the "randomness". As we have described, that problem has been addressed more-or-less satisfactorily for uniform numbers.

If we have "random" uniform numbers, we can get random numbers from other distributions by various transformations. We will describe only two here. In the first one we discuss, one uniform number yields one nonuniform number. In the second one we discuss, an indefinite number of uniform numbers are required to get one nonuniform number.

The Inverse CDF Method

If F_X is the CDF of a continuous random variable X, and we define the random variable U by the transformation $U = F_X(X)$, then U has a U(0, 1) distribution.

For a continuous random variable, this transformation can be inverted. If we start with U that has a U(0, 1) distribution, then $X = F_X^{-1}(U)$ is a random variable with CDF F_X (see equation (3.25) on page 254).

This leads to a technique called the *inverse CDF method*.

We can illustrate the inverse CDF method with the exponential distribution, which has CDF $1 - \exp(x)$ (see Table 3.2). The R code below produces the histograms in Figure 3.13.

```
u <- runif(n)
x <- -log(u)
par(mfrow=c(1,2))
hist(u, freq=FALSE, xlab="u", main="Uniform")
hist(x, freq=FALSE, xlab="x", main="Exponential")
```

Note the use of the transformation; it is equivalent to $-\log(1 - u)$.

The definition of the generalized lambda distribution in terms of the inverse of the CDF in equation (3.98) makes it particularly simple to generate random numbers from that distribution.

The inverse CDF method can also be adapted for discrete distributions by replacing the inverse with the generalized inverse of the CDF (see Gentle, 2003, page 104, for example).

Acceptance/Rejection

A very general way to transform a random variable from one distribution to another distribution is to make a random decision whether to accept or reject

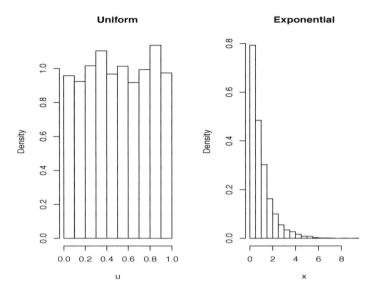

FIGURE 3.13
Histogram of a Simulated Uniform Sample and the Same Sample Transformed
by $-\log$

a given value of the random variable based on the relative probability densities
at the given value.

This technique is important in many statistical methods. For example,
acceptance/rejection is the basic idea in what is called Markov chain Monte
Carlo (MCMC), see page 388.

Figure 3.14 illustrates the idea. We want to collect a random sample from
a distribution with PDF $f(x)$, which has a range from 0 to 1, as shown in the
graph.

Here, we start with a uniform distribution, which we call the "proposal
distribution", and we call the distribution with PDF $f(x)$ the "target distribution".

We show four points somewhat uniformly spread over $(0, 1)$, as if they are
a sample from a $U(0, 1)$ distribution. We will randomly accept or reject these
points in our sample from the target distribution based on the ratio of the
PDF $f(x)$ to the uniform PDF, $g(x) = 1$. We will use the ratio of $f(x)$ to a
scalar multiple of $g(x)$, $cg(x)$.

Looking at the acceptance/rejection ratios for these points shown in the
figure, we see that we are more likely to accept a point where the probability density of the target distribution is large. After randomly accepting or
rejecting the points, those remaining will have a frequency distribution more
in line with that of the target distribution than of the proposal distribution.

We can show mathematically that the values accepted in this way will have a probability density exactly equal to the target density, $f(x)$ in this case. (See Gentle, 2003, page 114, for example.)

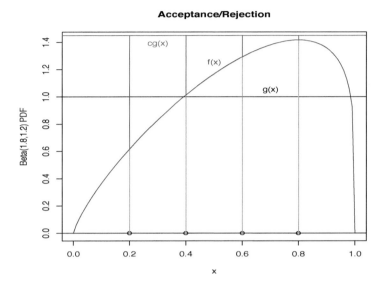

FIGURE 3.14
Acceptance/Rejection Showing Four Points with Uniform Majorizing Function

The starting distribution in the acceptance/rejection procedure, the proposal distribution, can be any distribution whose PDF can be scaled to "majorize" the PDF of the target distribution. Algorithm 3.1, where g_Y represents the PDF of the proposal distribution, describes the process.

Algorithm 3.1 The Acceptance/Rejection Method to Convert Uniform Random Numbers

    1. Generate x from the distribution with density function g_Y.

    2. Generate u from a U$(0, 1)$ distribution.

    3. If $u \leq f_X(x)/cg_Y(x)$, then
        3.a. accept x as the desired realization;
    otherwise
        3.b. discard x and return to step 1. ∎

Multivariate Distributions

The inverse CDF method does not apply directly to a multivariate distribution, although marginal and conditional univariate distributions can be used in an inverse CDF method to generate multivariate random variates. The acceptance/rejection method can be used for multivariate distributions. The proposal distribution would be chosen as a multivariate distribution whose PDF could be scaled to majorize the PDF of the target distribution.

For multivariate distributions, the main interest is often in the relationships among the elements. Measures of association, such as the variance-covariance matrix, model these relationships. We start with a multivariate vector of variates that have 0 correlations (covariances) with each other, and scale the vector to achieve a desired variance-covariance matrix. In a simple case where all variates in the multivariate vector have the same variance, say σ^2, and have 0 covariances, the variance-covariance matrix is σ^2 times the identity, $\sigma^2 I$.

Given a random vector Y whose variance-covariance matrix is the identity matrix, we can obtain a random vector X with any given variance-covariance matrix Σ_X, by use of equation (3.62). We first determine a matrix A such that $\Sigma_X = AA^{\mathrm{T}}$. We then form the random vector by the linear transformation

$$X = AY.$$

With $\mathrm{V}(Y) = I$, we have $\mathrm{V}(X) = \Sigma_X$. As discussed on page 268, A can be chosen as the Cholesky factor Σ_X. The R function `chol` computes the Cholesky factor.

This type of transformation applies for any distribution, but the meaning of covariances is not so clear in other distributions as in the multivariate normal distribution.

If we have a multivariate random vector Y from one distribution, we can generate a random vector X from another distribution with dependencies among its elements that are induced by the dependencies in Y, by use of a copula, equation (3.53) on page 265.

As an example of the use of a copula, assume X is a d-vector distributed as $\mathrm{N}_d(0, \Sigma_X)$, and let Φ be the CDF of a univariate $\mathrm{N}(0,1)$ distribution. Now, because the ith element of X is distributed as $\mathrm{N}(0, \sigma_{ii})$, where σ_{ii} is the ith diagonal element of Σ_X, $\Phi(X_i/\sqrt{\sigma_{ii}})$ is distributed as $\mathrm{U}(0,1)$ (see equation (3.25)). Now, let $F_1, \ldots, F_d$ be the CDFs of univariate distributions. Then the vector

$$Y = (F_1^{\leftarrow}(\Phi(X_1/\sqrt{\sigma_{11}})), \ldots, F_d^{\leftarrow}(\Phi(X_d/\sqrt{\sigma_{dd}}))) \qquad (3.114)$$

has a CDF F_Y of the form of the copula (3.53). Notice that each element of Y has a marginal distribution with CDF F_i. (You are asked to prove this for $d = 2$ in Exercise 3.15.)

The elements of Y will be correlated, but not with correlations equal to or easily determined by the correlations of the elements of X. The correlations of Y depend on the CDFs F_i, which ultimately determine the CDF F_Y.

In applications, the marginal distributions of Y are generally chosen to be in the same family, for example, they may be t distributions with different degrees of freedom.

The R `copula` package provides functions for random number generation with various classes of copulas. In Exercise 3.19c, you are asked to go through the computations using an expression of the form of (3.114).

3.3.3 Simulating Data in R

We have mentioned the useful collection of built-in functions involving several common probability distributions that compute the value of the density or probability function at a specified point, the value of the CDF at a specified point, or the quantile for a given probability. These functions have names that begin with "d", "p", or "q", and have another part that is a mnemonic referring to the name of the distribution. In addition to these three functions for a given distribution, there is another function that begins with "r" that indicates that a "random sample" of data from that distribution is to be simulated.

The sample is not actually random, as we have discussed, and two samples with the same starting point are exactly the same. The R random number generator functions choose a "random" starting point based on the system clock, so each time a random number function is called, a different random sample is generated (within limits; the default R generator has a period of $2^{19937} - 1$). Sometimes for testing purposes or to obtain reproducible examples, it is desirable to obtain the same "random" sample. This can be ensured by use of the R function `set.seed` prior to calling a random number generator function.

As we discussed in Section 3.3.2, the general starting point for simulating data is to generate numbers that seem to be distributed uniformly and independently over the interval $(0, 1)$. The R function `runif` does this. There are several ways that pseudorandom uniform numbers can be generated, and R provides a function `RNGkind` that allows the user to choose the type of basic generator.

Instead of pseudorandom numbers, quasirandom numbers can be used as the underlying sequence to transform to a sequence simulating a sample from any distribution. The `randtoolbox` package contains the functions `halton`, `sobol`, and `torus` that generate quasirandom uniform numbers. By default, these functions generate the same sequence each time. The package also has functions to perform statistical tests on samples of random numbers.

For a given probability distribution, the arguments to the "r" function are the same as in the other functions for that distribution, except instead of the point at which the function value is to be computed, the first argument is a positive integer or a vector.

For the Poisson family of distributions for which we defined the three functions that evaluate the density, probability, and quantile in equation (3.111), we have the function

```
rpois(n, lambda)
```

that generates a vector of n values (where n is some positive integer) that will be similar in their statistical properties as the corresponding properties of a sample of the same size from a Poisson distribution with parameter `lambda`. For example, we have the results shown in Figure 3.15. The sample mean

```
> set.seed(12345)
> lambda <- 5
> n <- 1000
> x <- rpois(n, lambda=lambda)
> mean(x)
[1] 5.077
> sd(x)
[1] 2.17389
```

FIGURE 3.15
Simulating a Sample of Poisson Data in R

and sample standard deviation in Figure 3.15 correspond closely to the mean and standard deviation of a random variable with a Poisson distribution with parameter λ, which are λ and $\sqrt{\lambda}$.

For the univariate normal distribution for which we defined the three functions that evaluate the density, probability, and quantile in equation (3.112), we also have the function

```
rnorm(n, mean, sd)
```

that generates a vector of n (where n is some positive integer) values that will be similar in their statistical properties as the corresponding properties of a sample of the same size from a normal distribution with mean **mean** and standard deviation **sd**. For example, we have the results shown in Figure 3.16.

The discrete uniform distribution arises often in statistical data analysis.

```
> set.seed(12345)
> m <- 100
> s <- 10
> n <- 1000
> x <- rnorm(n, mean=m, sd=s)
> mean(x)
[1] 100.462
> sd(x)
[1] 9.987476
```

FIGURE 3.16
Simulating a Sample of Normal Data in R

This is the distribution that has has k possible outcomes, each equally probable. The R function `sample` can be used to generate random numbers from this distribution. To generate n random integers from the set $1, \ldots, k$, use

```
sample(k, n, replace=TRUE)
```

The `sample` function can also be used to generate random permutations and to generate random variables from other discrete distributions with any specified probability function.

R has functions to generate random numbers from the heavy-tailed distributions we discussed in Section 3.2.4. These are the "r" functions corresponding to the R functions shown in Table 3.4.

Simulation of a time series in some cases is essentially the same as simulation of finite samples. Some time series models, however, are based on infinite sequences, and hence, present special issues. We will discuss simulation of time series data in R in Section 5.3.5.

Notes and Further Reading

Statistical inference depends on probability models. There are a number of "standard" probability distributions that serve well as models for a variety of different data. We have described a number of families of probability distributions. Further discussions of many common families of distributions are available in Krishnamoorthy (2015). There are many interesting relationships among various probability distributions, such as the relationship between a t distribution and a Cauchy and the relationship between a t and a normal. Leemis and McQueston (2008) present a graph of relationships among various univariate probability distributions.

As we have seen, financial data often exhibit heavy tails. Despite this fact, the normal distribution has often been used to model this data. Many authors including Fama and Roll (1968, 1971), for example, have emphasized the need to use heavy-tailed distributions in financial modeling. The t and stable general classes of distributions have tails of varying weights, each depending on parameters that can be chosen to match more closely the observed data. Mixture distributions can also be used to model the distribution of heavy-tailed data. Nolan (1997), Adler, Feldman, and Taqqu (1998), Jondeau, Poon, and Rockinger (2007), and Novak (2012) provide extensive discussions of the properties of various heavy-tailed distributions and their uses in financial modeling. Heavy-tailed distributions especially useful in modeling the inequalities of income and wealth among households are discussed in the articles in the book edited by Chotikapanich (2008).

Novak (2012) discusses extreme value theory and its applications to analysis of financial data. McNeil, Frey, and Embrechts (2015) discuss threshold exceedance and ways of measuring it.

In statistical modeling, there is often a need to go beyond the standard distributions to find a mathematical probability distribution that matches the shape and other characteristics of observational data. There are several general systems of distributions, each of which covers a wide range of shapes and other characteristics. Karl Pearson in the 1890s developed a general family of distributions based on a differential equation with five parameters (the *Pearson family*). Different relationships of the parameters led to solutions that corresponded to various families of probability distributions. Irving Burr in the early 1940s defined a flexible family of distributions over the positive real numbers (the *Burr distributions*). The Burr distributions include a number of common distributions, such as the Pareto, as special cases. Norman Johnson, beginning in 1949 with a simple transformation of a normal random variable, extended the transformations and developed a general system of distributions that corresponds to a variety of shapes (the *Johnson S_U distributions*). The lambda family of distributions was introduced by John Tukey, and extended to the generalized lambda family by Ramberg and Schmeiser (1974). Karian and Dudewicz (2000) describe ways of fitting the four parameters in the generalized lambda distribution to observed data, and they also discuss various applications of the distribution in modeling. Discussions of these general families of distributions are available in Stuart and Ord (1994).

The need to have a flexible general family of distributions also arises in choosing a Bayesian prior distribution (see Section 4.4.4). Two of the general methods for forming skewed normal distributions arose as methods of forming Bayesian priors that more accurately reflected prior beliefs. O'Hagan and Leonard (1976) suggested a prior as the maximum of two independent normals. The PDF of this distribution is the product of a PDF and a CDF.

Azzalini (1985) generalized this approach by adding a scaling in the CDF. Azzalini (2005) later extended the idea to multivariate families. Also motivated by the need to choose a Bayesian prior distribution, Fernandez and Steel (1998) suggested a scaling on one side of the mean combined with an inverse scaling on the other side to yield a skewed distribution from a symmetric one.

Copulas have become popular in financial modeling applications in recent years. Joe (2015), Cherubini et al. (2012), and Okhrin (2012) discuss various aspects of copulas in multivariate modeling.

Glasserman (2004) discusses Monte Carlo simulation methods, with a particular emphasis on financial applications. Gentle (2003) discusses random number generation and Monte Carlo methods in general. Niederreiter (2012) describes use of quasirandom simulation in financial applications.

Pricing of derivatives is a major topic in finance. Although we mentioned some approaches, the topic is beyond the scope of this book. Hull (2017) discusses methods for pricing options including use of binary trees, use of the Black-Scholes-Merton differential equation, and Monte Carlo methods. He gives the derivations of equations (3.80), (3.81), and (3.82). Fouque, Papanicolaou, and Sircar (2000) and Fouque et al. (2011) discuss the problems stochastic volatility present in pricing options. Haug (2007) provides an extensive compendium of various approaches to options pricing.

Exercises

3.1. **Transformations.**

Consider the simple discrete distribution of the random variable X with probability function as given in equation (3.18). Make two transformations,

$$Y = 2X + 1$$

and

$$Z = X^2 + 1.$$

(a) *Using the respective probability functions*, compute $E(X)$, $V(X)$, $E(Y)$, $V(Y)$, $E(Z)$, and $V(Z)$.

(b) Now, using $E(X)$ and $V(X)$, and *using the transformations only*, compute $E(Y)$, $V(Y)$, $E(Z)$, and $V(Z)$.

3.2. **Shrunken estimator** (see page 369).

Let X be a random variable and let $E(X) = \theta \neq 0$ and $V(X) = \tau^2 > 0$.

Let b be a constant such that $0 < b < 1$. (All of the salient results are unchanged if b is such that $|b| < 1$.)

(a) Show that $E(bX) \neq \theta$; that is, the shrunken estimator bX is not unbiased for θ.

(b) Show that $V(bX) < \tau^2$.

(c) Now "shrink" X toward the constant c; that is, let
$Y = c + b(X - c)$.
Show that $V(Y) < \tau^2$.

3.3. Simple exercises with probability distributions.

(a) Show that the variance of the Poisson distribution whose probability function is shown in Table 3.1 is λ, the same as the mean of the distribution.

(b) Show that the mean of the $U(0,1)$ distribution is $1/2$ and the variance is $1/12$.

(c) Show that the mean and variance of the gamma distribution with the PDF shown in Table 3.2 are $\alpha\beta$ and $\alpha\beta^2$.

(d) Use the PDF of the gamma distribution to evaluate the integral

$$\int_0^\infty \frac{1}{1728} \lambda^{13} e^{-5\lambda} \, d\lambda.$$

(This is the likelihood function for the sample $\{3, 1, 4, 3, 2\}$ from a Poisson distribution, as shown in Section 4.2.2.)

(e) Let X have the triangular distribution. The triangular distribution has PDF

$$f_X(x) = \begin{cases} 1 - |x| & \text{for } -1 < x < 1 \\ 0 & \text{otherwise.} \end{cases}$$

Show that $E(X) = 0$ and $V(X) = \frac{1}{6}$.

3.4. Simple exercises with probability distributions.

(a) Determine the survival function and the hazard function for the Weibull distribution with parameters α and β.

(b) Determine the hazard function for the exponential distribution with parameter λ.

What are the implications of this; that is, in what kinds of situations might the exponential be a useful model?

3.5. Transformations.

(a) Let X have the triangular distribution (see Exercise 3.3e.) Let $Y = 2X + 3$. Determine the PDF of Y, show that your function is a PDF, and compute $E(Y)$ and $V(Y)$.

(b) Let X have the triangular distribution.

Let $Z = X^2$. Determine the PDF of Z. Show that your function is a PDF, and compute $E(Z)$ and $V(Z)$.

(c) Let $V \sim N(0,1)$ and let $W = V^2$. Determine the PDF of W. (The distribution is the chi-squared with 1 degree of freedom.)

(d) Let $U \sim N(\mu, \sigma^2)$ and let $Y = e^U$. Determine the PDF of Y. (The distribution is the lognormal.)

(e) Let $U \sim N(\mu, \sigma^2)$ and let $Y = e^U$ as before. Determine $E(Y^2)$. *Hint:* Complete the square, as on page 288.

3.6. **Binomial trees.**

Suppose the price of stock XYZ is $100, suppose that the standard deviation of the daily returns of XYZ is 0.002, and suppose that the risk-free interest rate r_F is 1%.

(a) What is the (annualized) volatility σ of XYZ?

(b) Assume a unit of time as one year and a constant length of time intervals

$$\Delta t = 1/4,$$

a constant ratio of new value to old value following an up move of

$$u = e^{\sigma \sqrt{\Delta t}},$$

a constant ratio of new value to old value following a down move of $1/u$, and the probability of an up move of

$$p = \frac{e^{r_F \Delta t} - 1/u}{u - 1/u}.$$

Construct a binomial tree representing the possible trajectories of XYZ for one unit of time (four periods). Recall that under this model, the vertical scale in a tree representation is not linear.

(c) Suppose you owned a call option for XYZ at 103 for 251 days out (essentially one year) on the beginning date in this scenario.

What percentage of the time would your option have been in the money on the date of expiration, according to the binary tree model?

(d) What is the fair value of a European call option for XYZ at 103 for 251 days out on the beginning date in this scenario? (See also Exercise 3.20d.)

3.7. **Heavy-tailed distributions.**

(a) Plot on the same set of axes, using different line types and/or different colors, the PDFs of a standard normal distribution, a t with 10 degrees of freedom, a t with 2 degrees of freedom, and a Cauchy.

Briefly discuss the plots.

(b) Show that the mean and variance of the Cauchy distribution do not exist (or are infinite, depending on your definition of the improper integrals).

3.8. **Threshold exceedances.**

(a) Consider a normal distribution with mean 100 and variance 100. Write out the CDF of the threshold exceedance distribution with respect to 120 (two standard deviations above the mean). Just use symbols for functions that cannot be evaluated in closed form.

(b) Use R or another computer program to produce a plot of the CDF of the normal exceedance distribution in the previous question.

What is the probability that the random variable exceeds three standard deviations above the mean (that is, 130), given that it has exceeded the threshold?

(c) Consider a gamma distribution with shape parameter 4 and scale parameter 10. Write out the CDF of the threshold exceedance distribution with respect to 80 (two standard deviations above the mean). Just use symbols for functions that cannot be evaluated in closed form.

(d) Use R or another computer program to produce a plot of the CDF of the gamma exceedance distribution in the previous question.

3.9. **Skewed distributions.**

(a) For the standard normal distribution, plot the PDF in equation (3.101) for four values of α, $\alpha = -1, \frac{1}{2}, 1, 2$, along with the PDF of the standard normal.

(b) Generate 1,000 realizations of the maximum of two standard normal random variables. (For each, you could use `max(rnorm(2))`, but you should try to use vectorized operations, possibly with `ifelse`.)

Plot a histogram of the sample and superimpose a plot of the standard normal PDF.

(c) Write out the PDF of a skewed double exponential (Laplace) distribution using the method of equation (3.103).

For the standard double exponential, (mean $= 0$ and scale $= 1$), plot the PDF of the CDF skewed distribution for four values of α, $\alpha = -1, \frac{1}{2}, 1, 2$, along with the PDF of the standard double exponential.

3.10. **General families of probability distributions.**

(a) Generate 10,000 random variables following the beta$(5, 5)$, the beta(0.8, 0.8), the beta(0.8, 5), the beta(5, 2), the beta(2, 5), and the beta(5, 0.8) distributions.

Fit kernel densities to each one, and produce six plots of the empirical densities, similar to the mathematical densities in Figure 3.6.

(b) Generate 10,000 random variables following the generalized error distribution, with $\nu = 2$ (normal), $\nu = 1$ (double exponential), $\nu = 0.8$, and $\nu = 3$. (A function to do this is in the R package fGarch.)

Fit kernel densities to each one, and produce four graphs of the empirical densities on the same plot, similar to the mathematical densities in Figure 3.8.

Produce a legend to identify the separate densities.

(c) Generate 10,000 random variables following the skewed normal distribution, with $\xi = 1$ (normal), $\xi = 1.5$, and $\xi = 2$.

Fit kernel densities to each one, and produce three graphs of the empirical densities on the same plot, similar to the mathematical densities in Figure 3.9.

Produce a legend to identify the separate densities.

(d) Generate 10,000 random variables following each of the normal mixture distributions:

$$0.6N(0, 1) + 0.4N(0, 9)$$

$$0.6N(0, 1) + 0.4N(0, 0.64)$$

$$0.5N(0, 1) + 0.5N(3, 1)$$

$$0.9N(0, 1) + 0.1N(3, 1)$$

Fit kernel densities to each one, and produce six graphs of the empirical densities on the same plot, similar to the mathematical densities in Figure 3.10.

Produce a legend to identify the separate densities.

3.11. **Extreme value distributions.**

Consider a random sample of size n from the unit Pareto distribution with PDF

$$f(x) = 2x^{-2} \quad \text{for } 0 \le x;$$

that is, for the Pareto with $\alpha = 2$ and $\gamma = 1$.

(a) Write the cumulative distribution function for the maximum order statistic $X_{(n:n)}$ in the sample.

(b) Now form the normalized sequence $Z_n = c_n X_{(n:n)} + d_n$, where $\{c_n\}$ is the sequence $\{1/n\}$ and $\{d_n\}$ is the constant sequence $\{0\}$, and write the CDF of Z_n.

(c) Now take the limit as $n \to \infty$ of the CDF of Z_n.

This is the CDF of an extreme value distribution, and is in the form of the generalized extreme value distribution CDF given in equation (3.110) on page 309.

3.12. **Lognormal distribution.**

(a) Show formally that the median of a lognormal distribution with parameters μ and σ is e^{μ}.

Now generate 1,000 numbers that simulate a lognormal distribution with parameters 3 and 2. (Note that these parameters, while they are often denoted as μ and σ, are not the mean and standard deviation.)

(b) Compute the mean, variance, and median of your sample. What are the mean, variance, and median of the lognormal distribution with those parameters?

(c) Now, make a log transformation on the 1,000 observations to make a new dataset, $X_1, \ldots, X_{1000}$. Make a q-q plot with a normal distribution as the reference distribution.

(d) What are the theoretical mean and variance of the transformed data in Exercise 3.12c?

(e) Now, generate 1,000 numbers from a normal distribution with the same mean and variance as the sample mean and sample variance of the log-transformed data and make a graph of the histograms of the normal sample and the log-transformed data superimposed. Also make a q-q plot of the two samples.

3.13. **Location-scale t distribution.**

(a) Let X be a random variable with a location-scale t distribution with 3 degrees of freedom, location 100, and scale 10.

i. What is the probability that X is less than or equal to 110?
ii. What is the 0.25 quantile of X?

(b) Generate a sample of 1,000 random numbers following the (3,100,10) location-scale t distribution as above.

Make two q-q plots of the sample side by side, one with a t with 3 degrees of freedom as the reference distribution, and one with a standard normal as the reference distribution.

Comment on the plots.

(c) Generate a sample of 1,000 random numbers following the (6,100,10) location-scale t distribution.

Make two q-q plots of the sample side by side, one with a t with 6 degrees of freedom as the reference distribution, and one with a standard normal as the reference distribution.

Comment on the plots.

3.14. Linear multivariate transformations.

(a) Write a function to generate multivariate normal random numbers of a specified dimension and with a specified covariance matrix. The first statement for an R function should be

```
rmulvnorm <- function(n, d=1, mu=0, Sigma=diag(rep(1,d))) {
```

The R function `chol` computes the transformation matrix from `Sigma`. (The function `rmulvnorm` has the same functionality as the function `rmvnorm` in the `mvtnorm` package.)

(b) Use your function from the previous step to generate a sample of 1,000 trivariate normal random variables with mean of 0 and variance-covariance matrix

$$\Sigma = \begin{bmatrix} 9 & -3 & 2 \\ -3 & 4 & -1 \\ 2 & -1 & 1 \end{bmatrix}.$$

Now, produce three bivariate contour plots of each pair of variates. This is similar to the two plots in Figure 2.8.

3.15. Multivariate transformations; copulas.

Suppose X has a bivariate normal distribution, with mean 0, $V(X_1) = \sigma_1^2$, and $V(X_2) = \sigma_2^2$. Let F_1 and F_2 be the CDFs of two continuous univariate distributions.

Now, let

$$Y = (Y_1, Y_2) = \left(F_1^{-1}(\Phi(X_1/\sigma_1)), F_2^{-1}(\Phi(X_2/\sigma_2)) \right),$$

where Φ is the CDF of a univariate $N(0,1)$ distribution.

Show that Y_1 has a marginal distribution with CDF F_1.

3.16. "Random numbers".

Let $m = 2^{31} - 1$, $j = 5$, and $k = 17$. Let

$$\{x_1, x_2, \ldots, x_{17}\} = \{m/17 - 1, 2m/17 - 1, \ldots, m - 1\}.$$

Now for $i = 18, 19, \ldots, 1017$. Compute a "random sample" of size 1,000 in this manner.

```
x[i] <- mod(x[i-j]+x[i-k], m)
u[i] <- x[i]/m
```

Produce a histogram of your sample, and compute some summary statistics. Does the sample appear to have come from a $U(0,1)$ distribution?

3.17. Random number generation.

(a) Generate 1,000 random numbers from the triangular distribution. (The PDF is given in equation (3.28).) Produce a histogram and compute some summary statistics.

(b) Let $Y = 8X^3$, where X has the triangular distribution. (The PDF of Y is given in equation (3.29).)

Generate 1,000 random numbers corresponding to the random variable Y. (There are different ways of doing this. Which way do you think is better, and why?) Produce a histogram and compute some summary statistics.

3.18. Quasirandom number generation.

Use a Sobol' sequence to generate 1,000 quasirandom numbers from the triangular distribution, as in Exercise 3.17a. (A Sobol' generator is in the R package `randtoolbox`.)

Produce a histogram and compute some summary statistics. Can you tell any differences in the pseudorandom numbers of the other exercise?

3.19. Monte Carlo simulation of diffusion processes.

Suppose the price of stock XYZ is $100. Suppose that the daily log returns follow a location-scale t distribution with mean 0, scale 0.002, and degrees of freedom 6. Assume that the daily log returns are sequentially independent. (A large assumption!)

In this exercise you may wish to use functions developed in Exercise 3.13.

(a) Generate a random sequence of 253 daily log returns of XYZ. (This is approximately a year of trading days.)

Plot a histogram of the returns.

What is the standard deviation of the distribution of the daily log returns?

What is the observed standard deviation of your sample?

What is the statistical volatility of XYZ?

(b) Suppose the beginning price of stock XYZ is $100.

Using the random log returns generated in Exercise 3.19a, form a random sequence of 253 daily prices of XYZ, and plot them as a time series. (This is a diffusion process over approximately one year of trading days.)

(c) Now randomly permute the random log returns generated in Exercise 3.19a (use the R function `sample`), and again form a random sequence of 253 daily prices of XYZ starting at $100. Add a plot of these prices to the plot produced in Exercise 3.19b, using a different color or line type.

Now, again randomly permute the log returns generated in Exercise 3.19a, and plot the sequence of prices on the same plot with a different color or line type.

Do this two more times (for a total of 5 plots) on the same set of axes.

What is remarkable about the plots?

3.20. **Monte Carlo evaluation of European option values.**

Consider the daily prices of XYZ, as in Exercise 3.19, and assume the same location-scale t distribution for the returns.

Generate 1,000 random sequences of 253 daily prices of XYZ, each starting at $100.

Before answering any of the following questions, read all questions so as to accumulate the appropriate quantities to answer these questions.

(a) Plot the first 5 sequences as time series on the same graph.

(b) Compute the mean and sample standard deviation of the prices at the end of the period.

(c) Suppose you owned a call option for XYZ at 103 for 251 days out on the beginning date in this scenario.

What percentage of the time would your option have been in the money on the date of expiration?

(d) What is the fair value of a European call option for XYZ at 103 for 251 days out on the beginning date in this scenario if the risk-free interest rate is 1%?

Compare this with the results of the binomial tree model used in Exercise 3.6d. How do you account for any differences?

(e) Consider again the question in Exercise 3.20d except now assume that the t distribution has 3 degrees of freedom and a scale of 0.010. That is, suppose that the price of stock XYZ is $100, and that the daily log returns follow a location-scale t distribution with mean 0, scale 0.010, and degrees of freedom 3, and assume that the daily log returns are sequentially independent.

Using the Monte Carlo procedure as before, what is the fair value of a call option for XYZ at 103 for 251 days out on the beginning date in this scenario if the risk-free interest rate is 1%?

3.21. **Monte Carlo simulation of diffusion processes with Poisson jumps.**

Suppose that the daily log returns of an asset follow a location-scale t distribution with mean 0, scale 0.002, and degrees of freedom 6, and assume that the daily log returns are sequentially independent. Suppose further that the prices of the asset are subject to random shocks that follow a Poisson process. Assume the perturbations occur on average 4 times per year.

(a) What value of the Poisson parameter will yield a mean of 4 events per year (253 trading days)?

(b) Assume that each shock results in a 4% positive jump in the price of the stock with probability 0.4, and a 7% negative jump in the price of the stock with probability 0.6.

Generate and plot (on the same graph) 5 random sequences of 253 daily prices of the asset, each starting at $100.

3.22. **Monte Carlo simulation of two stock prices with a copula.**

We form a portfolio consisting of two stocks, XYZ, as in Exercise 3.19, and ABC whose daily log returns of the stock ABC have a location-scale t distribution with mean 0, scale 0.005, and degrees of freedom 3.

Assume that the returns are sequentially independent, and suppose further that the returns of XYZ and ABC tend to be positively related.

We use a copula to model the bivariate distribution of the returns. To model the relationship of the returns of XYZ and ABC, we begin with random variables X_1 and X_2 that have a bivariate normal distribution with mean 0, variance 1, and correlation 0.75, and let the 2-vector of returns be

$$R = \left(F_1^{-1}(\Phi(X_1)), F_2^{-1}(\Phi(X_2)) \right),$$

as in equation (3.114), where Φ is the CDF of a univariate $N(0,1)$ distribution, and F_1 and F_2 are the CDFs of the location-scale t distributions that model the individual returns of XYZ and ABC.

Consider a portfolio consisting of portions p_1 of XYZ and p_2 of ABC, where $p_1 + p_2 = 1$.

(a) Let $p_1 = 0.5$ and generate 1,000 random sequences of 253 daily prices of the portfolio where the starting with the prices of both XYZ and ABC are $100.

Compute the sample standard deviation of the prices of the portfolio at the end of that period.

(b) Now let $p_1 = 0.9$ and generate 1,000 random sequences of 253 daily prices of the portfolio, again starting with the price of each at $100.

Compute the sample standard deviation of the prices of the portfolio at the end of that period.

Comment on the results.

4

Statistical Models and Methods of Inference

A statistical model expresses a probability distribution for an observable variable and/or the model describes relationships among variables, both observable and latent. The model may include a special class of unobservable variables called parameters. As in the models of families of probability distributions discussed in Chapter 3, parameters allow the model specification to represent a *family* of models, each characterized by specific values of the parameters. In a general sense, statistical inference involves choosing a family of models and then, using observed data, choosing specific values or ranges of values of the parameters, that is, refining the model.

Data analysis generally involves *fitting* a model, that is, determining specific values of the parameters so that the observed data seem to correspond to the model.

Statistical *inference* involves not only fitting models, that is *estimating* parameters, but also providing statements of *confidence* regarding the fits. These statements may be in terms of "significance", or the accompanying statements that comprise statistical inference may address the distribution of estimators, particularly their variance.

Parameters either may not fully specify the model or they may be of infinite dimension. In either case, we may say the model is "nonparametric".

The model is at best an approximate description of the data-generating process. The form and level of detail of the model may depend on the specific application. Models in exploratory data analysis may be very different from models used to decide the amount of riskless reserves a bank must maintain, for example. A model that is adequate in one case may be inadequate in another.

In this chapter, we will consider the forms and basic components of statistical models, and then discuss and illustrate

- how to fit models using observed data;

- how to make statistical inferences about data-generating processes using data and models;

- how to compare different models;

- use of R software in fitting and analysis of statistical models.

4.1   Models

A model can be just a description of the frequency distribution of an observable random variable, or it can be an equation representing the relationships among many observable and unobservables variables.

A common reason that statistical models used in financial applications are inadequate is

- one or more aspects of the model change over time.

In the following, we describe some general types of statistical models.

Probability Models

The probability models discussed in Chapter 3 are useful for understanding the frequency distribution of a single random variable or a set of random variables. For example, we might express a model for the distribution of the random variable X as

$$X \sim N(\mu, \sigma^2), \tag{4.1}$$

or a model of a random d-vector X as

$$X \sim N_d(\mu, \Sigma), \tag{4.2}$$

where μ is a d-vector and Σ is a $d \times d$ matrix.

We fit these models by using observed data to estimate the parameters, and knowing that a data-generating process follows this model, we know generally what kinds of observations to expect.

The following are two common reasons that probability models used in financial applications are inadequate

- the probability distribution does not accurately reflect the tail behavior (normal distributions are assumed, but the tails are heavier);

- the model does not adequately account for correlations among all variables in the model (the model treats variables as independent, but they are not).

Models of Relationships and Dependencies

Often a more interesting problem is to fit a model that describes how different variables relate to each other by an equation involving the variables. The equation often is asymmetric, in the sense that there is just one quantity on the left side of the equation, the quantity of interest; and the right side of the equation consists of a *systematic component* that expresses the way in which the other variables may affect the quantity of interest and an *additive random component*, as, for example, in the model

$$y = g(x, \theta) + \epsilon, \tag{4.3}$$

where y and x are observable variables, g is a function, usually of some known fixed form, θ is an unknown and unobservable *parameter*, and ϵ is an unobservable adjustment often called the *error*, and treated as a random variable.

We assume some probability model for the distribution of the random error.

We often assume that the random error ϵ has a symmetric distribution and that $E(\epsilon) = 0$ so that the systematic component $g(x, \theta)$ is the conditional mean of y.

The variable of interest, y in this case, is called the "response" or the "dependent variable"; and the explanatory variable, x in this case, is called the "regressor", or the "independent variable". Often x is a vector, so we speak of the regressors or independent variable. ("Independent" in this context does not imply distributional independence, as the term is used on page 259.) Variables that are related to or dependent on other variables in the model are called "endogenous" variables. Some independent variables may be related to other variables or the error term in the model. Such variables are often called "instrumental" variables. Variables that are independent of other model terms are often called "exogenous".

Regression Models

Equation (4.3) is the general form of a *regression equation*, and it can have many different specific forms. A common form is a linear regression model,

$$y_i = \beta_0 + \beta_1 x_{i1} + \cdots + \beta_m x_{im} + \epsilon_i,$$

which is the subject of Sections 4.1.3 and 4.5.2. Market models of various forms, such as equation (1.35) on page 61, are specific instances of equation (4.3).

A common problem with regression models in financial applications, in addition to the probability model being wrong as note above, is that there are variables not included in the model that affect those variables that are in the model.

The regression model is one of the most common statistical models. We will encounter several examples in this chapter. The function g may be linear, but it can involve nonlinear terms or even derivatives or other transforms of the independent variables.

Classification Models

If the dependent variable y in equation (4.3) takes on only a finite set of values, the model is a *classification model*. Classification models are a special class of regression models.

Classification models are used in statistical machine learning to assign a category or class to an observational unit, given values of "predictors" or "features" (that is, values of x). In financial applications, classification models may be used to assign asset classes to "investment grade" or "junk"; they may be used to assign individuals to "high risk" or "low risk"; or they may be used to assign a stock as "up" or "down" at a specified short future interval.

In classification models, the function g in equation (4.3) is often nonlinear. Often, in fact, the function is very complicated, such as a sequence of rules in a decision tree, or even a "black box" in which there are many layers of variables that individually have no meaning.

One of its simpler forms of a classification model is a "generalized linear model" of the form

$$\Pr(Y = 1) = \frac{e^{\beta_0 + \beta_1 x_{i1} + \cdots + \beta_m x_{im}}}{1 + e^{\beta_0 + \beta_1 x_{i1} + \cdots + \beta_m x_{im}}}.$$

This particular generalized linear model is a *logistic model*. We will discuss this and other generalized linear models in Section 4.5.7.

Generalized linear regression models have a different form from the usual regression models with additive error terms. The models do not have explicit error terms, as the ϵ in other regression models.

Autoregressive Models

In some cases, especially in financial applications, the relationships of interest are those between the current value of a variable and its value at previous times. Equation (1.9), $x_t = f(\ldots, x_{t-2}, x_{t-1}) + \epsilon_t$, on page 6 is an example of such a model that is a specific case of equation (4.3). The present value of x is a function of past values, plus a random error component.

If the dependent variable is effectively the same as the independent variables but just represents the same variable later in time, the model is an *autoregressive* model (although we usually use that term only if the model is linear). In other forms, in which the g function in equation (4.3) includes derivatives of functions of the variable of interest, the model may represent various forms of what is called "geometric Brownian motion". These kinds of models are often used in financial modeling, especially of prices of derivative assets.

Autoregressive models and other time series models used in financial applications may be inadequate for all of the reasons noted above for other models. We defer discussion of time series models to Chapter 5.

Multivariate Models

In the model (4.3), $y_i = g(x_i, \theta) + \epsilon_i$, x_i is often a vector; that is, there are more than one regressor. (Recall that I do not use different fonts to distinguish vectors and scalars.) This is called "multiple regression".

Even in the case of multiple regression, if y_i is a scalar, then ϵ_i is also a scalar, and the analysis involves only a single random variable.

On the other hand, if the dependent variable y_i is a vector, then ϵ_i is also a vector. In this case, the model is a *multivariate model*, and we must consider the multivariate distribution of ϵ_i. This has major implications for the statistical analysis.

When y_i is a vector, the model is called a "multivariate regression" model. Such models are used often in financial and econometric applications. Some specific instances of these models are called in the literature "seemingly unrelated regressions", "simultaneous equations models", and "vector autoregressive" models.

Nonstationary Models

One of the biggest problems in application of statistical models is that the data-generating process changes over time. Often this requires use of different models in difference time regimes. Sometimes, however, the changes in some aspect of the model can themselves be modeled. If the basic functional relationships among the variables are persistent over time, changes in the error distribution can possibly be modeled separately.

A generalization of the model (4.3) is

$$y = g(x, \theta) + \delta h(x, \phi), \qquad (4.4)$$

with the g, x, and θ in the systematic component the same as in equation (4.1), but with a random error term that is also composed of a systematic component and an additional purely random component. The purely random component, δ, is a *multiplicative error* in the additive general error term.

There are many possible variations of the general statistical model (4.4). If $h(x, \phi) \equiv 1$, the error is purely additive, as ϵ before; if $g(x, \theta) \equiv 0$, the error is purely multiplicative. A model with a purely multiplicative error may be more appropriate if the relationships between y and the covariates x are highly nonlinear. In a purely multiplicative error model, we may assume that the error δ is such that $\log(\delta)$ has the distributional properties usually assumed for ϵ in the model with the simple additive error, equation (4.3).

In financial applications, the model (4.4) can be useful to account for stochastic volatility. In Section 5.4, we will discuss some specific instances of this model, called GARCH models.

Other Variations

There are many variations of the general models mentioned above. Models of various types can be combined. Individual terms in the models can be interpreted in various ways, for example, parameters in the model may be interpreted as random variables (a "Bayesian" approach). The individual explanatory variables or regressors may be related to each other in such a way that a hierarchical model may be appropriate. Some explanatory variables may not be observable, but may nevertheless be included in the model indirectly as "latent" variables.

4.1.1 Fitting Statistical Models

The data consist of pairs of observations on y and x, $(y_1, x_1), \ldots, (y_n, x_n)$, so the model equation (4.3) becomes

$$y_i = g(x_i, \theta) + \epsilon_i, \quad \text{for } i = 1, \ldots, n. \tag{4.5}$$

Statistical analysis uses the data to "fit" the model, that is, to estimate the parameters in the model, and then to assess the adequacy of the model for understanding the data-generating process and for predicting future data from the process. A common notation is to denote an estimate of a parameter θ as $\widehat{\theta}$, so the fitted model is $g(x_i, \widehat{\theta})$, without the error term.

We denote the value of y that corresponds to the fitted model as $\hat{y}$:

$$\hat{y}_i = g\left(x_i, \widehat{\theta}\right).$$

This value is called the "predicted value", the "estimated value", or the "fitted value".

Some other value of x, say x_0, not necessarily included in the dataset, can also be used in the fitted model,

$$\hat{y}(x_0) = \left(x_0, \widehat{\theta}\right). \tag{4.6}$$

We call $\hat{y}(x_0)$ the predicted, estimated, or fitted value of y at x_0.

Statistical inference goes beyond just the fitting of a model to assess the distributional properties of the estimators of the parameters, and then to make statements of confidence regarding the true values of the parameters or of predicted responses.

In the following sections we will discuss methods of fitting various types of statistical models and the statistical analyses that follow the fitting.

4.1.2 Measuring and Partitioning Observed Variation

A basic objective in statistical analysis is to model and to understand *variation*, that is, how and why a variable takes on different values at different times or in different situations.

In the model $y = g(x, \theta) + \epsilon$, we are interested in the variation in y. The form of the function g and the values of the parameter θ provide a systematic explanation of how y varies with x, and a probability distribution for ϵ describes random variation over and above the systematic variation.

We first develop a measure of variation. Then, in a given set of data, we partition the variation into a component that is explained by a systematic model and a component that is attributable to a separate probability distribution.

There are several ways of measuring variation (see Sections 1.3.1 and 1.3.8). One of the simplest and most useful measures is the sum of the squares of the deviations of the individual values from the mean of those values. For a given set of pairs of observations on y and x, $(y_1, x_1), \ldots, (y_n, x_n)$, the *total variation* in y is

$$\sum_{i=1}^{n} (y_i - \bar{y})^2.$$

This of course is the same as the sample variance except for the factor of $1/(n-1)$, and we could form a similar measure of the variation in x.

Using this same measure for the fitted ys in the model $\hat{y}_i = g(x_i, \hat{\theta})$, we have

$$\sum_{i=1}^{n} \left(\hat{y}_i - \bar{\hat{y}}\right)^2.$$

The fitted ys vary according to the model, so we could call this the *variation due to the model*.

Another set of values of interest are the *residuals*,

$$r_i = y_i - g\left(x_i, \hat{\theta}\right)$$
$$= y_i - \hat{y}_i. \tag{4.7}$$

The *residual variation* or *error variation* is, similar to the other expressions above, $\sum_{i=1}^{n}(r_i - \bar{r})^2$. In most common methods of fitting a model, the residuals sum to 0, so the residual variation is

$$\sum_{i=1}^{n} (y_i - \hat{y}_i)^2.$$

These three sources of variation, total, model, and residual, are fundamental in all statistical modeling and in other areas of statistical analysis. The relative amounts of variation due to these three sources indicate how well the fitted model does in "explaining" the variation in the observed data. The ratio of the model variation to the total variation, for example, may be taken as a measure of *goodness-of-fit* of the model. It is called R-squared.

There are other ways of measuring the variation due to these sources. The measures we have defined above are called *sums of squares*. The common method of fitting models called "least squares" is based directly on these measures.

We will revisit this concept of partitioning the variation in the response variable beginning on page 415.

Data Reduction, Dimension Reduction, and Parsimonious Models

Much of statistical analysis can be characterized as *data reduction*. In general, data reduction is the development of a model that is simpler than the given dataset, yet that captures the relevant characteristics of the full dataset. Note

the subjectivity in the previous statement: "simpler" and "relevant characteristics" may mean different things to different people or in different contexts.
We will not attempt to be more precise here, but continue to use these terms
with nontechnical, but intuitive meanings.

Given a univariate dataset $y_1, \ldots, y_n$, a simple model could be, for example,
that the data are a random sample of size n from a $N(\mu, \sigma^2)$ distribution. If the
population follows that model, then having estimates for μ and σ^2 provides
a good summary of all of the information about the population or the data-
generating process that is contained in the full dataset.

A dataset may also have many variables. We refer to the number of variables as the "dimension" of the data. Often in data analysis we seek to reduce
the dimension by forming some kind of linear combination of the variables
that summarizes the overall behavior of the individual variables. Stock indexes, such as the Dow Jones Industrial Average or the S&P 500 Index, serve
as summaries that reduce the dimension. A classic statistical method of dimension reduction is to form *principal components*, which we will discuss on
page 409.

A model that is a simple summary of a dataset cannot contain all of
the information in the data; hence, in modeling, there must be some lost
information. In both data reduction and dimension reduction, an objective
is to minimize the amount of information lost while capturing the salient
characteristics.

In general, more complicated models are able to explain more of the variation in a given set of data than simpler models can; thus, there is a general
tradeoff between the goodness-of-fit and the simplicity of the model. For two
given models, one may be "simpler" than the other because the number of regressors is fewer (lower dimension) and/or because the form of the function in
the regression equation, $g(\cdot, \cdot)$, is simpler. The simpler model reduces the data
and/or the dimension more. We say the simpler model is more *parsimonious*.

Comparing Models

We seek a model that fits the data well, yet one that is parsimonious. An
overall measure to compare two models contains a term for the goodness-
of-fit and a term or factor that reduces the measure by the complexity of
the model. (Of course, this overall measure could be defined in terms of the
negative of these two components.) In Section 4.6.3 beginning on page 467,
we will discuss some specific measures for comparing models.

The important principle is that we seek models that fit better *and* models
that are simpler. We seek a good balance in the tradeoff between these two
qualities.

4.1.3 Linear Models

A specific form of model (4.3) is a linear regression model,

$$y_i = \beta_0 + \beta_1 x_{i1} + \cdots + \beta_m x_{im} + \epsilon_i, \tag{4.8}$$

or in simpler notation,

$$y_i = \beta_0 + \beta^{\mathrm{T}} x_i + \epsilon_i. \tag{4.9}$$

We assume that the ϵs are uncorrelated random variables that have the same distribution.

We fit the model by estimating the βs. Denoting the estimates as $\widehat{\beta}_0, \widehat{\beta}_1, \ldots \widehat{\beta}_m$, we have

$$\hat{y}_i = \widehat{\beta}_0 + \widehat{\beta}_1 x_{i1} + \cdots + \widehat{\beta}_m x_{im}. \tag{4.10}$$

Matrix/Vector Notation

For n observations on y_i and on the corresponding $x_{i1}, \ldots, x_{im}$, we can organize the observations into an n-vector y and an $n \times m + 1$ matrix X whose first column consists of all 1s, and write the linear regression model as

$$y = X\beta + \epsilon, \tag{4.11}$$

where β is an $(m+1)$-vector and ϵ is an n-vector. (In a vector-matrix equation like this one, there is often some ambiguity about whether the matrix X includes a column of 1s. If the model is fit by least squares, that column is often omitted, and $\widehat{\beta}_0$ is obtained by use of the fact that the least squares solution goes through the means of the data. If the model is fit by some other method, the fitted model may not go through the means of the data, and generally β_0 must be estimated directly.) In the expression (4.11), the "X" includes a column of 1s, and hence, the matrix formulation (4.11) corresponds directly to the individual equation representation (4.8). In this representation, however "X" does not correspond directly to the "x" variables.

If "X" corresponds directly to the "x" variables, as in the formulation (4.9), we could write the model in matrix notation as

$$y = \beta_0 + X\beta + \epsilon,$$

where β_0 here is a vector of constants.

If the linear model is fit by least squares, there are some simple relationships that allow the model to be written in various forms.

We are not introducing these ambiguities; they exist in the statistical literature. By pointing out the ambiguities, however, we hope to prevent confusion that could result from them. In further discussions, we will generally state precisely whether or not "X" includes a column of 1s or is just the observations on the variables themselves.

The centering transformation $X_c = X - \overline{X}$ (see page 210) is often useful. In least squares fits, this transformation has the effect of removing the intercept term if the vector y is likewise centered.

Multivariate Linear Models

For d different dependent variables, with the same regressors, we may form d separate regression models,

$$
\begin{aligned}
y_{i1} &= \beta_{01} + \beta_{11}x_{i1} + \cdots + \beta_{m1}x_{im} + \epsilon_{i1} \\
\vdots\ \ \vdots\ & \qquad\qquad\qquad \vdots \\
y_{id} &= \beta_{0d} + \beta_{1d}x_{i1} + \cdots + \beta_{md}x_{im} + \epsilon_{id}.
\end{aligned}
\tag{4.12}
$$

The subscript notation may seem at first rather complicated, but for each value of the second subscript on the ys, the βs, and the ϵs, the regression equation is essentially the same as the linear regression model (4.8), and the separate analyses could proceed as for the multiple linear regression model.

For a given value of j, the ϵ_{ij}s are uncorrelated as before.

For a given value of i, however, the d ϵ_{ij}s *are likely to be correlated*. This means that separate analyses are likely to be misleading. This is a common problem in financial analyses, and the incorrect conclusions resulting from separate analyses have led to financial crises.

The model (4.12) is a *multivariate* multiple linear regression model. We can write it in a compact form similar to (4.11) as

$$
Y = XB + E,
\tag{4.13}
$$

where Y is a matrix whose columns each correspond to y in equation (4.11), X is the same matrix as in (4.11), B is a matrix whose columns each correspond to β in (4.11), and E is a matrix whose columns each correspond to ϵ in (4.11).

The important characteristic of the multivariate regression model (4.13) is that elements of the rows of E, which we can think of as d-vectors, are correlated. We may assume for example that, for a given i, the d-vector $(\epsilon_{i1}, \ldots, \epsilon_{id})$ has a multivariate normal distribution $N_d(0, \Sigma)$. This regression model is sometimes called *seemingly unrelated regressions* in the financial literature.

4.1.4   Nonlinear Variance-Stabilizing Transformations

It is often the case that as a variable increases in magnitude, the amounts to which it is likely to vary (that is, its variance) increases. The variance of the price of the stock of Apple when the price was in the range of \$30, for example, was less than when the price was in the range of \$160. (The percentage variation may be about the same.) When dealing with a set of data that varies over a wide range, it may be useful to apply a transformation so that the variances within subsets in various ranges are approximately equal. Such transformations are called *variance-stabilizing transformations*.

A common objective in transforming data is to make the frequency distribution of the data appear more like a normal distribution. Given a sample of positive numbers, a simple class of transformations that can yield a sample that is more like a normal sample are the power transformations with a given parameter λ,

$$
\tilde{x} = \begin{cases} \dfrac{x^\lambda - 1}{\lambda}, & \lambda \neq 0 \\[2mm] \log(x), & \lambda = 0. \end{cases}
\tag{4.14}
$$

These are called *Box-Cox transformations*. In regression analysis, the residuals often indicate the need for a variance-stabilizing transformation, and the Box-Cox transformations are often used for this purpose. The appropriate value of λ can be determined using a statistical criterion (maximum likelihood). We have seen an example of the special transformation for $\lambda = 0$, that is $\tilde{x} = \log(x)$, in Figure 1.13 on page 55. Constant geometric growth yields linear growth in the log, and the log transformation is often made for this reason rather than to change the statistical properties. We will not go into further details of the Box-Cox transformations here, but rather refer the interested reader to the statistical literature, for example, Kutner et al. (2004).

Transformations to change statistical properties of data, such as skewness, variance, and so on, are rather subjective, and it is generally not clear how to choose an appropriate transformation to improve the statistical properties. For a particular underlying probability distribution, however, it may be possible to determine analytically a useful transformation. For example, for a Poisson distribution, a square-root transformation is often used as a variance-stabilizing transformation (see Exercise 4.1). There are not many simple guidelines, however.

It is important to understand the effects of any transformations made. We discussed the effects of various transformations on the distribution of random variables in Section 3.1. For example, if $\tilde{X} = X^2$, then $E(\tilde{X}) = (E(X))^2 + V(X)$. If T is an unbiased estimator of θ, then T^2 is not an unbiased estimator of θ^2.

4.1.5 Parametric and Nonparametric Models

A statistical model often is actually a family of models. For example, in the model of the family of normal probability distributions for a single variable, denoted as $N(\mu, \sigma^2)$, μ and σ^2 are parameters which specify the specific normal distribution. Likewise, for a given g in the model (4.3), $g(x, \theta)$ specifies a *family of models*, each specific one depending on the value of the parameter θ.

Given observations $x_1, x_2, \ldots$ at times $t_1, t_2, \ldots$, we can form a simple family of models of trend as

$$
x_i = \alpha + \beta t_i + \epsilon_i.
\tag{4.15}
$$

The α and β in this equation are parameters. We refer to models of these types as parametric models.

In a standard formulation of a model, such as a single sample, $X \sim$

$N(\mu, \sigma^2)$, or the regression model (4.15), we distinguish three types of quantities: observable variables (usually realizations of random variables), unobservable or latent random variables (the ϵ_i), and unknown, unobservable parameters. The model is fit by assigning specific values to the parameters.

We refer to the process of fitting a model as *smoothing*, and we refer to the model as a smoothing model. The moving averages discussed on page 39, and the histograms and kernel densities discussed in Section 2.3 were smoothing models. A smoothing model, such as a histogram or a kernel density, often has a parameter that controls the extent of the smoothing. We call the parameter a smoothing parameter.

Instead of using a model such as equation (4.15), there are many other ways we could smooth the observed data. The moving averages and other trend curves we discussed and illustrated beginning on page 39 are examples. These are generally considered "nonparametric", but we will use "parametric" and "nonparametric" somewhat loosely. (Although a moving average requires specification of a window width, and other smoothing methods have similar parameters, these are different in fundamental ways from the α and β in model (4.15).)

4.1.6   Bayesian Models

Generally, we consider the unknown parameters to have fixed values. An objective of statistical inference is to estimate those parameters, or to test hypotheses concerning them.

In a different formulation of the model, we assume that the parameters are random variables. Such a model is called a *Bayesian model*. In that case, an objective of statistical inference is to refine any previous beliefs about the distribution of the random parameters.

We will discuss Bayesian data analysis further in Section 4.4.4 beginning on page 386, along with an example using R.

4.1.7   Models for Time Series

As we emphasized in the examples of Chapter 1, financial data often come in the form of a time series. In the general discussion of statistical models in this chapter, we mention the effect of time in models, but we will generally defer detailed discussion and analysis of time series to Chapter 5.

As we stated in Section 2.4.1, in the exploratory analysis of any data, it is appropriate to make some time series plots, and to do a general assessment of the presence of autocorrelations in the data.

Sometimes it is appropriate to go further in determining whether there is a meaningful, but hidden, time relationship. In regression analysis with financial data, for example, it is often advisable to perform a Durbin-Watson test on the residuals. (See page 434.)

4.2   Criteria and Methods for Statistical Modeling

We use statistical models to study and make inferences about a data-generating process. A statistical model may just describe a probability distribution, such as the model $N(\mu, \sigma^2)$ for a simple univariate data-generating process, or it may express a stochastic relationship between variables, such as $y = g(x, \theta) + \epsilon$, for the variables y and x. These models often include parameters that determine the specific properties of the model. An important step in parametric data analysis is to fit a model using sampled data; that is, to *estimate the parameters* in the model.

4.2.1   Estimators and Their Properties

We might decide that a good estimate of the mean μ in a model is the sample mean $\bar{x}$, and a good estimate of the variance σ^2 is the sample variance s^2. These are *point estimates*. They are not the only possibilities. In Section 1.3.8 for example, we mentioned the sample median, the trimmed mean, and the Winsorized mean, all of which could also serve as point estimates of μ.

As mentioned above, in a Bayesian approach, the parameters are random variables, so a "point estimate" of a parameter does not make much sense. Rather, we seek to develop (or "estimate") a conditional distribution for the parameters given the data. In the following, we will continue to refer to point estimates, but the basic approaches apply also to developing a conditional distribution. (There are clearly many subtleties here, and the interested reader is invited to pursue some of the relevant references cited at the end of the chapter.)

In an application with a given sample of observed data, an estimate is just a fixed value or a fixed function. We develop theory to help us understand estimates and to choose good estimates. The theory depends on a probability distribution, and we study "estimators", which are random variables. An *estimate* is a realization of an *estimator*.

Treating an estimator as a random variable, if the expected value of an estimator $\widehat{\theta}$ is the value of the parameter being estimated, θ, we say the estimator is *unbiased*. The *bias* is

$$E\left(\widehat{\theta}\right) - \theta. \tag{4.16}$$

We also call an estimate unbiased if it is a realization of an unbiased estimator. (Notice the "-e" and the "-or" in the foregoing.)

Given assumptions about the distributions of the random variables, we can analytically determine the expected value and, hence, the bias, of an estimator. For a given sample, however, in general, we have no direct way of assessing the bias in an estimate.

A desirable property of an estimator is that the estimator have small variance, because in that case, it would not vary much from sample to sample. In many simple cases, we have good methods of estimating the variance of an estimator in a given sample.

The *mean squared error*, or *MSE*, of an estimator combines the variance and the bias in one measure. The mean squared error is

$$ \mathrm{E}\left(\left(\widehat{\theta} - \theta \right)^2 \right) = \mathrm{V}\left(\widehat{\theta} \right) + \left(\mathrm{E}\left(\widehat{\theta} \right) - \theta \right)^2. \tag{4.17} $$

The bias of an estimator may decrease as the sample size increases. This is the case with many useful estimators, so we often focus on the asymptotic properties of estimators. Because the sample size is important, we may include it in our notation for an estimator, for example, "$\widehat{\theta}_n$" for an estimator of θ based on a sample of size n.

We refer to some of these asymptotic properties as *consistency*. There are different kinds of consistency, which we will not discuss here. The most important type of consistency for our purposes is consistency in mean squared error. As an estimator of the scalar parameter θ, we say $\widehat{\theta}_n$ is *consistent in mean squared error* if

$$ \lim_{n \to \infty} \mathrm{E}\left((\widehat{\theta} - \theta)^2 \right) = 0. \tag{4.18} $$

(See Gentle, 2019, Section 3.8.1 for discussion of different kinds of consistency and their properties.)

The concepts of unbiasedness, mean squared error, and consistency also apply to other types of statistical inference, not just to point estimation.

Predictors and Their Properties

A special kind of estimator is one that "estimates" or *predicts* a random variable. Since a random variable is a function that may take on a range of values, we must clarify the meaning of estimation or prediction in this sense. We refer to a realization of the random variable.

If X is a random variable, we will denote a predictor of X by $\widehat{X}$. Just as we sometimes treat estimators as random variables when we discuss their properties and other times treat estimates as fixed values in applications, we also may treat a predictor $\widehat{X}$ as a random variable and discuss its properties, or we may consider a prediction to be a fixed value.

The same properties we have discussed for estimators are also relevant for predictors. If, for example, X is a random variable with finite expectation $\mathrm{E}(X)$, an unbiased predictor of X is an $\widehat{X}$ such that $\mathrm{E}(\widehat{X}) = \mathrm{E}(X)$.

Just as with estimators, an important criterion for a "good" predictor is that the mean squared error be small.

Predictors are usually constructed from probability models that involve covariates for the variable of interest. Hence, the various properties such as bias and MSE are usually conditional properties, given the covariates. In regression modeling, for example, we may predict the response Y based on given values of the regressors X, and so the property of interest is $E\left(\widehat{Y}\,\middle|\,X\right)$.

In analysis of a time series, $\ldots, x_{t-1}, x_t, x_{t+1}, \ldots$, at time t, when we have observed $\ldots, x_{t-2}, x_{t-1}, x_t$, we may wish to predict x_{t+1}. We denote the predictor $\hat{x}_t^{(1)}$. Our interest obviously is in $E\left(x_t^{(1)}\,\middle|\,x_t, x_{t-1}, x_{t-2}, \ldots\right)$. Observe the notation here, in which, for the positive integer h, $\hat{x}_t^{(h)}$ denotes the forecast value of x_{t+h} based on the information from $x_t, x_{t-1}, x_{t-2}, \ldots$.. (We use this notation, which is not standard, in Chapter 5, particularly in Section 5.3.8.)

For given X, we see that the best predictor, $\widehat{Y}$, of Y in terms of conditional MSE is $E(Y|X)$; that is,

$$\widehat{Y} = E(Y|X).$$

This predictor is also conditionally unbiased.

4.2.2 Methods of Statistical Modeling

One of the first steps in statistical modeling is to consider possible *families* of models. If we choose a parametric family, the next step is to fit the model by using data to estimate the parameter θ (which is in general, a vector, so there are "parameters" to be estimated). We will denote the estimator of θ by $\hat{\theta}$.

The problem of statistical estimation has no meaning until we state some criteria for the estimator. (Any arbitrary value could be an "estimate".) We have alluded to some criteria above, such as unbiasedness, small variance, and small mean squared error.

While the criteria may identify desirable properties, they do not indicate how to determine the estimators.

There are several ways of defining and calculating an estimator. Two of the most popular of these methods are least squares and maximum likelihood. We will first briefly mention some other general methods, and then describe the methods of least squares and maximum likelihood in more detail in the sections below.

Method of Moments

A simple criterion for an estimator is that the moments of the fitted distribution match the sample moments. Use of this criterion is called the *method of moments*.

For example, in the Poisson distribution with parameter λ, the mean of the Poisson population is λ. Hence, the sample mean $\bar{x}$ is a method-of-moments estimator for λ.

As another example, consider the gamma(α, β) distribution. Because the mean of the gamma is $\alpha\beta$ and the variance is $\alpha\beta^2$, with a sample $x_1, \ldots, x_n$, a method of moments would lead to two equations in two unknowns:

$$\alpha\beta = \bar{x}$$
$$\alpha\beta^2 = s^2.$$

(An alternative formulation may use n instead of $n-1$ in the sample variance.)

Matching Quantiles

A very general method for nonparametric univariate estimation is use of the ECDF. The ECDF is an efficient estimator of the CDF of the population. It matches the sample quantiles to the population quantiles.

The CDF completely determines all properties of a population, so having an estimate of it provides estimates of all of those properties, including any parameters characterizing the distribution. The ECDF defines a discrete, finite probability distribution whose mean, variance, or other properties can be easily computed. These quantities serve as estimates of the unknown corresponding population quantities.

Matching quantiles is a useful method only for univariate distributions. The method can also be used on marginal distributions in a multivariate setting, however.

Fitting Copulas

Approaches such as use of the ECDF, a histogram, or a kernel density estimator model the entire distribution; that is, they show not only the location and spread, but also the general shape of a frequency distribution. These estimators work well with univariate distributions.

Another nonparametric way of modeling the entire distribution is by use of a copula (see Section 3.1.7, page 264). Copula models are especially useful for multivariate distributions. Copulas provide an alternative to other measures of association such as covariances and correlations, and so are useful in simulation modeling.

A d-variate copula is based on the d marginal distributions, transformations of the marginal random variables into uniform random variables, and finally transformations of those uniform variates into a multivariate random variable. The copula model is a specification of the forms of these various components. Thus, to fit a copula model to data, the marginal distributions are fit. This can be done by maximum likelihood or by other estimation methods, such as matching univariate quantiles. The function to transform the marginal variates to uniform variates at which to evaluate the copula is then fit by some estimation method, such as maximum likelihood.

There are various R packages that perform computations for copulas. Two of the most widely-used are `copula` and `fCopula`. These packages have functions, such as `fitCopula` in `copula`, for fitting copulas of a specified type to data.

Minimizing the Residuals; Least Squares

Another approach to fitting a statistical model is to compare data points to their expected values in the fitted model.

Consider the general model (4.3) with data $(y_1, x_1), \ldots, (y_n, x_n)$,

$$y_i = g(x_i, \theta) + \epsilon_i. \tag{4.19}$$

If ϵ_i is a random variable such that $E(\epsilon_i) = 0$, then $E(y_i) = g(x_i, \theta)$.

We cannot observe $\epsilon_i = y_i - g(x_i, \theta)$, but for any given value of the unknown parameter θ, say t, we have the *residual*,

$$r_i = y_i - g(x_i, t). \tag{4.20}$$

Given the data (y_i, x_i) for $i = 1, \ldots, n$, we want to determine a value of t, say $\hat{\theta}$, that makes the r_i small. Of course, some r_i may be small and others large, so we want to make some overall measure of the sizes of the r_i small.

There are several possible measures. One is the sum of the absolute values,

$$\sum_{i=1}^{n} |y_i - g(x_i, t)|. \tag{4.21}$$

This measure is the L_1 norm of the vector of residuals

$$r = (y_1 - g(x_1, t), \ldots, y_n - g(x_n, t)).$$

We denote this norm as $\|r\|_1$. Fitting the model (4.19) by minimizing the L_1 norm is called *least absolute values* estimation or L_1 estimation.

Another measure is the sum of squares of the residuals,

$$\sum_{i=1}^{n} (y_i - g(x_i, t))^2. \tag{4.22}$$

This measure is the square of the L_2 norm of the vector of residuals. We denote the L_2 norm of the residuals as $\|r\|_2$, or, because it is so common, just as $\|r\|$.

The sum of squares is the most commonly used measure. Finding the value of $\hat{\theta}$ that minimizes (4.22) is called a *least squares problem*, and the solution in the statistical modeling problem is called the *least squares estimator*. The least squares estimator for θ in the model (4.19) is

$$\hat{\theta} = \arg\min_{t} \sum_{i=1}^{n} (y_i - g(x_i, t))^2. \tag{4.23}$$

Least squares estimators are often unbiased and have small variances relative to other estimators. There are several other reasons the method of least squares is used for estimation. One is the mathematical tractability of the optimization problem. Another is the ease with which statistical properties of least squares estimators can be worked out. An additional reason is that if the underlying distribution is normal, the least squares estimators have desirable statistical properties (over and above the ease of working with them).

The Likelihood Function and Maximum Likelihood

"Probability" refers to a random variable and the values it may take on. "*Like-lihood*" refers to a somewhat similar concept that pertains to the probability model for given values of the random variable.

We can illustrate this with the Poisson family of distributions. If the random variable X is distributed as a Poisson random variable with parameter λ, then the *probability* that X has the value x, where $x = 0, 1, 2, \ldots$, is

$$f_X(x; \lambda) = \lambda^x e^{-\lambda} / x!, \qquad (4.24)$$

for a given $\lambda > 0$. Alternatively, for a given value of x, this quantity is the *likelihood* that the parameter has a value of λ. It is not the *probability* that the parameter has a value of λ.

Suppose we have a random sample of n observations from the Poisson distribution. Because we assume the observations are independent, the joint probability function is just the product

$$f(x_1, \ldots, x_n; \lambda) = \lambda^{\sum_{i=1}^{n} x_i} e^{-n\lambda} \prod_{i=1}^{n} \frac{1}{x_i!}. \qquad (4.25)$$

Given observations $x_1, \ldots, x_n$, we can form a function of λ at those fixed values of the xs:

$$L(\lambda; x_1, \ldots, x_n) = \lambda^{\sum_{i=1}^{n} x_i} e^{-n\lambda} \prod_{i=1}^{n} \frac{1}{x_i!}. \qquad (4.26)$$

A function of the parameters formed in this way from the joint probability function or joint PDF is called a *likelihood function*.

The probability function (4.24) or (4.25) depends on a specific value of the parameter. It is a function of the realizations of the random variable.

The likelihood function (4.26) depends on a sample of observations. It is a function of values of the parameter.

We generally denote a likelihood function by an expression of the form $L(\theta; X)$, where θ is the parameter of the distribution, often a vector, and X is a given sample from the distribution.

The probability function for a sample of size 5 with $\lambda = 3$ is

$$f(x_1, x_2, x_3, x_4, x_5; \ 3) = 3^{x_1 + x_2 + x_3 + x_4 + x_5} e^{-15} \frac{1}{x_1! x_2! x_3! x_4! x_5!}. \qquad (4.27)$$

The likelihood function is

$$L(\lambda; \ 3, 1, 4, 3, 2) = \frac{1}{1728} \lambda^{13} e^{-5\lambda}. \qquad (4.28)$$

This obviously does not integrate to 1 over the range of λ, $(0, \infty)$ (see Exercise 3.3d), so it cannot be a probability or probability density function.

Figure 4.1 shows the Poisson probability function for $\lambda = 3$, and the Poisson likelihood function for a given sample of size 5: $\{3, 1, 4, 3, 2\}$. (The sample was randomly generated in R using `rpois(5,3)`.)

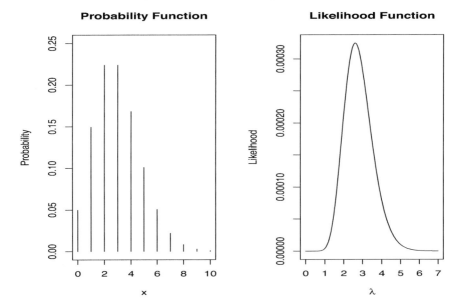

FIGURE 4.1
Poisson Probability Function for $\lambda = 3$, and Poisson Likelihood Function for
the Sample $\{3, 1, 4, 3, 2\}$

If there are more than one parameter, the likelihood function is defined
in the same way. Suppose, for example, $X \sim \mathrm{N}(\mu, \sigma^2)$. Given data $x_1, \ldots, x_n$,
the likelihood of μ and σ^2 for those data is

$$L(\mu, \sigma^2 \; ; \; x_1, \ldots, x_n) = \frac{1}{(\sqrt{2\pi\sigma^2})^n} \mathrm{e}^{-\sum_{i=1}^n (x_i - \mu)^2 / 2\sigma^2}. \qquad (4.29)$$

A likelihood function is not a probability function, and probabilities cannot
be computed from it. It is, however, formed from a probability function or
a probability density function. Intuitively, therefore, we may say that the
particular value of the parameters at the maximum is "most likely" to be the
"true" value.

These ideas lead to a formal method of statistical estimation called *maximum likelihood*.

Given a likelihood function $L(\theta; X)$ and a given sample X, the *maximum
likelihood estimator* or *MLE* for the parameter θ is

$$\widehat{\theta} = \arg\max_{\theta \in \overline{\Theta}} L(\theta; X), \qquad (4.30)$$

where $\overline{\Theta}$ is the closure of the parameter space of θ.

Notice that MLE depends on the assumption of a specific family of probability distributions. Least squares estimation, on the other hand, does not require that a family of probability distributions be assumed. For the linear regression model, such as $y = X\beta + \epsilon$ in equation (4.11), we can obtain the least-squares estimator of β without any reference to the distribution of ϵ. To obtain the MLE of β, however, we first need a likelihood function, which in turn, requires a probability function or a PDF.

For some distributions, such as the normal and the exponential, the MLEs are simple and well known. Thus fitting such distributions to data is straightforward. For other distributions, the fitting by MLE is not so simple.

The R package `MASS` has a function `fitdistr` for computing MLEs of the parameters in the PDF or probability function. The R function has a character argument that allows the user to specify one of many standard distributions. The function returns a list whose components depend on the distribution being fitted. One component of the list is `estimates`. For example, `estimates` in the list returned by `fitdistr(x, densfun="t")` are `m`, `s`, and `df`, corresponding to the parameters related to the "standardized" t distribution (see page 298).

Maximum likelihood methods can be used in nonparametric estimation also.

Because both least squares and MLE require us to find a minimum or a maximum, that is, to find an *optimum* of a function, in Section 4.3 below, we will interrupt the narrative briefly to consider the general problem of optimization.

Tuned Inference

Statistical inference is based on data, and it is performed in the context of a model of an underlying data-generating process. Fitting a statistical model occasionally requires a "tuning parameter". The nonparametric estimators of a probability density function discussed in Section 2.3 all were based on a tuning parameter, often called a smoothing parameter in those cases.

Inference about extreme events, such as is necessary in risk management, requires some quantification of "extreme". The specification of what is extreme is a tuning parameter. In a quantitative study of value at risk (VaR), we typically specify either an amount of loss or a probability of loss exceeding a given amount. These are tuning parameters. Study of outliers also depends on a specification of extreme.

The tail behavior of financial data is important in any analysis. There are various measures to describe the tail behavior, as discussed in Section 3.1.11. To make statistical inferences concerning these measures, such as the tail index, requires identifying the portion of a sample that corresponds to the tail. We may do this by specifying a quantile, say, declaring the tail to be the upper 5% of the distribution. The quantile, or how we specify it in the data, is a tuning parameter. Estimates or predictions based on that parameter is called *tuned inference*.

There are several types of tuned inference that we will encounter. The regularized least squares estimators, discussed beginning on page 369, are tuned estimators.

For tuned estimators, it is often not possible to work out the traditional exact statistical properties of estimators, such as bias and variance. We generally focus on asymptotic properties. Assessment of the statistical properties is often done by computational inference, such as the bootstrap, which we discuss in Section 4.4.5.

Simulation Methods

Simulation methods, also called Monte Carlo methods, involve the use of artificially-generated random data to study a data-generation process. The model of the data-generating process may be very simple and may be fully-specified, or it may be quite complicated, involving several interacting variables and the distributions of the individual variables may be quite different from one another. (The term "simulation" is also often used in economics to refer to the use of a model to study a financial process without the use of randomly-generated data. I will sometimes use the phrase "Monte Carlo simulation" to refer to simulation using artificially-generated random data.)

In applications of Monte Carlo simulation involving more than one variable, a suitable multivariate distribution must correspond not only to the univariate marginal frequencies, but it must also exhibit the same kinds of relationships among the variables. We can express these relationships through correlations or in the form of copulas, as we discussed in Sections 3.1.6 and 3.1.7. We discussed methods of modeling multivariate relationships on page 320. One of the main errors made in simulation models in practice is failure to account for the relationships among variables.

Monte Carlo Estimation of an Expected Value

In a simple application of Monte Carlo simulation, the objective is to estimate an expected value. We are interested in some function of a random variable X, which may be a multivariate random variable. Given a function g, we wish to estimate its expected value $E(g(X))$. If the PDF of X is f_X, and if g and f_X satisfy certain conditions, we have

$$E(g(X)) = \int_{\mathcal{D}} g(x) f_X(x) \, dx,$$

where $\mathcal{D}$ is the range of X. Now, suppose we can generate random variables to simulate X, in the manner discussed in Section 3.3. If we have a pseudorandom sample $x_1, \ldots, x_m$, then the *Monte Carlo estimate* of $E(X)$ is

$$\widehat{E(g(X))} = \frac{1}{m} \sum_{i=1}^{m} g(x_i). \tag{4.31}$$

This is a method-of-moments estimator; it is the sample mean of the estimand. This estimator is unbiased, as you are asked to show in Exercise 4.4a.

Another way of looking at the Monte Carlo estimator (4.31) is to consider the probability function of the elements x_i in the random sample, $x_1, \ldots, x_m$. Each element has the same distribution, which is a discrete uniform distribution with probability function $f_{x_i}(x_i) = 1/m$. Over this discrete distribution, the expected value of $g(x_i)$ is $\sum_{i=1}^{m} g(x_i) f_{x_i}(x_i)$, which is $\frac{1}{m} \sum_{i=1}^{m} g(x_i)$ as in equation (4.31).

The Monte Carlo estimator can easily be modified to estimate a conditional expectation, such as, for example, the expected shortfall. A conditional expectation is an expectation restricted to some subset of the range of X and perhaps another random variable Y, and so a very general expression for a conditional expectation of $g(X)$ is $E(g(X)|(X, Y) \in C)$ where C is the subdomain of $\mathcal{D}$ in which the conditions are satisfied. The Monte Carlo estimate is based on the pseudorandom sample $x_1, \ldots, x_m$, or on the pseudorandom multivariate sample $(x_1, y_1), \ldots, (x_m, y_m)$ if another random variable is involved. The estimate is as in equation (4.31), except that the summation is restricted to the values of x_i and y_i such that $(x_i, y_i) \in C$.

The estimator in equation (4.31) has a variance due to the variance in the distribution of the random numbers. In a practical application, this variance should be estimated, which can be done very simply as the Monte Carlo generation proceeds to accumulate the sum $\sum g(x_i)$. The variance is proportional to $1/m$, and the standard deviation, which is usually more easily interpretable, is proportional to $1/\sqrt{m}$.

Because of the variance in the Monte Carlo estimator, if the quantity $\int_{\mathcal{D}} g(x) f_X(x) \, dx$ can be computed analytically, it should be. Monte Carlo simulation should be resorted to only when a problem is mathematically intractable, or when there are other reasons to use Monte Carlo, such as to study distributions of associated quantities.

Monte Carlo Estimation of a Quantile

Monte Carlo simulation yields a sample or, perhaps, multiple samples. This may be a reason to use Monte Carlo even when some aspects of the problem are mathematically tractable.

The Monte Carlo sample, as a surrogate for the distribution of the model values, can be used to estimate other properties of the underlying distribution, such as quantiles. An estimate of the α-quantile of the distribution is the α sample quantile, which, as we have mentioned, can be defined in various ways (see page 215).

Value at risk, for example, corresponds to a quantile of the distribution of returns. VaR, as well as associated statistical properties of VaR, can be easily estimated by Monte Carlo. As mentioned above, unless some aspect of the computations are intractable, or unless there are associated properties of interest, Monte Carlo methods should not be used.

Monte Carlo simulation can also be used to estimate conditional quantiles using the same procedure outlined above.

Monte Carlo Estimation in a Stochastic Process

Monte Carlo simulation can be particularly useful in studying stochastic processes, such as a time series.

The model for a stochastic process is generally of the form

$$x_t = h(x_{t-1}, \ldots, x_{t-k}) + \epsilon_t.$$

If we can simulate random variables corresponding to the assumed distribution of ϵ, starting with fixed values of $x_{t-1}, \ldots, x_{t-k}$, it is an easy process to generate any number of observations in the process.

Different starting points yield different sequences in the simulation of stochastic processes. This of course suggests running multiple simulations with different starting points, and doing so is a standard way of dealing with the problem. The results of the multiple sequences may be averaged to obtain a more stable estimate. The multiple sequences also provide an assessment of the variance of the estimator.

Many models of stochastic processes do not assume a starting point. Some stochastic processes, such as some Markov chains, have a *limiting distribution* that will be reached, at least asymptotically, regardless of the starting point. In simulating stochastic processes, it is often desirable to have a *burn-in period*, in which the values generated are not used in the analysis. The R function `arima.sim` for simulation of a time series, for example, allows the user to specify the length of a burn-in; otherwise a "reasonable" value is computed by the program.

One of the most interesting classes of models of stochastic processes are *agent-based models*. In such a model, there are a number of "agents" whose interrelated actions affect a larger system, such as, for example, the actions of traders affecting the market. There can be different kinds of agents, such as individuals and institutions. The marginal stochastic processes that the individual agents follow can all be different, or some can be of the same family with parameters to allow different agents to have different characteristics, such as risk aversion. A famous agent-based model in finance is the Santa Fe Institute agent-based model of the stock market.

Resampling Methods

We use a sample to obtain information about a population. Sometimes we can get a better picture of the population by looking at the sample in different ways, and comparing the various views. One way of doing this is to "resample" the given sample. A straightforward way of doing this is to consider the sample to be a finite-sized population and to take multiple samples of the same size from that population. (This obviously would be sampling "with replacement".) These separate samples using only values contained in the original sample can provide information about the variance or the bias of statistical estimators. This kind of approach is called a "bootstrap", and we will discuss it further in Section 4.4.5.

4.3 Optimization in Statistical Modeling; Least Squares and Maximum Likelihood

Finding a minimum or a maximum of a function is a problem in applied mathematics called *optimization*. It turns out that many statistical methods, not just least squares and maximum likelihood, are based on optimization. Applications of optimization in statistics include methods of statistical inference such as estimation, testing, and prediction. Optimization is also used in other statistical techniques such as sampling and design of experiments. In financial applications, methods of optimization are used in determining an optimal portfolio or in determining an optimal hedge ratio.

In this section we first briefly describe optimization problems and general methods of solving them. We omit many of the details. We then describe some specific optimization problems and methods for applications in statistical inference.

4.3.1 The General Optimization Problem

In optimization, we have an *objective function* that we want to minimize or maximize. The problem is to find values of the arguments of the objective function, sometimes called the *decision variables*, that correspond to its minimum or maximum value. In parametric statistical estimation, the decision variables are the parameters to be estimated.

In the following, we will generally write the objective function as f and the decision variables as θ. There may be associated variables, which we will write as x. The objective function therefore is a function of both θ and x.

To simplify the notation in the general discussion, we will consider only minimization problems. The techniques of minimization for f apply immediately to maximization of $-f$.

There may also be restrictions or constraints on the decision variables. If there are no constraints, we write the general unconstrained problem as

$$\min_{\theta} f(\theta\,;\,x). \tag{4.32}$$

Often, there are constraints on the decision variables, such as requiring that they be nonnegative or that they sum to some specific constant. Constraints involving equalities are somewhat easier to work with, but we often write the general constrained optimization problem with inequality constraints as

$$\min_{\theta} f(\theta\,;\,x)$$
$$\text{s.t. } g(\theta) \leq c,$$

(4.33)

for some function g and constant c, usually a vector. An equality constraint can be expressed as two inequality constraints; one the negative of the other. An optimization problem with restrictions on the decision variables is called a *constrained optimization* problem.

Some functions do not have a minimum or maximum value, in which case the optimization problem is not well-defined. In other cases there may be multiple minima or maxima; we may have *local* optima and a *global* optimum.

Optimization is a large and important area of applied mathematics. There are many specialized areas and methods such as linear programming, quadratic programming, stochastic optimization including simulated annealing, use of genetic algorithms, and so on.

The decision variables may be discrete or continuous. They may also be more complicated; for example, the variables may represent permutations, as in the "traveling salesperson" problem in which the objective is to find the shortest route among a given set of cities such that each city is visited once. The decision variables are the permutations of the cities.

If the decision variables are continuous, the next consideration is whether or not the objective function is differentiable. There are often more than one decision variable; that is, θ is a vector. The differentiation is taken with respect to each variable; that is, the derivative is a vector, which we may write as $\nabla f(\theta, x)$. This is the *gradient* vector. At a minimum or maximum point in the interior of the domain of the function, the gradient is zero. Any point where the first derivative is zero is a stationary point.

The second derivative of f with respect to θ is called the *Hessian* and we denote it as H_f. If θ is a scalar, H_f is a scalar (a 1×1 matrix); if θ is a d-vector, H_f is a $d \times d$ matrix. The Hessian is a function of the same arguments as f, so we may write the matrix function as $\mathrm{H}_f(\theta, x)$.

If the second derivative is positive or positive definite in the multivariate case at a stationary point, then that stationary point is a minimum; if the second derivative is negative or negative definite, then the stationary point is a maximum; otherwise, it is an inflection point or a saddle point.

If the objective function is differentiable, using the properties of the first and second derivatives mentioned above, we have a direct method to find an optimum:

1. differentiate the objective function with respect to the decision variables, set the derivative to zero, and solve the equation to obtain a stationary point;

2. compute the second derivative to determine the type of the stationary point.

For a minimum or maximum found by this direct method, the question of whether it is a global optimum or only a local optimum remains.

Iterative Methods

Most of the interesting optimization problems cannot be solved in a closed form just by solving equations using the first derivative. A general approach is called an *iterative algorithm*, that is, one that proceeds through a sequence of possible solutions and eventually stops at a "close" solution.

An iterative algorithm is basically a rule that moves from one possible solution to another potential solution in such a way that by continuing using the rule to move, a point near to the true solution is encountered:

1. at step k with a possible solution, $\theta^{(k)}$, decide whether $\theta^{(k)}$ is an acceptable solution; if so, stop; if not, then determine another potential solution, $\theta^{(k+1)}$ and go to the next step;

2. determine whether to replace $\theta^{(k)}$ with $\theta^{(k+1)}$; whether or not $\theta^{(k+1)}$ replaces $\theta^{(k)}$, increment return to step 1.

An iterative algorithm begins with a possible solution, $\theta^{(0)}$, which may be chosen based on knowledge of the shape of the objective function or may just be some arbitrary point in the domain of the problem. Choice of the starting point may affect the performance of the iterative method. In the case of multiple optima mentioned above, different starting points can lead to different local optima. There is no general method that will lead to a global optimum.

If the objective function is differentiable, one way of choosing the direction of movement is to take the direction of the negative derivative. (If there is no such direction, then the current point is a minimum or an inflection point.) Methods that use the direction of the first derivative are called *steepest descent* algorithms. In a steepest descent algorithm,

$$\theta^{(k+1)} = \theta^{(k)} - \alpha_k \nabla f\left(\theta^{(k)}, x\right), \tag{4.34}$$

where α_k is a positive scalar that determines how far to go in the direction of the derivative.

Newton's Method and Variants

The path of a steepest descent algorithm may contain many zigzags. If the objective function is twice differentiable, the matrix of second derivatives may be used to modify the direction at each step in a way to reduce the zigzags. The use of the second derivatives in this way is called *Newton's method*.

How the second derivative can be used is seen in the Taylor series expansion truncated at the second order. The truncated expansion of f (ignoring the x

variable) about the point $\theta^{(k)}$ is

$$f(\theta) = f\left(\theta^{(k)} + \delta\right) \approx f\left(\theta^{(k)}\right) + \nabla f\left(\theta^{(k)}\right)\delta + \frac{1}{2}\delta^{\mathrm{T}}\mathrm{H}_f\left(\theta^{(k)}\right)\delta. \quad (4.35)$$

Now, we seek a stationary point of $f\left(\theta^{(k)} + \delta\right)$ with respect to δ; that is, a point where the derivative is 0. Differentiating both sides, we have

$$0 \approx \nabla f\left(\theta^{(k)}\right) + \mathrm{H}_f\left(\theta^{(k)}\right)\delta, \quad (4.36)$$

which, for exact equality, has the solution

$$\delta = -\left(\mathrm{H}_f\left(\theta^{(k)}\right)\right)^{-1}\nabla f\left(\theta^{(k)}\right). \quad (4.37)$$

This is the *Newton update*, and the next point in the iterations is

$$\theta^{(k+1)} = \theta^{(k)} + \delta.$$

The path of δ is generally in the direction of a stationary point.

A computationally expensive task in Newton's method is computing the Hessian at each step. An approximation to the Hessian often will work almost as well, especially in the early stages. Even using the Hessian from a previous step may be almost as good as using the exact Hessian. There are various ways of forming approximations and rules for reuse of Hessians over multiple steps. These variations are called *quasi-Newton methods*.

Other Iterative Methods

If the objective function is not twice-differentiable, Newton's method and quasi-Newton methods cannot be used. Also, if the function is not differentiable, steepest descent methods cannot be used. Even if the objective function is differentiable, however, it may be desirable to use a method that does not require a derivative. One good method that does not use a derivative is the *Nelder-Mead method*.

In each step in the Nelder-Mead method, three points on the surface of the objective function are identified, forming a triangle. At each iteration, the point with the largest value is replaced by a point with a smaller value. This is done by projecting the point with the largest value through the line segment formed by the other two points. The replacement point is chosen along that line of projection. In this way, the original triangle formed by the three points is "flipped" downhill into a new triangle.

Simulated annealing is a method that models the thermodynamic process in which a metal is heated beyond its melting temperature and then is allowed to cool slowly so that its structure is frozen at the molecular configuration of lowest energy. In this process the molecules go through continuous rearrangements, moving toward a lower energy level as they gradually lose mobility due to the cooling. As a method of optimization, simulated annealing moves through states in the space of the decision variables to arrive at a "state", that is, a point, of "lowest energy", that is, of minimum value of the objective function.

The rearrangements of molecules do not result in a monotonic decrease in energy. Likewise, simulated annealing does not follow a sequence of successively improving points, unlike steepest descent and Newton's method.

Even if the potential point is "uphill", that is, if it is not a "better" point, we may choose to move to that point anyway. This feature of simulated annealing helps the method to avoid converging to a local minimum. Converging to a local minimum rather than finding the global minimum is a problem when there are more than one minimum in the objective function.

At the kth step in the iterative process, we choose a potential point. If the point is better, we move to that point, but if it is not, we may base the probability of moving to that point on the change in the objective function and on a "temperature" parameter. If the potential point is uphill, we may still choose to move there with a nonzero probability.

Constrained Optimization; Lagrange Multipliers

When there are constraints on the decision variables, there are various ways that the basic methods can be modified so as to satisfy the constraints. The set of all points that satisfy the constraints is called the *feasible set*. The iterative methods we have described above can often be modified so that each successive potential point is restricted to the feasible set.

One way of handling constraints is to include them in the objective function. This can be done in various ways. In some cases, the extent to which the constraints are not satisfied is added to the objective function, and the sum is a new objective function to be minimized. In this case, the constraints may not necessarily be satisfied; but in some applications, these "soft constraints" correspond to a more appropriate formulation of the problem.

Consider a simple but common optimization problem in which the objective function is twice-differentiable with equality constraints. The formulation, as in equation (4.33) is

$$\min_{\theta} f(\theta\,;\,x)$$
$$\text{s.t. } g(\theta) = c.$$

We form the *Lagrangian function*,

$$L(\theta, \lambda\,;\,x) = f(\theta\,;\,x) + \lambda^{\text{T}}(g(\theta) - c), \qquad (4.38)$$

in which λ is a vector of *Lagrange multipliers* whose length is the same as the number of equations in the set of constraints.

In this case, any solution to the constrained optimization problem satisfies the equations

$$\frac{\partial L(\theta, \lambda; x)}{\partial \theta} = 0$$

$$\frac{\partial L(\theta, \lambda; x)}{\partial \lambda} = 0.$$

4.3.2 Least Squares

In the use of least squares to fit a statistical model $y = g(x, \theta)$, the objective function is the sum of squared residuals of observations y_i from the fitted model $g(x_i, \widehat{\theta})$,

$$\sum (y_i - g(x_i, \widehat{\theta}))^2.$$

The method of solving the optimization problem to minimize that sum depends on the form of $g(x, \theta)$. In any event, the sum of squares of the residuals, $\sum (y_i - g(x_i, \widehat{\theta}))^2$, is the key quantity. We call it *SSE*, or *RSS* for "residual sum of squares". It is the sum of the squared *residuals*, but it measures the effect of the *error* in the model.

Linear Least Squares

Least-squares estimation is very often used in fitting a linear model of the form (4.11),

$$y = X\beta + \epsilon, \tag{4.39}$$

where the vector y and the matrix X contain the data. In a fitted model, the unknown β is replaced by a computed value b. Because we assume $\mathrm{E}(\epsilon) = 0$, the expected values (or "predicted") of y in a fitted model are

$$\hat{y} = Xb. \tag{4.40}$$

Comparing the expected values with the actual observed values, we have the vector of residuals

$$r = y - Xb. \tag{4.41}$$

In least squares estimation, we minimize the sum of squares of the elements of r, which can be written as the vector inner product, $r^{\mathrm{T}} r$. It is also the sum of the L_2 norm of the vector of residuals. The objective function, therefore, is

$$f(b; y, X) = (y - Xb)^{\mathrm{T}}(y - Xb) = \|y - Xb\|_2^2. \tag{4.42}$$

Taking the derivative with respect to the decision variable b, setting it equal to zero, and dividing by 2 leads to the so-called *normal equations* involving the optimal value of b, which we call $\widehat{\beta}$,

$$X^{\mathrm{T}} X \widehat{\beta} = X^{\mathrm{T}} y. \tag{4.43}$$

Hence, in this case, if X is of full column rank (that is, the columns of X are linearly independent), then the least squares estimator is

$$\widehat{\beta} = (X^T X)^{-1} X^T y. \tag{4.44}$$

(This expression is mathematically correct, but it does not specify a good way to compute $\widehat{\beta}$.)

Estimation of σ^2

We often assume that the variance of ϵ in the model (4.39) is constant and finite; that is, $V(\epsilon) = \sigma^2$. The parameter σ^2 does not appear in the objective function above, and it is generally not estimated by direct use of least squares; rather, the least squares estimators of the other parameters are used in an expression that unbiasedly estimates σ^2. This unbiased expression can be derived without any distributional assumptions other than $E(\epsilon) = 0$, $E(\epsilon^2) = \sigma^2$, and $E(\epsilon_i, \epsilon_j) = 0$ for $i \neq j$. An unbiased estimator is the residual mean squared error, or MSE, as we will see in Section 4.5.2.

Properties of Least-Squares Estimators in Linear Models

An interesting property of the residuals in the least-squares fit of a linear model is that they are orthogonal to all columns of X:

$$r^T X = 0. \tag{4.45}$$

This follows from the normal equations, and is, in fact, a characteristic property of linear least squares.

We have often noted the common ambiguity concerning an intercept term in a linear regression model $y = X\beta + \epsilon$. If the first column of X consists of all 1s, then the model includes an intercept. Here, let us assume X is an $n \times m + 1$ matrix and that the first column contains only 1s.

Let $\bar{x} = (1, \bar{x}_1, \ldots, \bar{x}_m)$, that is, the vector consisting of 1 together with means of the regressors. Note that $\bar{x} = X^T 1_n / n$, where 1_n is the n-vector consisting of all 1s.

The first thing we note from equation (4.45) is that the sum of the residuals in a *model that contains an intercept and is fitted by least squares* is 0:

$$\sum_{i=1}^{n} r_i = r^T 1_n = 0. \tag{4.46}$$

A second result from equation (4.45) for a model with m regressors and an intercept fitted by least squares is that the fitted model goes through the point $(\bar{y}, 1, \bar{x}_1, \ldots, \bar{x}_m)$; that is,

$$\bar{y} = \bar{x}^T \widehat{\beta}. \tag{4.47}$$

We can see this easily using equation (4.45). Because 1_n is a column of X,

$$
\begin{aligned}
0 &= \left(y - X(X^{\mathrm{T}}X)^{-1}X^{\mathrm{T}}y\right)^{\mathrm{T}} 1_n/n \\
&= y^{\mathrm{T}}1_n/n - \left(X(X^{\mathrm{T}}X)^{-1}X^{\mathrm{T}}y\right)^{\mathrm{T}} 1_n/n \\
&= \bar{y} - (1_n/n)^{\mathrm{T}}X(X^{\mathrm{T}}X)^{-1}X^{\mathrm{T}}y \\
&= \bar{y} - \bar{x}^{\mathrm{T}}\widehat{\beta}.
\end{aligned}
$$

Another result that follows directly from the normal equations is that the vector of residuals is perpendicular to the vector of predicted values:

$$
r^{\mathrm{T}}\hat{y} = 0. \tag{4.48}
$$

This fact can be put into the broader geometrical context of linear least squares; the fitted values $\hat{y}$ are the projections onto a space spanned by the columns of X and the residuals are in the orthogonal complement space.

Equation (4.47) justifies the common practice of centering all variables and using the model

$$
y_{\mathrm{c}} = \widetilde{X}_c\widetilde{\beta} + \epsilon, \tag{4.49}
$$

where y_c is the vector $y - \bar{y}$, $\widetilde{X}_c$ is the last m columns of X with the mean of each column subtracted, and $\widetilde{\beta}$ is the last m elements of β, that is, all except β_0. Fitting the model (4.49) by least squares yields the same coefficient estimates as in the full model, and the intercept term β_0 is fit using equation (4.47).

The properties of least-squares estimation above are algebraic; they do not depend on any assumptions about probability distributions. These properties are so well known and widely used that statisticians sometimes forget the premises; for these properties to hold, **the model is linear with an intercept, and the fitting is done by least squares.**

On page 419, we will show that the least-squares estimators of β in the linear model are unbiased, and in Section 4.5.2, we will discuss various other statistical properties and applications of least squares estimators in linear regression models.

Nonlinear Models

Many processes, especially growth processes, do not follow a linear model. Growth of a business enterprise, or growth of assets, because of compounding, may follow an exponential trajectory. A simple exponential growth model may be of the form

$$
x_i = \alpha e^{\beta t_i} + \epsilon_i. \tag{4.50}
$$

Such a model may fit financial data such as revenue growth fairly well over limited time periods.

The model (4.50) represents exponential growth. Business enterprises often may grow exponentially and then begin to grow more slowly. Growth of many types of things follow an "S shape"; that is, growth is slow at first, then grows rapidly, and then flattens out. For example, the growth of new users of mobile phones in any given country is generally slow at first, then accelerates to rapid growth and then levels off as near market saturation is reached. A useful model for this kind of growth is the Gompertz model:

$$x_i = \alpha e^{\beta e^{\gamma t_i}} + \epsilon_i. \tag{4.51}$$

This type of model finds use in economic and financial applications in modeling growth of such things as the number of customers of a retail establishment, or growth of number of users of an app. The model is also widely used in biological studies of population growth of organisms as well as systemic growth of organs or cell masses.

Nonlinear Least Squares

Suppose we have data $(x_1, t_1), \ldots, (x_n, t_n)$, and we wish to fit the model (4.50) by least squares.

Because the model is nonlinear, the solution in general cannot be obtained in closed form. Newton's method or a quasi-Newton method is usually the best approach, although they may perform very poorly if a poor starting point is chosen or if the data do not fit the model well.

Letting x and t represent the vectors of the data, as in the development on page 363, for given values $\alpha = a$ and $\beta = b$, we form the residual vector,

$$r = x - ae^{bt}. \tag{4.52}$$

(We interpret e^{bt} as the vector $(e^{bt_1}, \ldots, e^{bt_n})$.)

The objective function to be minimized, analogous to expression (4.42), is

$$f(a, b; x, t) = \left(x - ae^{bt}\right)^{\mathrm{T}} \left(x - ae^{bt}\right). \tag{4.53}$$

Taking the derivative with respect to a and b, we get the gradient,

$$\nabla f(a, b; x, t) = 2(\nabla r)r, \tag{4.54}$$

where ∇r is the $2 \times n$ matrix,

$$\begin{bmatrix} -e^{bt_1} & \cdots & -e^{bt_n} \\ -at_1 e^{bt_1} & \cdots & -at_n e^{bt_n} \end{bmatrix}.$$

(This is the same as the Jacobian defined on page 268.) The gradient $\nabla f(a, b; x, t)$ can also be written as

$$2\nabla \left(x - ae^{bt}\right)^{\mathrm{T}} \left(x - ae^{bt}\right).$$

Unlike in the case of linear least squares, when this expression is equated to 0, the equation cannot be solved in closed form. (Notice that this expression is a 2-vector, each term of which is an inner product; it is a $2 \times n$ matrix times an n-vector.)

Since a direct method of optimization does not work here, as it did in the linear least squares problem, we must use an iterative method. We now compute the Hessian by taking the derivative of the expression for the gradient (4.54):

$$H_f = (\nabla r)(\nabla r)^{\mathrm{T}} + \sum_{i=1}^{n} r_i H_{r_i}, \qquad (4.55)$$

where H_{r_i} is the 2×2 matrix of second derivatives of the ith residual with respect to a and b;

$$H_{r_i} = \left[\begin{array}{cc} 0 & -t_i e^{bt_i} \\ -t_i e^{bt_i} & -at_i^2 e^{bt_i} \end{array} \right].$$

Now, with the gradient (4.54) and the Hessian (4.55) we can apply a Newton or quasi-Newton method.

Gauss-Newton Method

We observe an interesting feature of the gradient and the Hessian of a sum of squares function (4.53). At the minimum and in the vicinity of the minimum, the residuals should be small, in which case,

$$H_f \approx (\nabla r)(\nabla r)^{\mathrm{T}}. \qquad (4.56)$$

This suggests a quasi-Newton method using this approximation to the Hessian, because the first derivatives are easier to compute and form their outer product than it is to compute the matrix of second derivatives. Use of this approximation in the Newton updates is called the *Gauss-Newton method*. The Gauss-Newton method is the same as the Newton method in a linear least squares problem (Exercise 4.6).

A Gauss-Newton method is the default method in the R function for nonlinear least squares, `nls`. For given values of x and t, to use R to fit the model (4.50) we can use the following statement.

```
fitexp <- nls(formula=x~-1+a*exp(b*t), start=list(a=1,b=1))
```

Note that the formula in R assumes an intercept term, so if one is not present in the model, the formula must indicate that. One way is by use of "-1" in the formula, as above. (See Section 4.5.8 beginning on page 454 for descriptions of the `formula` keyword in R.)

The Gauss-Newton method can perform poorly if some of the residuals are large, because in that case the approximation to the Hessian is not very good. This may happen if the initial starting point in the iterations is not good, or if the dataset contains outliers, or just does not fit the model well.

Other Quasi-Newton Methods

We can attempt to find another quasi-Newton method in which we use a better approximation to H_f in equation (4.55) but still using only the first derivatives. One way is just to add some well-conditioned matrix to $(\nabla r)(\nabla r)^T$. This could be something as simple as a scalar multiple of the identity I,

$$H_f \approx (\nabla r)(\nabla r)^T + \lambda I,$$

where λ is some small positive number. This approach is called the Levenberg-Marquardt method, and it is probably the most widely used method for non-linear least squares.

Weighted Least Squares

Intuitively, minimizing the sum of the squared residuals $r^T r$ is a good way of fitting a regression model

$$y = g(x, \theta) + \epsilon.$$

Under simple assumptions about the variance of ϵ, the least squares estimators also have desirable statistical properties, as we will see in Section 4.5.2.

The assumptions about the variance of ϵ are that it is constant and finite and the ϵs have zero covariances.

If the variance of ϵ is not constant, estimators with better statistical properties can be obtained by weighting the residuals by the reciprocals of their variances. Instead of the sum of squares in expression (4.22), we have

$$\sum_{i=1}^{n} (y_i - g(x_i, t))^2 / \sigma_i^2, \tag{4.57}$$

where σ_i^2 is the variance of ϵ_i. If W is the $n \times n$ diagonal matrix whose (i, i) entry is $1/\sigma_i^2$ and r is the n-vector of residuals, the expression (4.57) can be written as the quadratic form $r^T W r$.

If the ϵs do not have zero covariances, their covariances can also be incorporated into the sum of squares. If

$$V(\epsilon) = \Sigma, \tag{4.58}$$

the quadratic form of interest is $r^T W r$, where $W = \Sigma^{-1}$. The minimum value of this weighted sum of squares, analogous to equation (4.44), is

$$\widehat{\beta}_W = (X^T W X)^{-1} X^T W y. \tag{4.59}$$

This is obtained by minimizing the weighted sum of squares of residuals, $r^T W r$. You are asked to show this in Exercise 4.8.

The estimator in equation (4.59) is called the *weighted least squares estimator* or *generalized least squares* estimator. The estimator with $W = I$, the identity (or the estimator in equation (4.44)) is called the *ordinary least squares* or *OLS* estimator.

The foregoing development involves $1/\sigma_i^2$ or W. In applications, these are likely to be unknown, but estimates of them may be available.

A more complicated form of generalized least squares estimation involves an iterative procedure in which residuals are fitted to a model that corresponds to an assumed structure for their variance-covariance. The W matrix in equation (4.59) is developed iteratively in accordance with an assumed model of the variance-covariance matrix, including one with autocorrelations.

Penalized Least Squares; Shrunken Estimators

The least squares estimator, $\widehat{\beta}$, in the linear model

$$y = X\beta + \epsilon,$$

is unbiased for β. (We show this on page 419.) This unbiased estimator, however, can have very large variance, especially if the columns of X are highly correlated. The mean squared error or MSE, that is, the variance plus the square of the bias (which is 0), may therefore be quite large.

There may be a biased estimator with smaller MSE. One way of forming a biased estimator with smaller MSE is to shrink the least squares estimator toward 0. (Shrinking any random variable toward a constant reduces its variance; see Exercise 3.2.) This is called "regularizing" the fit.

We can shrink the least squares estimators in the linear model by adding to the objective function a penalty term that increases in a norm of the estimators. Thus, we modify expression (4.42) and form the optimization problem

$$\min_{b} \|y - Xb\|_2^2 + g(\|b\|), \qquad (4.60)$$

where g is a nonnegative increasing function and $\|\cdot\|$ is a norm.

If the model has an intercept, that is, if the model is of the form

$$y = b_0 + b_1 x_1 + \cdots + b_m x_m + \epsilon,$$

then the term corresponding to the intercept is generally not included in $\|b\|$ in the penalty.

In a common case, the penalty is just a multiple of the square of the Euclidean norm:

$$\min_{b} \|y - Xb\|_2^2 + \lambda\|b\|_2^2, \qquad (4.61)$$

where $\lambda \geq 0$. As λ increases, the optimal values of b go toward 0. This method of fitting is called *ridge regression*. An informative graph to assess the effect of the tuning parameter is a plot of each regression coefficient estimate versus λ. This kind of graph is called a *ridge trace*. The coefficient estimates go toward zero, fairly rapidly at first as λ increases, and then slowly as λ continues to increase.

In another common case, the penalty is a multiple of the L_1 norm:

$$\min_b \|y - Xb\|_2^2 + \lambda\|b\|_1. \tag{4.62}$$

This method of fitting is called *lasso regression*. In lasso regression, as λ increases, some elements of the optimal b become exactly 0.

In ridge and lasso regression, the weight of the penalty, λ, is a *tuning parameter*. In this case, as in many instances of tuned statistical inference, the tuning parameter controls a tradeoff between bias and variance. We have seen this tradeoff, for example, in the smoothing parameter in nonparametric probability density estimation. In regularized least squares, as the penalty weight increases, the bias increases but the variance decreases.

Note the similarity of the penalized objective functions (4.61) and (4.62) to the Lagrangian function, equation (4.38). The penalty is conceptually similar to constraints.

The MSE, which is the sum of the variance and the square of the bias, typically decreases at first as λ increases from 0 and then increases as λ increases further. We can show mathematically that this happens, but we must be very clear: in practice, we do not know the MSE. We can estimate the variance by standard means, but we can only estimate the bias in nonstandard ways, such as by resampling and using bootstrap or cross-validation methods, as described in Section 4.6.2.

The penalties in ridge and lasso can be combined into a weighted average:

$$\lambda\left((1 - \alpha)\|b\|_2^2 + \alpha\|b\|_1\right). \tag{4.63}$$

This method of combining the penalties is called the *elastic net*. The elastic net has two tuning parameters, α and λ.

The R function `lm.ridge` in the `MASS` package performs the computations for ridge regression. The R package `glmnet` provides a function for fitting by the elastic net, in which the user selects α and can select λ or request that the function fit over a range of values of λ. The package also allows other regression models in addition to a linear model. (The actual objective function used in `glmnet` is slightly different from what is shown above.)

Constrained Least Squares; Quadratic Programming

In a regression model of the general form $y = g(x, \theta) + \epsilon$, it may be known that some or all of the parameters θ are in a certain range; for example, it may be known that they are nonnegative.

With a given set of observations $(y_1, x_1), \ldots, (y_n, x_n)$, it may be the case that the values of $\widehat{\theta}$ that yield the least squares $(y - g(x, \theta))^T(y - g(x, \theta))$ do not satisfy the restrictions on the model.

In order to ensure that the fitted parameters do satisfy the properties of the model, the least-squares optimization problem is constrained. The constraints can be handled in the computations in various ways, such as by Lagrange multipliers.

A common problem is fitting a linear regression problem

$$y = X\beta + \epsilon$$

with $\beta \geq 0$.

This nonnegative linear least squares problem can be solved by iteratively adjusting any negative $\widehat{\beta}_j$ to 0.

Notice that the least squares estimators without the restrictions may be unbiased; hence, the constrained least squares estimators may be biased.

The objective function in the nonnegative linear least squares problem above is $r^{\mathrm{T}} r$, where $r = y - X\widehat{\beta}$. This objective function can be generalized to be of the quadratic form $r^{\mathrm{T}} A r$, where A is a square matrix. (This is the form of a weighted least squares objective function.) The constraints can also be generalized to be of the form of more general linear constraints.

Using the general notation of this section, we can write a special form of the constrained optimization problem (4.33) where the objective function is a quadratic form in the decision variables θ plus a linear combination of them, and the constraints are linear:

$$\min_{\theta} \theta^{\mathrm{T}} A \theta + b^{\mathrm{T}} \theta \tag{4.64}$$
$$\text{s.t. } G\theta \leq c,$$

where A is a nonnegative definite matrix, b is a vector of the same order as the vector θ, G is a matrix with the same number of columns as the order of θ, and c is a vector of order equal to the number of rows in the matrix G.

The problem (4.64) is called a *quadratic programming problem*, and there are special algorithms for its efficient solution.

Quadratic programming problems arise in various areas of applied mathematics. The problem of determining an optimal portfolio (see page 67) is a quadratic programming problem. The decision variables are the weights of the individual assets in the portfolio and the matrix of the quadratic form is the correlation matrix of the returns of the assets in the portfolio.

The R function nnls in the nnls package performs the computations for fitting a linear model by least squares under the restriction that the coefficients are nonnegative. The R function solve.QP in the quadprog package performs the computations for solving a quadratic programming problem.

4.3.3 Maximum Likelihood

Given a likelihood function $L(\theta; X)$ and a sample X, the *maximum likelihood estimator*, or *MLE*, for the parameter θ is the value that maximizes the likelihood function for that given set of data; that is, it is a solution to the constrained optimization problem in equation (4.30) on page 353.

For a given probability distribution and a given set of data, the problem may have no solution; that is, the MLE may not exist. In most common data analysis problems, however, the MLE exists, and has desirable statistical properties.

Example: MLE in a Poisson Distribution

Consider again the Poisson distribution model with parameter λ and the data $X = \{3, 1, 4, 3, 2\}$. Looking at the plot of the likelihood $L(\lambda\,;\,X)$ on the right side of Figure 4.1 on page 353, it appears that the MLE of λ is slightly less than 3.

We can work this out analytically. Since $L(\lambda\,;\,x_1, \ldots, x_n)$ in equation (4.26) is differentiable, we obtain the derivative:

$$
\frac{\mathrm{d}}{\mathrm{d}\lambda} L(\lambda\,;\,x_1, \ldots, x_n) = \sum_{i=1}^{n} x_i \lambda^{\sum_{i=1}^{n} x_i - 1} \mathrm{e}^{-n\lambda} \prod_{i=1}^{n} \frac{1}{x_i!} - n\lambda^{\sum_{i=1}^{n} x_i} \mathrm{e}^{-n\lambda} \prod_{i=1}^{n} \frac{1}{x_i!}
$$

$$
= \lambda^{\sum_{i=1}^{n} x_i - 1} \left(\sum_{i=1}^{n} x_i - n\lambda \right) \left(\mathrm{e}^{-n\lambda} \prod_{i=1}^{n} \frac{1}{x_i!} \right).
$$

Now, we set the derivative equal to 0 and collect terms:

$$
\sum_{i=1}^{n} x_i = n\lambda, \tag{4.65}
$$

implying a stationary point at $\lambda = \bar{x}$. We get the second derivative by differentiating the first derivative. Letting $s = \sum_{i=1}^{n} x_i$ and $p = \prod_{i=1}^{n} \frac{1}{x_i!}$ to simplify the notation, we get as the second derivative

$$
(s(s-1) - sn\lambda - sn\lambda + n^2\lambda^2)\lambda^{s-2}\mathrm{e}^{-n\lambda}p,
$$

which is negative at $\lambda = \bar{x}$, if $\bar{x} > 0$. Hence, if $\bar{x} > 0$, the stationary point is a maximum, so in that case the MLE of λ is

$$
\hat{\lambda} = \bar{x}.
$$

In the example shown in Figure 4.1, the MLE of λ is $(3+1+4+3+2)/5 = 2.6$.

If $\bar{x} = 0$ (note that $\bar{x}$ cannot be negative in a Poisson distribution), the likelihood is $\mathrm{e}^{-n\lambda}$, which is maximized over the closure of the parameter space $\lambda \geq 0$ at $\lambda = 0$.

Note that the form of the MLE for the Poisson parameter does not depend on the particular sample; for any given sample, the MLE of the Poisson parameter is the sample mean. This is often the case for MLEs, they can be written as a general statistic computed in the same way from any sample.

Example: MLE in a Normal Distribution

Consider now a normal distribution, $N(\mu, \sigma^2)$, and a random sample $x_1, \ldots, x_n$. We have the likelihood function of μ and σ^2, $L(\mu, \sigma^2\,;\,x_1, \ldots, x_n)$, as shown in equation (4.29) on page 353. We could find the MLEs for μ and σ^2 by differentiating the expression and setting the two derivatives to zero.

Instead of working with $L(\mu, \sigma^2 \; ; \; x_1, \ldots, x_n)$, however, we might consider working with the log of the likelihood, which I denote as l_L:

$$l_L(\mu, \sigma^2 \; ; \; x_1, \ldots, x_n) = -\frac{n}{2}\log(2\pi) - n\log(\sigma) - \sum_{i=1}^{n}(x_i - \mu)^2/2\sigma^2. \quad (4.66)$$

The log of the likelihood function is called the *log-likelihood function*. Because the log is a one-to-one and strictly increasing function, the maximum point in the log of a function occurs at the same point as the maximum in the original function. From the log-likelihood (4.66), it is easy to compute stationary points of μ and σ^2 by differentiation,

$$\widehat{\mu} = \bar{x}$$

$$\widehat{\sigma^2} = \frac{1}{n}\sum_{i=1}^{n}(x_i - \bar{x})^2,$$

and then to show that these are MLEs using the second derivatives.

The log of the likelihood function is often a more convenient function to maximize, as in this case. Products in the likelihood function become sums in the log-likelihood, which are easier to work with.

Deviance

In least-squares modeling, for any given model or specific value of the parameters, the important quantity is the sum of the squares of the residuals, SSE, or RSS. In maximum likelihood approaches, the important quantity is some form of the likelihood function.

For computational purposes, we often work with the log of the likelihood, and the log-likelihood also has useful theoretical properties, such as an asymptotic relationship to a chi-squared distribution. The theoretical properties relate directly to twice the negative of the log-likelihood,

$$- 2\log(L(\theta; x)), \quad (4.67)$$

which we define as the *deviance*, which, as the likelihood itself, is a function of the parameter(s) of the model, θ. (The term "deviance" is also used in different ways by other authors in the statistical literature. This is how the term is used in R and this is the way I use it.)

In maximum likelihood approaches, the deviance is the important quantity, and it plays a similar role to the SSE in least-squares modeling.

Fitting the model by maximum likelihood is done by minimizing the deviance, analogous to minimizing SSE in least squares.

Instead of the ordinary residual for each observation, we define the *deviance residual* of an observation as the contribution that observation makes to the sum

$$-2\log(L(\theta; x)) = \sum_{i=1}^{n} -2\log(L(\theta; , x_i));$$

that is, the ith deviance residual is $-2\log(L(\theta; , x_i))$. (The deviance itself is sometimes called the "residual deviance"; so the residual deviance is the sum of the deviance residuals!)

Maximum Likelihood in Linear Models with Normal Errors

Consider again the linear regression model in the vector-matrix form of equation (4.11). Without any further assumptions, we obtained the least squares estimates as the solution to the normal equations (4.44).

For the MLE, however, we need an additional assumption; we need to assume a distribution.

Let us assume that the ϵs are iid $N(0, \sigma^2)$; that is, the vector ϵ has the multivariate normal distribution $N_n(0, \sigma^2 I)$, and the vector y has the multivariate normal distribution $N_n(X\beta, \sigma^2 I)$.

With this assumption, we can form the log-likelihood function, similar to the log of the likelihood (4.29):

$$l_L(\beta, \sigma^2 \; ; \; (y_1, x_1), \ldots, (y_n, x_n)) = \frac{-n\log(2\pi\sigma^2)}{2} - \frac{(y - X\beta)^{\mathrm{T}}(y - X\beta)}{2\sigma^2}.$$
(4.68)

Notice that to obtain the MLE of β, the only relevant term is the same inner product as in the least squares problem on page 363; hence, we see that the MLE of β is the same as the least-squares estimator of β.

This is a characteristic of the normal distribution; the least-squares estimator of the mean is the same as the MLE of the mean. This fact does not depend on the model being linear; we would arrive at similar expressions for the general regression model $y = g(x, \theta) + \epsilon$.

Notice another thing about the maximum likelihood approach to the problem in the case of a normal distribution. We also have an MLE of σ^2. Recall that we did not formulate a direct least squares estimator of σ^2.

Differentiating equation (4.68) with respect to σ^2, equating to zero, using the solution for the maximum with respect to β, and checking the second derivative, we get the MLE

$$\widehat{\sigma}^2 = (y - X\widehat{\beta})^{\mathrm{T}}(y - X\widehat{\beta})/n.$$
(4.69)

This estimator is biased for σ^2, but it is consistent in mean squared error. Notice, again, this result is true for the general regression model $y = g(x, \theta) + \epsilon$, as well, so long as the additive errors are iid $N(0, \sigma^2)$.

In the general maximum likelihood problem, we cannot treat the parameters separately, as in the case of a normal distribution.

Not all likelihood functions are as simple as the Poisson and the normal; for example, the derivatives obtained from the likelihood function for the gamma(α, β) distribution cannot be solved in closed form. Another example is the likelihood function in logistic regression (page 449). The distribution itself in that case is very simple, but the terms in the model make the optimization problem difficult to solve. Newton's method is the preferred method for solving that maximization problem.

Use of Newton's method on the log-likelihood function has given rise to two terms in the statistics literature, the "score function" and "Fisher scoring". The gradient of the log-likelihood is called the *score function* in statistics. Because the Hessian of the log-likelihood is the sample version of what is called the Fisher information, use of the Newton step with the Hessian and the gradient of the log-likelihood (equation (4.37), page 361) is called *Fisher scoring*.

4.3.4 R Functions for Optimization

R provides several functions both for general optimization problems and for special applications. It also has functions for fitting models using various criteria.

Some General Purpose R Functions for Optimization

- `nlm`
  solves a general differentiable unconstrained problem using Newton's method

- `nlminb`
  solves a general differentiable constrained problem using a Newton method (uses `nlm`)

- `optim`
  solves a general unconstrained problem using a Nelder-Mead method

- `constrOptim`
  solves a general constrained problem using a Nelder-Mead method with barriers (uses `optim`)

- `optim_sa{optimization}`
  solves a general unconstrained problem using a simulated annealing method

- `optimize`
  performs a one-dimensional search for a minimum

- `solveLP{linprog}`
  solves a linear programming problem

- `solve.QP{quadprog}`
  solves a quadratic programming problem

Some R Functions for Fitting Models

- `lm` or `lsfit`
  fits a linear model with least squares

- `nnls{nnls}`
  fits a linear model with least squares under the restriction that the coefficients are nonnegative

- `gls{nlme}`
  fits a linear model with generalized least squares

- `glm`
  fits a generalized linear model with maximum likelihood

- `lm.ridge{MASS}`
  fits a linear model using ridge regression

- `glmnet{glmnet}`
  fits a linear or generalized linear model using the elastic net

- `l1fit{L1pack}`
  fits a linear model with least absolute values

- `nls`
  fits a nonlinear model with least squares.

Optimization problems can present many computational difficulties. Although the formulation of a maximum likelihood problem may be straightforward, for many probability distributions, a sample size of 100 can generate values that vary so widely in magnitude that the software cannot cope with the relative sizes. The use of the log-likelihood usually will result in a better-conditioned numerical problem, but even so, the software will often fail even on apparently simple problems.

Many optimization problems require ad hoc simplifications in order to obtain a solution.

4.4 Statistical Inference

Statistical inference is the process of using a sample of data to infer properties of a larger population of which the sample is representative.

In the simplest and most common instances of statistical inference, we have a model of the aspects of the population of interest, and we collect a random sample from that population. With the sample in hand, we do some preliminary explorations as general checks on the model and on the data we have collected. Methods for doing this are described in Chapter 2. Statistical inference then involves either making more specific statements about the model or else predicting future observations based on the model and the data.

There are two different approaches to statistical inference that depend on fundamental differences in the models. For lack of better names, we call the two approaches "frequentist" and "Bayesian".

In frequentist inference, the objective is to narrow down the family of probability distributions of the observable data by estimating values or ranges of values of unknown model parameters, or otherwise to refine the family of models describing the observable. Statements about values of parameters are accompanied by statements about variances of estimators or levels of "confidence".

In Bayesian inference, we begin with a Bayesian model, as described in Section 4.1.6. Parameters in the model are considered to be random variables. First, as part of the model, we express our current "beliefs" about the probability distribution of the parameters. The objective of the analysis is to determine the conditional distribution of the parameters, given the prior beliefs and given the data.

These two different models lead to fundamental differences in what may be essentially the same conclusion. The frequentist approach may lead to the conclusion that "a 95% confidence interval of μ is $(93, 110)$", for example, while the Bayesian approach may yield the conclusion that "the probability that μ is in the interval $(93, 110)$ is 95%".

Types of Statistical Inference

Statistical inference may be couched in various forms. It may be a point estimator, that is, a specific value of a parameter, along with a statement of estimated variance. It may be an interval or subregion of the parameter space in which a parameter is believed to lie, along with a statement of confidence or credibility. It may be a decision on the truth of some statement about the distribution of the observables (a "hypothesis"), together with a statement about the confidence or probability levels associated with the decision. Yet another form of statistical inference may be a prediction about outcomes. In time series analysis, prediction is often called "forecasting". We will discuss prediction in Section 4.4.3, and forecasting in Chapter 5.

"Standard Errors"

A point estimator or a predictor is a random variable that has a probability distribution, often not known exactly. For an estimate or a prediction to be useful, some knowledge of its distribution is necessary. An important feature of the distribution is the variance or standard deviation; that is, some measure that indicates the precision of the statistic.

Along with any point estimate or prediction computed from data, we should compute an estimate of the variance or standard deviation of the underlying random variable. The estimate of the variance or standard deviation depends on the form of the statistic of course. As we develop estimators and predictors in statistical models, we also consider their distributions, or at least their variances and standard deviations.

An estimate of the standard deviation of an estimator or predictor is called the *"standard error"*. In statistical inference, the standard error is the most

commonly used measure of precision. The term is not only somewhat misleading, it is generally used indiscriminately with regard to the nature of the
estimator of the standard deviation. The estimator of the standard deviation
is likely to be consistent in mean squared error, but it is almost certainly
biased. Its properties as an estimator depend on the underlying distribution
and the nature of the estimator or predictor whose standard deviation is being estimated. In most cases, what is called the standard error is based on
approximations for a normal distribution.

The standard error is often used as a rule of thumb for rough limits above
and below a given point estimate or prediction. This interval is not a confidence interval, which has a more precise interpretation, as we describe in
Section 4.4.1. (In terms of a confidence interval the plus-or-minus interval
defined by a standard error under some fairly restrictive assumptions would
correspond to a 67% confidence interval.)

Methods of Statistical Inference

If we use the assumed underlying probability distribution to work out properties of our estimators or test statistics, we call the inference *exact inference*.
For many probability distributions, however, derivations of exact confidence
intervals or the significance of test statistics are mathematically intractable.

Asymptotic Inference

Sometimes, even if we cannot work out the exact distribution, we know that
the estimators or test statistics have a simpler asymptotic distribution. This
is often based on the central limit theorem (page 272). We might therefore use
the asymptotic distribution to approximate critical values or other properties
of the estimators. This is called *asymptotic inference*.

Computational Inference

Another approach when the probability distributions are mathematically intractable, is to simulate the distributions. This is called *computational inference*, and there are many forms of it. A common instance is the use of Markov
chain Monte Carlo (MCMC) methods, which we will describe beginning on
page 388. In Section 4.4.5, we will discuss another method of computational
inference, the bootstrap.

Tuned Inference

In some cases there may not be a direct relationship between the available data
and the model for the data. Examples of this are models of probability density,
because density is an infinitesimal quantity, and volatility of asset returns,
because volatility is an instantaneous quantity. Estimates of density and of
volatility require a tuning parameter, which of course can take different forms,
as we have seen. Inferences about tail behavior require a tuning parameter to

define "tail". There are several other instances in which a tuning parameter is necessary for the kind of inference to be made. Such inference is called *tuned inference.* It is related to two-stage or conditional inference, in which inference on the property of interest is preceded by statistical tests or estimates that determine the method of inference. Statistical properties of estimators involving a tuning parameter are generally very difficult to determine exactly, and so most instances of tuned inference also involve either asymptotic or computational inference.

In many instances of tuned inference, the tuning parameter controls a tradeoff between bias and variance. We see this, for example, in the smoothing parameter in nonparametric probability density estimation. As the bin width increases or the kernel density becomes more spread out, the variance decreases but the bias increases. Likewise, in regularized least squares (ridge or lasso), as the penalty weight increases, the bias increases but the variance decreases. Often the balance in the bias-variance tradeoff is chosen so as to minimize the MSE. In Exercise 4.16b, you are asked to investigate this tradeoff in the context of estimation of the tail index (discussed in Section 4.4.7).

4.4.1 Confidence Intervals

Given a sample of data, we can make more informative statements about the population mean than just giving a simple point estimate. One way of doing this is just to use the sample standard deviation as a measure of the precision of the mean estimate; in other words, to give some indication of the likelihood that the population mean is close to the estimated value.

Another way of approaching this problem is to choose some relatively large probability, say 95%, and determine some region on either side of the point estimate such that a random interval constructed in that way would have a 95% probability of enclosing the population mean.

This may sound rather complicated, but it is often quite easy if we assume that we know the family of distributions. For example, for fixed values of μ and σ^2, assume that the conditional distribution of the observable random variable X is $N(\mu, \sigma^2)$. Let $X_1, \ldots, X_n$ be iid as $N(\mu, \sigma^2)$, and let $\overline{X}$ be the mean of the X_i, and let

$$S^2 = \frac{1}{n-1} \sum_{i=1}^{n} (X_i - \overline{X})^2.$$

Now $(\overline{X} - \mu)/\sqrt{\sigma^2/n}$ has a $N(0, 1)$ distribution, and from the discussion beginning on page 289, we know that $(n-1)S^2/\sigma^2$ has a chi-squared distribution with $n-1$ degrees of freedom and is independent of $\overline{X}$. Hence,

$$t = \frac{\overline{X} - \mu}{\sqrt{S^2/n}}$$

has a t distribution with $n - 1$ degrees of freedom.

Now let $t_{n-1,0.025}$ be the value such that for a random variable t with a t distribution with $n-1$ degrees of freedom, $\Pr(t \leq t_{n-1,0.025}) = 0.025$. Because the t distribution is symmetric about 0, $\Pr(t \geq -t_{n-1,0.025}) = 0.025$. Finally, form the interval

$$\left(\overline{X} + t_{n-1,0.025}\sqrt{S^2/n}, \quad \overline{X} - t_{n-1,0.025}\sqrt{S^2/n}\right). \tag{4.70}$$

The probability that this interval includes μ is 95%.

Given n observations on X, $x_1, \ldots, x_n$, an interval constructed as above is called a *95% confidence interval*. The *confidence level* associated with the region is 95%.

For example, suppose for a sample of size $n = 250$, we compute $\bar{x} = 0.0667$, and $s^2 = 0.1759$. Now, $t_{249,0.025} = -1.970$ (from qt(0.025,249) in R), and so the 95% confidence interval for μ is

$$(0.0145, \quad 0.119). \tag{4.71}$$

In general, we speak of a $(1 - \alpha)100\%$ confidence level. The "critical t value", $t_{n-1,0.25}$ above, would be substituted with another quantile from the appropriate distribution. The quantile in the confidence interval is chosen so that the probability that the interval includes the parameter of interest is $1 - \alpha$, or alternatively, that the interval does not include the parameter is α. For a two-sided symmetric interval as in (4.70), we would use a value $t_{n-1,\alpha/2}$ such that $\Pr(t \leq t_{n-1,\alpha/2}) = \alpha/2$.

Under the same setup, there are different intervals we could choose that would have the same probability of including μ. One interval may be better than another because it is shorter, or because it has some other desirable property. A one-sided upper $(1 - \alpha)100\%$ confidence level would be

$$\left(\overline{X} + t_{n-1,\alpha}\sqrt{S^2/n}, \quad \infty\right). \tag{4.72}$$

We could also choose another confidence level and compute different intervals.

In the case of a normal distribution, confidence intervals are easily formed because of the relationships between the normal, chi-squared, and t distributions as described on page 289.

General Method of Forming Confidence Intervals

The discussion above focused on confidence intervals for the mean. For other parameters, we follow a similar development.

A general method to form a $(1-\alpha)100\%$ confidence interval for a parameter θ is to use a statistic, say T, that relates to θ and to find a *pivotal quantity* that involves θ and the statistic T. A pivotal quantity is a function of T and θ, $f(T, \theta)$, in which T and θ can be separated, and whose distribution is known.

We then find two quantiles f_{α_1} and f_{α_2} of $f(T, \theta)$ (either of which may be infinite), such that

$$\Pr\left(f_{\alpha_1} \leq f(T, \theta) \leq f_{\alpha_2}\right) = 1 - \alpha.$$

We then rearrange the terms in that probability statement.

In the confidence intervals for the mean of a normal distribution that we discussed above, the pivotal quantity is $f(T, \mu) = (\overline{X} - \mu)/\sqrt{S^2/n}$, which has a t distribution with $n - 1$ degrees of freedom. The statistic T in this case is the doubleton $(\overline{X}, \sqrt{S^2/n})$. The 95% confidence interval (4.70) is formed by use of the equation

$$\Pr\left(t_{n-1,0.025} \leq (\overline{X} - \mu)/\sqrt{S^2/n} \leq -t_{n-1,0.025}\right) = 95\%. \qquad (4.73)$$

4.4.2 Testing Statistical Hypotheses

A concept closely related to confidence is *significance*. The confidence level associated with the interior of a region determines the *significance level* associated with the exterior of the region. A $(1 - \alpha)100\%$ confidence region has a significance of $\alpha 100\%$.

The term "significance" is most often used in hypothesis testing.

Statistical Hypotheses

Statistical hypotheses state that the probability distribution of a given population has certain properties. Given an *assumed* general form of the distribution, a *statistical hypothesis* states that the distribution is some subfamily of the assumed distribution. We designate different hypotheses as "H_0", "H_1", "H_a" and so on. We may assume that the distribution is $N(\mu, \sigma^2)$, for example, and then we may formulate the hypothesis that the mean is some specific value, μ_0. We state this hypothesis as

$$H_0 : \mu = \mu_0. \qquad (4.74)$$

A *simple hypothesis* completely specifies the distribution (within the assumed family). For example, the hypothesis $H_0 : \mu = \mu_0$ is simple, but the hypothesis $H_0 : \mu \geq \mu_0$ is not. A hypothesis that is not simple is called *composite*.

A hypothesis to be *tested* is called a *null hypothesis*. The hypothesis is tested by collecting data and comparing an observed statistic with the distribution of that statistic under the null hypothesis. If the test statistic has an "extreme" value relative to the null distribution, the hypothesis is rejected. An extreme value is said to be "significant". The level of significance is the probability associated with the extreme region under the null hypothesis. This level of significance defining the *rejection region* is set before carrying out the procedure. Alternatively, for a given value of the test statistic, the probability of a value that extreme or more extreme under the null hypothesis is determined. This value is called the "p-value".

The same statistics used in setting confidence intervals for a parameter may be used in testing hypotheses concerning that parameter. For example, in testing the hypothesis above that $\mu = \mu_0$, we may form the statistic

$$t = \frac{\overline{X} - \mu_0}{\sqrt{S^2/n}}, \tag{4.75}$$

which is just a rearrangement of the terms in the confidence intervals (4.70) or (4.72). Testing the hypothesis by comparing t with the quantile $t_{n-1,\alpha/2}$ or $t_{n-1,\alpha}$ is equivalent to determining whether the appropriate confidence interval (4.70) or (4.72) includes μ_0. Also, equivalently, these procedures are equivalent to determining the p-value of the computed t, and rejecting the null hypothesis if the p-value is less than α.

The ideas and methods of statistical hypothesis testing is can be developed by starting with the problem of testing one simple hypothesis, called the "null", against another simple hypothesis, called the "alternative":

$$H_0 : \mu = \mu_0$$

$$vs$$

$$H_1 : \mu = \mu_1.$$

The *Neyman-Pearson* approach defines a procedure for testing these hypotheses that is *optimal* in the sense that for a stated limit on the probability of rejecting a true H_0, the probability of rejecting a false H_0 is maximized. This approach is extended to identify optimal tests for composite hypotheses in a variety of situations.

The basic idea in a statistical hypothesis test is to determine a "test statistic" and compare its value with the probability distribution of a random variable under the null hypothesis. If the value of the test statistic is "extreme", as determined by a fixed level of significance, then the hypothesis is rejected.

Errors in Hypothesis Testing

Erroneous decisions are made if a true hypothesis is rejected, or if a false hypothesis is not rejected. These decisions are called respectively "type I" and "type II" errors. The significance level of a test is the largest probability of a type I error for that test, but the probability of both type I and type II errors depends on the true state of the data-generating process.

Power of Statistical Tests

The power of a statistical test is the conditional probability of rejecting a null hypothesis, whether the hypothesis is false or not. For a test to be *valid*, the power when the null hypothesis is true cannot be greater than the significance level. Obviously we want the power of a test to be as large as possible when the null hypothesis is false.

The power of a test of a hypothesis concerning a parameter depends on the actual value of the parameter. For hypotheses concerning a given parameter, we call the power as a function of the parameter the *power curve*.

If the alternative hypothesis is basically "everything else", the test is called an *omnibus* test. The power of an omnibus test can vary widely, depending on the specific alternative.

Some Common Statistical Hypothesis Tests

Some of the most familiar hypothesis tests involve tests about parameters in a normal distribution. For example, we may have a sample $x_1, \ldots, x_n$ from an assumed $N(\mu, \sigma^2)$ population, and we want to test

$$H_0 : \mu = \mu_0$$

vs

$$H_1 : \mu \neq \mu_0.$$

This is a similar problem to that of setting a confidence interval for μ that we discussed on page 380.

The standard procedure is to compute the sample mean $\bar{x}$ and the sample variance s^2 and to compute the test statistic

$$t_c = \frac{\bar{x} - \mu_0}{\sqrt{s^2/n}}. \tag{4.76}$$

(This is the same as in equation (4.75), except there we used random variables instead of realizations of random variables, as here.) Now, under the assumptions, if the null hypothesis is true, t_c in equation (4.76) is a realization of a t random variable with $n - 1$ df. The test is completed by comparing the value of t_c with a critical value from the t distribution, or else by determining its p-value, that is, the probability that a t random variable is more extreme.

Consider, for example, the daily simple returns of the S&P 500, expressed as percentages, for the year 2017. (The log returns of these data were used in the scatterplots of Figure 1.22.) Now, using these data, we want to test

$$H_0 : \mu = 0.10 \quad \text{(that is, the mean daily simple return is 0.1\%)}$$

vs

$$H_1 : \mu \neq 0.10$$

at the 5% significance level.

The sample size is $n = 250$, the mean is $\bar{x} = 0.0667$, and the sample variance is $s^2 = 0.1759$. Then $t_c = -1.255$. Now we compute the two-sided p-value: $\Pr(t_{249} \leq t_c) + \Pr(t_{249} \geq -t_c) = 0.211$ (from `pt(-abs(t_c),249)+1-pt(abs(t_c),249)` in R). This p-value is greater than the 5% significance level set for the test, so it does not indicate rejection.

Alternatively, we can compute the critical values for the test statistic; they are $t_{249,0.025} = -1.970$ and $t_{249,0.9755} = 1.970$ (from `qt(0.025,249)` and `qt(0.975,249)` in R). Again, since $t_c > -1.970$ and $t_c < 1.970$, we do not reject.

On the other hand, had the null hypothesis been $H_0 : \mu = 0.15$, the computed value t_c would have been -3.14, yielding a p-value of 0.00189. In this case, we would reject the null hypothesis.

One-sided tests are handled in the same way as one-sided confidence intervals. The idea is to assign the relevant amount of probability to each alternative. Had the null hypothesis above been $H_0 : \mu \leq 0.10$, the p-value of the computed value t_c, which is the same -1.255 as before, would have been $\Pr(t_{249} \leq t_c) = 0.105$. In this case, we would also not reject the null hypothesis, but a slightly smaller value of t_c would result in rejection.

A test at the 5% level is equivalent to determining whether the test statistic lies within the appropriate 95% confidence interval. The example confidence interval in (4.71) was computed using these same data. The test statistic for $H_0 : \mu = 0.10$ is inside that confidence interval, so we do not reject; on the other hand, the test statistic for $H_0 : \mu = 0.15$ is outside the confidence interval, so we do reject in that case.

For tests concerning σ^2 in a normal distribution, we use a chi-squared test statistic. For tests relating to the means of two independent normal distributions, we use a t test statistic, and for tests relating to the variances of two independent normal distributions, we use an F test statistic.

While t, chi-squared, and F tests are exact only in the case of normal distributions, they may be approximately or asymptotically correct for other distributions, so they are widely used.

A common type of statistical hypothesis test is a goodness-of-fit test, in which the null hypothesis states that the population has a specific distribution or comes from a specific family of distributions, such as the normal family. Most goodness-of-fit tests are omnibus tests. We will discuss goodness-of-fit tests in Section 4.6.1, beginning on page 456.

Simultaneous Hypothesis Tests

If the same model and dataset are used to test more than one hypothesis, the correlations among the test statistics cause problems in the interpretation of the significance levels. (This is not the same situation as a test of a composite hypothesis.)

The significance level of each of a set of multiple tests is the probability of a type I error for that test. The probability of at least one type I error in the set of tests is higher. Depending on the context and the application, this problem can be dealt with in various ways, generally by reducing the significance level at which each separate test is conducted. We will not go into the details of the procedures here; but it is important that the analyst be aware of the issues of dependence in simultaneous hypothesis tests.

Evidence-Based Decisions

Statistical hypothesis testing treats the two possibilities asymmetrically. The null hypothesis is accepted as the default, and rejection of the null hypothesis results in a stronger level of confidence in the decision that it is not true than the level of confidence in deciding that it is true if it is not rejected. The traditional terms used to describe the results of a statistical hypothesis are "reject" or "fail to reject". Not all scientists or data analysts accept this asymmetric treatment of evidence as appropriate, and advocate other ways of incorporating evidence into the decision-making process. Royall (1997) gives several examples in which decisions based on p-values or confidence intervals are clearly not supportable from the available empirical evidence. Royall treats the decision-making process as a binary choice similar to choosing between "null" and "alternative" hypotheses, but he advocates basing any decision between two alternatives on the likelihood ratio, which is the ratio of the maximum of the likelihood functions under the two alternatives. Wasserstein, Schirm, and Lazar (2019) also decry the reliance on a simple measure of "significance" in making statistical decisions.

4.4.3 Prediction

One purpose of data reduction and dimension reduction, and generally of using models such as equation (4.15), is to help in understanding the data-generating process. In addition to understanding the process, however, the models can be used to predict future values. The prediction is based on the fitted model.

The confidence associated with a prediction must include not only the uncertainty arising from using a sample to fit the model, but also the uncertainty arising from the error term in the model. For example, in the general model (4.3), $y = g(x, \theta) + \epsilon$, using the fitted model, we predict the response y for a given value $x = x_0$, as $\hat{y} = g(x_0, \hat{\theta})$. A confidence interval for the mean of all y at the point x_0 can be constructed just based on x_0 and the distribution of $\hat{\theta}$. A *prediction interval*, however, must also include a term for the variation in the latent variable ϵ. A prediction interval for a point, therefore, is always larger than a confidence interval for a mean.

Nonparametric smoothing methods also yield predictions, of course. Prediction intervals in those cases, however, are generally more difficult to construct, and often cannot be formed analytically.

Some of the most effective methods of prediction are based on a "black box"; that is, there is no simple equation that is fitted. Some examples of black box methods are neural nets, support vector machines, and random forests. We will not describe these methods in this book.

4.4.4   Inference in Bayesian Models

Statistical inference in Bayesian models uses data to refine any previous beliefs about the distribution of the random parameters.

Formal Bayesian analysis begins with a model such as (4.1), $X \sim \mathrm{N}(\mu, \sigma^2)$, for a probability distribution of an observable random variable, or (4.3), $y = g(x, \theta) + \epsilon$, for a relationship among variables. In the Bayesian approach, however, the unobservable parameters such as μ, σ^2, and θ in the models are considered to be random variables. The Bayesian model, therefore, also includes components that model the probability distributions of the random parameters.

In the initial setup, the distributions of the parameters in the model are called *prior distributions*, and the probability distributions for observable random variables are conditional distributions; they are conditional on the values of the random parameters. From the marginal distribution of the parameters (the prior) and the conditional distribution of the observables given the parameters, we can form the joint distribution of the parameters and the observables as in equation (3.47) on page 260. Given the joint distribution, we can determine the marginal distribution of the observables (by integrating out the parameters) as in equation (3.39). Finally, from the joint distribution and the marginal distribution of the observations, we can determine the conditional distribution of the parameters given the observables, again, as in equation (3.47). This conditional distribution is called the *posterior distribution*. Instead of just a point estimate or a confidence interval, as is a major objective in ordinary statistical inference, in Bayesian inference, estimating the posterior distribution in the model is a primary objective.

All of these steps can be completed without having any data, in which case, the posterior distribution of the parameters involves unknown values of observables.

Notation and Steps

In terms of the notation we used on page 260, if we let X represent the observable random variables and Θ represent the parameters (both X and Θ may be vectors), and if we assume that these are continuous random variables whose PDFs exist, we start with the conditional distribution of the observables,

$$f_{X|\Theta}(x|\theta), \tag{4.77}$$

and the prior (marginal) distribution of the parameters,

$$f_{\Theta}(\theta). \tag{4.78}$$

From these, we have the joint distribution of the parameters and the observations (equation (3.47),

$$f_{X\Theta}(x, \theta) = f_{X|\Theta}(x|\theta) f_{\Theta}(\theta), \tag{4.79}$$

and from this we get the marginal of the observables (equation (3.39),

$$f_X(x) = \int f_{X\Theta}(x,\theta)\mathrm{d}\theta. \tag{4.80}$$

Finally, we have the posterior conditional distribution of the parameters given the observations (again, equation (3.47),

$$f_{\Theta|X}(\theta|x) = \frac{f_{X\Theta}(x,\theta)}{f_X(x)}. \tag{4.81}$$

In Bayesian analysis of data, after deriving the model for the posterior conditional distribution, we merely substitute the observed values in the appropriate places, and we have a distribution of the parameters.

All of the distributions above can be multivariate distributions, and the same steps apply. In an application to estimate both the mean and the variance, for example, θ may be the vector (μ, σ^2).

The shape of the prior $f_\Theta(\theta)$ depends on how strong are the analyst's prior beliefs. If the probability in a prior distribution is very concentrated around a single point, then for whatever sample of data is used, the Bayesian estimate of the conditional distribution for that parameter $f_{\Theta|X}(\theta|x)$ will be concentrated near to the same point of the prior. A "flat" prior distribution, on the other hand, is "non-informative", and the corresponding Bayesian estimate will depend more on the data sample.

Model Dependencies

The basic model is one of dependencies among variables, both observable variables and unobservable parameters or other latent variables. In the setup described above, the prior marginal distribution of the unobservable Θ with PDF given in equation (4.78) may be a conditional distribution given Ψ (called *hyperparameters*) with PDF $f_{\Theta|\Psi}(\theta|\psi)$. The parameters in general are vectors.

If each component of the parameter has an independent prior with its own vector of hyperparameters, we have a simple hierarchical setup with two separate inputs to a single conditional (the prior), which then provides input to the distribution of the observable X. For example, if the prior in equation (4.78) is replaced by a conditional prior with two components Θ_1 and Θ_2, where Θ_1 has a distribution conditioned on Ψ and Θ_2 has a distribution conditioned on Φ, the model can be represented as a simple directed acyclic graph (DAG), as shown below.

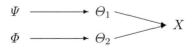

This DAG, for example, could represent a Bayesian model for a normal distribution of X, in which the mean (say, Θ_1) has a normal prior distribution with mean and variance having a joint distribution with hyperparameter Ψ and the precision (say, Θ_2) has a chi-squared prior distribution with hyperparameter (degrees of freedom) Φ (see Exercise 4.11). In many cases the hyperparameters are just chosen as constants; that is, the prior distributions of the hyperparameters are degenerate.

An advantage of a DAG representation of a Bayesian model is that it is compact and can be extended to any number of levels and variables. The simple DAG representation of the model is useful in specifying the model in computer software, and it is used in the model description in the BUGS and JAGS software systems (see below).

Notation and Terminology

The literature on Bayesian analysis often uses slightly different terminology from other areas of statistics. For example, the "inverse Gaussian distribution" in the Bayesian literature is not what other statisticians and probabilists mean by that term. And, of course, "probability" does not refer to a measure in the usual mathematical sense of the word, as it does in other areas of statistics. In much of the literature on Bayesian methods, instead of variance or standard deviation, "precision" is used. Precision is the reciprocal of the variance. (The term "precision" is used in various ways in statistics, often in a nontechnical, general sense. It is quantified in two different ways in the literature. In one definition, it is the reciprocal of the standard deviation; in the other, it is the reciprocal of the variance. In most literature on Bayesian methods, it is the reciprocal of the variance.)

Fitting Bayesian Models

If all of the expressions in equations (4.77) through (4.81) can be obtained, the Bayesian analysis is completed by inserting the observed data into the PDF in equation (4.81). This is the conditional distribution of the parameters of interest and it embodies all of the relevant information in the data about the model.

The step in this process that is potentially difficult is the evaluation of the integral in equation (4.80). This integration is trivial for certain forms of the prior distribution and the conditional distribution of the observables. Often in Bayesian analysis, special priors (called conjugate priors) are chosen so as to facilitate this computation. For example, in the simple case of a normal distribution with a mean μ, another normal distribution for μ is a conjugate prior, and the integral in equation (4.80) immediately reduces to the marginal PDF.

Another approach to obtaining the conditional distribution of the parameters given the data (that is, the posterior distribution) is to simulate samples from it. We do not form the expression in equation (4.81), rather, we obtain samples of realizations of that distribution. Having samples of any size provides the information about the distribution itself.

There are various forms these simulations can take. They are collectively known as "Markov chain Monte Carlo", or MCMC. The simplest type of MCMC method is called "Gibbs sampling". This method is implemented in the BUGS and JAGS software. The simulations proceed through a sequence of conditional distributions that ultimately converge to a stationary distribution in a Markov chain. We will not discuss the details of the simulations here.

Because of questions of convergence of the process, much of the effort in a Bayesian analysis is devoted to analysis of the *method*, rather than analysis of the data and model.

Software

A practical problem in performing a Bayesian analysis is how to specify the model to a general-purpose computer program. A widely-used program for Bayesian analysis is BUGS, which is a standalone computer program that has its own keywords for defining the model. There are various versions of BUGS, such as WinBUGS and OpenBUGS. A BUGS executable program can be invoked from R.

Another system with a similar design and similar keywords is JAGS. JAGS is open source software, and an executable program can be obtained from SourceForge. *This program must be installed in a directory accessible by R.* Once the executable program is installed, it can be invoked from R using the `rjags` package.

An Example

We will illustrate the use of `rjags` to perform a simple analysis of the INTC daily returns for the first three quarters of 2017, discussed in Chapter 1. (These data are shown in the histogram of Figure 1.20 and elsewhere.)

We wish to estimate the population mean and variance of the daily returns. The sample mean is 0.000251 and the sample standard deviation is 0.00887 (see page 80). These sample statistics can be considered as estimates of the population parameters, with no further assumptions about the distributions. Of course, we could also use other sample statistics that we discussed in Chapter 1, such as a trimmed mean or a Winsorized mean, as estimates (see also Section 4.4.6 below).

Bayesian estimates require prior information or beliefs. For this example, we model the observables with a normal distribution and we assume a normal prior for its mean and a chi-squared prior for its precision (reciprocal of the variance). For the mean, we choose a flat or noninformative prior normal distribution. We choose a mean of 0 and a variance of 1,000,000. The large variance means that we put more weight on the mean of the sample than on the mean of prior, 0. For the precision, we choose a chi-squared distribution with 5 degrees of freedom. (The choice of chi-squared is motivated by the distribution of the sample variance, which involves the precision; see page 289. The choice of the degrees of freedom is made arbitrarily just for the example.)

In terms of the directed acyclic graph on page 387, Θ_1 represents the mean daily return and Θ_2 represents the standard deviation of the daily returns. The hyperparameter Ψ is a vector of the two parameters in the prior distribution of Θ_1. In this case, the hyperparameter is the constant (0, 1000000), or in a different form, (0, 1.0e-6). The hyperparameter Φ is a single parameter in the prior distribution of Θ_2. In this case, this hyperparameter is also a constant. It is 5.

Defining the Bayesian Model

JAGS and `rjags` require an external text file that specifies the model. The external text file specifying the model is saved in an accessible directory with a filename extension of "`bug`", as shown in Figure 4.2 below.

The external file uses its own notation, which is used in other BUGS software. The BUGS/JAGS language is somewhat similar to the R language, but it has some functions with the same or similar names that have different meanings for the arguments. There is also a class of functions that *specify a distribution*. The names of these functions all begin with "`d`", and the other component is a mnemonic similar to the R functions for probability distributions (Table 3.4). For example, in the BUGS/JAGS language, `dnorm` just means "the normal distribution", not the CDF of the normal distribution, as in R. BUGS/JAGS provides many common distributions. These distributions are used in the model statement to specify how a particular variable is distributed.

Another important difference between the BUGS/JAGS functions that specify distributions and the R functions that evaluate distributional functions is that arguments of some BUGS/JAGS functions may specify a "precision", which in the BUGS/JAGS language, as in most Bayesian literature, is interpreted as the reciprocal of the *variance*. The corresponding R functions specify a *standard deviation*. For the normal distribution, for example, we have

$$\text{BUGS/JAGS} : \texttt{dnorm}(\texttt{mu}, \texttt{tau}) \quad \Leftrightarrow \quad \text{R} : prenorm(arg, \texttt{mu}, 1/\texttt{sqrt}(\texttt{tau}))$$

The file to define the model, therefore has the form shown in Figure 4.2. We store it in an external file that we name `INTC173Q.bug`.

In Figure 4.2, the observables `x[i]` have a normal distribution, `mu` has a prior normal distribution, and `tau` has a prior chi-squared distribution.

Using the Model to Analyze Data

After defining the model using the BUGS/JAGS language and storing the file in the R working directory, we can process it, and use it to analyze data using R commands. The R function `jags.model` in the `rjags` package performs the usual BUGS/JAGS operations using the appropriate executable file. Notice that `jags.model` requires starting values (`inits`), and these are obtained by ordinary R functions `rnorm` and `rchisq`. The R function `coda.samples` in the

```
model{
# observables
   for(i in 1:N) {
       x[i] ~ dnorm(mu,tau)
   }
# priors on parameters
   mu ~ dnorm(0, 1.0e-6)
   tau ~ dchisq(5)
   sigma <- 1/sqrt(tau)   # the standard deviation
}
```

FIGURE 4.2
JAGS/BUGS Model Setup to be Stored in `INTC173Q.bug`

`rjags` package organizes the output into an object of class `mcmc.list`, and the standard R function `summary` displays the output.

Figure 4.3 illustrates the use of **rjags**. All statements in that figure are regular R statements.

```
library(rjags)
r <- as.numeric(INTCd20173QSimpleReturns)
N <- length(r)
INTCRet <- list(x = r, N = N)
inits <- function(){rwlist(mu = rnorm(1,0,1), tau = rchisq(1,5))}
INTC173Q <- jags.model("INTC173Q.bug", data=INTCRet)
INTC173Q.coda <- coda.samples(INTC173Q, c("mu", "tau", "sigma"),
               n.iter = 1000)
summary(INTC173Q.coda)
```

FIGURE 4.3
R Statements to Perform Analysis

CODA (Convergence Diagnostic and Output Analysis) is a software system that monitors and reports on the MCMC performance in the analysis. There are separate versions of CODA that are incorporated into JAGS and the various versions of BUGS. The `coda.samples` function in Figure 4.3 produces simple output, a portion of which is shown in Figure 4.4.

In Bayesian analysis, since the parameters are assumed to be random variables, we do not just estimate the parameters themselves; rather, we estimate their distributions. The estimated distributions can be specified in terms of their quantiles. Some estimated quantiles of the conditional distributions of the mean ("`mu`"), the standard deviation ("`sigma`"), and the precision ("`tau`") are shown in Figure 4.4.

```
1. Empirical mean and standard deviation for each variable,
   plus standard error of the mean:

           Mean        SD  Naive SE Time-series SE
mu     1.667e-04  0.005186 0.0001640       0.0001640
sigma  7.337e-02  0.003908 0.0001236       0.0001519
tau    1.874e+02 19.809482 0.6264308       0.7416085

2. Quantiles for each variable:

            2.5%        25%       50%       75%      97.5%
mu      -0.009336  -0.003325 1.723e-05 3.744e-03    0.01066
sigma    0.066222   0.070924 7.324e-02 7.581e-02    0.08243
tau    147.180765 174.016964 1.864e+02 1.988e+02  228.03284
```

FIGURE 4.4
Output from `rjags`

The results in Figure 4.4 will vary on different runs because of the random starting points and the randomness of the Monte Carlo procedure itself. We recall from page 80 that the mean of the sample was 0.000251 and the standard deviation was 0.00887, compared to the Bayes fits on this run (0.0001667 and 0.07337 in Figure 4.4). Notice that because the prior distribution of the mean was flat or noninformative, the posterior distribution of the mean is concentrated near the sample value of 0.000251. On the other hand, the posterior distribution of the standard deviation is not concentrated near the sample standard deviation. The prior distribution of the standard deviation (which is a transformed inverse chi-squared distribution) is somewhat more concentrated, so the posterior distribution will not be as affected by the sample value.

There are several additional arguments to both the `jags.model` function `coda.samples` function that control the *computational aspects* of the analysis. As we mentioned above, much of the effort in a Bayesian analysis is devoted to analysis of the *method*, rather than analysis of the data and model. There are many types of diagnostic output available about the *performance of the method of analysis*, which is not the same as the analysis itself.

4.4.5 Resampling Methods; The Bootstrap

Statistical inference involves statements about the confidence or significance level of estimates or test statistics. These statements are based on probability distributions in the model and distributions of various transformations made in the analysis. This is called *exact inference*. In some cases, the distributions can be derived from the basic underlying probability distribution in the model. In other cases, a tractable asymptotic distribution can be derived, and variances or other properties of quantities in the analysis can be computed. This provides the basis for approximate statements of confidence or significance. This is called *asymptotic inference*.

In many cases, neither exact nor relevant asymptotic distributions can be worked out. Even if such distributions are available, there may be questions about whether the data really arose from the assumed distribution. Simulation of probability distributions or random sampling within the given dataset can be used to estimate variance, to estimate bias, and to establish levels of confidence or significance. This is called *computational inference*. The MCMC methods referred to above, as well as other Monte Carlo simulation methods are examples of computational inference. Resampling methods of various types are also examples of computational inference.

Resampling

Resampling is the reuse of the same dataset in the same analysis. Multiple subsets of the original dataset are repeatedly used. In some cases, this is done randomly and in some cases it is systematic. These methods are very general and can be used in various statistical methods.

Reuse of the sample does not mean that we are getting something for nothing. The inferences based on resampling are based on the probability distributions of subsamples.

The two main types of resampling are cross-validation, which we will discuss in Section 4.6.2, and the bootstrap.

The Bootstrap

A given sample of data $x_1, \ldots, x_n$ defines a discrete population. If each element is distinct, and we assign a probability of $1/n$ to each element, we have the discrete uniform distribution, which we have referred to previously. *Bootstrap methods* repeatedly draw samples (with replacement) from that discrete uniform distribution; that is, from the given sample.

A *bootstrap sample* is a simple random sample from the finite, discrete population that consists of the points $x_1, \ldots, x_n$, each with a probability of $1/n$. Let us designate the elements of the bootstrap sample as $x_1^*, \ldots, x_n^*$. Any

x_i^* may be any x_j, and the same x_j may occur multiple times in the bootstrap sample $x_1^*, \ldots, x_n^*$.

In all aspects of data analysis, we need estimates of the variances and biases of the statistics we compute. For simple statistical procedures, such as estimating a mean, we can easily compute an estimate of the variance of the statistic used as the estimator. In more complicated situations, however, we cannot determine an estimate of the variance of the statistic. Likewise, we may not know anything about the bias of the statistic. Bootstrap methods can be used to estimate the variance as well as to estimate, and to correct for, bias.

Suppose T is a statistic that we compute from the original data, $x_1, \ldots, x_n$, that is, it is a function of those data, $T(x_1, \ldots, x_n)$. We can generate a bootstrap sample $x_1^*, \ldots, x_n^*$, and compute the same statistic, which we will call T^*; that is, $T^*(x_1^*, \ldots, x_n^*)$. We can do this many times. We generate B bootstrap samples, and for each, we compute the statistic; call them $T^{*1}, T^{*2}, \ldots, T^{*B}$. The basic ideas of the bootstrap is that $T^{*1}, T^{*2}, \ldots, T^{*B}$ provides information about the distribution of the original T, such as its variance and bias. The variance of the $T^{*1}, T^{*2}, \ldots, T^{*B}$, for example, is an estimate of the variance of the original T.

We know a lot about the $T^{*1}, T^{*2}, \ldots, T^{*B}$ because they are computed from a finite, discrete distribution.

In applications, B is of the order of 1,000 or so, depending on n and on the nature of the statistical problem and the underlying distributions.

If the statistic T is something simple, there is no reason to do this; we already know the distribution of T, or at least we can estimate its variance easily. The bootstrap can be used in more complicated situations, such as estimating the tail index, discussed later in this chapter.

Bootstrap Estimate of the Variance of an Estimator

As mentioned above the bootstrap estimator of the variance of T is $V(T^*)$. We possibly could work this out analytically because T^* is computed from a simple, finite, discrete uniform distribution, but we can easily estimate it from the observed values $T^{*1}, T^{*2}, \ldots, T^{*B}$. Hence, an estimator of $V(T^*)$ is the sample variance of the $T^{*1}, T^{*2}, \ldots, T^{*B}$.

The basic idea is that the variation of the individual T^{*j} about their mean $\overline{T}^*$ is similar to the variation in T about its mean.

The finite-sample exact properties of this estimator of the variance depend on the underlying distribution and on the nature of the statistic T. Under mild assumptions, the estimator is consistent.

In a simple case, such as if T is the sample mean, the bootstrap estimate of the variance is the same as what we would get from a direct approach. For the mean, it is always $\widehat{\sigma}/n$, where $\widehat{\sigma}$ is an estimate of the variance of the underlying sample.

Bootstrap Estimate of the Bias of an Estimator

If the statistic T is an estimator of the parameter θ, its bias is $\mathrm{E}(T) - \theta$. Without knowing the underlying distribution, both $\mathrm{E}(T)$ and θ are unknown. The bootstrap can be used to estimate the difference in $\mathrm{E}(T)$ and θ, however.

The idea is similar to the idea underlying the bootstrap estimator of the variance. An estimator of the bias is the difference in T computed from the original sample and the mean of the bootstrap samples, that is

$$\overline{T}^* - T.$$

The finite-sample exact properties of this estimator of bias depend on the underlying distribution and on the nature of the statistic T. In many cases, all we can conclude is that it provides an indication of the direction of the bias.

Bootstrap Confidence Intervals

As we discussed on page 381, a method of forming a confidence interval for a parameter θ is to find a pivotal quantity that involves θ and a statistic T, whose distribution is known, to form a probability statement relating to quantiles of the distribution of the pivotal quantity, and finally rearrange terms so that the expression is in the form of a confidence interval. If we let $f(T, \theta)$ be the pivotal quantity, the probability statement is of the form

$$\Pr\left(f_{(\alpha/2)} \leq f(T, \theta) \leq f_{(1-\alpha/2)} \right) = 1 - \alpha.$$

In simple cases, such as the example of the normal mean in Section 4.4.1, we can easily determine the quantiles $f_{(\alpha/2)}$ and $f_{(1-\alpha/2)}$, but in more complicated situations, we do not know these. If we could simulate $f(T, \theta)$, we could estimate $f_{(\alpha/2)}$ and $f_{(1-\alpha/2)}$ by Monte Carlo. This is not possible, since we do not know θ.

A bootstrap method for estimating $f_{(\alpha/2)}$ and $f_{(1-\alpha/2)}$ that requires few assumptions of the distribution is to simulate the distribution of $f(T^*, T_0)$, where T_0 is the value of T in the given sample; that is, we use $f(T^*, T_0)$ in place of $f(T, \theta)$. The confidence interval formed in this way is called a *basic bootstrap confidence interval*.

Consider again the example of the daily simple returns of the S&P 500, expressed as percentages, for the year 2017. We are interested in the mean return. We computed the mean of that sample of size $n = 250$ as 0.0667. Using the sample mean and the sample variance, under the assumption of a normal distribution, on page 380, we computed a 95% confidence interval for the population mean as

$$(0.0145, \quad 0.119).$$

We know the frequency distribution of the returns has heavier tails than the normal distribution, so that confidence interval is not exact.

We now compute a 95% basic bootstrap confidence interval, with $B = 1000$. In this case, the statistic T is the sample mean $T_0 = 0.0667$ (`xbar` in the R code below). This can be done in R with the following code.

```
set.seed(1)
for (i in 1:B) fB[i] <- mean(sample(x,n,replace=TRUE))
fB <- fB- xbar
quantile(fB,probs=c(0.025,0.975))
```

We get
$$(-0.0527, \quad 0.0541). \tag{4.82}$$

This is considerably wider than the parametric confidence interval based on the normal distribution, but the fact is that the means of a substantial number of resamples of the same data (more than 5%) lie outside the parametric interval (4.71). Exactly 5% of the means of the bootstrap samples lie outside of the interval (4.82).

The basic bootstrap confidence interval depends on very few assumptions about the underlying distribution; consequently, it tends to be conservatively wide. There are various other ways of using bootstrap methods to form confidence intervals. These other methods tend to give tighter (smaller) confidence intervals. Several methods and other applications are discussed by Efron and Tibshirani (1993).

Computational Issues

There are several R packages that perform bootstrap computations. Two of the most useful packages are `boot` and `bootstrap`. Many bootstrap applications require ad hoc code, however.

Many bootstrap applications are computationally intensive, but computations can often be performed in parallel. Serious applications require parallel processing, and in R, computations can often be speeded up by use of the `foreach` directive in the `foreach` package.

4.4.6   Robust Statistical Methods

Although many statistical procedures are based on a normal distribution, this does not mean that it is assumed that the underlying population is actually a normal population. It means that it is assumed that the characteristics of the distribution are not sufficiently different from normal to invalidate the statistical procedure. Simple graphs of financial data, for example, such as the log density plot in Figure 1.26 on page 91 or the q-q plot in Figure 1.27, indicate that the distribution of the returns is quite different from a normal distribution. The question is how much do these differences affect statistical inference.

The removal of just a few outliers might make the dataset sufficiently normal so that procedures based on normal populations would not lead to incorrect decisions. An alternative to removing outlying observations is the use of statistical procedures that are *robust* to departures from the distributional assumptions. Either of these approaches has drawbacks; the financial analyst must be aware of the consequences of removing and/or ignoring potentially informative data.

For inferences about a population mean, the sample mean is generally a good estimator, but extreme outliers can have a large effect on the sample mean. A robust estimator may be better. The sample median, for example, is a robust estimator. It is unaffected by a single observation larger than the median being made ever larger, or by a single observation smaller than the median being made ever smaller. Just as the sample mean is the least squares estimator of the population mean, the sample median is the least absolute values estimator of the population mean.

The sample median is not so affected by outliers, but if the population is skewed, the median is not a good estimator of the population mean. The α-trimmed mean, which we discussed on page 95, and the α-Winsorized mean (equation (1.60)) are compromises between the mean and the median. For $\alpha = 0$, they are the mean, and for $\alpha \approx 0.5$, they are essentially the median. For $\alpha > 0$, they are robust to outliers, but yet have good properties as estimators of the population mean for normal data.

Outliers can have a very large effect on the sample variance because it is an average of squared deviations. Instead of an average of squared deviations to measure the spread of the data, an average of the absolute deviations may be more appropriate. Also, instead of using deviations from the mean, which itself is subject to the effect of outliers, we may use the median. Finally, as an average of these deviations, instead of their mean, we may use the median of the deviations; that is, for a numeric vector x, we take the median absolute deviation (MAD) as in equation (1.62) on page 96.

Another robust measure of the spread of the data, is the interquartile range (IQR), as in equation (1.61).

Both the MAD and the IQR are more similar to the standard deviation than to the variance. The relationships to the standard deviation depend on the sample size and on the underlying probability distribution. Asymptotic approximations based on a theoretical normal distribution are often used as estimates of the standard deviation (see equations (1.63) and (1.64) and Exercise 2.2).

Use of quantiles-based measures such as MAD and IQR may make a normal distribution more similar to a frequency distribution of financial data. Figure 1.24 on page 89 showed a histogram of the daily simple returns of INTC for the first three quarters of 2017 with a normal probability density function superimposed. The normal distribution in that figure used the sample mean and the sample standard deviation. Figure 4.5 shows that same graph on the left side, and on the right side shows the same histogram but with a superimposed normal distribution that is fit using the sample median and the MAD.

While the standard deviation based on the MAD seems to yield better fits in the tails, the kurtosis of the sample is not fit well using either normal distribution. This highlights the main difference in the sample frequency distribution and the normal distribution.

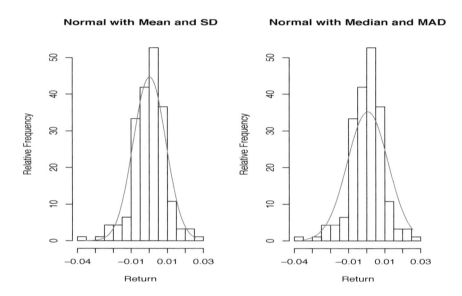

FIGURE 4.5
Frequency of Daily Simple Rates of Return for INTC Compared to Idealized Normal Frequency Distributions

The fits in Figure 4.5 are poor, and the problems of the heavy tails seen in the q-q plots of Chapter 1 remain.

Use of the ranks of data, rank(x), rather than the actual data x is more robust to outliers and heavy-tailed distributions. For the set of order statistics $x_{(1)}, \ldots, x_{(n)}$, recall that the rank transformation is

$$\tilde{x}_j = \text{rank}(x_j) = i \quad \text{where } x_j = x_{(i)}.$$

(This is equation (1.66) on page 97, where we introduced and illustrated ranks using Spearman's rank correlations for daily returns of DJIA, S&P 500, Nasdaq, and Intel.)

Relationships between two sets of rank-transformed data are similar to relationships between the two sets of raw data, and are sometimes more informative because they are not obscured by outlying observations. We recall that Spearman's rank correlation coefficient, or "Spearman's rho" (equation (1.67)), is based on rank data, and it is a more robust measure of association than the covariance or correlation.

Another measure of association that is more robust than the covariance or correlation, is Kendall's tau or Kendall's rank coefficient (equation (1.68)). It is the difference in the numbers of concordant and discordant pairs normalized by the total number of pairs.

Which Population Measures Are Relevant?

Our objective in statistical analysis is to understand a population, using data. We use sample statistics to estimate population measures, such as a sample mean to estimate a population mean. When we are concerned about the presence of erroneous or misleading outliers in the data, we may resort to robust sample statistics. If the population is skewed, however, use of a trimmed mean or the median, which is an extreme trimmed mean, may not be a good estimate of the population mean. We might consider whether the population mean is the most relevant characteristic of the population. Perhaps the population median is more relevant.

Similar considerations arise for other population measures. The population correlation, for example, may not be the relevant descriptor of the association between variables. (Recall the example of X and Y on page 262, where $Y \approx X^2$.)

4.4.7 Estimation of the Tail Index

In financial applications, the tails of probability distributions are of interest. This is because the tails correspond to extreme gains or, more seriously, to extreme losses. In a normal probability distribution, the probability density in the tails decreases at a rate of e^{-x^2}, but the probability density in heavier-tailed distributions may decrease at a polynomial rate, $|x|^{-(\alpha+1)}$.

In distributions with polynomial tails, either left or right, the important measure is the tail index, α in the expression above (see page 275). Larger values of the tail index indicate lighter tails (hence, as we stated earlier, sometimes the relevant measure is taken as the reciprocal of the tail index as we have defined it).

In this section we will consider only the lower (left) tail, although the methods of estimation for either tail are the same with the appropriate changes (see Exercise 4.14d).

We discussed general methods of estimation of the probability density in Section 2.3, but an estimate of the overall probability density may not yield a very good estimate of the tail index.

The PDF of a distribution with a polynomial left tail is asymptotically

$$f_X(x) = c|x|^{-(\alpha+1)},\qquad(4.83)$$

for some $c > 0$, as $x \to -\infty$.

The question is how small must x be for this to be an adequate approximation to the PDF. Let us assume that it is a good approximation whenever $x \le x_u$ for some $x_u < 0$.

If we assume that this is adequate approximation for $x \le x_u < 0$, then the conditional PDF for $x \le x_u$ is

$$f_{X \le x_u}(x) = \alpha |x_u|^{-\alpha} |x|^{-(\alpha+1)}\qquad(4.84)$$

(Exercise 4.14a). Given a random sample of size n with order statistics $x_{(1:n)}, x_{(2:n)}, \dots, x_{(n:n)}$, in which n_{x_u} sample values are less than x_u, the log-likelihood given the smallest n_{x_u} order statistics is

$$l_L(\alpha) = n_{x_u} \log(\alpha) - n_{x_u} \alpha \log(|x_u|) - (\alpha + 1) \sum_{i=1}^{n_{x_u}} \log(|x_{(i:n)}|).\qquad(4.85)$$

Differentiating the log-likelihood with respect to α and equating the result to 0, we have the stationary point

$$\widehat{\alpha} = \frac{n_{x_u}}{\sum_{i=1}^{n_{x_u}} \log(x_{(i:n)}/x_u)}.\qquad(4.86)$$

The second derivative of the log-likelihood at this point is negative, indicating that this is the MLE of α. (In Exercise 4.14b, you are asked to confirm the derivatives and the MLE above.)

The development above depends on the value x_u below which the PDF is exactly proportional to $|x|^{-(\alpha+1)}$. Given a random sample, this value determines how many order statistics to use in the estimate of α.

An alternate approach, instead of using x_u to determine the number of order statistics n_{x_u}, would be to choose the number of order statistics, perhaps as some fraction of the overall sample size n. This would be equivalent to taking $x_u = x_{(k:n)}$, for some k, so long as $x_{(k:n)} < 0$. The resulting estimator from equation (4.86) is called *Hill's estimator*.

$$\widehat{\alpha}_{\mathrm{H}} = \frac{k-1}{\sum_{i=1}^{k-1} \log(x_{(i:n)}/x_{(k:n)})}.\qquad(4.87)$$

In this form, the estimator is the reciprocal of the mean of the logs of the absolute values of the first $k-1$ order statistics minus the log of the absolute value of the kth order statistic. While the empirical properties of the two estimators $\widehat{\alpha}$ and $\widehat{\alpha}_{\mathrm{H}}$ are essentially the same, Hill's estimator is an exact MLE only if the PDF is proportional to $|x|^{-(\alpha+1)}$ over its full support.

Hill's estimator is a tuned estimator. The tuning parameter is k. (See the discussions beginning on page 354 and on page 378.) As in most cases of tuned inference, the value of the tuning parameter affects the results, but its choice is somewhat subjective.

Consider the daily log returns for the S&P 500 from January 1, 1990, through December 31, 2017. (These are the data shown in Figure 1.27 on page 92.) A q-q plot of those data indicated that the returns had heavier tails than the normal distribution.

As we have mentioned regarding this dataset, because the distribution of returns is not stationary (see the plot of these data shown in Figure 1.29 on page 99), we may question the relevance of any analysis that uses data over a period of 28 years. Nevertheless, we compute $\widehat{\alpha}_H$ for k ranging from 11 to the maximum value of 500, and show a plot of the estimator versus the tuning parameter in Figure 4.6.

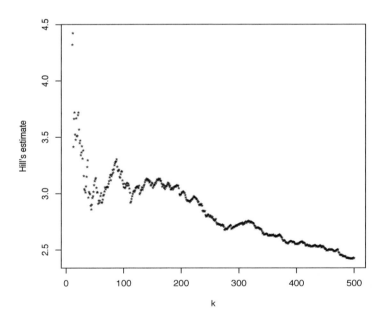

FIGURE 4.6
Hill's Tail Index Estimates for the Daily Log Returns of the S&P 500 from 1990 through 2017 for Different Values of k

The plot in Figure 4.6 showing the effect of the tuning parameter on the estimator is called a *Hill plot*. The tuning parameter, as in many cases of tuned inference affects the bias and the variance in opposite ways. There is a bias-variance tradeoff. As the tuning parameter gets smaller, less of the sample is used and the variance increases; on the other hand, as more order statistics are used, it is likely that the PDF behaves less like $|x|^{-\gamma}$, and hence the estimator is more biased. In Exercise 4.16b, you are asked to investigate this empirically using bootstrap estimates of the bias and variance. Hill plots can be highly variable, and it is sometimes difficult to obtain any useful information from the plot. You will see some various shapes of Hill plots in Exercise 4.15. Other similar plots that may be more informative have been suggested (see Drees, de Haan, and Resnick, 2000, for example).

The Hill plot can be useful to identify ranges of the tuning parameter in which the estimate is relatively stable. In Figure 4.6, this *interval of stability* appears to be between about 100 and 200. The estimates are approximately 3 in this range. A practical method of estimating the tail index is first to identify an interval of stability in a Hill plot, and then to take the estimate as the mean of Hill's estimates over that interval.

Many of the statistical distributions useful in finance have polynomial tails, and so the tail index is an important characteristic of the distributions. The tail index may indicate the need for robust statistical methods, and may even be useful in deciding on the type of method.

Most statistical distributions in finance vary over time. The stochastic volatility of returns shown in Figure 1.29 in Chapter 1 illustrates this dramatically. Certain aspects of the distribution of the returns are essentially the same during periods of either high or low volatility, but others vary over time. In Exercise 4.15, you are asked to investigate how the tail index changes over different periods of high and low volatility.

The R package `evir` has functions for computing the Hill estimator of the tail index, for producing plots, and performing other computations related to the generalized Pareto distribution, as mentioned previously. The `hill` function produces a Hill plot along with confidence intervals computed using the bootstrap. The `gpd` function computes a fit of the generalized Pareto distribution to a given dataset.

Estimating Tail Dependence

The tail index is a *local measure* of a probability distribution for a single random variable. Just as the distribution of a single random variable in the tails is of interest, the relationship between two random variables in the tails is also of interest. This would require a *local measure of association*.

One obvious measure of the relationship would be the correlation conditional on each random variable being greater than specified values or specified quantiles. Another measure is based on the "tail dependence" (see page 276). For two random variables, Y and X with CDFs F_Y and F_X, the *upper tail dependence function* $\lambda_u(p)$, equation (3.76) on page 276, and the *lower tail dependence function* $\lambda_l(p)$, equation (3.77), are local measures of association in the upper or lower p proportion of the distribution. The lower tail dependence function is

$$\lambda_u(p) = \Pr(Y > F_Y^\leftarrow(p) \mid X > F_X^\leftarrow(p)) \tag{4.88}$$

The limits at the extremes of the distribution are called the *coefficient of upper tail dependence* and the *coefficient of lower tail dependence* (equations (3.78) and (3.79)). The coefficient of lower tail dependence is

$$\lambda_l = \lim_{p \to 0^+} \lambda_l(p). \tag{4.89}$$

These measures are conditional probabilities, and hence, are not symmetric in Y on X; nevertheless, they are widely used in financial applications. (In some of the literature, they are denoted as "$\chi_{XY}(p)$" and "χ_{XY}". This notation is preferable, because the "X" and "Y" indicate that the measure is directional. This is the terminology used in the extRemes R package.) The extRemes R package provides various functions for inference about extreme values, including estimators of tail dependence.

Although the tail dependency is defined in terms of a conditional distribution, which is asymmetric, we generally use it as a symmetric measure. A symmetric estimator of $\lambda_l(p)$ for $0 < p < 1$, based on a bivariate sample $(x_1, y_1), \ldots, (x_n, y_n)$, uses the univariate sample quantiles x_p and y_p. The estimator is

$$\widehat{\lambda_l}(p) = \#\{(x_i, y_i) < (x_p, y_p)\}/pn, \tag{4.90}$$

where "#" is the number of items satisfying the condition, and $(x_i, y_i) < (x_p, y_p)$ means that both $x_i < x_p$ and $y_i < y_p$. This is the quantity $\widehat{\chi_{XY}}(p)$ chi.hat(x, y, p) computed in the taildep function in the R extRemes package.

Tail dependencies can be seen qualitatively in scatterplots. Consider the scatterplot matrix on the 2017 daily returns of the three major indexes and INTC and GLD on page 86. In each scatterplot, large tail dependencies would result in a relatively large proportion of points being in the lower left and the upper right of the plot. We see that this is the case for all pairs of indexes, and it is largely true for INTC with each of the indexes. (The outlier in INTC distorts the view somewhat.) The tail dependency of GLD, the gold bullion ETF, and any of the indexes and the tail dependency of GLD and INTC, however, all appear to be rather small. In Exercise 4.17, you are asked to compute the sample tail dependencies at the $p = 0.05$ level for INTC and the Nasdaq Composite and for GLD and the Nasdaq Composite for the year 2017, corresponding to the data in the scatterplots in Figure 1.22.

In portfolio construction, assets with small tail dependencies are often selected to diversify the portfolio.

There are various other local measures of association, including a local or conditional version of Kendall's tau. (The unconditional Kendall's tau is defined in equation (1.68) on page 97.)

4.4.8   Estimation of VaR and Expected Shortfall

Risk management is one of the most important purposes of financial analysis. Risk is the variation in prices of assets. It is measured by the volatility, which is the standard deviation of returns over some period. In the management of risk, we are concerned primarily with negative returns. We are interested in the possible amounts of losses.

In risk management, there are three components: a time period, an amount of loss at the end of that period, and a probability of that amount of loss or of a greater amount. All of these components depend on the returns. Returns used in risk management are usually simple one-period returns.

Beginning on page 114 in Chapter 1, we defined various measures useful in risk management, the most important of which were value at risk or VaR, and expected shortfall or conditional value at risk, CVaR.

The value of a risky asset t time units in the future is modeled as a random variable. For that time period t, and a given probability α, VaR is the α-quantile of that distribution of changes in values. Rather than modeling the distribution of the values of the asset directly, we model the distribution of returns, and then model the changes in asset value using the return.

For a given value of a risky asset P, VaR is defined for a given probability α and for a given time forward t as

$$\text{VaR}(t, \alpha) = -Pr_{t,\alpha}, \tag{4.91}$$

where $r_{t,\alpha}$ is the α-quantile of the distribution of the returns for the next t time units; that is, $r_{t,\alpha}$ is such that

$$\Pr\left(R_t \leq r_{t,\alpha}\right) = \alpha,$$

where R_t is the random variable representing the simple return t time units forward. The minus sign is used just to make the quantity positive, because α is small enough that $r_{t,\alpha} < 0$.

Since the initial principal P is known, all uncertainty in VaR is due to uncertainty in the distribution of the returns.

We estimate VaR for a given α and t by first estimating $r_{t,\alpha}$, and then applying it to the initial value of the asset:

$$\widehat{\text{VaR}}(t, \alpha) = -P\widehat{r}_{t,\alpha}. \tag{4.92}$$

Another useful measure for risk management is the *expected shortfall*, discussed on page 119. Here, we focus on the amount of *loss* in the value of the asset, $L(t)$, at t time units in the future. By the definition of VaR, we see that for given α and t,

$$\Pr\left(L(t) \geq \text{VaR}(t, \alpha)\right) = \alpha.$$

(Here, we have assumed that L is a continuous random variable, and we have written the probability of "$\geq$", rather than of "$>$", which would result directly from the definition of VaR.) The expected shortfall for α and t is the conditional expected loss, given that the loss is at least as great as $\text{VaR}(t, \alpha)$:

$$S(t, \alpha) = \text{E}(L(t) \,|\, L(t) \geq \text{VaR}(t, \alpha)). \tag{4.93}$$

An estimate of the expected shortfall is obtained by averaging all losses greater than or equal to $\widehat{\text{VaR}}_{t,\alpha}$. Once the estimate $\widehat{r}_{t,\alpha}$ is available, however, the estimate of the expected shortfall is obtained more easily by averaging all returns less than or equal to $\widehat{r}_{t,\alpha}$ and then multiplying by the initial value P.

The Time Horizon

The quantile $r_{t,\alpha}$ of course depends on the distribution of the rates of return, and that distribution depends on the length of time over which the return is computed. The shapes of the distributions of returns are different for different intervals (see the q-q plots in Figures 1.27 and 1.28 of samples of daily, weekly, and monthly rates of return). This means that to estimate $r_{t,\alpha}$, we need a sample of simple rates of return for the specific length of the time horizon in the VaR.

Estimation of the α-Quantile of the Returns

Both $\text{VaR}(t, \alpha)$ and $S(t, \alpha)$ depend on $r_{t,\alpha}$, the α-quantile of the returns for a time period of t, and so estimates of those quantities can be based on an estimate of $r_{t,\alpha}$.

Estimation of distributional properties of returns is not a simple process. "Historical volatility", which is the standard deviation of recently observed returns, is not a good measure; that is, it does not fit well with financial models of returns. Nevertheless, for estimating $r_{t,\alpha}$, we do use recent historical returns. Deciding how far back in time to include returns in the historical sample is not straightforward. The farther back in time, the more data, which is good, but also the less likely that the distribution is the same, which is bad. On the other hand, if the distribution is changing, the farther back in time provides more information about the nature of the changing distribution. In practice VaR computations may use as little as six months of data to as much as ten years of data.

Given a set of historical returns, there are two ways we can estimate the quantile $r_{t,\alpha}$. A simple way is *nonparametric*. One nonparametric method uses the frequency distribution of a sample of returns. This is the method we used on page 116 (see the figure).

Another nonparametric estimator is merely the sample quantile of the observed returns. As we mentioned on page 216, there are various ways empirical quantiles are defined. The R function `quantile` provides nine different sample quantiles (and there are others). Fortunately, which form of the sample quantile is used does not make much difference in the estimated VaR or shortfall.

Figure 1.37 on page 116 illustrates another nonparametric approach, in which case the estimate of the quantile is taken as the point on a histogram corresponding to a probability of α. This approach depends on how the bins in the histogram are defined.

Another way of estimating the quantile $r_{t,\alpha}$ from historical data is *parametric*. It involves fitting a parametric probability distribution to the returns, and then from the fitted PDF, evaluating the α-quantile. The normal distribution is easy to fit to data, and although it is sometimes used in VaR and shortfall computations, it is not very good for this purpose, because its tails do not fit the returns very well in the areas where we are most interested. There are several other more appropriate families of probability models, such as those we discussed in Section 3.2.4 beginning on page 295. The location-scale t distribution discussed in that section is useful for this purpose. The R function `fitdistr{MASS}`, discussed on page 354, fits the three parameters of the location-scale t distribution. Then, using the relationship between the location-scale t distribution and the ordinary t distribution, the R function `qt` can be used to compute the α-quantile.

A parametric model can also be chosen based on past experience, and rather than the model being fitted to the given data, that data is used to set the values of some aspects of the model. Figure 1.38 illustrates this simple parametric approach. A t distribution with 4 degrees of freedom was chosen as an idealized model. It was not fitted to the data in that example; rather, the quantile in the data was taken as the quantile of the t distribution times the sample standard deviation.

Another method of estimating a quantile is to start with the normal quantile and make an approximate adjustment based on the sample skewness and kurtosis. A good approximation is based on the *Cornish-Fisher expansion* of the unknown CDF, based on Hermite polynomials. The simple form of the expansion is based on cumulants instead of moments, so there are some complicated computations involving cumulants as well as the Hermite polynomials. The R package `PDQutils` has functions to convert cumulants to moments or vice versa and to evaluate the Hermite polynomials necessary for the cumulant approximation. We will not discuss this method further here. (Gentle, 2019, Chapters 0 and 1, has a discussion of orthogonal polynomials and various expansions.)

A parametric method may perform better than a nonparametric approach for small sample sizes and/or for small values of α, of course assuming the parametric model provides a good fit.

Once the estimate $\widehat{r}_{t,\alpha}$ is available, whether it is parametric or nonparametric, the estimate of $\mathrm{VaR}(t, \alpha)$ is obtained merely by multiplying by -1 and the initial value P, as in equation (4.92). Likewise, an estimate of the expected shortfall is obtained by averaging all returns less than or equal to $\widehat{r}_{t,\alpha}$ and then multiplying by the initial value P.

VaR and the expected shortfall are both defined with respect to a specific future time. In the case of many financial institutions, such as banks, the time interval is usually one day. Individuals and portfolio managers may be interested in other time horizons. Given a set of historical daily returns, the returns over other intervals can be obtained by adding the appropriate number of daily returns, as we showed on page 31. This aggregation is done before any sorting of the data; quantiles cannot be aggregated in this way.

VaR and Expected Shortfall in Portfolios

We usually define a portfolio of assets by specifying the proportionate values, $w_1, \ldots, w_N$, of the individual assets. Over time, these weights change because the returns of the individual assets, $r_1, \ldots, r_N$, are not equal. Portfolio management often consists of rebalancing the assets in order to preserve a target set of weights that achieve an optimal combination of risk and return, as we discussed in Section 1.2.9.

The rate of return of a portfolio depends on the relative proportions of the individual assets and their rates of returns. As we have seen, the single-period simple return of a portfolio is a linear combination of individual simple returns. Because the relative proportions change, however, a simple formula such as equation (1.39) is not adequate.

For the historical rates of returns needed to estimate the VaR or expected shortfall, we often compute the daily or weekly portfolio returns using a constant set of weights that correspond to the correct weights at some point during the relevant time interval. (see Exercise 4.18f.)

Another issue in measuring the risk of a portfolio is the fact that quantiles of linear combinations do not behave coherently; that is, the quantile of a linear combination of two variables may not be between the quantiles of of the two variables individually (see Section 3.1.3).

Confidence Intervals for VaR(t, α) and for the Expected Shortfall

Knowing the distribution of the estimator $\widehat{r}_{t,\alpha}$ allows us to derive a confidence interval for VaR(t, α). In practice, we generally just use an estimate of the standard deviation of $\widehat{r}_{t,\alpha}$ and then form a normal approximation for the confidence interval,

$$- P \left(\widehat{r}_{t,\alpha} \pm z_\gamma \widehat{\sigma}_{\widehat{r}_{t,\alpha}} \right), \tag{4.94}$$

where, for a $(1-\gamma)100\%$ confidence interval, z_γ is the γ-quantile of a standard normal distribution, and $\widehat{\sigma}_{\widehat{r}_{t,\alpha}}$ is the estimated standard deviation of $\widehat{r}_{t,\alpha}$.

The lower bound in a confidence interval for the expected shortfall would be computed by averaging all returns less than or equal to the lower confidence bound on $r_{t,\alpha}$ and then multiplying by the initial value P. The upper bound in a confidence interval for the expected shortfall would be computed in a similar manner.

Another way of forming a confidence interval for VaR(t, α) or for the expected shortfall is to use the bootstrap method described on page 395. This involves first resampling the returns and obtaining bootstrap sample estimates $\widehat{r}_{t,\alpha}^{*j}$, for $j = 1, \ldots, B$.

4.5   Models of Relationships among Variables

Many statistical datasets can be organized into a matrix ("flat file") in which the columns correspond to variables or features, and each row corresponds to an observation, which is a vector consisting of one value for each variable. We will call this matrix the "data matrix", and we will often represent it as X.

An important objective is to understand the relationships among the variables, that is, the columns of the data matrix.

First, we may compute the mean of each variable. The means are important, of course, but they may obscure other interesting relationships, so we may center the data by subtracting the mean of each column from all elements in the column. We will let $\overline{X}$ be a matrix whose columns are the means of the variables; that is, they are constant. The centered matrix is $X_c = X - \overline{X}$, in the notation on page 210. The ideas and operations here are similar to those discussed in Section 2.1.2 for variables, except here we are operating on a column of a data matrix, so the notation may be slightly different.

One further step may be to standardize the variables. To do this, we divide each column in X_c by the sample standard deviation of the column.

Covariances and Correlations

Let n be the number of rows of the data matrix X; that is, n is the number of observations. The matrix S, given by

$$S = \frac{1}{n-1}(X - \overline{X})^{\mathrm{T}}(X - \overline{X}) = \frac{1}{n-1}X_c^{\mathrm{T}}X_c, \qquad (4.95)$$

contains important information about the relationships among the variables.

Note that this matrix is symmetric and that the element in the ith row and j^{th} column of S, call it s_{ij}, is the sample covariance between the ith and j^{th} variables. (This is the same value as given in equation (1.54) on page 82.) The diagonal elements of S are the sample variances.

We will call S the *sample variance-covariance* matrix. We often drop one or more of these words, and refer to i as the variance-covariance matrix, the variance matrix, or the covariance matrix. The *sample* matrix is computed from data; it corresponds to the *population* or *model* variance-covariance matrix we defined in equation (3.48) on page 261.

Correlations are often more useful because they are all on the same scale, ranging from -1 to 1. The sample correlation between the ith and j^{th} variables is

$$r_{ij} = \frac{s_{ij}}{\sqrt{s_{ii}}\sqrt{s_{jj}}}.$$

The *sample* correlation matrix formed in this way corresponds to the *population* or *model* correlation matrix we defined in equation (3.49).

The correlation matrix would be the result in equation (4.95), if the matrix

X_c were standardized; that is, if for each i, the ith column is divided by $\sqrt{s_{ii}}$. The correlation matrix is the variance-covariance matrix of standardized data.

It may be of interest to test the hypothesis that the correlation, ρ_{ij}, between the ith and j^{th} random variables is 0. The obvious test statistic is the sample correlation, r_{ij}. The distribution of r_{ij} depends on the joint distribution of the underlying random variables X_i and X_j. Even if $\rho_{ij} = 0$, the distribution of r_{ij} is rather complicated. Without working out the distribution, we could test hypotheses concerning ρ_{ij} or set confidence intervals using various nonparametric techniques, including bootstrap methods.

If r is the sample correlation from a sample of size n from two independent normal distributions (that is, with $\rho = 0$), the quantity

$$t = r\sqrt{\left(\frac{n-2}{1-r^2}\right)} \tag{4.96}$$

is distributed approximately as t with $n-2$ degrees of freedom. This provides an approximate test of the null hypothesis that $\rho = 0$ and it provides approximate confidence intervals for ρ, based on critical values of the t distribution. (The significance levels of this test and of the confidence intervals are good approximations for samples from normal distributions. If the underlying samples are not normal, or if the elements of each sample are not independently distributed within the sample, then the distribution of t in equation (4.96) may be very different from a t distribution; see page 585 for an example.)

In Section 4.5.1 below we consider symmetric relationships. This leads to principal component analysis. In Sections 4.5.2 through 4.5.8 we consider various forms of asymmetric relationships in which one or more variables are considered "dependent" on the others. This is regression analysis.

4.5.1   Principal Components

Covariances and correlations provide useful information about *bivariate linear* relationships among the variables. The covariance or correlation matrix can also be used to identify other linear relations among more than just two variables at a time.

In *principal components analysis (PCA)*, we use S or a correlation matrix to identify a single linear combination of the variables that contains as much of the variation in all the individual variables as possible. We call that linear combination the *first principal component* or just the *principal component*. We can then identify a second linear combination of the variables that contains the most information remaining. This linear combination is linearly independent of the first one. We call this latter linear combination the *second principal component*. We can proceed to identify as many linearly independent principal components as there are linearly independent variables. Each linear combination of the original variables is effectively a new variable that is linearly independent of the other new variables; that is, the new variables have 0 correlations with each other.

A purpose in PCA is data reduction and dimension reduction. Instead of using all of the variables in a model, we may use only a few of the new variables, that is, the linear combinations. In a financial application, for example, a stock index or an ETF could be defined in terms of a principal component.

The data reduction is accomplished by forming new variables as linear transformations, which are the principal components.

Given a set of variables, $x_1, x_2, \ldots, x_m$, we seek to find a linear combination of them, $z = a_1 x_1 + \cdots + a_m x_m$, that has maximum variability. If $z = a^T x$ is a linear combination of the xs, then, as discussed on page 269, the sample variance of z is $a^T S a$. Obviously, by choosing larger values of the as, $a^T S a$ can be increased without bound. (Recall that S is nonnegative definite, so for any a, $a^T S a \geq 0$.) We are interested in the relative values in the linear combination, so if we restrict the size of the values in a, there is a maximum value.

Let us restrict a so that its Euclidean norm is 1, that is, $\sqrt{\sum a_i^2} = 1$. Among all such vectors, the one that maximizes $a^T S a$ is the normalized eigenvector of S that corresponds to the maximum eigenvalue. (This can be shown using the spectral decomposition of S. A proof of this is beyond the scope of this book. See Gentle, 2017, Section 3.8.10, for definitions and a careful development of this decomposition. It is not necessary for our purposes here to understand all of the mathematics.)

The elements of the vector that forms the linear combination are called "loadings", because they represent the contribution of each of the original variables to the principal component.

The main use of PCA is for data reduction. The first principal component accounts for the most variation in the data, the second principal component accounts for the second most, and so on. The relative amounts of variance in the original data that are "explained" by the principal components are proportional to the eigenvalues of the correlation matrix. A small number of principal components may account for a relatively large proportion of the variation in the individual variables. These few principal components may contain much of the information in the larger set of variables.

The idea of seeking a linear combination of a set of variables only makes sense if the number of variables is large. In financial applications, these may be tens or hundreds of stocks, interest-bearing securities, or other economic variables.

An Example

Here, we will just illustrate the concept with two variables we will call x and y. PCA is usually applied to a larger number of variables, of course, but two variables can be represented more easily.

A scatterplot of observations on the two variables is shown in Figure 4.7. (The data are actually the weekly log returns of the Dow Jones Industrial Average and the Nasdaq Composite for 2017.)

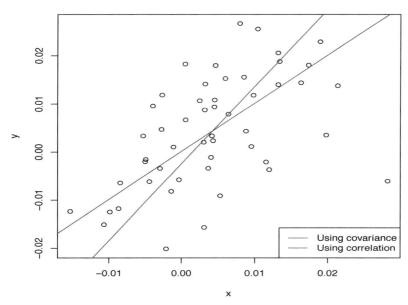

FIGURE 4.7
Scatterplot and Principal Component

The first principal component (that is, the vector $a = (a_1, a_2)$) is computed such that the sample variance of $a^T(x, y)$ is maximum. As indicated above, this is done by determining the eigenvector corresponding to the maximum eigenvector of the variance-covariance matrix or of the correlation matrix. Either matrix can be used, but there are differences. The correlation matrix has effectively been scaled, so that the values are comparable. In a covariance matrix a very large variance of a single variable may obscure the relationships among the other variables. It is generally better to standardize the variables or to use the correlation matrix. (As we mentioned above, the correlation matrix is the variance-covariance matrix of standardized data.)

In Figure 4.7, we show the first principal components computed using both the variance-covariance matrix and the correlation matrix. Both were computed using the R function `princomp`. We see that the first principal component (in both cases) seems to be in a direction of maximum variation of the data. (See the additional comments about the correlation case below, however.)

The R code for computing and displaying the first principal component using both the variance-covariance matrix and the correlation matrix is shown in Figure 4.8. (The result using the correlation matrix is the same as if we had standardized the data.) Figure 4.8 also shows the eigenvector of each that corresponds to the maximum eigenvalue, and we see that they are the same.

```
> princomp(XY)$loadings[ ,1]
        x         y
0.5293222 0.8484209
> eigen(var(XY))$vectors[ ,1]
[1] 0.5293222 0.8484209
> princomp(XY, cor=TRUE)$loadings[ ,1]
        x         y
0.7071068 0.7071068
> eigen(cor(XY))$vectors[ ,1]
[1] 0.7071068 0.7071068
```

FIGURE 4.8
Computing a Principal Component and an Eigenvector; R Code and Output

This is just a toy example to show the properties of principal components. It only gives a hint as to how PCA may be used in statistical analysis. As we mentioned above, PCA is often used to reduce the dimensions of a large number of variables. Although this is not an important point, our toy example illustrates that the PCA is actually meaningless for two standardized variables, or when using their correlation matrix: the loadings will always be equal.

Transformations prior to PCA

Unlike in the case of fitting a linear model with least squares, where the results are not affected by centering the data, centering the variables prior to extracting the principal components does make a difference. The results obtained after centering the data are more meaningful. The data should always be centered prior to PCA; that is, given a data matrix X, we should form X_c, in the notation on page 210.

Standardizing the data also affects the principal components. In many cases, it is best to standardize the data prior to extracting the principal components; that is, from the X data matrix, we form X_{cs}, in the notation used on page 210. Doing this equalizes the variances in the data, when otherwise, differences in the variances may just depend on the different units of measurement.

Applications in Finance

Data reduction is an important aspect of analysis of financial data. When there is a large number of economic measures to be considered, a single linear combination of them may serve just as well.

A stock index, for example, is a reduction of the data of several individual stocks. For the simple purpose of an index to track the variation in market prices, the first principal component would perform best. In Exercise 4.20, you are asked to form an index of stocks based on the amount of variation in the prices of the individual stocks that is accounted for in the index.

The principal component, however, may contain negative loadings, which would represent short positions in the corresponding stocks. This would be an undesirable property for a stock index.

Constrained Principal Component

For a given sample variance-covariance matrix or correlation matrix S, the first principal component is the solution to a constrained optimization problem: maximize $a^{\mathrm{T}} S a$ subject to $\|a\| = 1$.

The optimal value of a, that is, the vector of loadings, may contain negative elements. By adding a nonnegativity constraint, we have the constrained principal component problem:

$$\max_a a^{\mathrm{T}} S a$$

$$\text{s.t.} \quad \|a\| = 1$$
$$a \geq 0.$$

(4.97)

This optimization problem is a quadratic programming problem except for the additional constraint $\|a\| = 1$. The $\|a\| = 1$ constraint is usually handled by decomposing the matrix S into a weighted sum of outer products of its eigenvectors, as we indicated earlier. The constraint $a \geq 0$ is of the standard form in quadratic programming problems.

4.5.2 Regression Models

One of the most important types of models of relationships is a *regression model*, which is an asymmetric model relating a "dependent" variable (or a vector of variables) to another set of variables, called "regressors", "explanatory" variables, or "independent" variables. In the financial literature, regressors are often called "factors", and the regression model is called a "factor model".

The most common form of the regression model includes an additive "error" term, as in

$$y_i = g(x_i, \theta) + \epsilon_i,$$

(4.98)

where y_i and x_i are the ith values of observable variables, g is a function, usually of some known fixed form, θ is an unknown and unobservable *parameter*, and ϵ_i is an error term. The error term is an unobservable adjustment for the ith observation.

We use data to *fit the model*, that is, to estimate parameters in the model. We arrive at a model such that for the observed data,

$$y_i \approx \hat{y}_i = g(x_i, \widehat{\theta}).$$

The *residual* $y_i - \hat{y}_i$ is an important quantity in the fit.

Note that the residual is the *vertical distance* from y_i to $\hat{y}_i$. An alternative measure of the difference between the observed value and the fitted value would be the Euclidean distance from y_i to the fitted line $g(x_i, \widehat{\theta})$. Minimizing the sum of squares of these distances is called *orthogonal regression*.

Assumptions

We make a few important assumptions in regression models. First of all, there is a basic assumption of constancy in the data-generating process over the time in which the data are collected. As we have mentioned, most financial data have a time component, and the appropriate analysis often requires an explicit time component. (Time series models account for this directly. We will discuss time series models in Chapter 5.)

Secondly, we assume that the model relationships hold for all values of x, at least within a range of values.

We assume that the errors, as random variables, have 0 expectation (otherwise, we would just add their expectation into an intercept term).

We also assume that the errors, the ϵ_i, have a constant variance; that is, they are *homoscedastic*. Failure of this assumption, *heteroscedasticity*, is a common problem in financial applications, however. There are two main reasons for this. For one, many random quantities have variances that depend generally on their magnitudes. A second reason is time dependency, as mentioned above. Data-generating processes change over time.

Another important assumption about the errors is that they have 0 correlation with each other.

Sometimes we make a very specific assumption about the joint distribution of the errors. We may assume that they are iid $N(0, \sigma^2)$.

In the following, we will discuss various properties of the quantities in a regression analysis. For most statements we will give the mathematical reasons, but some are based on theory beyond the scope of this book.

Many of the statements depend on whether or not the model is linear; others depend only on the 0 expectation, constant covariance, and 0 correlation of the errors; and some depend on the errors having a normal distribution. For each property or statistical technique, it is important to recognize what assumptions are made.

Many formal aspects of statistical inference (tests of hypotheses, confidence intervals, and so on) in regression analysis are based on an assumption of a normal distribution of the errors. Even if the errors are not normally distributed, these procedures are often approximately correct.

Model Errors and Residuals

In regression analysis, as in other analyses in which we fit a model to data, it is necessary to assess how well the model fits the data. In general, we must inspect and analyze the residuals

$$r_i = y_i - \hat{y}_i. \tag{4.99}$$

If they are too large or if they do not appear to be randomly distributed, the model may not be appropriate.

The residuals r_i in equation (4.99) are not the same as the the errors e_i in equation (4.98). The errors, that is, the ϵ_is, are unobservable random variables. The residuals are not observed realizations of the corresponding random variables. To repeat, the errors are not observable.

The "hat" notation is often used to denote an estimate, such as $\widehat{\theta}$ for θ and $\hat{y}_i$ for the mean response at x_i. With this meaning of the hat notation, it is not appropriate to write "$\hat{\epsilon}_i$" for the ith residual r_i, because r_i is not an estimate of ϵ_i.

The residuals are just the result of fitting the regression model to a set of data. They have some relation to the random errors, of course, but even if the errors are independent of each other, the residuals, in general, are not independent of each other.

Qualitative Regressors

Some regressors in a regression model may be qualitative; that is, they may represent one of a set of discrete states, such as homeowner or not, employed or not, and so on. Binary discrete variables, such as these, present no problem; they can be arbitrarily coded as 0 and 1.

If a qualitative regressor can take on k different values, one way of incorporating it into a standard regression model is by forming a dummy variable for each of $k - 1$ (arbitrary) values of the qualitative regressor. Each dummy variable takes only the values 0 and 1, corresponding to the level of the qualitative variable in a given case or observation. If the qualitative variable takes its kth value, all dummy variables corresponding to that variable are 0.

Partitioning the Variation

As mentioned in Section 4.1.2, a useful way of thinking about fitting a regression model to data is that the model partitions the total variation of the dependent variable in the data into two sources of variation: that accounted for in the model and the residual variation.

Either source of variation is measured by a sum of squares. In the case of the total variation, it is

$$\text{SST} = \sum_{i=1}^{n} (y_i - \bar{y})^2. \tag{4.100}$$

(This is sometimes called the *adjusted* total variation. For other models, the total variation may just be $\sum y_i^2$; that is, there is no adjustment for the mean $\bar{y}$.) The variation accounted for by the fitted regression model is measured by

$$\text{SSR} = \sum_{i=1}^{n} (\hat{y}_i - \bar{y})^2. \tag{4.101}$$

This is sometimes denoted as SSReg, for "sum of squares due to regression". What is left over, the sum of squares of the residuals is

$$\text{SSE} = \sum_{i=1}^{n} (y_i - \hat{y}_i)^2. \tag{4.102}$$

If $\sum_{i=1}^{n} (y_i - \hat{y}_i) = 0$, then the equation

$$\text{SST} = \text{SSR} + \text{SSE} \tag{4.103}$$

is an algebraic identity if $\sum_{i=1}^{n} (y_i - \hat{y}_i) = 0$, which can be shown by expansion, as you are asked to do in Exercise 4.21. If the model is fitted by least squares, then $\sum_{i=1}^{n} (y_i - \hat{y}_i) = 0$, and so equation (4.103) holds; however, if the model is fitted by some other criterion, it may not hold.

The idea of partitioning the variation in a variable of interest into different sources of variation is one of the most fundamental principles in statistical inference. The partitioning of the observed variation and the analysis of variance address the question of whether or not the given data seem to follow the model.

These sums of squares for measuring the variations are called by different names in the literature. Often the "SS" is placed at the end, for example, what I call "SSE" is often called "RSS", for "residual sum of squares". R uses that term.

The variation is measured by the sums of squares of deviations of the individual terms from their mean values or from the values in the fitted model. The partitioned variation, that is, the sums of squares representing the different sources of variation, is compactly presented in an *analysis of variance* (ANOVA or AOV) table:

Source	df	SS
Model	p	SSR
Error	$n - 1 - p$	SSE
Adjusted total	$n - 1$	SST

A simple measure of the extent to which the model accounts for the observed variation is called R-squared, or R^2:

$$R^2 = 1 - \frac{\text{SSE}}{\text{SST}}. \tag{4.104}$$

R-squared can also be written as SSR/SST, and it is clear that $0 \leq R^2 \leq 1$. R-squared is often expressed as a percentage.

R-squared is also called the *coefficient of multiple determination*. The non-negative square root of R-squared is called the *coefficient of multiple correlation*.

Estimation of σ^2

Another parameter in the model is σ^2, the variance of the ϵs. (In the matrix/vector formulation, we assume the variance-covariance matrix of the vector ϵ is $\sigma^2 I$.)

In many cases, we can show that the expected value of SSE is $(n-p-1)\sigma^2$, so we can form an unbiased estimator is

$$\widehat{\sigma}^2 = \text{SSE}/(n - p - 1), \tag{4.105}$$

where p is the number of regressors in the model, not including the constant term. The quantity $\text{SSE}/(n - p - 1)$ is denoted as MSE. (Recall that "MSE" for "mean squared error" also refers to population variance plus squared bias, which cannot be computed from a sample.) The MSE in the sense above is a quantity computed from the sample, and we often call it the "mean squared residual". These quantities may have simpler distributional properties in linear regression models. We will discuss some of their applications on page 421.

Individual Regressors and Individual Observations

In fitting a regression model to data, there are two general types of considerations: one involves the regressors, and the other involves the individual observations.

We must consider whether the individual regressors belong in the model and how they affect or are affected by the other regressors. This may involve selection of which meaningful regressor variables should be in the model. Inclusion of irrelevant regressors results in *overfitting*, which aside from misleading results, increases the variances of the estimators.

We will address these issues in linear regression models in Section 4.5.4, and give some examples using R beginning on page 444.

Some of the observations may not fit the model very well, or some observations may yield an undue influence on the fitted model. The individual observations are studied by their residuals in the fitted model.

If the data fit the model well, then we would expect the residuals to be randomly distributed, with no obvious patterns. On the other hand, if the model is not appropriate for the observed data, we might expect that some of the residuals would be very large or that there would be patterns in the residuals. Inspection and formal analysis of the residuals is an important part of regression analysis.

If a few residuals are outliers, it may indicate that the corresponding observations are outliers, possibly because they were recorded erroneously or under differing circumstances. On the other hand, if many residuals appear anomalous, or if there are patterns in the residuals, it is likely that the model is inappropriate or that it does not include relevant terms.

Outliers may also indicate that the model should be fit using a different method.

We discuss methods of analysis of residuals in linear regression models in Section 4.5.5, and give some examples using R beginning on page 440.

4.5.3   Linear Regression Models

In this and the following couple of sections, we briefly describe some of the relevant theory for linear regression models. We mention many details and definitions. In Section 4.5.6, beginning on page 435, we give an extensive example to illustrate the concepts. The reader who is familiar with linear regression may wish to skip to that section, or at least to look at some of the specific techniques that we discuss in the context of that example.

The most common form of the general regression model is a *linear regression* model. If there are m "x" variables, that is, m predictors or regressors, the model has the form

$$y_i = \beta_0 + \beta_1 x_{i1} + \cdots + \beta_m x_{im} + \epsilon_i. \tag{4.106}$$

If there is only one regressor, the model in that case is called a "simple" linear regression model. The simple linear regression model is sometimes called a "one-factor model" in financial applications, and it is useful to study how one financial asset (or rate or index) relates to another one. The market model (1.35) on page 61 is an example of a simple linear regression model. The market model relates returns of an asset to the returns of another asset or index. In Figure 1.17, we show the fitted regression of the excess daily returns of INTC on the excess returns of the S&P 500 Index for 2017. In some exercises in this chapter, you are asked to fit simple linear regression models of prices and of returns of stocks or of ETFs and of the S&P 500 Index.

Least-squares Fit of the Linear Model

For a given set of data, we often write the linear regression model in vector/matrix form as

$$y = X\beta + \epsilon, \tag{4.107}$$

where y is an n-vector of observations on the dependent variable, X is an n by $p+1$ matrix whose columns are observations on the regressors, β is an $(p+1)$-vector of unknown parameters, and ϵ is an unobservable vector of errors with

$$\mathrm{E}(\epsilon) = 0 \quad \text{and} \quad \mathrm{V}(\epsilon) = \sigma^2 I_n. \tag{4.108}$$

(Note that there is some ambiguity in a matrix/vector expression similar to (4.107) regarding the constant intercept β_0 in equation (4.106) above. Here we are assuming that the first column of X is a column of all 1s. Another interpretation, of course, is that the values in X are centered. In any specific application, we will be clear on what the individual elements are.)

We recall that the ordinary least squares estimate of β is

$$\widehat{\beta} = (X^{\mathrm{T}}X)^{-1}X^{\mathrm{T}}y. \tag{4.109}$$

Under the assumption that the errors are random variables with mean 0, that is, $E(\epsilon) = 0$, we see an immediate property of the least squares estimator $\widehat{\beta}$. It is *unbiased*:

$$
\begin{aligned}
E(\widehat{\beta}) &= E\left((X^{\mathrm{T}}X)^{-1}X^{\mathrm{T}}y\right) \\
&= (X^{\mathrm{T}}X)^{-1}X^{\mathrm{T}}E(y) \\
&= (X^{\mathrm{T}}X)^{-1}X^{\mathrm{T}}X\beta \\
&= \beta.
\end{aligned} \tag{4.110}
$$

The estimator $\widehat{\beta}$ is a random vector, of course. It is a linear transformation of the random vector y, as seen in equation (4.109). Given that the variance of y is $I\sigma^2$, using equations (3.60) and (3.62) on page 268, we can work out the variance of $\widehat{\beta}$:

$$
\begin{aligned}
V(\widehat{\beta}) &= V\left((X^{\mathrm{T}}X)^{-1}X^{\mathrm{T}}y\right) \\
&= (X^{\mathrm{T}}X)^{-1}X^{\mathrm{T}}V(y)\left((X^{\mathrm{T}}X)^{-1}X^{\mathrm{T}}\right)^{\mathrm{T}} \\
&= (X^{\mathrm{T}}X)^{-1}X^{\mathrm{T}}I_n X(X^{\mathrm{T}}X)^{-1}\sigma^2 \\
&= (X^{\mathrm{T}}X)^{-1}\sigma^2.
\end{aligned} \tag{4.111}
$$

We will not prove it here, but the Gauss-Markov theorem states that if the errors have constant variance and if they have 0 covariances, then any linear combination of the elements of $\widehat{\beta}$ has the smallest variance for that linear combination of the elements of all unbiased estimators of β of the form Ay, for a matrix A such as $(X^{\mathrm{T}}X)^{-1}X^{\mathrm{T}}$ above.

If we also assume that the errors have a normal distribution, then y has a normal distribution, and so $\widehat{\beta}$ also has a normal distribution:

$$\widehat{\beta} \sim N_p\left(\beta,\ (X^{\mathrm{T}}X)^{-1}\sigma^2\right).$$

(Any linear combination of normal random variables has a normal distribution.) The normal distribution also allows us to form exact confidence intervals and tests of hypotheses concerning β.

Also if the errors have a normal distribution, then $\widehat{\beta}$ is also the maximum likelihood estimator of β, as we have seen on page 374.

Qualitative Regressors in Linear Models

Qualitative variables in linear regression models can be viewed as forming multiple parallel regression lines or planes. Binary discrete variables are generally handled by arbitrarily coding them as 0 and 1. The effect on the regression model (4.106) is that the binary variable is essentially absent when its value is 0, and when its value is 1, it is also essentially absent, but the intercept term is $\beta_0 + \beta_b$, where β_b is the coefficient of the binary variable. The binary variable is a dummy variable; it is either present as an addition to the intercept (with a value of 1) or absent (with a value of 0). The least-squares estimators are the same as those below.

If a qualitative regressor can take on k different values, a simple way of incorporating it into a standard regression model is by forming a dummy variable for each of $k - 1$ (arbitrary) values of the qualitative regressor. The computations and analysis proceed in the same way as before.

Projecting y; The Hat Matrix

Fitting a linear regression model $y = X\beta + \epsilon$ by least squares determines a *projection* of the vector y onto a vector space of lower dimension. The projection is the fitted vector, $\hat{y}$. We have

$$\hat{y} = X(X^{\mathrm{T}}X)^{-1}X^{\mathrm{T}}y,$$

or

$$\hat{y} = Hy \tag{4.112}$$

where

$$H = X(X^{\mathrm{T}}X)^{-1}X^{\mathrm{T}}. \tag{4.113}$$

This fitting matrix H is called the "hat matrix" for obvious reasons. It is a smoothing matrix that smooths the elements of y onto a linear space.

The hat matrix is a *projection matrix*; that is, it is symmetric and idempotent:

$$HH = X(X^{\mathrm{T}}X)^{-1}X^{\mathrm{T}}X(X^{\mathrm{T}}X)^{-1}X^{\mathrm{T}} = X(X^{\mathrm{T}}X)^{-1}X^{\mathrm{T}} = H.$$

This fact implies a number of useful properties of H. (Equation (4.48) on page 365, for example, is a direct result of H being idempotent.) We will state and use some of these properties in the following. Several facts are just results from matrix algebra, and do not depend on statistical or distributional properties.

Some important facts about H are that it depends only on X (not on y); its diagonal element h_{ii} is a quadratic form involving the ith row of X; and h_{ii} determines the extent to which the projected value $\hat{y}_i$ differs from y_i. The diagonal element h_{ii} is a measure of the leverage of the ith point.

One of the most useful facts about H involves the residuals from fitting the model $y = X\beta + \epsilon$ with all the data and the residuals from fitting the model with all of the same observations except one. We will use these latter facts in the development beginning on page 429.

F Tests in Linear Regression

The quantities such as SSR and SSE, which we discussed on page 415, have simpler distributional properties in linear regression models.

If, in addition to the standard assumptions about the errors ϵ, we also assume that they have a normal distribution, then the residuals, r, also have a normal distribution, and SSE/σ^2 has a chi-squared distribution with $n - p - 1$ degrees of freedom. Furthermore, if $\text{E}(y) = 0$ (recall, according to the regression model, that $\text{E}(y) = X\beta$), then SSR/σ^2 has an independent chi-squared distribution with m degrees of freedom. (These facts are part of the conclusion of Cochran's theorem, which we will not prove here.)

The ratio SSR/p is denoted as MSR, or sometimes as MSreg. The ratio of these two quantities, MSR/MSE, in which the unknown σ^2 has canceled out, has an F distribution with p and $n - p - 1$ degrees of freedom; that is,

$$F = \frac{\text{MSR}}{\text{MSE}} \tag{4.114}$$

has an F distribution if $\beta = 0$. (See page 289 for the relationship of chi-squared and F distributions.)

The F statistic in equation (4.114) is the test statistic for an omnibus F test in the regression model $y = X\beta + \epsilon$ for the null hypothesis that $\beta = 0$. If F exceeds an F critical value, the null hypothesis is rejected at the level of significance associated with the critical value.

t Tests

A hypothesis regarding any of the β coefficients can be performed using a t test with the t statistic similar to equation (4.75). To test the null hypothesis $\beta_j = \beta_{j0}$, for example, the t statistic, with $n - p - 1$ degrees of freedom, is

$$t = \frac{\widehat{\beta}_j - \beta_{j0}}{\sqrt{\text{MSE}}}. \tag{4.115}$$

Transformed Linear Regression Models

As we have indicated, there may be some ambiguity in the model $y = X\beta + \epsilon$ as to whether or not X includes a column of 1s. The centering transformation $X_{\text{c}} = X - \overline{X}$ (see page 210) is often useful. In least squares fits, this transformation has the effect of removing the intercept if the vector y is likewise centered.

It is often helpful to make other linear transformations on the data matrix X in linear models. Transformations may also be made on the dependent variable. If the transformations are linear and invertible, it is easy to transform statistics using one model to corresponding statistics of another model. Suppose, for example, we transform the data in the model (4.11)

$$y = U\theta + \tilde{\epsilon}, \tag{4.116}$$

where
$$U = XC,$$

and C is an invertible matrix. Then the least squares fit for θ has a very simple relationship to the least squares fit of β:

$$
\begin{aligned}
\widehat{\beta} &= (X^\mathrm{T}X)^{-1}X^\mathrm{T}y \\
&= ((UC^{-1})^\mathrm{T}(UC^{-1}))^{-1}(UC^{-1})^\mathrm{T}y \\
&= ((C^{-1})^\mathrm{T}U^\mathrm{T}UC^{-1})^{-1}(C^{-1})^\mathrm{T}U^\mathrm{T}y \\
&= C(U^\mathrm{T}U)^{-1}C^\mathrm{T}(C^{-1})^\mathrm{T}U^\mathrm{T}y \\
&= C(U^\mathrm{T}U)^{-1}U^\mathrm{T}y \\
&= C\widehat{\theta}
\end{aligned}
\tag{4.117}
$$

A useful linear transformation is to standardize the independent variables; that is, let $C = D^{-1}$, as in equation (2.5), where D is an $m \times m$ diagonal matrix whose nonzero entries on the diagonal are the sample standard deviations $s_{x_1}, \ldots, s_{x_m}$. In this case, if U is also centered, $U = X_{cs}$, the centered and standardized matrix (see page 210).

While a basic assumption in linear regression models is that the variance of the errors, and hence of the dependent variable, is constant, in some applications, the variance is different over different ranges of the dependent variable. The Box-Cox transformations, which are nonlinear depending on a tuning parameter λ (see page 345), can often stabilize the variance. The `boxcox` function in the `MASS` R package accepts a fitted linear model in an `lm` object or an `aov` object and computes values of the likelihood corresponding to a range of values of λ. The value of λ corresponding to the maximum of the likelihood could be chosen as the value for the Box-Cox transformations to be used.

4.5.4   Linear Regression Models: The Regressors

The variation in each regressor or "independent variable" in the model contributes to the fit of the overall model. Some regressors are more useful than others in accounting for the variation in the response. There may be interactions among the regressors that affect how they fit into the model.

Observations that correspond to unusually large or small values of the regressor often have a greater effect on the fit of the model than other observations. We will discuss this on page 429.

From equation (4.111) and the estimate of σ^2 given in equation (4.105), we have an estimate of the variance-covariance matrix of $\widehat{\beta}$. An estimate of the variance of the individual estimator of β_j is just the jth diagonal element of $(X^\mathrm{T}X)^{-1}\widehat{\sigma}^2$.

Variable Selection

For a given set of observations, we may consider the linear model (4.106)

$$y_i = \beta_0 + \beta_1 x_{i1} + \cdots + \beta_m x_{im} + \epsilon_i \tag{4.118}$$

to be a *potential* model, with a potential set of regressors, $x_1, \ldots, x_m$. Different sets of regressors used in the model form different partitions of the variation. In regression analysis, it is often instructive to fit a model with only one or two regressors, and then to add regressors one at a time to the model.

As more regressors are included in the model, the SSE will generally decrease and the R-squared will increase. The R-squared will not in any event decrease. This is a general algebraic result of least squares fitting. The stronger the linear relationship of a regressor to the dependent variable, the greater reduction in SSE in general. The reduction depends on the other regressors in the model, however, and if the newly included regressor is linearly related to them, the reduction in SSE may be small. The fact that the R-squared increases may lead to including irrelevant regressors, that is, to overfitting, which, as we will see, inflates the variances of other estimators.

The increase in R-squared resulting from including another regressor in the model indicates the existence of a linear relationship over and above the linear relationships of the regressors already in the model. Including another regressor in the model, however, results in a more complex model.

The selection of regressors to include in the model must involve consideration not only of the goodness of the fit, but also of the complexity of the model.

Adjusted R-Squared

R-squared is a good measure of how well the model fits the data, but it does not account for model complexity. The *adjusted R-squared* is a variation that adjusts R-squared downward to account for the number of variables included in the model. It is

$$R_a^2 = 1 - \frac{n-1}{n-p} \frac{SSE_p}{SST}, \tag{4.119}$$

where SSE_p denotes SSE based on p regressors in the model. (The adjusted R-squared is sometimes defined as above but using the fraction $(n-1)/(n-p-1)$.)

Unlike the R-squared, the adjusted R-squared does not necessarily increase as more regressors are added to the model. The model with the largest adjusted R-squared may be considered to be the "best" model.

Mallows's C_p

Another measure of how well the model fits, but that also accounts for the complexity of the model is Mallows's C_p. This measure uses the fitted model that includes all of the *potential* variables as a basis.

If there are m potential variables as in the model (4.118), and if s^2 is the mean squared residual using the model with all potential regressors, that is, SSE_m divided by the appropriate quantity as in equation (4.105), then Mallows's C_p is defined as

$$\text{Mallows } C_p = \frac{\text{SSE}_p}{s^2} - (n - 2p - 2). \tag{4.120}$$

The model with the smallest Mallows's C_p may be considered to be the "best" model.

The idea in adjusted R-squared and in Mallows's C_p is that they account for the tradeoff between how well the model fits the data and how complex the model is.

Mallows's C_p is based on the SSE_m of the best potential model, that is, the model that includes all of the variables in the dataset. On the other hand, it is not necessary to identify potential models for the adjusted R-squared.

Mallows's C_p can easily be modified so as to be based on a maximum likelihood approach, using the deviance of the model under consideration in place of SSE_p, and using the deviance of the "biggest" model in place of s^2. In a maximum likelihood fit assuming a normal distribution, SSE_p is the same as the deviance (possibly missing an additive constant, see page 373).

There are other measures that trade off goodness of fit with model complexity. We will discuss two of them, AIC and BIC, in Section 4.6.3. Both of those are based on a maximum likelihood approach, and they also involve the concept of all "potential" variables.

Sequential Linear Models; Partial Regressions

For the n-vectors $y, x_1, \ldots, x_p$, consider a sequence of linear regression models, each building on the previous one by the addition of another regressor:

$$
\begin{aligned}
y &= \beta_{0,0} + \epsilon_0 \\
y &= \beta_{0,1} + \beta_{1,1} x_1 + \epsilon_1 \\
&\vdots \\
y &= \beta_{0,p} + \beta_{1,p} x_1 + \cdots + \beta_{p,p} x_p + \epsilon_p
\end{aligned}
\tag{4.121}
$$

The subscripts on the coefficients indicate that they are different depending on which regressors are in the model; the constant term in the first model in the equations (4.121) $\beta_{0,0}$ (which is just the mean) is different from the constant term $\beta_{0,1}$ in the second equation, which is the intercept in that simple linear regression model. Likewise, the error terms have subscripts to distinguish them, although we make the same standard assumptions about the distribution of each.

Now consider combining the equations sequentially, for example, substituting the first equation into the second one and rearranging terms. This yields the regression equation

$$\epsilon_0 = (\beta_{0,1} - \beta_{0,0}) + \beta_{1,1} x_1 + \epsilon_1; \tag{4.122}$$

and, in general, substituting the kth equation into the $(k+1)^{\text{st}}$ yields the regression equation

$$\epsilon_{k-1} = \begin{aligned}[t] &(\beta_{0,k} - \beta_{0,k-1}) + (\beta_{1,k} - \beta_{1,k-1})x_1 + \cdots \\ &\cdots + (\beta_{k,k} - \beta_{k,k-1})x_k + \beta_{k+1,k+1}x_{k+1} + \epsilon_k. \end{aligned} \tag{4.123}$$

The coefficient $\beta_{k+1,k+1}$ in equation (4.123) is exactly the same as that in the $(k+1)^{\text{st}}$ equation of equations (4.121).

Each equation in equations (4.121) can be considered to be a *partial regression* model of the succeeding equation. Each regression such as equation (4.123) can be considered to be a *conditional* regression given a previous equation in the set of equations (4.121).

If the equations (4.121) are fit unbiasedly, say by least squares, then the results above hold for the residuals in successive fits; that is, for example, $\widehat{\beta}_{k+1,k+1}$ obtained by fitting the regression model

$$r = \theta_0 + \theta_1 x_1 + \cdots + \theta_k x_k + \beta_{k+1,k+1}x_{k+1} + \epsilon$$

where

$$r = y - \widehat{\beta}_{0,k} - \widehat{\beta}_{1,k}x_1 - \cdots - \widehat{\beta}_{k,k}x_k$$

yields the same least squares estimate of $\beta_{k+1,k+1}$ as would be obtained by fitting the model

$$y = \beta_{0,k+1} + \beta_{1,k+1}x_1 + \cdots + \beta_{k,k+1}x_k + \beta_{k+1,k+1}x_{k+1} + \epsilon.$$

The relevance of these results in developing regression models is that at each stage, the residuals from a previous fit can be inspected for patterns or anomalies.

Another main point in the results above is that the additional reduction in the sum of squares for error, SSE, can be the basis for a "partial" F test for the significance of the regressor that is added at each step. You are asked to perform the computations to illustrate these relationships in Exercise 4.22. We discuss the partial F tests in the next section, and we show an example of them on page 440.

These results are also the basis for the use of the partial autocorrelation function (PACF) in identification of the order of an autoregressive model, see page 547.

Partial F Test

In addition to the F in equation (4.114), we can develop other F tests for β in the linear regression model under the assumption of normality. The idea in the F test is to form a ratio of two independent chi-squared statistics that represent a partitioning of the observed variation into a part due to the model or portion of the model under consideration and the residual variation due to the remaining error ("unexplained" variation) in the model.

The issue is, given some variables already included in the model, how much improvement (reduction in unexplained variation) is achieved by adding some more variables. Let us denote the variables already in the model by "X_{in}", those considered for adding to the model by "X_{add}", and those already in the model together with the additional ones by "X_{all}".

A measure of the variation accounted for by the additional variables is the reduction in the error sum of squares,

$$\text{SSE}(X_{\text{in}}) - \text{SSE}(X_{\text{all}}),$$

where $\text{SSE}(X_{\text{in}})$ is the residual sum of squares when the X_{in} variables are the only ones in the model, and $\text{SSE}(X_{\text{all}})$ is the residual sum of squares when the X_{add} variables as well as the X_{in} variables are all included in the model. This difference is the *partial sum of squares* associated with the additional variables, which we denote as

$$\text{SSR}(X_{\text{add}}|X_{\text{in}}).$$

The degrees of freedom associated with $\text{SSR}(X_{\text{add}}|X_{\text{in}})$ is the number of variables in X_{add}, say m_a. Often, of course, $m_a = 1$, as we consider adding just one new variable to the model.

The relevant sum of squares for error is $\text{SSE}(X_{\text{all}})$. Let m represent its degrees of freedom; that is, let m be the total number of variables in X_{in} and X_{add}.

Again, we form independent chi-squared random variables, divide by their degrees of freedom, and form a ratio in which the unknown σ^2 cancels out. We therefore have the *partial* F statistic,

$$F = \frac{\text{MSR}(X_{\text{add}}|X_{\text{in}})}{\text{MSE}(X_{\text{all}})}, \qquad (4.124)$$

which, if the β associated with X_{add} is 0, has an F distribution with m_a and m degrees of freedom.

Correlations among the Regressors; Multicollinearity

If the regressors have nonzero correlations among themselves, their value in the model is reduced. Consider the extreme case of two regressors having a sample correlation of 1 or -1. In this case, the two variables are linearly related, and two terms in the model (4.8) can be combined. Also, the matrix $X^{\text{T}}X$ in the normal equations (4.43) is singular, and it does not have an inverse so as to yield $\widehat{\beta}$ in equation (4.44). Such an extreme case rarely exists, but it is often the case that the correlations are large or that one or more regressors are "nearly" a linear combination of the others. The matrix $X^{\text{T}}X$ is "nearly" singular, and computational problems can arise in solving the normal equations. We often use the term *multicollinearity* to describe this general situation.

There are various quantitative measures of near-singularity or multi-collinearity. For numerical computations, we use a *condition number*, the most relevant one of which is the ratio of the largest singular value of X to its small-est nonzero singular value. (If and only if a singular value is 0, the matrix $X^{\mathrm{T}}X$ is singular.)

In statistical applications, the effects of multicollinearity generally involve increased variances of the various estimators, resulting in instability of estimates from one analysis to another.

Procedures to reduce the effects of multicollinearity include removing some variables from the model or regularizing the least squares fit so as to shrink the estimators, such as by ridge or lasso regression, as discussed on page 369.

Variance Inflation Factors

The variance-covariance matrix for the least-squares estimator of the vector β, as given in equation (4.111) is $V(\widehat{\beta}) = (X^{\mathrm{T}}X)^{-1}\sigma^2$. The diagonal elements of this matrix are the variances of the individual coefficients when all of the other regressor variables are included in the model.

The effect of a single regressor over and above the effects of the other regressor variables depends on how the regressor is related to the others. A measure of the strength of this relationship can be measured by the coefficient of multiple correlation between this variable and the other variables; that is, the R-squared of the linear regression of the variable of interest on the other regressors.

Assume that we have m regressors as in equation (4.118). Consider the jth regressor, x_j. Let s_{x_j} be the sample standard deviation of this variable.

The variance of the least squares estimator of the associated coefficient, $V(\widehat{\beta}_j)$ is the jth diagonal element of $(X^{\mathrm{T}}X)^{-1}\sigma^2$.

Let us consider the regression of x_j on the other regressors. The model for this regression is

$$x_{ij} = \alpha_0 + \sum_{k \neq j} \alpha_k x_{ik} + \tilde{\epsilon}_i.$$

Let R_j^2 be the R-squared associated with this regression.

It can be shown by some tedious algebra that

$$V(\widehat{\beta}_j) = \frac{1}{1 - R_j^2} \frac{\sigma^2}{(n-1)s_{x_j}}. \qquad (4.125)$$

The factor $1/(1 - R_j^2)$ is called the *variance inflation factor (VIF)* of the regressor x_j, for obvious reasons.

If x_j is strongly linearly related to the other regressors, then the R-squared R_j^2 is large (close to 1) and the variance inflation factor is large.

We can see that the variance inflation factor is always greater than or equal to 1. As a general rule of thumb, a variance inflation factor greater than 5 may warrant further consideration of the relationships of the regressors to each other, and a value greater than 10 may be taken as an indicator of serious multicollinearity.

The fact that the variance inflation factor is always greater than 1 means that the inclusion of any regressor in the model increases the variance of the other estimators. This supports the statement we made earlier about overfitting; inclusion of additional regressors increases the variance, and inclusion of irrelevant regressors, needlessly so. Variance inflation factors, however, are more effective in identifying collinear regressors than irrelevant ones.

We will illustrate the variance inflation factor in the example on page 446.

In ridge regression (see page 369), as the tuning parameter λ begins to increase from 0, the variance inflation factors of all regressors decrease fairly rapidly at first, and then slowly (toward 1) as λ continues to increase. Thus the ridge trace in terms of the variance inflation factors may also help to determine an appropriate value of the tuning parameter.

4.5.5   Linear Regression Models: Individual Observations and Residuals

When a model is fit to data, the best way of assessing the adequacy of the fit is by analyzing the residuals.

If the model is appropriate and if it has been fitted appropriately, then the frequency distribution of the sample of residuals should be similar to the assumed distribution of errors.

The residuals are affected by the patterns among the regressors. The values of the regressor values may give some observations more "leverage", which we discuss below.

Each observation affects the fit of the model, and thus affects its own residual. We may fit the model without a given observation, and then obtain a "deleted residual", as we discuss below.

Outliers

Outliers are anomalous observations that do not seem to follow the same data-generating process as the other observations. Sometimes this is a result of an erroneous measurement or recording. Other times it may be a consequence of a heavy-tailed frequency distribution, which, as we have observed, often characterizes financial data.

Outliers in regression modeling usually result in large residuals, and observations with extraordinarily large residuals should be inspected. If the observations are erroneous, they should be corrected or removed from the dataset. Otherwise, the effect of outliers on estimates can be explored by computing estimates without the outlying observations or by use of a robust method for fitting the model. As mentioned previously, fitting by least absolute values instead of by least squares is less affected by outliers. This is illustrated in Figure 4.9. The L_1 fit is unchanged by the presence of either outlier in the two panels. This is not to suggest, however, that L_1 estimation would be appropriate whenever there are outliers. Outliers always demand further consideration;

possibly correcting them, possibly exploring the data with and without them, or possibly using other methods of estimation.

Outliers among the Regressors; Leverage

The effect that each observation has on the least squares estimate of β depends on how far the values of the regressors in that observation are from their means. This may not be so obvious in the multiple linear regression model, but we can see it easily in the case of simple linear regression,

$$y_i = \beta_0 + \beta_1 x_i + \epsilon_i.$$

As we have shown, the least squares fit goes through the means of the observations, so we can write the simple linear regression model as

$$y_i - \bar{y} = \beta_1(x_i - \bar{x}) + \epsilon_i,$$

and the least squares estimate of β_1 is

$$\widehat{\beta}_1 = \arg\min_b \sum_{i=1}^n \left((y_i - \bar{y}) - b(x_i - \bar{x})\right)^2. \tag{4.126}$$

The amount that each term in this summation is affected by b depends on the difference $x_i - \bar{x}$, either positively or negatively. The effect is due to the *leverage* exerted on $\widehat{\beta}_1$ by the ith observation. The corresponding value of y_i in relation to $\bar{y}$ also of course affects $\widehat{\beta}_1$.

This is illustrated in Figure 4.9, which shows a simple dataset fitted with a least squares line, then augmented with an additional data point, and then again fitted with a least squares line. In each case, the new point is the same distance below the original fitted regression line, but in the plot on the left side the additional point is near the mean of the regressors and in the plot on the right side it is far away from the mean of the regressors, giving it more leverage. Notice that the high leverage point (right plot) has more effect on both $\widehat{\beta}_0$ and $\widehat{\beta}_1$. Notice also that the magnitude of the new residual of the high-leverage point (the dotted line) is considerably less than that of the new point near the middle.

Larger leverage results in a smaller residual because the high-leverage point exerts a larger effect on the fitted model.

The Deleted Residual

Suppose we have n observations and fit the model $y = X\beta + \epsilon$ using all n observations. Let $r_1, \ldots, r_n$ be the residuals, and let H be the hat matrix. Now, suppose we delete the ith observation and fit the model again. With

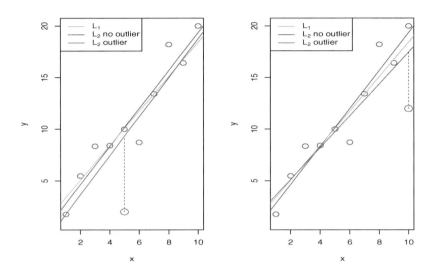

FIGURE 4.9
Outliers; Effect of Leverage; Least Squares and Least Absolute Values

the new estimated coefficients, we predict the response corresponding to the observation that was deleted. Let $\hat{y}_{i(-i)}$ be that predicted value, and consider the difference in that predicted value and the actual value y_i, which was not used in the fit without the ith observation. We call that difference the "deleted residual", and denote it by d_i.

We have the useful fact

$$d_i = y_i - \hat{y}_{i(-i)} = \frac{r_i}{1 - h_{ii}}, \qquad (4.127)$$

which can be shown by some rather tedious algebra.

Frequency Distribution of Residuals in Linear Models

In the linear model $y - X\beta + \epsilon$, we assume that $E(\epsilon) = 0$ and $V(\epsilon) = \sigma^2 I$; that is, all the errors have the same expected values and variances. This is not true in general for the residuals, however.

The expected values of the residuals are not all the same. High leverage points have residuals with smaller expected absolute values (recall Figure 4.9). If the linear model includes an intercept and is fit by least squares, however, the mean of all of the residuals is always 0. (This is just a fact from linear algebra; it does not depend on any statistical or distributional properties.)

Now, consider the variances of the residuals:

$$
\begin{aligned}
V(r) &= V(y - X\widehat{\beta}) \\
&= V((I - (X^TX)^{-1}X^T)y) \\
&= (I - X(X^TX)^{-1}X^T)V(y)(I - X(X^TX)^{-1}X^T) \\
&= (I - X(X^TX)^{-1}X^T)(I - X(X^TX)^{-1}X^T)\sigma^2 \\
&= (I - X(X^TX)^{-1}X^T)\sigma^2 \\
&= (I - H)\sigma^2.
\end{aligned}
\tag{4.128}
$$

From this we see that the residuals do not have a constant variance (unless all diagonal elements of the hat matrix are equal). This is also related to the leverage of the residuals, as we observe heuristically in Figure 4.9.

Because the residuals have different variances, in order to compare their relative magnitudes, we need to stabilize them so that their variances are equal. From equation (4.128), we see that this can be done by dividing each by the square root of the corresponding diagonal element of $I - H$, that is by $\sqrt{1 - h_{ii}}$. We have

$$
V\left(r_i / \sqrt{1 - h_{ii}}\right) = \sigma^2.
\tag{4.129}
$$

Hence, dividing each residual by $\sqrt{1 - h_{ii}}$ yields a set of stabilized residuals that all have the same variance.

Equation (4.128) also means that even if the errors have 0 correlation, the residuals do not have 0 correlation (unless the hat matrix is a diagonal matrix).

If the errors are normally distributed, then both y and $\widehat{\beta}$ have normal distributions, and so the residuals have normal distributions because they are linear combinations of y and $\widehat{\beta}$.

Studentized Residuals

If the errors are normally distributed with a variance of σ^2, then the residuals are normally distributed and $(n-m)\widehat{\sigma}^2/\sigma^2$ has a chi-squared distribution with $n - m$ df because it is a quadratic form in a matrix of rank $n - m$. Hence, if the errors are normally distributed, we can form Student's t random variables from the residuals by studentization using $\widehat{\sigma}^2$ (see page 289 for definitions of t random variables and studentization).

We need to account for the variance itself, which from equation (4.129), is $(1 - h_{ii})\sigma^2$. Hence, we define the *studentized residual* as

$$
t_i = \frac{r_i}{\sqrt{\widehat{\sigma}^2(1 - h_{ii})}},
\tag{4.130}
$$

which, under the assumption of normality (and the assumption that the data follow the model), has a t distribution with $n - m$ df. The studentized residual in equation (4.130) is also sometimes called the *standardized residual* or the *internally studentized residual.* (Most documentation for R uses the phrase "standardized residual".)

The t_i **do not have 0 correlations** and they **are not "estimates"** of the ϵ_i.

Externally Studentized Residuals

To avoid the effects of leverage, we can form a more useful studentized random variable associated with the ith residual.

Instead of using $\hat{\sigma}^2$, which depends on all residuals including the ith one, we use $\hat{\sigma}^2_{(-i)}$, which is computed as in equation (4.105), except r_i is not used, and the divisor is $n - m - 1$; that is,

$$\hat{\sigma}^2_{(-i)} = \frac{\sum_{j \neq i} r_j^2}{n - m - 1}. \tag{4.131}$$

The scaled variable $(n - m - 1)\hat{\sigma}^2_{(-i)}/\sigma^2$ has a chi-squared distribution with $n - m - 1$ df.

We now define the *externally studentized residual*:

$$t_{i(-i)} = \frac{r_i}{\sqrt{\hat{\sigma}^2_{(-i)}(1 - h_{ii})}}, \tag{4.132}$$

which, under the assumption of normality (and the assumption that the data follow the model), has a t distribution with $n - m - 1$ df.

The externally studentized residual is the same as the corresponding deleted residual d_i (from equation (4.127)) divided by the square root of the estimate of σ^2 computed using the residuals from fitting the model without the ith observation. (This estimate is not the same as $\hat{\sigma}^2_{(-i)}$, which is based on $n - 1$ of the residuals from fitting the model using all n observations.)

The $t_{i(-i)}$ are "more similar" to the ϵ_i, but they **do not have 0 correlations** and they **are not "estimates"** of the ϵ_i.

Terminology and Notation

There are various terms used in the literature for the various types of residuals of interest in regression analysis. The unqualified word "residual" almost always means $y_i - \hat{y}_i$, as we have been using the word. The other types of residuals result from adjustments that make the distributions of the residuals be more similar to a t distribution. (Of course if the underlying distribution of the errors is not normal, a t distribution cannot be formed; it can often be approximated, however.)

One approach is just to divide the residuals by $\sqrt{1 - h_{ii}}$, as would be suggested by equation (4.129). This is sometimes done, and the resulting residuals are called "stabilized residuals".

It is more common, however, also to studentize the residuals (see page 289) by dividing by $\hat{\sigma}$ as in equation (4.130). The resulting residuals are called "standardized residuals", "studentized residuals", or "internally standardized residuals". Most documentation for R uses the term "standardized residuals".

Finally, the residuals in equation (4.132), which have been studentized using $\widehat{\sigma}_{(-i)}$ are called the "externally studentized residuals" or the "studentized deleted residuals". They are occasionally also called just "studentized residuals" or "standardized residuals". Various software packages may use different terminology.

Also, note some inconsistencies in the notation involving the subscript "$(-i)$". In $\widehat{\sigma}^2_{(-i)}$, the notation means that the ith value in the sum $\sum_{i=1}^n r_i^2$ has been omitted, but otherwise, the r_i are the regular residuals from the fit. On the other hand, in $\widehat{y}_{i(-i)}$ and $\widehat{\beta}_{k(-i)}$ below the notation is used to refer to values obtained from fitting the model using data with the ith observation deleted.

Leverage: DFFITS, DFBETAS, and Cook's Distance

There are various measures of the leverage exerted by a single observation in linear regression analysis. Three common ones, defined below, use the same idea as in externally studentized residuals; that is, *leave-out-one deletion*, removal of one observation in order to assess the effect of that observation in the full analysis. These measures can also be expressed in terms of the deleted residuals (from equation (4.127)). The names of some of these measures were given by economists, who like to use a string of upper-case letters as the name of a mathematical variable.

("DF" in the quantities defined below stands for "difference" in the "FITS" and in the "BETAS".)

DFFITS is a measure of the influence that the ith observation has on the ith predicted value, $\widehat{y}_i$.

$$\text{DFFITS}_i = \frac{\widehat{y}_i - \widehat{y}_{i(-i)}}{\sqrt{\widehat{\sigma}^2_{(-i)} h_{ii}}}. \tag{4.133}$$

The predicted value using the observations with the ith one deleted is $\widehat{y}_{i(-i)}$, as above.

This quantity is algebraically the same as

$$t_{i(-i)} \sqrt{\frac{h_{ii}}{1 - h_{ii}}}, \tag{4.134}$$

where $t_{i(-i)}$ is the ith externally studentized residual.

DFBETAS is a measure of the influence that the ith observation has on the kth fitted coefficient, $\widehat{\beta}_k$. We denote the fitted coefficient omitting that observation by $\widehat{\beta}_{k(-i)}$

$$\text{DFBETAS}_{ki} = \frac{\widehat{\beta}_k - \widehat{\beta}_{k(-i)}}{\sqrt{\widehat{\sigma}^2_{(-i)} c_{kk}}}, \tag{4.135}$$

where c_{kk} it the kth diagonal element of $(X^{\text{T}}X)^{-1}$.

Cook's distance, D_i is a measure of the influence that the ith observation has on all of the predicted values; that is, instead of $\hat{y}_i - \hat{y}_{i(-i)}$, as in DFFITS, we use the sum of the squared differences, which in vector notation we can write as the inner product $(\hat{y} - \hat{y}_{(-i)})^{\text{T}}(\hat{y} - \hat{y}_{(-i)})$. Cook's distance is

$$D_i = \frac{(\hat{y} - \hat{y}_{(-i)})^{\text{T}}(\hat{y} - \hat{y}_{(-i)})}{m\hat{\sigma}^2}. \qquad (4.136)$$

The numerator and denominator in Cook's distance are similar to scaled chi-squared random variables with m and $n - m$ degrees of freedom. The ratio of independent chi-squared random variables is an F random variable. Although D_i does not have an exact F distribution, the F distribution can be used to approximate the significance of D_i. There are general rules of thumb concerning the significance of D_i to suggest when the effect of the ith observation may be considered overly influential. One rule of thumb is that if D_i exceeds $F_{m,(n-m),0.50}$, the 50th percentile of the F distribution with m and $n - m$ degrees of freedom, then the influence of the ith observation is great enough to warrant concern.

Cook's distance can also be expressed as

$$D_i = \frac{r_i^2}{m\hat{\sigma}^2}\left(\frac{h_{ii}}{(1 - h_{ii})^2}\right). \qquad (4.137)$$

Serial Correlations; The Durbin-Watson Statistic

If the residuals have a sequential pattern, possibly related to the times at which the data were collected, we may observe increasing or decreasing subsequences in the residuals. A reasonable measure of the extent to which the sequence of residuals exhibit a pattern is

$$d = \frac{\sum_{i=2}^{n}(r_i - r_{i-1})^2}{\sum_{i=1}^{n}r_i^2}. \qquad (4.138)$$

This is called the Durbin-Watson statistic.

Note that this is the same as the sample autocorrelation $\hat{\rho}(1)$ (equation (1.69), page 100), because in this case, $\bar{r} = 0$.

By expanding the numerator into three sums involving r_{i-1}, r_i, and r_{i+1}, we can see that the Durbin-Watson statistic does indicate a pattern or lack thereof. We can also see that the value of d is always between 0 and 4, and a value close to 2 indicates essentially 0 correlation. Small values of d (near 0) indicate positive correlation and large values (near 4) indicate negative correlations.

The R function `dwtest` in the `lmtest` package performs a Durbin-Watson test. If the sequence of the data do not correspond to a time-order, a different order can be specified in the arguments to `dwtest`. The distribution of the

Durbin-Watson statistic can be computed exactly but also a simple normal approximation is available. Either can be requested in `dwtest`.

There are versions of the statistic based on larger lags. For lag $h \geq 1$, the numerator in equation (4.138) has terms of the form $(r_i - r_{i-h})^2$. The R function `durbinWatsonTest` in the `car` package performs a Durbin-Watson test using lags up to a specified maximum.

The distribution of the Durbin-Watson statistic for larger lags is quite complicated. For lags greater than 1, `durbinWatsonTest` uses an ad hoc bootstrap approximation.

If the regression model errors are serially correlated, the inferences we make may not be valid. If, however, the autocorrelations in the errors are stationary, use of a time series model combined with the regression model may lead to valid analyses. One way of addressing this problem is by use of generalized least squares, where the weight matrix (W in equation (4.59) on page 368) is chosen to approximate the autocovariance matrix. The R function `gls` in the `nlme` package implements this method. We will consider an example of the use of `gls` for fitting a regression model in Section 5.3.12, but we will not discuss the details of the methodology. We refer the interested reader to Pinheiro and Bates (2009), who also developed the R package.

4.5.6 Linear Regression Models: An Example

To illustrate some of the concepts in linear regression, let us consider an example in financial analysis. We use R functions to illustrate various steps in the regression analysis.

Our interest is in how the rates of B-rated corporate bonds relate to treasury rates and to actions of the FOMC (which sets a target for the effective fed funds rate).

We will use weekly data on the rates of Moody's Seasoned Baa Corporate Bonds, the effective fed funds rates, rates of 3-month T-Bills, rates of 2-year Treasury Notes, and rates of 30-year Treasury Bonds. The rates for all of these are available from FRED as `WBAA`, `FF`, `WTB3MS`, `WGS2YR`, and `WGS30YR`, respectively (see Table A1.17 on page 169). We will analyze data from January 2015 through 2017.

We have observed some general properties of these treasury rates in Chapter 1. We noted that the 3-month T-Bill rate is often used as the "risk-free" rate, and in Figure 1.4 on page 23 we plotted the daily rates (`TB3MS`) back to 1934. The rates of the 3-month T-Bills, 2-year Treasury Notes, and 30-year Treasury Bonds at a single point in time constitute part of the yield curve, and we plotted some of these same data in Figure 1.6.

Regressing any of these rates on any of the others would not be very revealing, but instead we will look at the weekly changes in the rates.

An important concern is the sequential nature of the data. We will do further analysis of these data in Chapter 5, but for now, we will ignore the time factor, other than performing a Durbin-Watson test for serial correlations in the residuals.

Data Preparation and Exploration

Using R and quantmod, we bring in the data this way.

```
getSymbols("WBAA", src = "FRED")
getSymbols("FF", src = "FRED")
getSymbols("WTB3MS", src = "FRED")
getSymbols("WGS2YR", src = "FRED")
getSymbols("WGS30YR", src = "FRED")
```

We will just form a simple R data frame instead of an xts data frame, because after merging the datasets, the dates are not exact anyway.

```
dBaa <- diff(as.numeric(WBAA["20150101/20171231"])[-1])
dfed <- diff(as.numeric(FF["20150101/20171231"]))
d3m  <- diff(as.numeric(WTB3MS["20150101/20171231"])[-1])
d2y  <- diff(as.numeric(WGS2YR["20150101/20171231"])[-1])
d30y <- diff(as.numeric(WGS30YR["20150101/20171231"])[-1])
YX   <- data.frame(cbind(dBaa, dfed, d3m, d2y, d30y))
```

The first thing is to obtain an overview of general properties of the data. We can print the first few observations using head, and determine if there are any missing values using sum and is.na. Doing this (not shown), the data appear to be what we would expect, and we find no missing values.

As mentioned previously, the dates associated with weekly data from FRED do not match (see discussion on cleansing data beginning on page 183). All the weekly data are for Fridays, except the fed funds rate, FF, which is tabulated as of Wednesdays. There are more Fridays in this period than there are Wednesdays, so we will just omit the first Friday in the dataset collected. Although the data are not reported on the same day, they are from the same week.

Now we look at a scatterplot matrix to get a quick overview of their variation.

```
pairs(YX)
```

The scatterplot matrix of weekly changes is shown in Figure 4.10.

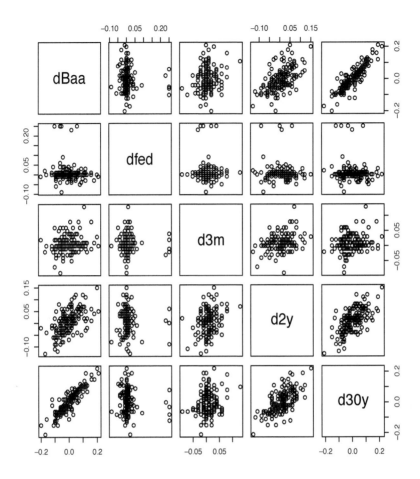

FIGURE 4.10
Weekly Differences in Baa Bond Rates, Fed Funds Rates, and Treasury Rates

All weekly changes are clustered around 0, as we would expect. We notice a strong positive correlation between the changes in the Baa corporates and those of the 30-year T bonds, and decreasingly strong positive correlations between the Baa corporates changes and changes in shorter term Treasuries. The changes of the 30-year and the 2-year Treasuries have a rather strong positive correlation; otherwise, the regressors do not exhibit strong correlations.

The only unusual thing in the scatterplot matrix in Figure 4.10 is the nature of the variation in dfed, the weekly differences in the effective fed funds rates. The weekly changes seem relatively constant centered at 0, except for a few points. It might be instructive to explore this further before using those data in our analysis. Figure 4.11 shows a time series plot of the effective funds rates (not the differences).

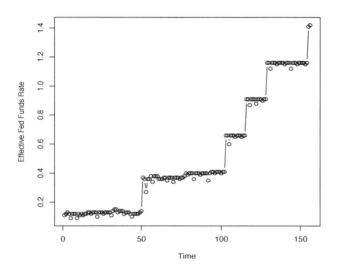

FIGURE 4.11
Weekly Effective Fed Funds Rates

Figure 4.11 reminds us of exactly what the effective fed funds rate is. It corresponds very closely to the target rate set by the FOMC (the fed funds rate, or discount rate). Following the 2008 financial crisis, the target rate was cut ("quantitative easing"), and remained very low until the end of 2015, when the FOMC began a series of 25 basis point increases. Thus, there were only a few nonzero weekly differences in the target rate. The effective fed fund rate changes very little on a weekly basis, and generally the change is approximately 25 basis points. This raises the question of whether the changes in the effective fed funds rate should even be used in the regression analysis. Except for the noise of the negotiations, it is essentially a discrete variable taking on a value of 0 most of the time, and values of ±25 basis points or maybe ±50 basis points occasionally. We will keep it in our regression model for now, however.

Least Squares Linear Regression

Now we are ready to do the regression. We use the R function lm.

```
regr <- lm(formula=dBaa~dfed+d3m+d2y+d30y, data=YX)
```

The keyword `formula` can be omitted, and because the model includes all of the variables in the `YX` data frame, the formula can be specified in shorthand form as `dBaa~..`. (Note the period character.) See Section 4.5.8 beginning on page 454 for descriptions of the `formula` keyword in R.

We next print a simple summary that shows information about the residuals, the coefficient estimates and sample statistics about the estimates, and some overall measures of the fitted model.

```
> summary(regr)

Residuals:
      Min        1Q     Median        3Q       Max
-0.136241 -0.013876 -0.000879  0.013094  0.092480

Coefficients:
              Estimate Std. Error t value Pr(>|t|)
(Intercept) -0.003300   0.002802  -1.178   0.2408
dfed         0.001432   0.054797   0.026   0.9792
d3m          0.128558   0.098534   1.305   0.1940
d2y         -0.146627   0.072924  -2.011   0.0462 *
d30y         0.903769   0.047153  19.167   <2e-16 ***
---
Signif. codes:  0  0.001 ** 0.01 * 0.05 . 0.1  1

Residual standard error: 0.03249 on 150 degrees of freedom
Multiple R-squared:  0.7962,    Adjusted R-squared:  0.7907
F-statistic: 146.5 on 4 and 150 DF,  p-value: < 2.2e-16
```

The R-squared of 79.6% indicates that a linear combination of the treasury rates provides a relatively good model for the Baa corporates rates. (We can never make a more definitive statement about R-squared than something qualified by "relatively".)

The coefficient estimates and associated statistics in the summary should be self-explanatory. We see that the 30-year T-Bonds have the strongest relationship to the Baa corporates, which is not surprising because the corporates generally have long terms. The effective fed funds rate is not significant in this regression model. We expect the regression coefficients to be positive, generally. The negative coefficient associated with the 2-year T-note is a result of the multicollinearity among the regressors; with the relationship of the 30-year T-bonds in the model, the conditional relationship of the 2-year note is negative. We would not necessarily expect this, but it is not surprising. We will consider the multicollinearity of the regressors further on page 446.

While the statistics shown in the summary above are marginal values assuming all four regressors are in the model, the `anova` function shows the individual sequential sums of squares and the partial or conditional F values for the variables added to the model in the order specified in the model. In the ANOVA table produced by `anova`, each partial F value has 1 degree of freedom for the numerator.

```
> anova(regr)
Analysis of Variance Table

Response: dBaa
            Df  Sum Sq Mean Sq  F value      Pr(>F)
dfed         1 0.00521 0.00521   4.9401     0.02774 *
d3m          1 0.02375 0.02375  22.5007 4.849e-06 ***
d2y          1 0.20163 0.20163 191.0560 < 2.2e-16 ***
d30y         1 0.38770 0.38770 367.3564 < 2.2e-16 ***
Residuals  150 0.15831 0.00106
---
Signif. codes:  0 *** 0.001 ** 0.01 * 0.05 . 0.1   1
```

Residual Analysis

The residuals can reveal several things about the model and the data.

Given the magnitudes of the data and the general spreads of their values observed in Figure 4.10, we do not expect any residuals to have unusual values. Nevertheless, we should *always* do some exploratory analysis of residuals in any regression analysis, and we will illustrate some of the methods using functions in R.

The summary of the results above show that the residuals range from -0.1362 to 0.0925, with a median of -0.000879. In this standard output from R, the "residuals" are the raw residuals; that is, their scale corresponds to the scale of the data, and their variances are different, depending on the diagonal elements of the hat matrix, h_{ii}. These residuals can be obtained by use of `$residuals` in an `lm` object, `regr$residuals` in this example.

In financial data, especially these data that are time series, it is often of interest to see if there are any obvious patterns in the data as a function of time; hence, we plot `regr$residuals`. The index of the vector is sequential in time. Figure 4.12 shows the residuals sequential in time.

We do not see any irregularities in Figure 4.12, other than a few of the residuals are rather large. There are no time patterns apparent in the residuals.

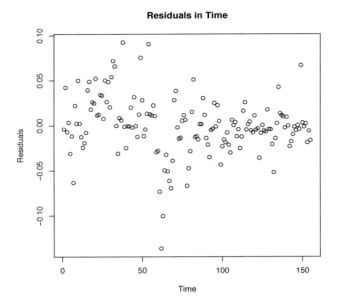

FIGURE 4.12
Residuals of Regression of Baa Bond Rates on Fed Funds Rates, and Treasury
Rates over Time

Residuals that are adjusted for their differing variances and that are studentized by dividing by the square root of an estimate of the model variance, that is, the internally studentized residuals of equation (4.130), can be obtained by the `rstandard` function. (In most R documentation, the internally studentized residuals are called "standardized residuals".)

The `plot` function in R applied to an object of class `lm` produces various plots of the raw residuals and the internally studentized residuals.

```
plot(regr, which=c(1:3, 5))
```

Each plot is displayed in the active plot area until a "return" is sent to the graphics device. The plots desired are specified by the `which` argument. Figure 4.13 shows four of these plots. Points associated with fairly large values of residuals or leverages are labeled.

The magnitudes of raw residuals often depends on the magnitude of the predicted values, so a plot similar to that in the upper left of Figure 4.13 often

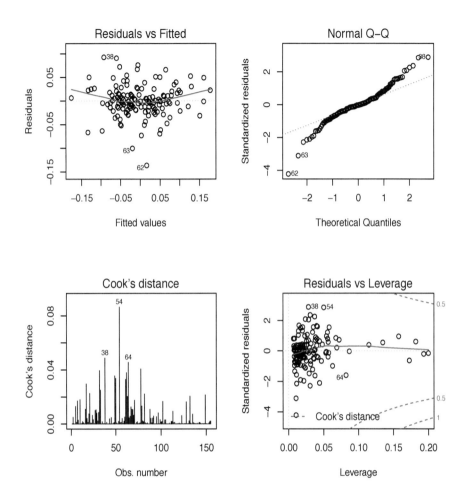

FIGURE 4.13
Various Plots of Residuals and Influence for Regression of Baa Bond Rates
Changes on Fed Funds Rate and Treasury Rate Changes

reveals concerns. The plot in this example does not, although observations
number 62 and 63 stand out as large.

The q-q plot in Figure 4.13 shows that the frequency distribution of the
residuals has a heavier tail than a normal probability distribution. Again,
observations number 62 and 63 stand out as large.

Cook's distance in the lower left plot of Figure 4.13 shows one fairly large
leverage point, observations number 54. In the plot in the lower right, we
see that the point is also associated with a fairly large residual, but it had

not been identified in any of the previous plots. The residuals for that point, both the raw and the standardized, are very similar in magnitude to those of observation number 38. The general rule of thumb suggests that a Cook's distance greater than pf(4, 150, 0.50) would be cause for concern. None of the computed values even come close to that critical point, which is 0.456.

Overall, the plots do not raise major concerns about any of the observations.

Some additional R functions for analyzing influence are hat, hatvalues, dfbetas, dffits, covratio, and cooks.distance. Some of these functions were used by R itself in the plot function for the lm object. The olsrr package has several functions for regression analysis, especially for producing diagnostic plots. We will not pursue them further here.

Residual Analysis: Autocorrelations in the Residuals

Financial and economic data are often in the form of a time series. As mentioned above, the data in each of the variables that we have been analyzing are time series. In models with variables that are time series, it is often the case that the residuals will be correlated with each other or that the residuals will exhibit patterns over time. Patterns, of course, can also be due to lack of stationarity in any of the series.

There are no patterns visible in Figure 4.12 showing a plot of the residuals versus time, but autocorrelations often do not exhibit obvious patterns. Because of the time series nature of financial data, and these data in particular, it is desirable after a regression fit to perform a statistical test for autocorrelations in the residuals. A simple test for nonzero autocorrelations in the residuals is the Durbin-Watson test.

The R function durbinWatsonTest in the car package performs a Durbin-Watson test for serial correlations up to a maximum lag specified.

```
> library(car)
> durbinWatsonTest(regr, max.lag=3)
 lag Autocorrelation D-W Statistic p-value
   1      0.52982469     0.9385733   0.000
   2      0.27848495     1.4297074   0.000
   3      0.09229081     1.7996013   0.272
 Alternative hypothesis: rho[lag] != 0
```

This test indicates that both the lag 1 and lag 2 serial correlations are nonzero. (The tests for different lags are not independent of each other.)

The R function dwtest in the lmtest package performs a Durbin-Watson test for serial correlations at lag 1. It is shown below just for illustration.

```
> library(lmtest)
> dwtest(regr)

        Durbin-Watson test

data:  regr
DW = 0.93857, p-value = 1.214e-11
alternative hypothesis: true autocorrelation is greater than 0
```

Autocorrelations in the residuals indicate autocorrelations in the errors (recall the difference in "residuals" and "errors"), which of course is a violation of the fundamental assumption in the regression model that the errors have zero correlation. Although it is important to recognize the autocorrelations, they rarely make major changes in the analysis. The effect is generally relatively small changes in p-values and significance levels.

If the autocorrelations in the errors are stationary, use of a time series model combined with the regression model may lead to valid analyses. One way of addressing this problem is to estimate the autocorrelations and then use generalized least squares, where the weight matrix (W in equation (4.59) on page 368) corresponds to the autocorrelations of the errors. The R function gls in the nlme package implements a weighted least squares fitting of linear models. Use of estimated autocorrelations in a GLS fit, however, does not yield exact p-values and significance levels.

We will consider an example of the use of gls for fitting a regression model with autocorrelated errors in Section 5.3.12, beginning on page 571.

Variable Selection

We have seen nothing unusual in our analysis of the individual observations, so now we will take another look at the regression variables in the model.

In the summary output shown on page 439, the significance codes indicate that some variables are much more important than others.

We now explore which combinations of regressors provide the best fit, at the lowest cost in model complexity (that is, smallest number of regressors). We do this in a systematic way. There are several R functions that are useful in doing this. We will discuss only three, step in the standard stats package, and two from the leaps package, the leaps function and the regsubsets function.

Variable Selection: Stepwise Selection

One of the simplest R functions for variable selection is step, which for a given set of regressors, by default begins with no regressors, that is, just a constant model, and then adds the single best regressor according to the Akaike information criterion or AIC. (The AIC is negatively proportional to Mallow's C_p; smaller values of AIC indicate a better fit after accounting for the number of variables in the model. We will discuss AIC and related measures in Section 4.6.3 below.)

After fitting the model with a single regressor, step proceeds to find the next best regressor, according to the AIC, and if the AIC is smaller than before, that variable is added. If that variable is added, then step proceeds to find the third best regressor, according to the AIC, and if the AIC is smaller than before, that variable is added. If that variable is added, then step determines the best variable among those in the model to remove from the model, again according to the AIC. If the AIC would be decreased by removal of that variable, the variable is removed. This two-step procedure now continues: find the best variable among the potential ones not in the model, and if it would decrease the AIC by doing so, add it to the model; find the best variable among the ones in the model, and if it would decrease the AIC by doing so, remove it from the model. The process terminates when there is no potential variable not already in the model that would decrease the AIC by adding it to the model.

Variable Selection: All Best Subsets

The stepwise procedure is not guaranteed to find the best subset of regressor variables larger than two. Another approach is to find the best subset of regressor variables of each size starting with one and continuing until all variables are included. For m potential variables, there are $2^m - 1$ possible subsets, and so this is obviously a computationally-intensive process. Once the best set of a given size is determined and the best subset of each smaller size is known, the number of potential variables that could constitute the best subset of the next larger size is reduced. There is an algorithm, called "leaps and bounds", based on this fact that greatly reduces the number of subsets that must be examined.

The R function leaps in the leaps package implements this algorithm. The input to leaps is a matrix containing the regressors and a vector containing the response. It allows use of either the Mallow's C_p or the adjusted R-squared criterion (or the simple R-squared, which would choose the same sets as the adjusted R-squared). The function will find the k best subsets of each size if there are at least k subsets of that size. An example of the use of leaps on the interest rate regression model is shown below. We use the adjusted R-squared criterion and we request the 2 best subsets of each size.

```
> library(leaps)
> X <- cbind(dfed, d3m, d2y, d30y)
> leapout <- leaps(X, dBaa, method="adjr2", nbest=2,
+              names=c("dfed","d3m","d2y","d30y"))
> leapout
```

```
$which
     dfed   d3m   d2y   d30y
1 FALSE FALSE FALSE   TRUE
1 FALSE FALSE  TRUE  FALSE
2 FALSE FALSE  TRUE   TRUE
2 FALSE  TRUE FALSE   TRUE
3 FALSE  TRUE  TRUE   TRUE
3  TRUE FALSE  TRUE   TRUE
4  TRUE  TRUE  TRUE   TRUE


$label
[1] "(Intercept)"  "dfed"   "d3m"    "d2y"    "d30y"


$size
[1] 2 2 3 3 4 4 5


$adjr2
[1] 0.7883313 0.2883811 0.7911289 0.7879048
[5] 0.7921041 0.7897458 0.7907191
```

The output is easy to interpret, the best single variable is the 30-year bond d30y, and the two best variables are the 30-year bond and the 2-year note d2y. The adjusted R-squared for the model with just d30y is 0.788, and the adjusted R-squared for the model with both d30y and d2y is 0.791. Adding a third variable, d3m increases the adjusted R-squared slightly to 0.792, but adding the fourth variable decreases the adjusted R-squared to 0.790.

The leaps package also includes a function, regsubsets, that allows several options for which subsets will be included. Under one set of inputs, regsubsets is the same as leaps, but the user can specify a specific set of regressors to include or a specific set to exclude in all subsets considered.

There are many issues in variable selection in regression analysis. They often entail practical considerations that go beyond the computations, such as ease of obtaining data (usually not an issue in financial applications, but extremely important in medical applications) and precision of the measurements. Variable selection is part of the more general process of model selection, which we discuss in Section 4.6.3.

Multicollinearity: Variance Inflation Factors

The value of including any specific regressors in the model depends on the correlations of those regressors with other regressors in the model.

We expect a high degree of multicollinearity in a set of interest rates. It is not surprising that only one or two series of rates would do as well as four or more.

With suspected large multicollinearity, it is desirable to compute the variance inflation factors for the regressors. The `vif` R function in the `faraway` package computes these:

```
> library(faraway)
> vif(regr)
    dfed       d3m       d2y      d30y
1.009564 1.094674 1.901146 1.800922
```

None of these variance inflation factors is noticeably large, which is perhaps surprising because we expect high multicollinearity in these regressors. The implication, however, is that the multicollinearity is not excessive.

Regularization: Ridge Regression

Although we did not observe a high degree of multicollinearity, before leaving this example, we may consider regularization of the fit.

We use the R function `lm.ridge` in the `MASS` package. We use four values of λ.

```
> library(MASS)
> lm.ridge(formula=dBaa~dfed+d3m+d2y+d30y, data=YX,
           lambda=c(0,0.25,0.5,1.0))
                       dfed          d3m        d2y      d30y
0.00 -0.003300489 0.0014322050 0.1285576 -0.1466267 0.9037692
0.25 -0.003314719 0.0011597893 0.1279668 -0.1435074 0.9009622
0.50 -0.003328801 0.0008903027 0.1273856 -0.1404238 0.8981795
1.00 -0.003356528 0.0003599466 0.1262518 -0.1343611 0.8926853
```

While the values of the coefficient estimates generally shrink (notice that the shrinkage is not monotonic in each coefficient), there does not appear to be any meaningful effect, so we will not pursue this approach here. (Note that the change in the intercept estimate was due to the changes in the other estimates; it was not included in the regularization penalty.)

The `lm.ridge` function produces a list with statistics relating to each regressor at the specified values of λ. A ridge trace for each regressor can be produced for a sequence of λs by

```
plot(lm.ridge(dBaa ~., data=YX, lambda=seq(0,1.0,0.001)))
```

(also, note how the formula is specified in this example). As noted above, the ridge regulation had little effect on the fit, so the plot is not shown.

A reason to use ridge or lasso regression is to reduce the MSE. Because the MSE includes the square of the bias, which in general is not known, we do not have a simple way of estimating it. We will discuss a method of estimating the MSE for any value of λ using cross-validation methods in Section 4.6.2.

Predictions

For a given set of values of the regressor variables, the fitted model yields a predicted value of the response. Assuming an underlying normal distribution along with the standard assumptions of the linear regression model, we can form a confidence interval or a prediction interval for the predicted response.

The R function `predict` uses the output of other R functions that fit various models. The first argument to `predict` is the output object from the fitting function. In this example, the output from fitting the full model above is `regr`, which contains the information about the model itself, the method of fitting (least squares), and the coefficient estimates and other output statistics. To predict, for example, the change in the rates of Baa corporates if the fed fund rate decreased by 25 basis points, we could use the fit from `lm`, `regr`, together with a data frame with a decrease in `dfed` and no change in the Treasury rates.

The `predict` function also computes an approximate 95% confidence interval or an approximate 95% prediction interval, if requested with the `interval` argument.

```
> predict(regr,
+          data.frame(dfed=-0.0025, d3m=0.0, d2y=0.0, d30y=0.0),
+          interval="confidence")
           fit            lwr          upr
1 -0.003304069 -0.008893432 0.002285293
> predict(regr,
+          data.frame(dfed=-0.0025, d3m=0.0, d2y=0.0, d30y=0.0),
+          interval="prediction")
           fit            lwr          upr
1 -0.003304069 -0.06773713 0.06112899
```

The predicted change in the Baa corporates rate is a decrease of 33 basis points. The intervals are fairly wide, however, indicating fairly large estimated variances of the estimators. We note that the prediction interval is wider, accounting for the variation in the individual values over and above the variation due to the model coefficient estimators.

4.5.7 Nonlinear Models

There are many interesting relationships among variables that are not linear. A simple example is a quadratic relationship, in which the dependent variable is related to the square of a regressor.

Polynomial Regression Models

A polynomial regression model in a single regressor variable takes the form

$$y_i = \beta_0 + \beta_1 x_i + \beta_2 x_i^2 + \cdots + \beta_m x_i^m + \epsilon_i. \qquad (4.139)$$

This model is very similar to the general linear regression model (4.8), with powers of the single regressor taking the place of other regressor variables. Fitting the model by least squares or other criteria would be exactly the same as fitting a multiple linear regression model. It is likely, however, that multicollinearity would be more of a problem in the polynomial model because of the relationships of the various powers of a single variable to each other.

A variable selection procedure may be used to decide on the appropriate degree of the polynomial. Higher degrees (more than, say, degree 3) are generally not useful, because higher degree polynomials can be very wiggly.

Logistic Regression

In many applications where regression may be useful, the linear models discussed above are not appropriate because the response variable is qualitative, often just binary. In such a case we might use the predictor variables similar to the way we do in a linear model, but modify the model so that it conforms to the problem addressed.

These models have application in classification problems, which are the basic problems in statistical machine learning. The main objective in statistical learning is to build a model to predict a qualitative response variable using a set of data on various features associated with the response. In medical applications, for example, the features may include various physiometric variables, such as blood pressure, serum triglyceride levels, and so on, and the response may be "heart disease", yes or no. In commercial applications, the response may be "credit risk", good or bad, and the predictive features may be income, outstanding debt, assets owned, debt service records, and so on. In financial applications, the features may be various measures of recent stock price movements, and the response may be "change in price over the next 10 days", up or down. (This is what is done in "technical analysis", although the classification models are generally not stated precisely.) In the use of classification models in statistical learning, one of the most important aspects is the identification of which measurable features are useful for prediction.

Consider a common decision problem faced by many financial institutions: whether or not to issue a credit card to an applicant for the card. The decision can be based on a number of variables that are entered in the application form.

These variables include such things as income, debt, and so on. The variables may also be qualitative, such as whether or not the applicant owns a home.

A standard linear regression model would have the form

$$y_i = \beta_0 + \beta_1 x_{i1} + \cdots + \beta_m x_{im} + \epsilon_i,$$

with these symbols taking their obvious meanings. The response variable in this model makes no sense, however. It can vary continuously over a wide range. The response of interest, however, is just "no" or "yes", which we might code as 0 or 1.

The right-hand side of the equation presumably captures the relationships of interest between the regressors and the dependent variable. One approach, therefore, is to form some function of the right-hand side of the equation that varies between 0 and 1. This approach is equivalent to modeling the probability that $Y = 1$. In the credit card example, the dependent variable is the probability that the applicant is a good credit risk.

While there are many ways we could transform the right-hand side to be between 0 and 1, a way that is mathematically tractable and that works well in practice is to use the *logistic function* of the right-hand side, and form the model

$$\Pr(Y = 1) = \frac{e^{\beta_0 + \beta_1 x_{i1} + \cdots + \beta_m x_{im}}}{1 + e^{\beta_0 + \beta_1 x_{i1} + \cdots + \beta_m x_{im}}}. \tag{4.140}$$

This *logistic model* looks very different from the usual regression models with additive error terms. This model has no explicit error term.

The randomness modeled here is a Bernoulli process (see page 283); each applicant equals 1 ("yes", "good credit risk") with the probability in equation (4.140) that depends on the values of the regressor variables. Recall that a Bernoulli distribution is a special case of the binomial distribution, and that the sum of n independent Bernoullis with the same probability is a binomial random variable with that probability and a number n.

Another way of writing the model (4.140) is by use of *odds*. We can represent the odds as the ratio $\Pr(Y = 1)/(1 - \Pr(Y = 1))$, and taking the log of the odds, we have the *logit* as a linear model,

$$\log\left(\frac{\Pr(Y = 1)}{1 - \Pr(Y = 1)}\right) = \beta_0 + \beta_1 x_{i1} + \cdots + \beta_m x_{im}. \tag{4.141}$$

The logit is the *link* function that linearizes the logistic model for the probability, equation (4.140). It is essentially the probability function of a binomial distribution.

We usually fit a logistic model by maximum likelihood.

To simplify the notation, let $p(x_i; \beta)$ represent $\Pr(Y = 1)$. Now, given n observations of regressors and corresponding values of Y, the likelihood function is

$$L(\beta; x, y) = \prod_{i:y_i=1} p(x_i; \beta) \prod_{i:y_i=0} (1 - p(x_i; \beta)). \tag{4.142}$$

The MLE of β is obtained by maximizing this likelihood.

The model (4.140) is a *logistic regression* model. It is also called a binary regression model, although there are different forms a binary regression model can take.

There are other models of this general form. They are called *generalized linear models* because of the linear terms in the exponential. The specific form of a generalized linear model depends on the probability model of the process and on the linearizing link function. In the case of logistic regression, it is a Bernoulli (or binomial) distribution with a logit link. There are other link functions that can be used with an underlying binomial distributions, and other types of generalized linear models can be formed with other distributions, such as the Poisson.

As mentioned in the discussion beginning on page 373, in maximum likelihood modeling, the important quantity is the deviance, $-2\log(L(\beta; x, y))$, more so than the sum of squares of the residuals as in least-squares modeling. Also, instead of the actual residuals, we generally work with the deviance residuals, $-2\log(L(\beta; x_i, y_i))$.

Example: Fitting a Logistic Model in R

The R function `glm` uses maximum likelihood to fit generalized linear models. The model are specified in `glm` by a formula similar to that used in `lm`. (See Section 4.5.8 beginning on page 454 for descriptions of the `formula` keyword in R.) We will briefly illustrate the use of `glm` to fit a logistic regression model to credit card default data. The dataset `Default` is from the `ISLR` package (built and described by James et al., 2013). It consists of 10,000 observations on four variables, the binary response `default`, a binary regressor `student`, and two continuous variables, `balance` and `income`. Of these credit card holders, there were 333 defaults in the year.

```
> require(ISLR)    The package is loaded only to get the dataset
> head(Default)
  default student   balance    income
1      No      No  729.5265 44361.625
2      No     Yes  817.1804 12106.135
3      No      No 1073.5492 31767.139
4      No      No  529.2506 35704.494
5      No      No  785.6559 38463.496
6      No     Yes  919.5885  7491.559
> glmfit <- glm(formula=default~student+balance+income,
+                data=Default, family=binomial)
> contrasts(default)
    Yes
No    0
Yes   1
```

```
> summary(glmfit)

Call:
glm(formula = default ~ student + balance + income,
    family = binomial, data = Default)

Deviance Residuals:
    Min       1Q    Median       3Q      Max
-2.4691  -0.1418  -0.0557  -0.0203   3.7383
Coefficients:
              Estimate Std. Error z value Pr(>|z|)
(Intercept) -1.087e+01  4.923e-01 -22.080  < 2e-16 ***
studentYes  -6.468e-01  2.363e-01  -2.738  0.00619 **
balance      5.737e-03  2.319e-04  24.738  < 2e-16 ***
income       3.033e-06  8.203e-06   0.370  0.71152
---
Signif. codes:  0 *** 0.001 ** 0.01 * 0.05 . 0.1  1

(Dispersion parameter for binomial family taken to be 1)
    Null deviance: 2920.6  on 9999  degrees of freedom
Residual deviance: 1571.5  on 9996  degrees of freedom
AIC: 1579.5

Number of Fisher Scoring iterations: 8
```

The interpretation of the `formula` keyword depends on the family of models. The family in this case is "binomial"; hence, `glm` interprets the right-hand side as the right-hand side of equation (4.141), using the same rules for interpreting a formula as in other R functions as described in Section 4.5.8.

For binary variables with "No" and "Yes", R codes "No" as 0 and "Yes" as 1, but just to be sure, use of the `contrasts` function shows the coding of the response variable.

The debt balance and whether or not the card holder is a student are highly significant in this model. For a fixed balance and income, the negative coefficient -0.6468 indicates that a student is less likely to default than a nonstudent.

The main use of a logistic regression model such as fitted model above is to make decisions about whether or not to issue credit cards to applicants. In practice, the credit card issuing company would probably ask for more information from an applicant than just the student status, debt balance, and income, as we included in this example model. For purposes of illustration, however, let us use this fitted model to predict the probability of default. Just as in the example with a linear regression model beginning on page 448, we can use the R function `predict` to make the prediction. We will do this for two students; one with a debt balance of $1,000 and one with a debt balance of $3,000, and both with income of $10,000. Note that in using `predict` for a fitted generalized model, we must specify `type="response"`.

```
> predict(glmfit,
+          data.frame(student="Yes",balance=1000,income=10000),
+          type="response")
          1
0.003175906
> predict(glmfit,
+          data.frame(student="Yes",balance=3000,income=10000),
+          type="response")
        1
0.9967441
```

While there is very small probability that the student applicant with $1,000 in debt and an income of $10,000 will default, the student with $3,000 debt and the same income almost surely will, according to our fitted model. The credit card issuing company, after using historical data to build the model, can use it to compute the estimated probability of default and can then decide whether or not to issue a credit card. The company may set credit limits if the card is issued, based on the predicted probability.

Other R functions for use in regression analysis, such as `step` and `glmnet`, can also be used with generalized linear models.

When financial institutions build classification models for purposes such as deciding whether or not and how much credit to issue, whether the models are logistic regression models or models of other forms, the data scientists generally begin with many potential predictive features. In the process of statistical learning, a number of variables are inspected in different models, and a good set of variables and good predictive model is selected. Variables and models are selected using methods described in Sections 4.6.2 and 4.6.3 below, often involving breaking the available data into "training sets" and "test sets".

Other Nonlinear Models

Many variables have nonlinear relationships with each other. This is often the case with models in which a regressor is time. Growth models, for example, have a regressor that is time. Growth often follows an "S shape"; that is, growth is slow at first, then grows rapidly, and then flattens out.

Growth regression models, such as the Gompertz model on page 366, can be fitted by least squares, maximum likelihood, or other methods, using the optimization techniques discussed in Section 4.3.

The distributions of statistics computed for nonlinear models are often far more complicated than those computed for linear models or even for generalized linear models. The distributions often must be approximated by asymptotic distributions.

4.5.8 Specifying Models in R

Statistical models in R are specified by a *formula*, which is an R object of class formula. A formula class object consists of three parts,

$$\text{response } \tilde{ } \text{ terms,} \tag{4.143}$$

where "response" is a numeric response vector and "terms" is a series of vectors or transformations of variables in a data frame specifying the predictors and their form in the model. The symbol $\tilde{ }$ separates the response from the predictor portion of the model.

The form of the predictor portion of the model is indicated by operators such as "+", "*", and ":". The model is assumed to contain an additive constant, unless a "no-intercept" model is specified by "0+" or "-1". For example, the R formula

$$\text{y } \tilde{ } \text{ x1+x2}$$

may specify the model

$$y_i = \beta_0 + \beta_1 x_{1i} + \beta_2 x_{2i} + \epsilon_i,$$

and

$$\text{y } \tilde{ } \text{ 0+x1+x2}$$

may specify the model

$$y_i = \beta_1 x_{1i} + \beta_2 x_{2i} + \epsilon_i.$$

On the other hand, in glm if family=binomial, these same formulas may specify the models

$$\log\left(\frac{\Pr(y_i = 1)}{1 - \Pr(y=1)}\right) = \beta_0 + \beta_1 x_{1i} + \beta_2 x_{2i},$$

and

$$\log\left(\frac{\Pr(y_i = 1)}{1 - \Pr(y=1)}\right) = \beta_1 x_{1i} + \beta_2 x_{2i}.$$

Mathematical functions and operators can be included in the formula, except for the exponentiation operator, which, because of its special meaning (see example below), must be escaped using the I function (see the example). Table 4.1 shows some examples, using the standard statistical notation, and

TABLE 4.1

Some Example Model Formulas in R

y ~ x1+x2-1	$\beta_1 x_1 + \beta_2 x_2$
y ~ x1+x2+x1:x2	$\beta_0 + \beta_1 x_1 + \beta_2 x_2 + \beta_3 x_1 x_2$
y ~ x1*x2	$\beta_0 + \beta_1 x_1 + \beta_2 x_2 + \beta_3 x_1 x_2$
y ~ (x1+x2+x3)^2	$\beta_0 + \beta_1 x_1 + \beta_2 x_2 + \beta_3 x_3 +$
	$\beta_4 x_1 x_2 + \beta_5 x_1 x_3 + \beta_6 x_2 x_3$
y ~ x1+I(x1^2)+log(x2)	$\beta_0 + \beta_1 x_1 + \beta_2 x_1^2 + \beta_3 \log(x_2)$

with the obvious interpretation of the R objects as statistical variables. The operators "*", ":", and "^" are used primarily in classification (AOV) models.

The full meaning of the models in Table 4.1 depends on the usage. In a linear model, in each case, the response y is interpreted as the variable on the left side of the model, and an error term ϵ is added to the right side of the model.

In a generalized linear model, the expression in the right column of Table 4.1 is the linear portion of the link function in the model.

A `formula` object can be constructed by the `formula` function, as shown in Figure 4.14.

```
> model1<-formula("y~x1+x1^2+log(x2)")
> model1
y ~ x1 + x1^2 + log(x2)
> class(model1)
[1] "formula"
> reg <- lm(model1)
...
```

FIGURE 4.14

A Formula in R

4.6 Assessing the Adequacy of Models

Statistical theory is developed for specific classes of models and families of distributions. For data that follow the models, the use of the models allows meaningful statistics to be computed and tests and confidence statements to be made. For real-world financial data, the accuracy of an analysis depends on the extent to which the data follow the model. If the match is not good, the results of the analysis will not be good.

In any data analysis, we must first consider whether the model is a good one for the data-generating process, and then after fitting the model with the data, we must assess the adequacy of the model.

4.6.1   Goodness-of-Fit Tests; Tests for Normality

In statistical applications, one of the simplest models is that the data follow a particular distribution or family of distributions. Given a sample of data, checking this assumption can be framed as a statistical hypothesis test. The null hypothesis is

$$H_0 \; : \; \text{the distribution from which the sample came is } H_0,$$

where "H_0" is the specific distribution of distribution family. For example, "H_0" may be "normal", that is, the normal family of distributions, or "H_0" may be "N(0,1)", that is, a very specific normal distribution. A hypothesis specifying a distribution family is a composite hypothesis; one specifying a specific distribution is simple.

A test of a hypothesis that a sample is from a specific probability model or class of models is called a *goodness-of-fit test*. Although this could apply to any distribution, such as Poisson, gamma, or so on, the normal distribution is the one that we test for most often.

The usual alternative hypothesis in a goodness-of-fit test is "any other distribution".

A goodness-of-fit statistical test is usually different from other statistical hypothesis tests, in that it is often not a main objective of the analysis.

A more general null hypothesis for goodness-of-fit considerations is

$$H_0 \; : \; \text{the model that fits the sample is } M_0,$$

where "M_0" is the specific model, which may include a distribution of some terms. For example, "M_0" may be "$y_i = \beta_0 + \beta_1 x_i + \epsilon_i$, for some β_0 and β_1 and ϵ_i are identically and normally distributed". We will discuss goodness-of-fit and, hence, selection of models in Section 4.6.3. In this section, we discuss goodness-of-fit of a sample to a probability distribution.

Tests for Normality

The normal distribution corresponds in many ways to our intuitive feelings of how things in nature and how financial and economic quantities vary. Our perceptions of "unusual" or "extreme" occurrences are in large part tempered by what is extreme in a normal probability distribution.

Mathematically, the normal probability distribution is easy to work with, and we have available in the literature a wealth of theory about the normal distribution. Many formal statistical tests of hypotheses and other procedures are based on the assumption that the random variables are normally distributed. We also have theoretical results that allow us to use the normal distribution as an approximation in situations when the underlying distribution is not normal. Many of these approximations depend on the asymptotic results embodied in the central limit theorems.

We wish to decide if a sample of data arose from a normal distribution, or from a distribution that is very similar to a normal distribution, and if not, how that underlying distribution differed from a normal distribution. We often do this visually, using various kinds of plots. In Chapter 1, we compared samples of various quantities, mostly returns, with a normal distribution. In the plots in Figures 1.24, 1.25, and 1.26 we superimposed a normal probability density function on a frequency display of a sample to facilitate the comparisons.

Q-q plots are effective visual methods for determining goodness-of-fit, and in Figures 1.27 and 1.28, we used q-q plots for a visual assessment of the extent that samples were similar to normal samples. These visual assessments are among the most important techniques of exploratory data analysis discussed in Chapter 2.

In addition to visual assessment of evidence that a sample did or did not arise from a normal distribution, there are several formal statistical tests of the null hypothesis that the distribution from which the sample came is normal. Some of these goodness-of-fit tests are for the normal distribution specifically, but some can be used for any distribution. Some tests require that the distribution be completely specified, and others can be used when only a family of distributions is hypothesized.

Two widely-used statistical goodness-of-fit tests for null hypotheses that fully specify the distribution are the chi-squared test and the Kolmogorov test, or Kolmogorov-Smirnov test.

The chi-squared goodness-of-fit test is based on the bins of a histogram, and the Kolmogorov-Smirnov test is based on the order statistics that define an ECDF. We will discuss these two general tests first.

Chi-Squared Goodness-of-Fit Tests

The computations for a chi-squared goodness-of-fit test are based on counts in bins, such as in a histogram. The counts are compared with the expected number in each bin under the hypothesized distribution.

For a hypothesized normal distribution, the histogram in Figure 1.24 on page 89 shown with the superimposed normal probability density illustrates the setup. The data are the simple daily returns of INTC for the first three quarters of 2017. Recall that the normal probability density used in this figure is the one with mean equal to the sample mean of the returns and with standard deviation equal to the sample standard deviation of the returns. This plot is reproduced below on the left side of Figure 4.15. Also shown are counts of observations in some bins, along with the expected number in that bin for the given sample size, $n = 188$.

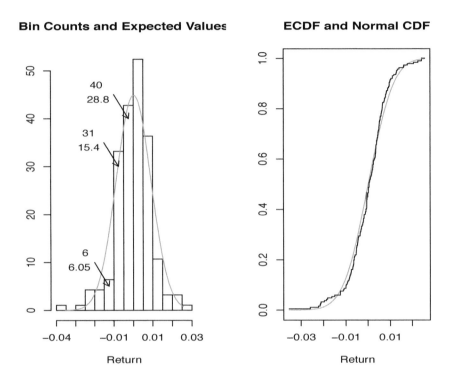

FIGURE 4.15
Statistics for Chi-Squared and Kolmogorov Goodness-of-Fit Tests for Daily Simple Rates of Return for INTC

In a formal chi-squared goodness-of-fit test, the range of the hypothesized distribution is divided into regions or bins. For k bins, the chi-squared test is based on the test statistic

$$c = \sum_{i=1}^{k} \frac{(o_i - e_i)^2}{e_i}, \qquad (4.144)$$

where o_i is the number of observations falling into the ith bin, and e_i is the expected number in the ith bin, computed as

$$n\Pr(X \in \text{bin}_i),$$

where the probability is computed from the hypothesized distribution. For the INTC returns shown in the left-side plot in Figure 1.24, the o_i and the e_i are (to one decimal place)

1	0	1	4	4	6	31	40	49	34	10	3	3	1
0.0	0.1	0.4	1.7	6.1	15.4	28.8	39.6	40.0	29.7	16.2	6.5	1.9	2.4

Under the null hypothesis, the chi-squared statistic (4.144) has an approximate chi-squared distribution. How good the approximation is depends on a number of things. If any e_i is very small, say less than 5, the approximation is not good. This can be remedied by combining bins with very small values of e_i. In our case, we combine the first five bins into one, and combine the last three bins into one. Doing this for the data above in the 8 resulting bins, we get

$$c_{\text{computed}} = 14.6,$$

where we have used the full precision in R (not the rounded values shown).

The approximation to the chi-squared distribution also depends on whether other computations from the sample are used to specify the hypothesized distribution. If the computations do not depend on any auxiliary statistics that estimate properties of the hypothesized distribution, the chi-squared distribution under the null hypothesis has degrees of freedom equal to the number of groups minus 1.

In this example of INTC returns, we had 8 groups, but we used the sample mean and the sample standard deviation of the returns to determine the specific member of the family of normal distributions. The chi-squared approximation can be improved by subtracting from the degrees of freedom in the approximate chi-squared distribution the number of parameters that we estimate from the sample. In this example, we estimated 2 parameters; therefore, we use a chi-squared distribution with 5 degrees of freedom.

Hence, to complete the chi-squared goodness-of-fit test that this sample of INTC returns came from a normal distribution, we compare the computed value of the test statistic with a critical value from the chi-squared distribution with 5 degrees of freedom. For a test at the 5% significance level, we have

```
> qchisq(0.95,5)
[1] 11.0705
```

and since the computed value is greater than this, the chi-squared test results in rejection of the null hypothesis that the returns follow a normal distribution.

There is also an R function `chisq.test` that facilitates performing this test.

Kolmogorov-Smirnov Goodness-of-Fit Tests

Another general goodness-of-fit test is based on the ECDF of the sample and the CDF of the hypothesized distribution. The test statistic is the supremum of the absolute difference between these two functions.

We refer to this supremum as the *Kolmogorov-Smirnov distance*, and we will refer to tests based directly on it as *Kolmogorov-Smirnov tests*. There is also a version of the test to compare the distributions of two samples, using two ECDFs instead of an ECDF and a CDF, and some people use different combinations of "Kolmogorov" and "Smirnov" for the names of the different tests; I will call them all "Kolmogorov-Smirnov tests". I will also call them "KS tests".

The main issue in using the KS test is that the null hypothesis is a simple hypothesis. Just as the case of a chi-squared goodness-of-fit test, a specific distribution must be specified. For chi-squared tests, we had a simple adjustment to account for sample estimates to specify the member of a family of distributions. The use of estimated values changes the distribution of the test statistic, and unfortunately, there is no simple adjustment for the KS test, as in the case of the chi-squared test.

As an exploratory technique, however, and with the understanding that exact significance levels cannot be stated, it is useful to compute the Kolmogorov-Smirnov distance between the ECDF of a sample and a CDF of interest.

Figure 2.1 on page 212 shows the ECDF of the 2017 INTC returns discussed above, along with the CDF of a normal distribution whose mean and standard deviation correspond to the sample mean and standard deviation. This plot is reproduced on the right side of Figure 4.15, along with a line that shows the supremum of the difference between the ECDF and the CDF. The supremum difference occurs close to the return value of 0. (It is difficult to see clearly in the plot, but that is not important.) The Kolmogorov-Smirnov distance is 0.065.

The R function `ks.test` performs the computations for a KS test on a given sample for a specified distribution. For a hypothesized normal distribution with given mean and standard deviation, for example, we can use

```
ks.test(ret, "pnorm", mean=m, sd = s)
```

The p-value reported by `ks.test` as above is not exact if m and s are estimated from the sample `ret`, as in the case here. The Lilliefors modification (see below) adjusts the p-value for the fact that the mean and the standard deviation are estimated from the same dataset.

Goodness-of-Fit Tests Specifically for the Normal Distribution

There are some (very complicated) approximations to the distribution of the KS test statistic for the normal distribution with estimated parameters. One test that uses approximations with the KS test statistic is called the *Lilliefors test*. The test statistic is the KS distance, but the p-value is adjusted for the estimation of the parameters, using Monte Carlo simulations.

The power of the Lilliefors test is not large for many classes of alternatives.

Other tests for normality use relationships between the sample order statistics as in the ECDF and the expected values of corresponding quantiles from a standard normal distribution. Two similar tests that use these differences are the Shapiro-Wilk test and the Shapiro-Francia test, and two others based on these differences are the Cramér-von Mises test and the Anderson-Darling test.

Another approach to testing for normality is to use the sample skewness and kurtosis. Any normal distribution has a skewness of 0 and a kurtosis of 3 (or excess kurtosis of 0). Hence, if the sample skewness and/or kurtosis deviate too much from these values, there is evidence that the sample did not come from a normal population. Two tests that are based on skewness and kurtosis are the D'Agostino-Pearson test, and the Jarque-Bera test. Both of these tests seem to be popular among financial data analysts.

We have mentioned several tests for normality without explicitly defining them, and we have said very little about their relative merits. Unfortunately, there is no best test for normality; the tests perform differently depending on how the properties of the given sample depart from the properties of the normal distribution. The issue, of course, is the power of the test, and this depends on the alternative hypothesis, that is, on how the given sample differs from what would be expected from a normal distribution. If the given sample differs from a normal distribution because it is skewed or because the tail is heavier (or lighter), the D'Agostino-Pearson test or the Jarque-Bera test may be better. If the sample is not skewed and/or the tails are not particularly heavy, the Shapiro-Wilk test, the Shapiro-Francia test, the Cramér-von Mises test, or the Anderson-Darling test may be better. The Lilliefors test generally has less power than any of these other tests. In Exercise 4.31, you are asked to perform these tests on several different artificially generated samples.

When testing for normality or any other distributional family, it is good practice to examine graphical displays, such as q-q plots. It also is important to recognize that when multiple tests are conducted, the significance levels must be interpreted with care.

R Functions for Testing for the Normal Distribution

There are R functions for performing all the tests for normality mentioned above. These functions are in various R packages:

- `stats`: Shapiro-Wilk, `shapiro.test`

- nortest: Lilliefors `lillie.test`; Shapiro-Francia, `sf.test`; Cramér-von Mises, `cvm.test`; Anderson-Darling, `ad.test`

- fbasics: D'Agostino-Pearson, `dagoTest`

- tseries: Jarque-Bera, `jarque.bera.test`

We illustrate below all of the tests on the INTC returns. Remember that the results of multiple tests on the same sample are not independent.

```
> shapiro.test(ret)
        Shapiro-Wilk normality test
data:  ret
W = 0.9705, p-value = 0.0005501

> require(nortest)
> lillie.test(ret)
        Lilliefors (Kolmogorov-Smirnov) normality test
data:  ret
D = 0.065333, p-value = 0.05016

> sf.test(ret)
        Shapiro-Francia normality test
data:  ret
W = 0.96717, p-value = 0.0004298
> cvm.test(ret)
        Cramer-von Mises normality test
data:  ret
W = 0.21225, p-value = 0.003716

> ad.test(ret)
        Anderson-Darling normality test
data:  ret
A = 1.435, p-value = 0.001014

> require(fBasics)
> dagoTest(ret)
Title:
 D'Agostino Normality Test
Test Results:
  STATISTIC:
    Chi2 | Omnibus: 16.8101
    Z3   | Skewness: -2.5903
    Z4   | Kurtosis: 3.1781
  P VALUE:
```

```
    Omnibus  Test: 0.0002237
    Skewness Test: 0.009588
    Kurtosis Test: 0.001482

> require(tseries)
> jarque.bera.test(ret)
        Jarque Bera Test
data:   ret
X-squared = 29.261, df = 2, p-value = 4.426e-07
```

The Lilliefors test yields ambivalent results, but all of the other tests suggest rejection of the null hypothesis that the INTC simple daily returns have a normal distribution.

4.6.2 Cross-Validation

Cross-validation methods are used with a variety of statistical procedures, such as regression, time series analysis, and density estimation, that yield a "fit" at each observation. Cross-validation is based on the ideas of predictive sample reuse similar to the general methods of resampling in Section 4.4.5.

Cross-validation means to hold out part of the data, apply the basic procedure to the remaining data, and use the results to fit the data that were held out and compare the fitted values with the observed values. This is the same idea that we used in linear regression to externally studentize the residuals and to assess the influence of the individual observations.

Cross-validation is one of the main methods of *model selection*. In selecting and fitting a model, our concern is with the mean squared error (MSE), which has both a variance and a bias component. In many fitting methods, especially methods of tuned inference, such as estimating a probability density with a histogram, there is a tradeoff between variance and bias.

"Validation" in the sense of cross-validation, is estimating bias and/or selecting a model with a small bias. We cannot compute either the variance or the bias, but estimation of variance is often easy; a good estimator can usually be computed as some function of the sample variance. It is not so easy to estimate bias, however. There is no bias term in the model, even if the model is indeed biased. (See the discussion in Section 4.4.5 concerning the bootstrap and its use in estimating bias.) The point is we often have good estimators of the variance using the sample variance, but there is no direct way of assessing the bias in an estimator.

The mean squared residual MSE, which we compute from a sample, is not a good estimator of the population mean squared error (also denoted as MSE), but it is the best one we can compute directly.

Validation Samples

One way of using the given data both to fit the model and to evaluate the fitted model is to divide the dataset into two subsets, a "training set" and a "test set" or a "validation set". The model is fitted (parameters are estimated and so on) using data from the training set. The data in the test set are then compared with their values predicted by the fitted model, and the residuals from the test set are used to estimate the population MSE. The MSE from the fitted training set is different from the MSE in the test set, but of course, there is only one population MSE.

The basic idea is that the sample used to fit a model, that is, the training set, cannot itself be used to evaluate the fit; after all, the method used to make the fit (least squares, maximum likelihood, whatever) is based on making the fit as good as possible for the given data.

Dividing the original sample into two sets, a training set and a validation set, however, means that we are not using all of our data for the basic purpose of fitting the model; that is, for getting good estimates of the parameters.

Because the fit from the training set is not as good as it would be if all data were used, and further because the fit is often made in such a way as to minimize the variance in the training set, the mean of the squared residuals from the validation set is biased upward, that is, it overestimates the error in the model.

These two problems of validation samples are worse for larger portions of the original data that are allocated to the validation sample.

Leave-One-Out Cross-Validation

Using the training-set/validation-set approach, if we choose the largest training set, we have just one observation in the validation sample. The training set is almost as large as possible. The bias in the error estimate should be smaller, because more observations are used to fit the model. The problem, however, is that the residual, $(y_i - \hat{y}_{i(-i)})^2$, will be highly variable. (This is the deleted residual, introduced on page 429.)

The obvious solution is to do this many times holding out different observations and to take an average. This approach is called *leave-one-out cross-validation* or LOOCV. In each of the n steps of LOOCV, we leave out one observation at a time and fit the model n times on the remaining $n-1$ observations.

Each time we do this, we get an estimate of the population MSE, which we denote as $\text{MSE}_{(i)}$ because it is based on the $(y_i - \hat{y}_{i(-i)})^2$. Averaging these, we get the LOOCV estimate for the population MSE,

$$\text{CV}_{(n)} = \frac{1}{n} \sum_{i=1}^{n} \text{MSE}_{(i)}. \tag{4.145}$$

This formula requires fitting a model n times. Refitting models n times is computationally intensive.

(As a side note, we may observe, using equation (4.127), that $\mathrm{CV}_{(n)}$ can be expressed as

$$\mathrm{CV}_{(n)} = \frac{1}{n} \sum_{i=1}^{n} \left(\frac{y_i - \hat{y}_i}{1 - h_{ii}} \right)^2 ; \tag{4.146}$$

hence, in the case of *least squares* fits of *linear* models, the LOOCV estimate of the MSE can be computed using just the original fit with all of the data. Cross-validation in this simple case, however, is not very useful.)

Another drawback to LOOCV is that the variance of $\mathrm{CV}_{(n)}$ is large because the individual terms have positive covariances.

k-Fold Cross-Validation

An approach that reduces the computational burden of LOOCV is to divide the sample into k groups, or "folds", of approximately equal size and then do cross-validation using them. We leave out one fold, fit the model using the rest of the data, and then compute the MSE from the held-out fold. This is k-fold cross-validation. In this terminology, LOOCV is n-fold cross-validation.

For the jth fold, the computed MSE is

$$\mathrm{MSE}_{(j)} = \frac{1}{n_j} \sum_{i \in j^{\text{th}} \text{ fold}} (y_i - \hat{y}_{i(-j)})^2, \tag{4.147}$$

where $\hat{y}_{i(-j)}$ is the predicted value of y_i using the fit from the data with the jth fold held out, and n_j is the number of observations in the jth fold.

Averaging these, we get the k-fold CV estimate for the population MSE,

$$\mathrm{CV}_{(k)} = \frac{1}{k} \sum_{j=1}^{k} \mathrm{MSE}_{(j)}. \tag{4.148}$$

As in other statistical procedures with tuning parameters, there is a bias-variance tradeoff in the choice of k in k-fold cross-validation. Based on the reasoning above, we see that as k increases (approaching LOOCV), the bias of the MSE estimator decreases, but the variance of the MSE estimator increases. The computational burden also increases as k increases. The computational issues with cross-validation can be mitigated by use of parallel processing. (Recall the `foreach` directive in the `foreach` package mentioned on page 396.)

Cross-validation is used in many statistical procedures, especially in probability density estimation and in classification and clustering. A common type of usage of cross-validation is to evaluate different statistical models, such as regression models of different forms. Another common type of usage of cross-validation is to evaluate different statistical fits of a given model, such as ridge regression criteria with different values of the penalty weight. A particular model or method is chosen that minimizes the cross-validation MSE.

Example: Use of Cross-Validation to Choose the Ridge Penalty Parameter

We will use cross-validation in the example regression model discussed on pages 435 through 448. On page 447 we fit the model with ridge regression using the R function `lm.ridge` in the `MASS` library. We used four values of λ, including $\lambda = 0$ which is ordinary least squares. As λ increased, the estimates shrank toward 0, but we had no measure of how good the fits were. We know that the MSE (which we did not compute) increased as λ increased.

Although the results did not seem to indicate that ridge regression would provide a substantially better fit, let us use CV estimates of the MSE to select a value of λ.

It is straightforward to write a computer program to do k-fold cross-validation. The setup is just to divide the sample into k folds of approximately equal size. After the folds have been identified, cross-validation is done by a two-part loop over the k folds in which the model is fit using all data except the selected fold, and then the MSE is computed on the excluded fold.

The division of the sample into folds should be done randomly. If the sequence in the sample is "random", then we can just identify the folds as the first approximately n/k observations, the next approximately n/k observations, and so on. If there is some possibility that the ordering in the sample is not random, as in the case of bond yields at different points in time, a random permutation of a sample of n observations can be made in R by use of `sample`:

```
set.seed(5)
index <- sample(1:n,n)
permuteddata <- data[index, ]
```

In this example, instead of writing code to do cross-validation on ridge regression, we will use an R function that does just that. The R function `cv.glmnet` in the `glmnet` package referred to earlier does these computations based on 10-fold cross-validation, unless a different value of k is specified. This function will accept a vector of λ values over which to fit ridge regressions, or if the user does not specify the values, the function will generate a set of values for λ. The generated set will not include 0, so if CV is to be performed on the OLS fit, then 0 must be specified when the function is invoked.

In the following, we specify the same set of λ values as in the R code on page 447. The `cv.glmnet` function does cross-validation for the elastic net with parameter α (see page 370), so for ridge regression we set the parameter `alpha` to 0. (For lasso regression we would set `alpha` to 1.)

The folds are formed randomly in `cv.glmnet`, so in order to ensure reproducible results, we must set the seed of the random number generators. Another thing to be aware of is that the arguments in `cv.glmnet` cannot be

data frames. Using the `YX` data frame from page 447, we may first set
`X<-as.matrix(YX[,2:5])`.

```
> library(glmnet)
> set.seed(5)
> cv.out <- cv.glmnet(X, dBaa, alpha=0, lambda=c(0,0.25,0.5,1.0))
> lam <- cv.out$lambda.min
> lam
[1] 0
> msr <- min(cv.out$cvm)
> msr
[1] 0.001083626
> sqrt(msr)
[1] 0.03291848
```

The output in `$lambda.min` indicates that the optimal λ of those tried is 0, that is, OLS yields the lowest estimated MSE of any of the λ values tried. (From the results on page 447, we would have already suspected that that might be the case; regularization does not make much difference for this dataset and model.) The output in `$cvm` is the estimated MSEs at the values of λ, but `$cvm` is not ordered the same as `lambda`, so we explicitly find its minimum, corresponding to the best value of λ. This is the cross-validated estimated MSE, $\text{CV}_{(k)}$ of equation (4.148), at the best value of λ.

How does the cross-validated estimated MSE compare with the computed MSE in the least-squares fit of the model? The MSE was not shown in any of the output on pages 435 through 448 where we used ordinary least squares on this data and model, but in the summary output on page 439, the "Residual standard error" was printed. It is 0.03249, which is the square root of the MSE. This is smaller than the square root of $\text{CV}_{(k)}$. A possible explanation is that the cross-validated estimated MSE includes an estimated bias. Under our assumptions for this model, however, the estimators are unbiased. The least-squares method of fitting minimizes the MSE for the given sample. In cross-validated estimation, when the estimated fit based on least squares in each given sample is used in other samples, the MSEs in those samples are likely to be larger just due to sample variation. The cross-validated MSE is an estimate of the *conditional* MSE (of any other sample) using a fit based on one given sample.

4.6.3 Model Selection and Model Complexity

The guiding principles for fitting statistical models discussed in Section 4.2 yield criteria for comparing models and for selection of the most appropriate model. The two most common criteria for fitting models, least norm of residuals and maximum likelihood, are also the most commonly used criteria in selecting models.

The choice of the best model, however, involves not only how well the model fits as determined by the residuals or by the likelihood, but also the complexity of the model. As an extreme case that we mentioned, we can form a linear regression model with a perfect fit for any given dataset (all residuals being zero) if we allow enough regressor variables (even random ones not among the observations). Hence, any measure of the goodness of the model must include a penalty term for the model complexity. Extra terms in a model not only increase the complexity of the model, in most cases, as we have described (see the discussion of VIF on page 427), they increase the variance of the estimators we use in fitting the model and in making predictions.

There are several ways of measuring model complexity. For parametric models, the simplest measure is the number of parameters. Of course, with parametric models of different forms, this measure may not be relevant.

Two measures for use in selecting variables to include in a linear regression model that we discussed on page 423 were adjusted R-squared and Mallows's C_p. Both of these measures included a component measuring goodness-of-fit based on sums of squares, and a component relating to the complexity of the model.

There are various ways that we may measure the goodness-of-fit of a general model. A common method of assessing goodness-of-fit of a model is by use of the likelihood, but, as we have pointed out before, *defining a likelihood means assuming a family of distributions.*

Unless we assume a specific family of distributions of the errors, the "likelihood" is only a vague concept. The relevant quantity is the likelihood function evaluated at the given fully-specified model. We use "$L\left(\widehat{M}\right)$" to represent this value; here, $\widehat{M}$ is the model with all parameters having assigned values (presumably MLEs from a sample). We will use "p" as a measure of the model complexity. In practice, p is often just the number of parameters. We then will use "$\widehat{M}_p$" to denote the "best" model with this number of parameters.

The specific function of the likelihood in some of the measures below is the deviance, $-2\log\left(L\left(\widehat{M}\right)\right)$, as defined on page 373.

Two of the most common measures of the goodness of a model based on the deviance are the *Akaike information criterion* or *AIC*,

$$\text{AIC} = -2\log\left(L\left(\widehat{M}_p\right)\right) + 2p, \tag{4.149}$$

and the *Bayes information criterion* or *BIC*,

$$\text{BIC} = -2\log\left(L\left(\widehat{M}_p\right)\right) + \log(n)p. \tag{4.150}$$

The Bayes information criterion was developed from a Bayesian perspective, but its form and use do not depend on a Bayesian approach to modeling. The BIC is also called the *Schwarz information criterion* or *SIC*.

In both cases, the measure decreases the better the fit (larger likelihood) and increases in model complexity; hence, smaller values of AIC or BIC indicate "better" models. Both of the measures are often negative.

The difference is trivial; $\log(n)p$ instead of p. Of course, for larger values of n, this difference can become large. Since the weighting of model complexity is greater in BIC, it places more emphasis on simpler models than does the AIC.

Because each of these measures is used to compare models, they can be modified by an additive constant or by a positive multiplicative constant, so some definitions of the AIC and BIC are slightly different from those above.

In the simple case of multiple linear regression in which the errors have independent and identical normal distributions, for a model with p parameters, $y = X\beta_p + \epsilon$, we get a simple form of the deviance using equations (4.68) and (4.69) We then readjust the deviance and the additive penalty term, so that for multiple linear regression with p regressors, we have

$$\text{AIC} = \log(\widehat{\sigma}_p^2) + \frac{2p}{n} \tag{4.151}$$

and

$$\text{BIC} = \log(\widehat{\sigma}_p^2) + \frac{\log(n)p}{n}, \tag{4.152}$$

where $\widehat{\sigma}_p^2$ is the residual mean square using the k regressors.

In the case of least squares fitting, AIC and BIC can be expressed in terms of the residual sum of squares, SSE. AIC is directly proportional to Mallows's C_p, which we wrote on page 424 in terms of SSE. Both AIC and BIC can be used to select the best set of variables to include in a linear regression model. They often yield the same models, but occasionally BIC chooses a model with fewer variables. In Chapter 5, we illustrate the use of AIC and BIC to select the orders in an ARIMA model.

There is also a "corrected" version of AIC, called AICc, that uses a different form of the additive penalty term. It is only relevant in small samples. Again, we remark that the specific formulas for AIC and BIC differ in the statistical literature from author to author. For comparative evaluation of models, the differences are not important so long as the same definition is used in the comparisons.

Notes and Further Reading

Statistical modeling and inference is a large area, and this chapter obviously has only covered a selection of the relevant topics. Any number of texts cover the theory and the applications, both in finance and in other areas. Section 3.2 in `mason.gmu.edu/~jgentle/books/MathStat.pdf`

has an extensive discussion of the theory underlying methods of estimation, including Bayesian procedures, least squares, maximum likelihood, and others. Davidson and MacKinnon (2004) discuss many aspects of the analysis of linear models with an emphasis on applications in economics and finance.

Many of the technical terms and concepts of statistical inference discussed in Section 4.4 are often misunderstood by scientists and data analysts. One technical term, in particular, has become so widely misunderstood and misused that some statisticians have argued that use of the term by statisticians and others should discontinued. The term is "significance", along with the related terms "significant p-value", "statistical significance", and so on. In this text, I have defined and used these terms correctly. It is unfortunate that these terms also have nontechnical meanings. One nontechnical meaning of "significant" is "important". Some people confuse "significant" with "important", or otherwise do not understand the technical statistical meaning of the term.

There are, of course, other technical terms in statistical inference that are also common everyday words, with general nontechnical meanings, and hence their statistical usage is misunderstood.

I will not attempt a list of these words here, but I will mention one: "likelihood". Despite the technical definition of "likelihood", even advanced statistics students sometimes incorrectly use it synonymously with "probability".

"Significant" does not mean "important", and "likelihood" does not mean "probability". I will continue to use both of those words correctly. (If I wrote for the popular media, I might not, however.)

Bayesian methods generally require a basic shift in the underlying statistical paradigm. Berger (1993) provides an extensive discussion of the theory and applications of Bayesian methods. Kruschke (2015) discusses practical issues in Bayesian analysis and provides an introduction to the use of JAGS and `rjags`.

Hull (2017) has an extensive discussion of models for the analysis of pricing of futures assets, such as stock options. The methods include the use of Monte Carlo simulations.

The Santa Fe Institute agent-based model of the stock market is described by Ehrentreich (2008). That book also discusses agent-based models in general.

Joe (2015) describes various methods of fitting copulas to data.

Optimization plays a major role in statistical estimation and other methods. Section 4.3 only touched the surface of the topic. Griva, Nash, and Sofer (2009) provides background material as well as discussion of specific methods for optimization.

Friedman, Hastie, and Tibshirani (2010) describe lasso and ridge regression and the elastic net, and James et al. (2013) place particular emphasis on the methods implemented in the R software in the `glmnet` package.

The tails of frequency distributions and extreme values in general are important topics in the analysis of financial data. A general discussion of extreme values is available in Novak (2012). Gilleland, Ribatet, and Stephenson (2013) review available R software for analysis of extreme values.

The method of estimating the tail index discussed in Section 4.4.7, which was originally described by Hill (1975), is commonly used. It is generally adequate, although, as mentioned, it can fail spectacularly, resulting in "Hill horror plots". Various methods have been suggested for improving the Hill estimator, such as use of the bootstrap described in Pictet, Dacorogna, and Müller (1998). There are other methods of estimating the tail index that have merit, and some other methods are presented in Politis (2002) and Bacro and Brito (1998), for example. Reiss and Thomas (2007) discuss other aspects of extreme value theory and methods of estimating the tail index and tail dependencies.

Buthkhunthong et al. (2015) describe a local version of Kendall's tau useful in measuring tail dependencies. Jones (1996) describes general issues in local statistical measures.

There are many books on regression analysis, including Kutner et al. (2004), Sheather (2009), Harrell (2015), and Fox and Weisberg (2018). The latter three books discuss the use of R in regression. Fox and Weisberg (2018) have an extensive discussion of regression diagnostics, using the terminology in the R documentation. Harrell (2015) discusses logistic models as well as linear regression models. Pinheiro and Bates (2009) discuss methods for computing generalized least squares estimates for various structures of the variance-covariance matrix of the errors in linear models. Gentle (2009), on pages 610 through 613, discusses orthogonal regression and describes a method for fitting a model by minimizing the orthogonal residuals.

Many of the ideas in leave-out-one deletion were first developed in Belsley, Kuh, and Welsch (1980). Proofs of the various facts about matrices and linear algebra referred to in Sections 4.5.2 and 4.5.5 are given in Gentle (2017).

The books on regression analysis above discuss model selection, at least among linear models. Harrell (2015) discusses regression model selection in a more general context. Some criteria are based on the heuristic underlying the fitting criterion, such as least squares. Several criteria are based on "information". This term in general refers to a function of the log-likelihood, and there are some specific definitions appropriate in different contexts, such as Fisher information, Kullback-Leibler information, and so on. We discussed the AIC and BIC in Section 4.6.3 and mentioned AICc, but there are other criteria based on information. Some more common ones are Watanabe-Akaike information criterion, Shibata's information criterion, and the Hannan-Quinn information criterion. These all involve an information measure and a penalty for the model complexity. The latter two are particularly popular in fitting some time series models. Claeskens and Hjort (2008) and Wang (2012) discuss use of various information-based criteria in model selection.

Model selection is an important consideration in Bayesian analysis, and while the basic principles are essentially the same, there are some additional considerations, and Wang (2012) discusses model selection using Bayes factors.

A related topic is *model averaging* or *multimodel inference* (instead of model selection). Although multimodel inference has many potential applications in finance, it it not widely used in that field, and we have not discussed it in this book. Burnham and Anderson (2002) discuss multimodel inference and model averaging from a frequentist perspective. Claeskens and Hjort (2008) discuss Bayesian model averaging, as well as frequentist model averaging.

Hastie, Tibshirani, and Friedman (2009) and James et al. (2013) discuss and illustrate methods of cross-validation.

Regression techniques that are often used in econometrics, but ones that we have not discussed, include least squares for seemingly unrelated regressions and two- and three-stage least squares for instrumental variables. The systemfit R package has functions for these kinds of statistical analysis.

The basic stats package in R has many functions for regression and other types of statistical analyses. Other R packages that we have mentioned or used in this chapter include MASS (variable selection, ridge, and many other computations), faraway (VIF and many other regression computations), olsrr (several functions, especially for diagnostic plots), leaps (variable selection), lmtest (ridge and lasso), glmnet (generalized linear models), and nlme (generalized least squares). These packages may also be useful in the exercises for this chapter.

Exercises

Many of the exercises in this chapter require you to use R (or another software system if you prefer) to acquire and explore real financial data.

Often the exercises do not state precisely *how* you are to do something or *what type* of analysis you are to do. That is part of the exercise for you to decide.

4.1. Transformations.

Consider a random variable X in the family of Poisson distributions with parameter λ. As λ varies from 1 to 9, both the mean and the variance of X increase from 1 to 9. Now consider the transformation $Y = \log(X+1)$. How do the mean and variance of Y vary as λ varies from 1 to 9?

Simulate 1,000 realizations of X for each $\lambda \in \{1, 2, \ldots, 9\}$, and for each value of λ, compute the sample mean and variance of your simulated sample. How do the mean and variance vary?

4.2. Maximum likelihood estimation.

(a) Write the log-likelihood function for the Poisson distribution, given data $x_1, \ldots, x_n$, corresponding to the likelihood function in equation (4.26).

(b) Determine the MLE of λ.

4.3. Method of moments and maximum likelihood estimation.

Given a sample $x_1, \ldots, x_n$ from a $N(\mu, \sigma^2)$ distribution, determine the method of moments estimate and the MLE of σ^2.

4.4. Monte Carlo estimation.

(a) Suppose we have a function g of a random variable X whose PDF is f_X. We wish to estimate $E(g(X))$. The first consideration is whether or not $E(g(X))$ exists. Let us assume that g is such that $E(g(X))$ exists and is finite. Another issue is whether we can generate random (or pseudorandom) variables that simulate observations on the random variable X.

Given a pseudorandom sample $x_1, \ldots, x_m$ from the distribution of X, show that Monte Carlo estimate of $E(g(X))$,

$$\widehat{E(g(X))} = \frac{1}{m} \sum_{i=1}^{m} g(x_i),$$

is unbiased.

(b) Let $X \sim N(0, 1)$. Use Monte Carlo to estimate $E(X^2)$.

(c) Use Monte Carlo simulation to estimate the conditional mean of a t random variable with 4 df given that the variable is greater than or equal to its 95th percentile.

   i. In this case, you may assume that the 95th percentile is known.

     What other issues should you consider in reporting your result?

   ii. For a general problem of this nature, we may not know the 95th percentile of the underlying distribution. If that is the case, how would you proceed?

4.5. MLE in the "standardized" t distribution.

Obtain the daily adjusted closing prices of Intel (INTC) for the period January 1, 2017, through September 30, 2017, and compute the returns. These are the data used in the VaR examples in Section 1.4.5, although because INTC pays dividends the adjusted closes may be slightly different.

Fit a location-scale ("standardized") t distribution to these data using maximum likelihood. The R function `fitdistr` in the package `MASS` performs this fit. Identify the parameter estimates you obtain.

What is the estimated standard deviation of the fitted t distribution?

4.6. Gauss-Newton method.

To fit the linear model $y = X\beta + \epsilon$, Newton's method leads to the normal equations in one step, and the solution to the normal equations is the solution to the optimization problem. Show that the Gauss-Newton method for the linear least squares problem is the same as Newton's method.

4.7. Gompertz model.

Write out the gradient and the Hessian for the least squares estimation of the parameters in the Gompertz model

$$x_i = \alpha e^{\beta e^{\gamma t_i}} + \epsilon_i.$$

4.8. Weighted least squares; GLS.

By minimizing the weighted sum of squared residuals, derive the weighted least squares estimator for the linear model with $V(\epsilon) = \Sigma$.

4.9. Confidence intervals and tests of hypotheses.

Show that the confidence interval (4.70)

$$\left(\overline{X} + t_{n-1,0.025}\sqrt{S^2/n}, \quad \overline{X} - t_{n-1,0.025}\sqrt{S^2/n}\right)$$

is formed from the pivotal quantity $(\overline{X}, \sqrt{S^2/n})$ as in equation (4.73).

4.10. Confidence intervals and tests of hypotheses.

This exercise requires computation of the individual quantities to test hypotheses and set confidence intervals for μ and σ^2 in an assumed $N(\mu, \sigma^2)$ distribution.

Given the small set of data $\{10, 8, 9, 13, 9, 11\}$, do the following.

(a) Set a 90% two-sided confidence interval for μ.
(b) Test at the 10% level the hypothesis $\mu = 11$ versus the alternative $\mu \neq 11$.
(c) Set a 90% lower one-sided confidence interval for σ^2 (that is, a confidence interval in which the lower bound is $-\infty$).
(d) Test at the 10% level the hypothesis $\sigma^2 \leq 3$ versus the alternative $\sigma^2 > 3$.

4.11. Bayesian models.

Consider a hierarchical Bayesian model for a univariate normal distribution $N(\mu, \sigma^2)$ in which the mean μ is a realization of a normal random variable M and σ^2 is a scaled realization of a random variable Σ^2 that has an inverted chi-squared distribution with ν_0 degrees of freedom. (For the scaling factor σ_0^2, this means that

$1/(\sigma_0^2 \Sigma^2)$ has a chi-squared distribution.) Now, given $\Sigma^2 = \sigma^2$, assume that M has a conditional normal distribution with mean μ_0 and variance σ^2/κ_0.

(a) Draw a directed acyclic graph that describes this model. Identify each variable with the corresponding variable in the DAG on page 387.

(b) Choose μ and σ^2 (say, 100 and 100) and generate 100 pseudorandom variates from a $N(\mu, \sigma^2)$ distribution.

Now set up the Bayesian formulation with chosen values of μ_0, σ_0^2, and κ_0 (say, 90, 90, and 1), and analyze the data. (In the Bayesian sense, that means to develop posterior distributions for μ and σ^2, given the simulated data.)

It is suggested that you use R and BUGS/JAGS (with rjags) to do this analysis. The JAGS executable program must be installed in a directory accessible by R. It is available from SourceForge.

4.12. **Bayesian analysis.**

Obtain the INTC adjusted closing prices for the year 2018 and compute the daily returns. Now determine Bayes estimates of the mean and the standard deviation of the returns, similar to the example illustrated in Figures 4.2 through 4.4. For the prior on the mean use $N(0, 100)$, and for the prior on the variance use the inverse chi-squared distribution with 10 degrees of freedom. (The "inverse chi-squared distribution" is a phrase often used in Bayesian analysis. If X has the inverse chi-squared distribution, then $1/X$ has the chi-squared distribution.)

The JAGS executable program must be installed in a directory accessible by R. It is available from SourceForge.

4.13. **Robust fitting.**

Obtain the daily closing prices of the S&P 500 Index for the year 2017 and compute the daily returns.

(a) Compute the standard deviation (the volatility), the MAD, and the IQR of the returns.

(b) Now produce two side-by-side histograms with normal probability density curves superimposed, where for one the normal density has mean equal to the sample mean and standard deviation equal to the sample standard deviation, and for the other one the normal density has mean equal to the sample median and standard deviation based on the MAD (see equation (1.64)). Your plots should be similar to those in Figure 4.5. Which one seems to fit better? Does this mean that the normal distribution may be a good model for returns if the appropriate mean and variance are used?

4.14. **The tail index.**

(a) Let $f_X(x)$ be the PDF of a random variable X. Suppose that for some $x_u < 0$, the form of $f_X(x)$ over $x \leq x_u$ is

$$f_X(x) = c(-x)^{-(\alpha+1)},$$

where $c > 0$ and $\alpha > 1$

Show that the conditional PDF for $x \leq x_u$ is $f_{X \leq x_u}(x) = \alpha |x_u|^{\alpha} |x|^{-(\alpha+1)}$.

(b) Consider a random sample of size n from the distribution with PDF $f_X(x) = c|x|^{-(\alpha+1)}$. Let the n_{x_u} be the number of order statistics in the sample that are are less than or equal to x_u. Show that the log-likelihood of α, given the smallest n_{x_u} order statistics is

$$l_L(\alpha) = n_{x_u} \log(\alpha) - n_{x_u} \alpha \log(|x_u|) - (\alpha + 1) \sum_{i=1}^{n_{x_u}} \log(|x_{(i:n)}|).$$

Show that the MLE of α is

$$\widehat{\alpha} = \frac{n_{x_u}}{\sum_{i=1}^{n_{x_u}} \log(x_{(i:n)}/x_u)}.$$

(c) Show that Hill's estimator of the tail index is the reciprocal of the mean of the logs of the absolute values of the first $k - 1$ order statistics minus the log of the absolute value of the kth order statistic. (The log of the absolute value of the kth order statistic is subtracted before computing the mean.)

(d) The discussion in Section 4.4.7 and Exercises 4.14a through 4.14c concerned the index of the lower tail of a distribution. That discussion also assumed that the lower tail was over the negative reals.

What modifications must be made for a similar method to estimate the index of the upper tail?

What modifications must be made for a similar method to estimate the tail index if that tail may be over both negative and positive reals?

4.15. **The tail index.**

Obtain the daily closes of the S&P 500 from January 1, 1990, through December 31, 2017, and compute the daily log returns. These are the data used in Figure 4.6.

As in Exercise 2.12b, divide the S&P 500 daily log returns into four periods,

January 1, 1992, through December 31, 1995;
January 1, 1997, through December 31, 2002;
January 1, 2003, through December 31, 2005;
January 1, 2006, through December 31, 2009.

In each of these periods, the volatility was relatively constant, but it differed markedly between the periods.

For the data in each period, produce a Hill plot for the lower tail. These plots will show the types of variation that we often see in Hill plots. You may use the `hill` function in the `evir` R package, or any other function to compute the Hill estimator, or you may just write code to compute the estimates. (The algorithm is just a simple direct coding of the definition.)

Identify an interval of stability in each plot. (This is rather subjective.) Compute an estimate of the lower tail index for the distribution in each period as the mean of Hill's estimate over the interval of stability.

How do the estimates of the tail index differ? How do the estimates compare to the tail index of a t distribution?

Do you see any relation to the differing levels of volatility over those different periods?

4.16. **Estimation of the tail index.**

Obtain the daily closing prices of the S&P 500 Index for the year 2017 and compute the log returns.

(a) Compute the Hill estimator of the lower tail index of the distribution of the daily returns of the index for that period, letting the tuning parameter range from 11 to 100.

You may use the same computations as used in Exercise 4.15. (That might have been the `hill` function in the `evir` R package, or some other function.) Your plot should look somewhat similar to that in Figure 4.6, but the tail index of the returns for that period may be different.

(b) Compute an estimate of the lower tail index for these data using tuning parameter values of 20, 30, 40, and 50.

How do they differ?

Now, use the bootstrap to estimate the bias and the variance for the estimator at each value of the tuning parameter.

What conclusions can you draw about the bias-variance trade-off?

(c) Compute a basic bootstrap 95% confidence interval for the the tail index of the distribution of the daily returns of the index for the year 2017.

For this, take the estimate as the mean of Hill's estimate over the same interval in each bootstrap sample, where this interval is chosen (subjectively) by inspection of the Hill plot for the full dataset.

4.17. Tail dependencies.

Obtain the daily unadjusted closing prices of INTC, GLD, and the Nasdaq Composite for the year 2017 and compute the log returns. These data were shown in the displays in Figure 1.22 on page 86.

(a) Produce a scatterplot of the returns of INTC and versus the returns of the Nasdaq Composite, and another scatterplot of the returns of GLD and versus the returns of the Nasdaq Composite, for the year 2017.

Our interest is in the relationships of the returns in the tails of the bivariate distributions.

On each scatterplot draw a vertical line at the $p = 0.05$ quantile of the sample of returns on the horizontal axis and a horizontal line at the $p = 0.05$ quantile of the sample of returns on the vertical axis.

(b) Compute the sample tail dependency at $p = 0.05$ for returns of INTC and the Nasdaq Composite, and for returns of GLD and the Nasdaq Composite for the year 2017.

Comment on the results.

4.18. Value at risk and expected shortfall.

Obtain the daily adjusted closing prices of Intel (INTC) for the period January 1, 2017, through September 30, 2017, and compute the returns. These are the data used in Exercise 4.5 and in the VaR examples in Section 1.4.5, although because INTC pays dividends the adjusted closes may be slightly different.

Assume a holding of $1,000 in INTC on October 10, 2017.

(a) Compute a nonparametric estimate of VaR for this holding with 95% confidence for October 11, 2017 (one day out).

(b) Compute a nonparametric estimate of the expected shortfall for this holding with 95% confidence for October 11, 2017 (one day out).

Which is larger, the VaR or the expected shortfall? Is this always the case?

(c) Compute a nonparametric estimate of VaR for this INTC holding with 95% confidence for October 17, 2017 (one week out). This computation should be based on the historical simple weekly returns over a recent time period.

(d) Compute a nonparametric estimate of the expected shortfall for this holding with 95% confidence for October 17, 2017 (one week out).

(e) Now, obtain the daily adjusted closing prices of Microsoft (MSFT) for the same period, January 1, 2017, through September 30, 2017, and compute the returns.

Assume a holding of $1,000 in MSFT on October 10, 2017.

i. Compute a nonparametric estimate of the VaR for this MSFT holding with 95% confidence for October 11, 2017 (one day out).

ii. Compute a nonparametric estimate of the expected shortfall for this MSFT holding with 95% confidence for October 11, 2017 (one day out).

(f) Now, consider a portfolio consisting of of $500 in INTC and $500 in MSFT on October 10, 2017.

For measures of risk, we need a model of the distribution of the returns, or at least of the returns no greater than the 0.05 quantile. Use historical frequency data as the model.

Assume that the portfolio is continuously rebalanced so that the proportions of INTC and MSFT are 50/50 at all times. (In practice, of course, portfolios would not be rebalanced this frequently, but it the assumption provides an approximate frequency distribution for the returns.

i. Compute a nonparametric estimate of VaR for this portfolio with 95% confidence for October 11, 2017 (one day out). Compare this with the VaRs for INTC and MSFT individually as computed above. Does this indicate whether or not VaR is coherent?

ii. Compute a nonparametric estimate of the expected shortfall for this portfolio with 95% confidence for October 11, 2017 (one day out). Compare this with the expected shortfalls for INTC and MSFT individually as computed above. Does this indicate whether or not expected shortfall is coherent?

4.19. **Sample correlations.**

Obtain the daily unadjusted closing prices of INTC and GLD and compute the log returns (Exercise 4.17).

Test the hypothesis that the correlation between the returns is 0.

4.20. **Principal components.**

Obtain the daily unadjusted closing prices of each of the 30 stocks in the Dow Jones Industrial Average for the year 2017. There were no stock splits during that period. (There is a text file linked on the website for the book that lists the stock symbols for the stocks in the DJIA during that period. As time goes on, some of these symbols may become invalid. In that case, the exercise can be modified in one of two ways, either of which will preserve the intent of the exercise. One way, is just to change the year 2017 to the most recent full year. In that case, the list of companies may not be the same. The other way is just to remove the invalid stock symbols the list, and to use fewer than 30 stocks.)

(a) Determine the first principal component of these 30 variables.
(b) Now compute the correlation between the daily closes of the first principal component and the daily closes of the DJIA for 2017.
(c) Form a market index based on the first principal component of the Dow 30 stocks as a market index. Normalize your index so that it was 100 on the first trading day of 2017. Plot the daily closes of your index and of the DJIA for the two years 2017 and 2018, on a single graph with the indexes normalized so as to have the same value on the first trading day of 2017.
(d) From a practical standpoint, why might a market index based on a principal component of the prices of a set of stocks not be desirable? (See page 413.)

4.21. **Partitioning the sum of squares.**

Show that for a model fit by least squares, the total sum of squares is equal to the model sum of squares plus the residual sum of squares. That is, show that

$$\sum_{i=1}^n (y_i - \bar{y})^2 = \sum_{i=1}^n (\hat{y}_i - \bar{y})^2 + \sum_{i=1}^n (y_i - \hat{y}_i)^2$$

is an algebraic identity if the residuals sum to 0; that is,

$$\sum_{i=1}^n (y_i - \hat{y}_i) = 0.$$

4.22. **Sequential regression and partial F statistics.**

Consider the models

$$\begin{aligned} y &= \beta_{0,0} + \epsilon_0 \\ y &= \beta_{0,1} + \beta_{1,1}x_1 + \epsilon_1 \\ y &= \beta_{0,2} + \beta_{1,2}x_1 + \beta_{2,2}x_2 + \epsilon_2 \\ y &= \beta_{0,3} + \beta_{1,3}x_1 + \beta_{2,3}x_2 + \beta_{3,3}x_3 + \epsilon_3 \end{aligned}$$

as in equation (4.121). Generate some random data for y, x_1, x_2, and x_3,

```
n <- 100
set.seed(12345)
x1 <- runif(n)
x2 <- runif(n)
x3 <- runif(n)
y <- 4 + 1*x1 + 2*x2 + 3*x3 + .05*rnorm(n)
```

(a) Fit the model $y = \beta_{0,0} + \epsilon_0$ and compute the residuals; call them r_0. Now fit the two models

$$
\begin{aligned}
r_0 &= \theta_{0,1} + \beta_{1,1}x_1 + \epsilon_{1p} \\
y &= \beta_{0,1} + \beta_{1,1}x_1 + \epsilon_1
\end{aligned}
$$

and compare the coefficient estimates as well as the partial F statistics.

(b) Let r_1 be the vector of residuals from the model $y = \beta_{0,1} + \beta_{1,1}x_1 + \epsilon_1$ in Exercise 4.22a.

Now fit the two models

$$
\begin{aligned}
r_1 &= \theta_{0,2} + \theta_{1,2}x_1 + \beta_{2,2}x_2 + \epsilon_{2p} \\
y &= \beta_{0,2} + \beta_{1,2}x_1 + \beta_{2,2}x_2 + \epsilon_2
\end{aligned}
$$

and compare the coefficient estimates as well as the partial F statistics.

(c) Let r_2 be the vector of residuals from the model $y = \beta_{0,2} + \beta_{1,2}x_1 + \beta_{2,2}x_2 + \epsilon_2$ in Exercise 4.22b.

Now fit the two models

$$
\begin{aligned}
r_2 &= \theta_{0,3} + \theta_{1,3}x_1 + \theta_{2,3}x_2 + \beta_{3,3}x_3 + \epsilon_{3p} \\
y &= \beta_{0,3} + \beta_{1,3}x_1 + \beta_{2,3}x_2 + \beta_{3,3}x_3 + \epsilon_3
\end{aligned}
$$

and compare the coefficient estimates as well as the partial F statistics.

4.23. **Regression and beta.**

Financial analysts using the common formulas such as the "market model", equation (1.35) on page 61, or the simple formula for beta, equation (1.36), must make decisions about how to assign values to the terms in the formulas.

To use historical data (is there any other way?) to assign values to the terms in these formulas requires decisions: (1) which market, M; (2) frequency of returns; (3) time period of the data; and (4) returns of adjusted or unadjusted prices. For some quantities, there may also be a choice of formulas. Simple returns are more commonly used in this kind of analysis, but log returns may also be used.

For the market model, we also must decide what values to use for $R_{F,t}$.

On January 11, 2019, TD Ameritrade, on its website linked from users' accounts, quoted a beta for INTC of 0.8. On the same day, E*Trade, on its website linked from users' accounts, quoted a beta for INTC of 1.3.

The difference (assuming each used a "correct" formula correctly) would be due to the formula used, which market is used, the frequency of returns, and the time period.

In Exercise A1.13, you explored the effects of these choices on calculation of beta for INTC, just using the simple correlation formula.

In this exercise, you are to consider the same issues using beta computed from the market model (1.35),

$$R_{i,t} - R_{F,t} = \alpha_i + \beta_i(R_{M,t} - R_{F,t}) + \epsilon_{i,t}.$$

For $R_{F,t}$, use the 3-Month US T-Bill rate, and recall that the data from FRED may contain missing values (see Exercise A1.7).

Compute betas for INTC based on the 18 combinations of the following

- 3 benchmarks: S&P 500 (^GSPC), Nasdaq composite (^IXIC), and VGT (an information technology ETF owned by Vanguard)
- 2 frequencies: daily returns and weekly returns;
- 3 time periods: 2018-07-01 through 2018-12-31, 2018-01-01 through 2018-12-31, and 2017-01-01 through 2018-12-31.

Compare these results with those obtained in Exercise A1.13. Note, of course, that different definitions of beta were used, different types of returns were used, and further, since adjusted closes were used, the results can differ slightly due to changes in the adjustments over time.

4.24. **Regression.**

The SPY ETF is designed to track the S&P 500 Index. Obtain the unadjusted monthly closing prices of SPY and the corresponding monthly closes of the S&P 500 Index for the year 2017. (Note that we should use either the adjusted closes of SPY and the S&P 500 Total Return as in Figure 1.14 or the unadjusted prices of SPY and the S&P 500 Index itself; see Exercise 1.10.) Plot the prices and perform a regression analysis of the SPY closes on the S&P 500. (See also Exercise 2.3.)

Interpret your results.

4.25.**Regression.**

The SDS ETF is designed to produce returns twice the negative of the returns of the S&P 500 Index (see page 58, and see Exercise 2.3). Obtain the weekly closes of SDS and of the S&P 500 Index and compute the weekly returns for each. Note that for short ETFs, we focus on the *returns*, not on the prices. (See Figure 1.16 on page 59.)

Plot the returns.

Fit a simple linear regression model regressing the SDS weekly returns on the S&P 500 Index weekly returns.

Interpret your results.

4.26.**Regression residuals.**

A q-q plot of regression residuals, as in Figure 4.13, is often a part of regression analysis.

Why might a q-q plot of the residuals be of interest?

What kind of regression residuals should be used in a q-q plot? Why?

4.27.**Regression;** Corporate bond yields and yield changes on US Treasury yields and yield changes.

Obtain the weekly yields of Moody's seasoned Aaa corporate bonds and the weekly yields of 3-month Treasury bills for the years 2008 through 2018. (Obtain data from FRED.)

(a) Regress the weekly yields of Moody's seasoned Aaa corporate bonds on the weekly yields of 3-month Treasury bills for the period 2008 through 2010.

Are the results surprising? Discuss.

(b) Regress the change in weekly yields of Moody's seasoned Aaa corporate bonds on the weekly change in yields of 3-month Treasury bills for the period 2008 through 2010.

Are the results surprising? Discuss.

Analyze and discuss the residuals.

4.28.**Regression;** Corporate bond yield changes on US Treasury yield changes.

Obtain the weekly yields of Moody's seasoned Aaa corporate bonds, the yields of 3-month Treasury bills, the yields of 2-year Treasury notes, the yields of 10-year Treasury notes, and the yields of 30-year Treasury bonds for the years 2008 through 2018.

(Note that the 30-year Treasury was discontinued on February 18, 2002, but reintroduced on February 9, 2006.)

(a) Analyze the linear regression of the change in weekly yields of Moody's seasoned Aaa corporate bonds on the weekly changes in yields of 3-month treasury bills, of 2-year treasury notes, of 10-year treasury notes, and of 30-year treasury bonds.

Compute the variance-inflation factors.

Which variables seem most important, and which, if any, should be eliminated from the model?

(b) Produce three plots: raw residuals versus the predicted values; a q-q plot of internally studentized residuals with respect to a normal reference distribution; and Cook's distance for each observation.

Summarize any information gleaned from the plots.

(c) Perform a Durbin-Watson test of the null hypothesis that serial correlations of the errors in the linear regression model up to lag 3 are 0.

Summarize the results. See also Exercise 5.27.

4.29. **Logistic regression.**

Define a dependent variable that takes the values "up" or "down", representing the direction of the price move of the S&P 500 Index for the following day from the previous day. Now define regressors that represent the difference in the current (closing) price and the 50-day moving average and the difference in the current (closing) price and the 200-day moving average.

Obtain the daily closing of the S&P 500 for the year 2018, and as much of 2017 as necessary to compute 200-day moving averages. Use the 2018 data to fit a logistic regression model with the up or down closes of the following day as the response variable. (Notice that response variable is for the day following the corresponding regressor variables.)

Test the fit of your model for the first 10 trading days of 2019.

4.30. **Tests for normality of returns.**

Obtain the closing prices of the S&P 500 Index for 2017, and compute the simple returns.

Test the returns for a normal distribution using the Anderson-Darling test, the Jarque-Bera test, and the Shapiro-Wilk test.

4.31. **Tests for normality.**

Generate samples of size 100 from four different t distributions, one each with 100 df, 20 df, 5 df, and 3 df. (See Figure 3.7.) Also generate samples of size 100 from two different gamma distributions both with rate equal to 1, one with shape equal to 5 and one with shape equal to 100.

Now perform the Anderson-Darling test, the Cramér-von Mises test, the D'Agostino-Pearson test, the Jarque-Bera test, the Lilliefors test, the Shapiro-Francia test, and the Shapiro-Wilk test on each sample. (There are six different samples.)

Summarize your results, stating the resulting p-value for each test for each distribution.

4.32. **Cross-validation.**

(a) It was stated that the LOOCV estimator $CV_{(n)}$ may have large variance "because the individual terms have positive covariances".

Why do they have positive covariances and why does that cause the variance of $CV_{(n)}$ to be large?

(b) It was stated that in k-fold cross-validation estimation of the MSE, "as k increases (approaching LOOCV), the bias of the MSE estimator decreases, but the variance of the MSE estimator increases."

Explain why this is the case.

(c) Explain why the cross-validated estimated MSE, $CV_{(k)}$ is in general larger that the estimated variance (even if the estimator is unbiased.)

Relate your explanation specifically to a regression model in which the fitted "Residual standard error" is 0.03249, as on page 439, yet the square root of the cross-validated estimated MSE is 0.03331, as on page 467.

5

Discrete Time Series Models and Analysis

A discrete time series is a stochastic process, that is, a sequence of random variables, $\{X_t\}$, indexed by time. At a given time s, we may have access to $\{X_r : r \leq s\}$ but not to X_t for $t > s$.

Time is a continuous quantity, but since in many time series we have access to $\{X_t\}$ only at fixed points in time, we will treat t as a discrete quantity. There are continuous-time analogues of most of the models we discuss in this chapter. Although these models are useful in many analyses (pricing of options, for example), we will not discuss them here.

The points at which $\{X_t\}$ is observed may be equally spaced, or nearly so. Many time series models useful in finance assume equal spacing, and although the spacing may not be equal, often it is nearly so. (See the discussion in Section 1.1.3 concerning weekend and holiday effects.)

In this chapter we will consider some useful discrete linear time series models in which the times are considered to be equally spaced (or nearly so). For such models, we use a notation of the form x_{t-1}, x_t, x_{t+1}, or in general x_{t+k}, where t and k are integers. (Recall previous comments about the notational differences for random variables and realizations of random variables. Here, as before, we will not adhere strictly to precise notation for the differences.) We may use the notation $\{x_t\}$ or $\{X_t\}$ to represent the time series itself, and except occasionally, we will not distinguish between random variables and realizations of random variables.

The basic model is of the form

$$x_t \approx f(x_{t-1}, x_{t-2}, \ldots). \tag{5.1}$$

Often the approximation "$\approx$" is just addition of some noise, and as with the models of Section 4.1 with additive error terms, we can write the general time series model as

$$x_t = f(x_{t-1}, x_{t-2}, \ldots; \theta) + w_t, \tag{5.2}$$

where θ represents some constant, but possibly unknown, vector of parameters, and w_t represents an error or noise in the system. In specific models, we may make various assumptions about the error term w_t.

Although there is a temptation to rush to more advanced topics such as heteroscedasticity and cointegration, the emphasis in this chapter is on the fundamentals. If those are not understood well, the other topics can only be pursued mechanically.

We will begin with a few basic models, discuss general properties of time series and relevant statistical measures, and then, beginning in Section 5.3, we will describe the common ARMA models and their applications. The most common failure of these models in financial applications is due to heteroscedasticity, and in Section 5.4, we discuss extensions of ARMA models to allow nonconstant variance, or stochastic volatility in GARCH models.

In Section 5.5, we briefly discuss unit-root processes and cointegrated time series.

White Noise

A process $\{w_t\}$ that is a sequence of random variables with mean 0, constant finite variance, and 0 autocorrelations at all lags is called *white noise*. Because there is a similar model in continuous time, we sometimes call this *discrete white noise*. The 0 autocorrelation condition can be stated as

$$\mathrm{Cor}(w_t, w_{t+h}) = 0 \quad \text{for } h = 1, 2, \ldots$$

We will indicate that a sequence $\{w_t\}$ is a white noise by the notation

$$w_t \sim \mathrm{WN}(0, \sigma_w^2).$$

(A more appropriate notation would be "$\{W_t\} \sim \mathrm{WN}(0, \sigma_w^2)$", but in this book, I have generally chosen simpler notation unless it is incorrect or confusing.)

A white noise process does not need to have a "beginning", and for any t, $\mathrm{E}(w_t) = 0$ and $\mathrm{V}(w_t) = \sigma_w^2$.

The zero correlation does not imply independence.

A special type of white noise is *Gaussian white noise*, defined as white noise in which the random variables have a normal (Gaussian) distribution. In this case, the sequence is iid, as we have seen in Chapter 3.

A white noise process in which the sequence is iid is called a *strict white noise* whether Gaussian or not, and the white noise process without independence is sometimes called a *weak white noise*. (Note that other authors use these terms somewhat differently. Some may impose further conditions on a sequence in order to call it a white noise. Some require that each term in the sequence has the same distribution in addition to equal first two moments, and/or require that the sequence be independent, instead of just zero-correlated, and/or may allow the mean to be nonzero.)

A simple linear regression with time as the independent variable and a white noise error term,

$$x_t \approx \beta t + w_t,$$

is a *white noise with drift*.

Linear Processes

A *linear process* $\{x_t\}$ is a linear combination of white noise variates w_t of the form

$$x_t = \mu + \sum_{j=-\infty}^{\infty} \psi_j w_{t-j}, \tag{5.3}$$

where ψ_j are constants such that

$$\sum_{j=-\infty}^{\infty} \psi_j < \infty.$$

Linear processes that depend only on the past are of interest. A *one-sided linear process* is a linear process that can be written in the form

$$x_t = \mu + \sum_{j=0}^{\infty} \psi_j w_{t-j}. \tag{5.4}$$

In a *finite* linear process, $\psi_j = 0$ for $j < k_1$ or $j > k_2$ for some finite constants k_1 and k_2.

Moving Average of White Noise

A *moving average of a white noise* is a finite linear process of normalized sums of the form

$$x_t = \frac{1}{k} \sum_{j=1}^{k} w_{t-j}. \tag{5.5}$$

We call k the "window width" or the "band width".

We will also use the term "moving average" to refer to more general linear combinations of the form $x_t = \sum_j \psi_j w_{t-j}$, with various restrictions on ψ_j, but the word "average" in this case does not imply a weighted mean.

Notice that for a finite moving average of white noise,

$$x_t = \sum_{j=1}^{k} \psi_j w_{t-j},$$

we have

$$E(x_t) = \sum_{j=1}^{k} \psi_j E(w_{t-j}) = 0, \tag{5.6}$$

and

$$V(x_t) = \sum_{j=1}^{k} \psi_j^2 V(w_{t-j}) = \sum_{j=1}^{k} \psi_j^2 \sigma_w^2. \tag{5.7}$$

The covariance of two terms, $\text{Cov}(x_t, x_{t+h})$, depends on whether there is any overlap in the white noise terms in x_t and x_{t+h}. If there is no overlap, the covariance is 0.

Random Walk

A simple model of the general form of equation (5.2) is the *random walk*, in which f is just x_{t-1},

$$x_t = x_{t-1} + w_t, \tag{5.8}$$

and $\{w_t\}$ is a white noise. (Other authors may require w_t to be iid; that is, $\{w_t\}$ to be a strict white noise.)

Although many time series models do not have a "beginning", a random walk process needs to have a starting point, which we usually denote as x_0, and run t over the positive integers.

A random walk is a simple constant diffusion process, as discussed on page 280, but of course not all diffusion processes are random walks.

With a starting point of x_0, the random walk process in equation (5.8) can be expressed as

$$x_t = x_0 + \sum_{i=1}^{t} w_i. \tag{5.9}$$

Figure 5.1 shows a simulated random walk.

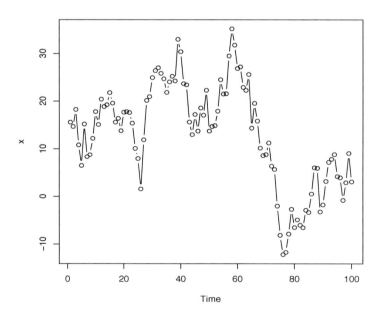

FIGURE 5.1
A Random Walk

The mean of a random variable following a random walk process starting at x_0 is

$$E(x_t) = E(x_0) + E\left(\sum_{i=1}^{t} w_i\right)$$

$$= x_0; \tag{5.10}$$

that is, the unconditional mean of a random walk is constant, depending only on x_0. The variance of a random walk variable, on the other hand, is

$$V(x_t) = V(x_0) + V\left(\sum_{i=1}^{t} w_i\right)$$

$$= 0 + \sum_{i=1}^{t} V(w_i)$$

$$= t\sigma_w^2, \tag{5.11}$$

that is, it depends on t and in fact grows linearly as t grows. We can see this in the random walk shown in Figure 5.1.

The R function cumsum together with a source of white noise can be used to generate a random walk. The random walk in Figure 5.1 was generated by the following R statements.

```
w <- rnorm(100, sd=4)
x <- cumsum(w) + 10
```

In financial applications, we often model log returns as random walks. We use the standard deviation more often than the variance. The standard deviation grows as $\sqrt{t}$. This is why we annualize daily volatility σ_d by $\sqrt{253}\sigma_d$, as on page 107.

Aggregated Log Returns

Because log returns are additive (see equation (1.26) on page 31), for any time period, the accumulated return of a sequence of daily returns $r_1, \ldots, r_t$ over that period form a walk:

$$r_t = \sum_{i=1}^{t} r_i. \tag{5.12}$$

This is a random walk if, for the individual sub-periods, the log returns correspond to a random variable with 0 mean and constant finite variance, and if they have 0 correlation with each other. One form of the Random Walk Hypothesis, referred to in Chapter 1, assumes log returns follow a white noise process; hence, the log returns are a random walk under that hypothesis.

Random Walk with Drift

A variation is a *random walk with drift*, given by

$$x_t = \delta + x_{t-1} + w_t, \tag{5.13}$$

or

$$x_t = x_0 + t\delta + \sum_{i=1}^{t} w_i. \tag{5.14}$$

In financial applications, fixed payments constitute a drift, for example.

The random walk with drift δ has $E(x_t) = t\delta + x_0$. This is an important difference; the mean of a random walk is independent of the time (it is constant), but the mean of the random walk with drift is dependent on the time.

Geometric Random Walk

A *geometric random walk* is a process whose logarithm is a random walk:

$$x_t = x_{t-1}e^{w_t}, \tag{5.15}$$

or, starting at x_0,

$$x_t = x_0 \exp(w_t + \cdots + w_1). \tag{5.16}$$

Under a form of the Random Walk Hypothesis, cumulative log returns are a random walk, and so asset prices follow a geometric random walk.

The additive nature of log returns allows us to relate a principal amount at time t, P_t, in terms of the initial amount P_0, as

$$P_t = P_0 \exp(r_t + \cdots + r_1), \tag{5.17}$$

where $r_1, \ldots, r_t$ are log returns. (This results from equation (1.20) on page 29 and equation (5.12).) If the log returns r_i are a white noise process, that is, if their sums form a random walk, the process $\{P_t\}$ forms a geometric random walk. Simple returns do not have this property.

If the log returns are from a Gaussian white noise process, that is, if the r_i in equation (5.17) are random variables distributed iid as $N(0, \sigma^2)$, then P_t has a lognormal distribution. The process is called a *lognormal geometric random walk* with parameter σ. (Note that P_t has a lognormal distribution with parameters 0 and $t\sigma$; going through the details to show this is part of Exercise 5.11. See also page 288 for general discussions of the lognormal family of distributions.)

General Autoregressive Processes

An extension of the random walk with drift model (5.13) is the autoregressive model,

$$x_t = \phi_0 + \phi_1 x_{t-1} + \cdots + \phi_p x_{t-p} + w_t, \tag{5.18}$$

which we mentioned briefly on page 338, and will discuss more fully in Section 5.3.2. In the random walk, $p = 1$, and in the simple random walk $\phi_0 = 0$.

Multivariate Processes

All of the models we have described above are univariate; that is, the individual observations and random variables are scalars. Each model has a multivariate generalization, however.

The multivariate generalization of any of these models retains the model's essential characteristics from one observation to another; that is, from one point in time to another point in time. The elements of the vector variables in the multivariate models at a given point in time, however, may have any kind of relationships to each other.

For example, the elements of a multivariate white noise process may have nonzero correlations among themselves, but the correlations of a given element over time are zero. If

$$\ldots, w_{-1}, w_0, w_1, \ldots$$

are random d-vectors with constant mean 0 and constant $d \times d$ variance-covariance matrix Σ_w, the process $\{w_t\}$ is a *multivariate white noise* process if the covariance matrix of w_s and w_t, $\mathrm{E}(w_s w_t^{\mathrm{T}})$, for $s \neq t$ is zero.

Likewise, we can define multivariate moving averages of white noise, multivariate linear processes, multivariate random walks, and so on. A multivariate random walk, for example, has the same form as the model (5.8),

$$x_t = x_{t-1} + w_t,$$

except that x_t and x_{t-1} are vectors, and $\{w_t\}$ is a multivariate white noise.

Linear Combinations of Univariate Processes

Linear models involving time series processes are of interest. The properties of the processes, however, may affect the distributions of computed statistics, and, hence, may lead to incorrect inference. The case of autocorrelated errors, referred to on page 434 in Chapter 4, is an example. Another example is the spurious regression between two unrelated processes, which we will discuss beginning on page 584.

Linear combinations of uncorrelated univariate processes may retain some of the salient properties of the individual processes; for examples, a linear combination of uncorrelated white noise processes is a white noise, and a linear combination of uncorrelated random walk processes is a random walk. (When we say the processes $\{x_t\}$ and $\{y_t\}$ are "uncorrelated", we mean $\mathrm{Cor}(x_s, y_t) = 0$ for all s and t.)

To see that a linear combination of uncorrelated white noise processes is a white noise, assume $\{u_t\}$ and $\{v_t\}$ are univariate white noise processes, and $\mathrm{Cor}(u_s, v_t) = 0$. For

$$w_t = \left(\begin{array}{c} u_t \\ v_t \end{array} \right),$$

$\{w_t\}$ is a bivariate white noise with

$$\Sigma_w = \begin{pmatrix} \sigma_u^2 & 0 \\ 0 & \sigma_v^2 \end{pmatrix}.$$

Now for a constant a, let $x_t = au_t + v_t$. We want to show that $\{x_t\}$ is a white noise process. First, since $\mathrm{E}(u_t) = 0$ and $\mathrm{E}(v_t) = 0$, $\mathrm{E}(x_t) = 0$. Now, since $\mathrm{V}(u_t)$ is the constant σ_u^2 and $\mathrm{V}(v_t)$ is the constant σ_v^2, $\mathrm{V}(x_t)$ is the constant $a^2\sigma_u^2 + \sigma_v^2$. Finally, since for any s and t with $s \neq t$, $\mathrm{Cov}(u_s, u_t) = 0$ and $\mathrm{Cov}(v_s, v_t) = 0$,

$$\mathrm{Cov}(x_s, x_t) = \mathrm{Cov}(au_s + v_s, au_t + v_t) = \\ \mathrm{Cov}(au_s, au_t) + \mathrm{Cov}(au_s, v_t) + \mathrm{Cov}(v_s, au_t) + \mathrm{Cov}(v_s, v_t) = 0.$$

Hence, $\{x_t\}$ is a white noise process, by the definition on page 488. (Here, we have used the fact that $\mathrm{Cov}(z_s, z_t) = 0 \Leftrightarrow \mathrm{Cor}(z_s, z_t) = 0$.)

You are asked to show that a linear combination of two uncorrelated random walks is a random walk in Exercise 5.1.

State Space Models

A time series model of the form (5.2) can be generalized. Suppose that we do not observe x_t directly, but rather observe y_t that depends on the *state* x_t:

$$y_t = g(x_t; \delta) + u_t,$$

where $\{u_t\}$ is a white noise process. The states may be observable or they may be latent variables or parameters. They may be random variables, especially in a Bayesian sense. (The term "space" has to do with early applications, and is not indicative of some aspect of the methodology, so "state-space" should not be interpreted literally.)

We model the states, similar to (5.2), as

$$x_t = f(x_{t-1}\,;\,\theta) + w_t,$$

where $\{w_t\}$ is a white noise process uncorrelated with $\{u_t\}$. The analysis of this model, however, depends on inferences from the corresponding y_t and y_{t-1}.

This formulation of separate entities allows for a broader focus on the time series. Emphasis on the two stages yields a "dynamic" approach, and computational methods such as the Kalman filter allow updating of states using observed data.

The R package KFAS (Helske, 2017) provides several computationally efficient functions for Kalman filtering, smoothing, forecasting, and simulation of state space models with a variety of probability distributions.

The text by Durbin and Koopman (2012) discusses the state space approach to time series analysis, and the interested reader is referred to that text. We will not pursue this topic further here.

5.1 Basic Linear Operations

Linear time series models, just as linear regression models, are useful in data analysis. Even if the underlying process is not linear, the linear model often serves as a good approximation.

Linear Functions

A function f is said to be a *linear function* or a *linear operator* on d-vectors if, for any real number a and real d-vectors u and v,

$$f(au + v) = af(u) + f(v). \tag{5.19}$$

For example, consider a function g of 2-vectors. Let $g(x) = 2x_1 + x_2$ for the 2-vector $x = (x_1, x_2)$. Then g is a linear function because, for any real number a and 2-vectors x and y,

$$g(ax + y) = 2ax_1 + ax_2 + 2y_1 + y_2 = a(2x_1 + x_2) + 2y_1 + y_2 = ag(x) + g(y).$$

The function $h(x) = x_1^2 + x_2$, for example, is not linear because it does not have this property.

A linear function of a linear function is linear. To see this, suppose f and g are linear functions of d-vectors, and a is a real number and u and v are real d-vectors. Then

$$\begin{aligned} f(g(au + v)) &= f(ag(u) + g(v)) \\ &= af(g(u)) + f(g(v)), \end{aligned}$$

which, of course, fits the definition of $f(g)$ being linear. This is called *composition* of the functions.

Linear functions have a number of useful properties, and there are some basic linear functions, which we will call linear operators, that have useful applications in discrete time series, as we discuss below.

A function of a time series that yields another time series over a subset of the same time domain is a *filter*. For example, the moving average of white noise is a filter of a white noise process. We will now consider some important linear filters.

5.1.1 The Backshift Operator

One of the most useful linear filters is the *backshift operator*, $\mathrm{B}(\cdot)$, which for the time series

$$\ldots, x_{i-1}, x_i, x_{i+1}, \ldots$$

is defined by

$$\mathrm{B}(x_t) \equiv x_{t-1}. \tag{5.20}$$

This is also called a *lag operator*, and it is sometimes denoted as $\mathrm{L}(\cdot)$.

It is easy to see that this is a linear operator:

$$\mathrm{B}(ax_s + y_t) = ax_{s-1} + y_{t-1} = a\mathrm{B}(x_s) + \mathrm{B}(y_t).$$

We also observe a nice property of the composition of the backshift operator with itself:

$$\mathrm{B}(\mathrm{B}(x_t)) = x_{t-2},$$

which we can generalize for the positive integer k, and introduce the notation B^k:

$$\mathrm{B}^k(x_t) = x_{t-k}. \tag{5.21}$$

The inverse of the backshift filter is the operator $\mathrm{B}^{-1}(\cdot)$, where

$$\mathrm{B}^{-1}(x_t) = x_{t+1}.$$

With this notation, we see that the expression in equation (5.21) applies to all integers as well, with $\mathrm{B}^0(\cdot)$ being the identity operator. The inverse of the backshift operator is also called the forward shift operator.

The simple properties of the backshift operator allow us to treat it like an ordinary numerical variable; for example, we can write

$$
\begin{aligned}
(1 - \mathrm{B})^2(x_t) &= (1 - 2\mathrm{B} + \mathrm{B}^2)(x_t) \\
&= x_t - 2x_{t-1} + x_{t-2}. \tag{5.22}
\end{aligned}
$$

The backshift operation $\mathrm{B}(x_t)$ is often written without the parentheses: $\mathrm{B}x_t$; I use either notation, generally making the choice to enhance clarity.

We sometimes use the notation $\mathrm{B}(\{x_t\})$ to represent the time series formed by shifting one time step ahead in the time series $\{x_t\}$. Applied to a finite time series, $x = (x_1, x_2, \ldots, x_T)$, the backshift operator just results in a shorter series:

$$\mathrm{B}(x) = (x_2, \ldots, x_T).$$

A backshift on a time series in R can be performed simply by manipulating the index of a vector (or matrix).

```
> x <- c(1, 2, 3, 3, 2, 1)
> n <- length(x)
> x
[1] 1 2 3 3 2 1
> c(NA,x[-n])                        #  backshift once
[1] 2 3 3 2 1
> c(NA,NA,x[-c(n-1,n)])              #  backshift twice
[1] NA NA  1  2  3  3
> k <- 2
> c(rep(NA,k),x[-c((n+1-k):n)])   #  backshift k times
```

```
[1] NA NA  1  2  3  3
> c(x[-c(1,2)],NA,NA)              # forward shift twice
[1]  3  3  2  1 NA NA
```

The `lag` function in the `dplyr` package of the `tidyverse` suite does backshifts as above, and the associated `lead` does forward shifts. The resulting vectors are the same length as the operand vector, so they contain some NA entries as above. (The `lag` function in the R `stats` package does not do a backshift.)

The Multivariate Backshift Operator

For operations on vectors, as in multivariate processes, the backshift operator has a simple generalization. If the elements of the process $\ldots, x_{-1}, x_0, x_1, \ldots$ are d vectors, then we define the d-variate backshift operator $\mathrm{B_d}$ as

$$\mathrm{B_d} \begin{pmatrix} x_{1,t} \\ \vdots \\ x_{d,t} \end{pmatrix} = \begin{pmatrix} x_{1,t-1} \\ \vdots \\ x_{d,t-1} \end{pmatrix}.$$

5.1.2 The Difference Operator

Another useful linear filter for working with a time series is the *difference operator*, $\triangle(\cdot)$, defined by

$$\triangle(x_t) \equiv x_t - x_{t-1}. \tag{5.23}$$

The difference operator is related to the backshift operator,

$$\triangle(x_t) = x_t - \mathrm{B}(x_t) = (1 - \mathrm{B})(x_t), \tag{5.24}$$

and hence, $(1-\mathrm{B})$ is also called the *difference operator*. The difference operator is also called the *backward difference operator*, for obvious reasons.

Some authors denote the difference operator as ∇ instead of $\triangle$. The difference operation $\triangle(x_t)$ is often written without the parentheses: $\triangle x_t$; I use either notation, generally making the choice to enhance clarity.

The difference operator is a linear operator:

$$\begin{aligned} \triangle(ax_s + y_t) &= ax_s + y_t - (ax_{s-1} + y_{t-1}) \\ &= a(x_s - x_{s-1}) + (y_t - y_{t-1}) \\ &= a\triangle(x_s) + \triangle(y_t). \end{aligned}$$

We sometimes use the notation $\triangle(\{x_t\})$ to represent the time series formed by differencing on the time series $\{x_t\}$. Applied to a finite time series, $x = (x_1, x_2, \ldots, x_T)$, the difference operator results in a shorter series:

$$\triangle(x) = (x_2 - x_1, \ldots, x_T - x_{T-1}).$$

The composition of the difference operator with itself has a simple relationship to the backshift operator:

$$\triangle^2(x_t) = x_t - 2x_{t-1} + x_{t-2} = (1 - B)^2(x_t).$$

We can easily see the extension to *differences of order k*:

$$\triangle^k(x_t) = (1 - B)^k(x_t), \tag{5.25}$$

where k is a positive integer. This also leads to another equation,

$$\triangle^k(x_t) = \sum_{h=0}^{k} (-1)^h \binom{k}{h} x_{t-h}, \tag{5.26}$$

where $\binom{k}{h}$ represents the binomial coefficient, $k!/(h!(k-h)!)$.

Differences of order k lead to expressions involving the backshift operator similar to real numbers in a binomial expansion, and equation (5.26) is equivalent to

$$(1 - B)^k = \sum_{h=0}^{k} \binom{k}{h} (-1)^h B^h, \tag{5.27}$$

as in equation (5.22) for $k = 2$.

Fractional Differencing

The binomial formula (5.27) has another form:

$$(1 - B)^k = \sum_{h=0}^{\infty} \prod_{0 < j \le h} \frac{j - 1 - k}{j} B^h, \tag{5.28}$$

in which $h = 0, 1, 2, \ldots$ and $j = 0, 1, \ldots, h$. Notice that for a positive integer k, the infinite summation is equivalent to the finite sum in (5.27) because the terms $\prod_{0 < j \le h} \frac{j-1-k}{j}$ are 0 for all $h > k$. For $k = 2$, for example, the formula (5.28) yields $(1 - B)^2 = 1 - 2B + B^2$, as expected.

If k is not an integer, we call this *fractional differencing*. In that case, the summation is an infinite series in B^h. The expansion in equation (5.28) has meaning for any value of k subject to the convergence of the series. Applying the B^h to x_t, we get a sequence $x_{t-1}, x_{t-2}, \ldots$.

Useful financial time series models can be built on fractional differencing $(1 - B)^k$ with $-0.5 < k < 0.5$. We will mention this application again later, but we will not pursue the development of those models in this book.

Lags

We call the difference operator defined in equation (5.23) a difference operator of *lag 1*. We can generalize the difference operator to a difference of *lag k*, for any integer k, denoted by $\triangle_k(\cdot)$, and defined by

$$\triangle_k(x_t) \equiv x_t - x_{t-k}. \tag{5.29}$$

Note that $\triangle_{k_1}(\triangle_{k_2}(x_t)) = \triangle_{k_1+k_2}(x_t)$, and $\triangle_0(x_t) = 0$. We also observe that $\triangle_k(x_t) = (1 - B^k)(x_t)$. (You are asked to go through the simple steps to prove this latter equivalence in Exercise 5.3c.)

In R, the `diff` function performs differencing on a time series. On a numeric vector or an object of class `ts`, the `diff` function yields an object of shorter length than the operand; that is, there are no NAs, as in the lag filter illustrated above. By default, the `diff` function performs differencing at lag 1. Differencing at a lag greater than 1 can be specified by a second argument.

```
> x <- c(1, 2, 3, 3, 2, 1)
> x
[1]  1  2  3  3  2  1
> diff(x)                       # difference of order 1
[1]  1  1  0 -1 -1
> x[-1] - x[-length(x)]  # difference of order 1
[1]  1  1  0 -1 -1
> diff(diff(x))                 # difference of order 2
[1]  0 -1 -1  0
> diff(diff(diff(x)))           # difference of order 3
[1] -1  0  1
> diff(x,2)                     # difference at lag 2
[1]  2  1 -1 -2
```

The `diff` function performs differently on an object of class `xts`. It yields NAs in that case, similar to the NAs in the lag filter illustrated on page 496. See page 174 and the example shown in Figure A1.51 for the results of `diff` applied to `xts` objects.

Returns

The backshift and difference operators are central to the computation of returns. A one-period simple return (equation (1.17)) is

$$R = \triangle(x_t)/\text{B}(x_t),$$

and the log return (equation (1.22)) is

$$r = \triangle(\log(x_t)).$$

The sequence of returns is a filter of the time series of prices.

In R, these returns are computed using the `diff` function:

```
xsimpret <- diff(x)/x[-length(x)]
xlogret <- diff(log(x))
```

(Again, recall the difference in operations on xts objects.)

For any filter on a finite vector that results in a shorter vector, such as computation of returns, differencing, or backshifting, the computer software designer must decide whether to return a shorter vector (as diff in the base package) or to return a vector of the same length with some meaningless values, usually NAs (as diff in the quantmod package or lag in the dplyr package). And, of course, the software user must be aware of the decision.

5.1.3   The Integration Operator

Equation (5.24) leads to an inverse of the backward difference operator:

$$
\begin{aligned}
\triangle^{-1}(x_t) &= (1 - \mathrm{B})^{-1}(x_t) \\
&= (1 + \mathrm{B} + \mathrm{B}^2 + \cdots)(x_t) \\
&= x_t + x_{t-1} + x_{t-2} + \cdots \\
&= \sum_{h=0}^{\infty} x_{t-h} \\
&= \mathrm{S}(x_t),
\end{aligned}
\tag{5.30}
$$

where, if the series is convergent, the inverse $\mathrm{S} = \triangle^{-1}$ is called the *summation operator*, the *infinite summation operator*, or the *integration operator*. If it exists, the integration operator is a linear filter.

The mathematics in the steps of equation (5.30) is not precise. We are treating the operator B as a real number. This leads to the infinite expansion of $(1 - \mathrm{B})^{-1}$, but that expansion does not hold for all real numbers. Secondly, after the infinite series of operators has been applied to x_t, yielding $x_t + x_{t-1} + x_{t-2} + \cdots$, we must address the convergence of this series. Its convergence clearly depends on the values of the x_i. If they are all larger than any positive constant, for example, the series does not converge, and so $\triangle^{-1}$ does not exist; that is, in that case, equation (5.30) is meaningless.

5.1.4   Summation of an Infinite Geometric Series

Both the integration operator and the expansion of $(1 - \mathrm{B})^k$ when k is not a nonnegative integer involve infinite sums in B^k. They are both *geometric series* in the operator B. Since they are infinite sums of operators, the question arises as to whether or not the series converges. Because the operators applied to real numbers yield real numbers, the question of convergence becomes the question of convergence of a geometric series of real numbers; that is, whether or not the sum is finite.

Geometric series in real numbers, which are also called power series, arise often in time series analysis, and so here we will recall a well-known fact about an infinite power series.

For real numbers a and b, the power series $\sum_{j=0}^{\infty} ab^j$ is convergent if and only if $|b| < 1$, and in that case,

$$\sum_{j=0}^{\infty} ab^j = \frac{a}{1-b}. \tag{5.31}$$

(We can prove the "if" part of this statement very easily by first writing

$$\sum_{j=0}^{k} ab^j (1-b) = a \left(1 + \sum_{j=1}^{k} b^j\right)(1-b) = a(1 - b^{k+1})$$

and then observing $\lim_{k \to \infty} b^{k+1} = 0$ if $|b| < 1$. The "only if" part is likewise easily proved following an expansion.)

5.1.5 Linear Difference Equations

Many useful models for time series analysis have the form of *linear difference equations*. A linear difference equation is formed from a function $f(t)$ by considering a discrete change in t, $\triangle t$, and the corresponding change in f

$$\triangle f = f(t + \triangle t) - f(t).$$

(This is the same operator as considered in Section 5.1.2.) If t is a discrete variable, that is, if it can be mapped one-to-one to the integers, we may write the function values as f_t and $f_{t+\triangle t}$.

The relative difference of order j is written as $\frac{\triangle^j f}{\triangle t^j}$ analogously to the derivative of order j of f with respect to t, $\frac{d^j f}{dt^j}$, in a *differential equation*. A difference equation in f and t, in general, involves $f, t, \frac{\triangle f}{\triangle t}, \ldots, \frac{\triangle^k f}{\triangle t^k}$.

There is a rich mathematical theory relating to difference equations, but we will only use some simple results from this theory.

Simpler forms of linear difference equations play a major role in discrete time series analysis. In these applications, f is a function of t, but t only takes integral values, so any $\triangle t$ is an integer and any $\triangle f$ is just the difference between f at one value of t and some other value of t. Hence, a pth order difference equation in this context has the general form

$$f_t - a_1 f_{t-1} - \cdots - a_p f_{t-p} - a_0 = 0, \tag{5.32}$$

or

$$(1 - a_1 B - \cdots - a_p B^p)(f_t) - a_0 = 0. \tag{5.33}$$

If $a_0 = 0$, the difference equation is said to be *homogeneous*. A linear time series model may have this form with added random noise and it may or may not be homogeneous.

If some elements in the difference equation are random variables, the equation is a *stochastic difference equation*. This is analogous to the continuous time models of financial processes that we have mentioned before, such as equation (3.80), which are stochastic differential equations.

The basic linear models occurring in time series analysis are of the form of stochastic difference equations. In the analysis of such models, we can sometimes express the ACF as a recursive deterministic difference equation (see page 543).

Solutions of Difference Equations

A *solution* of a difference equation is a function of t that does not involve differences (just as a solution to a differential equation is a function that does not involve differentials). A difference equation may have a *general solution* that involves undetermined constants. *Initial conditions* or *boundary conditions* may remove the indeterminacy in the constants. Solving a difference equation or a differential equation usually begins by finding a general solution.

Linear difference equations and linear differential equations may have general solutions of a simple form. Consider the difference equation (5.32), with $a_0 = 0$. Now, suppose a solution is

$$f_t = r^t.$$

Now consider a formal substitution into equation (5.32), yielding the polynomial equation,

$$r^t - a_1 r^{t-1} - \cdots - a_p r^{t-p} = 0. \tag{5.34}$$

Any root r_0 of the polynomial in r provides a solution to the difference equation, because with the formal substitution, $f_t = r_0^t$, it makes the equality in (5.34), and consequently in equations (5.32) and (5.33), true.

As an example, let $a_1 = -5$ and $a_2 = 6$, with $p = 2$. We have the difference equation

$$f_t - 5f_{t-1} + 6f_{t-2} = 0,$$

and the associated polynomial equation

$$r^2 - 5r + 6 = 0.$$

Since $(r-3)(r-2) = 0$, we have two roots $r_{01} = 3$ and $r_{02} = 2$, and so both 3^t and 2^t, are solutions to the difference equation. Consider the solution 3^t, for example:

$$3^t - 5(3)^{t-1} + +6(3)^{t-2}.$$

This expression is equal to 0 for any t.

The Characteristic Polynomial

A general approach to solving a *homogeneous difference equation*, just as in solving a homogeneous differential equation, is to identify and solve a *characteristic equation*, also called the *auxiliary equation*. For the simple difference equation (5.32) or (5.33), the characteristic equation is formed by a *characteristic polynomial* that has the same form as the "polynomial" in B.

Instead of $f_t = r^t$ leading to the polynomial (5.34), it is more common to consider $f_t = z^{-t}$, which for $a_0 = 0$, leads to

$$1 - a_1 z - \cdots - a_p z^p, \tag{5.35}$$

in the arbitrary variable z. The characteristic polynomial is a pth degree polynomial for a pth order difference equation.

Roots of the Characteristic Polynomial

A pth degree polynomial with real coefficients has p roots, which may contain imaginary components, and which may not all be distinct. A solution to the difference equation is a function of these roots. The properties of these roots (their magnitudes, real or non-real, distinct or not) determine important properties of the solution, which, in turn, determine properties of the time series process. Non-real roots occur only as complex conjugate pairs.

There are many interesting and useful results the mathematical theory relating to difference equations and the corresponding characteristic polynomials. We will consider the relationship of the roots of the characteristic polynomial to solutions of the difference equation in two simple cases. This relationship gives meaning to the phrase "unit root", which we will encounter in later sections.

Consider, for example, the first-order difference equation

$$(1 - a\mathrm{B})(f_t) = 0, \quad \text{for } a \neq 0 \text{ and } t = 1, 2, \ldots. \tag{5.36}$$

Note that if $a = 1$, this corresponds to a random walk without the additive noise.

Observing that $f_1 = af_0$, $f_2 = a^2 f_0$, $f_3 = a^3 f_0$, and so on, we see that a solution to that difference equation, for any t, is

$$f_t = a^t f_0. \tag{5.37}$$

For the simple difference equation (5.36), the characteristic polynomial is

$$1 - az. \tag{5.38}$$

This polynomial has one root, $z_1 = a^{-1}$. It is not just a coincidence that a solution to the difference equation, equation (5.37), is

$$f_t = z_1^{-t} f_0. \tag{5.39}$$

This result follows from the theory of difference equations, but we will not elaborate further on that theory here.

The solution in equation (5.39) is a *general solution* for equation (5.37) in the sense that f_0 could be replaced by f_s and f_t by f_{s+t}. If a specific value is known, it is called an *initial condition*, and when it is substituted at the proper place, the solution is the *particular solution* corresponding to the given initial condition. For instance, if $f_0 = c$, the particular solution is $f_t = z_1^{-t}c$.

Our interest is in the behavior of the solution as t increases; that is, as time moves forward.

Notice that if $|z_1| > 1$, the solution converges to 0; if $z_1 = 1$, the solution is constant; and if $|z_1| < 1$, the solution grows in magnitude without bound.

Example: Second Order Difference Equation

Consider, for example, a second order difference equation,

$$(1 - a_1 B - a_2 B^2)(f_t) = 0, \quad \text{for } a_2 \neq 0 \text{ and } t = 2, 3, \dots . \tag{5.40}$$

The characteristic polynomial is

$$1 - a_1 z - a_2 z^2. \tag{5.41}$$

This polynomial has two roots, z_1 and z_2, which we can determine by the quadratic formula,

$$\frac{1}{2a_2} \left(-a_1 \pm \sqrt{a_1^2 + 4a_2} \right). \tag{5.42}$$

These roots may have imaginary components, so a criterion that depends on the absolute value as above must be replaced by a criterion that depends on the *modulus*, which we also denote by $|z|$. If $z = x + iy$, where i is the imaginary element, then $|z| = \sqrt{x^2 + y^2}$. If z is real, $|z|$ is just the ordinary absolute value. Again, the criterion involves the relationship of $|z_1|$ and $|z_2|$ to one; that is whether they lie outside, on, or inside the *unit circle*, which is the locus of points in the complex plane, such that

$$|z| = \sqrt{x^2 + y^2} = 1. \tag{5.43}$$

If the discriminant of the quadratic formula, $a_1^2 + 4a_2$, is zero, then the two roots are equal, and in that case the solution is

$$f_t = (c_1 + c_2 t) z_1^{-t}, \tag{5.44}$$

where c_1 and c_2 are constants that depend on the initial conditions; that is, they depend on known values for two values of f_t, say f_0 and f_1. As before, if $|z_1| > 1$, the solution converges to 0; if $z_1 = 1$, the solution is constant; and if $|z_1| < 1$, the solution grows in magnitude without bound. The *unit root*, $z_1 = 1$, is the critical point between the two types of solution.

If the discriminant is not zero, there are two different roots z_1 and z_2, and in that case the solution is

$$f_t = c_1 z_1^{-t} + c_2 z_2^{-t}, \tag{5.45}$$

where, again, c_1 and c_2 are constants that depend on the initial conditions. Notice that if both $|z_1| > 1$ and $|z_2| > 1$, the solution converges to 0, but if either $|z_1| < 1$ or $|z_2| < 1$, the solution grows in magnitude without bound. Again, a *unit root* is the critical point between the two types of solution.

If the discriminant is negative, then z_1 and z_2 are complex conjugates of each other, $z_2 = \bar{z}_1$. (The overbar represents the complex conjugate; that is, if $z = x + iy$, then $\bar{z} = x - iy$. Note, therefore, $|\bar{z}| = |z|$.) The sum in equation (5.45), that is, f_t, is real, so it is also the case that $c_1 z_1^{-t}$ and $c_2 z_2^{-t}$ are complex conjugates of each other.

In the case of complex roots, again, whether or not the solution converges depends on $|z_1|$ and $|z_2|$ (where $|z_2| = |z_1|$). If $|z_1| > 1$, the solution converges to 0. Expressing z_1 as $|z_1| e^{i\psi}$ and, hence, z_2 as $|z_1| e^{-i\psi}$ for some real number ψ, and using the fact that $e^{i\theta} + e^{-i\theta} = 2\cos(\theta)$ for any θ, we have

$$f_t = a|z_1|^{-t} \cos(t\psi + b), \tag{5.46}$$

for some real numbers a and b that depend on the initial conditions. (Two degrees of freedom are necessary to span the space covered by c_1 and c_2 in equation (5.45).) This expression implies that if $|z_1| > 1$, the solution converges sinusoidally to 0.

We will mention the characteristic polynomial, the unit circle, and unit roots from time to time in connection with properties of time series. The roots, whether distinct or not, and their magnitudes, determine such things as how a process evolves over time and whether or not it is stationary. A *unit root* occurs when one or more of the roots of the characteristic polynomial is equal to 1 or -1. As a preview of properties that we will discuss later, consider the nature of the solutions (5.39), (5.44), and (5.45) if the roots z, z_1, or z_2 are ± 1. We have either the constant 1 or an alternation between 1 and -1.

5.1.6 Trends and Detrending

A time series model with a constant additive term,

$$x_t \approx \alpha + f(x_{t-1}, x_{t-2}, \ldots), \tag{5.47}$$

has a simple linear trend.

The constant additive term in equation (5.47) means that observations following that process will have a general linear trend, either positive or negative depending on the sign of α. The simple random walk model with drift is an example. Consider a random walk with drift, $x_t = \delta + x_{t-1} + w_t$. Figure 5.2 shows a random sample from this model with $x_0 = 10$ and $\delta = 2$. (I used rnorm in R with a seed of 12346.)

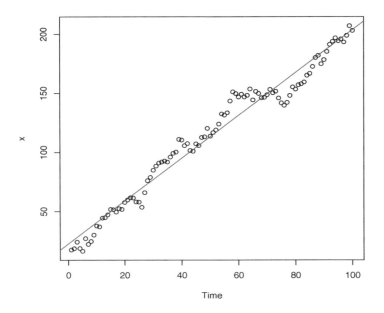

FIGURE 5.2
A Random Walk with Drift

The model for a random walk with drift is a simple linear regression model with time as the regressor, and $E(X_t) = x_0 + \delta t$. Figure 5.2 shows a least squares regression line through the random walk data. It has an intercept of 22.77, and a slope of 1.81. The points on the line correspond to $\hat{x}_t = \hat{x}_0 + \widehat{\delta} t$. (The individual steps with the drift are exactly the same as the numbers shown in Figure 5.1.) In Exercise 5.5 you are asked to generate a similar sequence, but with the white noise following a Student's t distribution.

The residuals from the regression model should be free of the trend. Hence, the process $r_t = x_t - \hat{x}_t$ does not exhibit the trend. The regression residuals do, however, retain a random walk characteristic. (They are not the same as the w_t in Figure 5.1; recall the discussion of residuals and errors on page 415.)

By forming the residuals, $r_t = x_t - \hat{x}_t$, we have "detrended" the data, but because of the patterns in the data, the residuals in the least squares fit inherit a random walk pattern without drift.

Another way of removing the trend, as we have indicated, is by use of the difference operator. This, however, yields a white noise process; hence, all information about the path of the realized walk is lost.

Figure 5.3 shows both detrended processes.

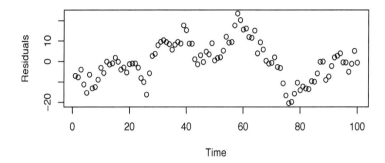

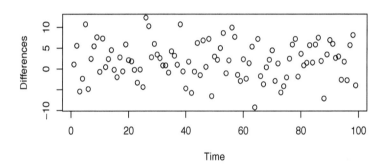

FIGURE 5.3
Detrended Processes

The Kwiatkowski-Phillips-Schmidt-Shin (KPSS) test can be used as a test for linear trend. (Both the null hypothesis and the alternative hypothesis are rather complicated; both are highly composite. The test is also used in time series as a test for a unit root, and we will encounter it again for that condition in Section 5.5.2.)

The KPSS test is implemented in the R function kpss.test in the tseries package. The function is designed around a paradigm of "reject at a significance level below 0.01, do not reject at a significance larger than 0.10, and report a p-value if it is between 0.01 and 0.10".

We illustrate an application of the test here on the random walk data ("Walk") and on both detrended series ("Residuals" and "Differences"). It rejects level stationarity in the random walk at a significance level below 0.01. It computes a significance of 0.069 in the residuals from a linear fit, and it does not reject stationarity of the differenced series at the 0.10 level. (The output below has been edited for compactness.)

```
> library(tseries)
> kpss.test(Walk)
        KPSS Test for Level Stationarity
data:  Walk
KPSS Level = 3.3156, p-value = 0.01
In kpss.test(Walk) : p-value smaller than printed p-value
> kpss.test(Residuals)
        KPSS Test for Level Stationarity
data:  Residuals
KPSS Level = 0.41942, p-value = 0.06878
> kpss.test(Differences)
        KPSS Test for Level Stationarity
data:  Differences
KPSS Level = 0.06181, p-value = 0.1
In kpss.test(Differences) : p-value greater than printed p-value
```

Although many time series of financial data follow a general trend, some for days, and others for years, it is not likely that any time series of financial data would follow such a regular pattern as we see in Figure 5.2 for very long. First of all, changes in values, such as prices of stocks, are generally proportional to their magnitude, that is, the changes are appropriately measured as *returns*, either simple or log, rather than as raw differences. A random walk with drift model would require a drift that varies proportionally to x_t. More importantly, however, is the nature of the random fluctuations; their distributions have heavy tails and furthermore, the distributions change over time. (Recall the plots of financial data in Chapter 1.)

5.1.7   Cycles and Seasonal Adjustment

"Cycle" is a generic term that indicates repetitive patterns in a process. There are various "cycles", such as business cycles, cycles of sunspots, cycles of the El Niño-Southern Oscillation, and so on. Cycles may or may not have *fixed periods*.

"Business cycles" exist, in the sense that almost any measure of economic activity goes up and down over time, but the lengths of the time intervals within the cycles vary.

Some economic and financial cycles do have fixed periods. We call the variations in cycles with fixed periods *seasonal effects*, where the "season" may be any fixed unit of time, such as quarter, month, week, and so on. The numbers of airline passengers are higher in the summer months, and candy sales are greater in October, for example, so these cycles have periods of one year.

The earnings of some companies also vary seasonally. To illustrate seasonal effects, we will use an old set of data that is an R built-in dataset named JohnsonJohnson, which contains the quarterly earnings of Johnson & Johnson Corporation (JNJ) for the period 1960 through 1980. We can clearly see seasonal effects in the earnings shown in the left-hand plot in Figure 5.4.

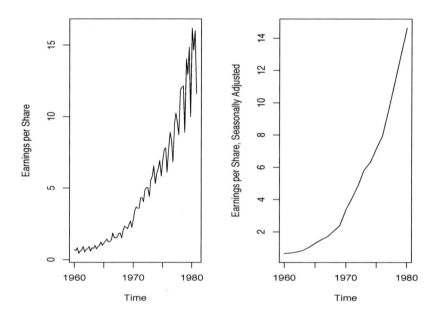

FIGURE 5.4
Quarterly Earnings of JNJ

The plots in Figure 5.4 were produced using R. The object containing the data, JohnsonJohnson is a ts object with a frequency of 4. (This can be determined by frequency(JohnsonJohnson).) The variation is from quarter to quarter within a period of one year.

Sometimes we wish to analyze the seasonal effects; other times, the interest is in longer trends, and the seasonal effects are a nuisance. There are various ways of adjusting for seasonal effects. A simple way is just to aggregate the observations of all seasons within the next higher time unit to that time unit; days into weeks or weeks into months, for example.

In the case of JNJ earnings, there are four quarters per year, so we may aggregate quarters to years and take the mean of the four quarters. The aggregated data are shown in the right-hand plot in Figure 5.4, and of course no seasonal effects are seen.

The R function `aggregate` was used to remove the seasonal effects. The `aggregate` function by default will sum the values within the seasons, and of course, that is correct in this case to get the annual earnings. In this example, however, we are interested in the adjusted quarterly earnings; hence, we want the quarterly averages for the years. To get this, we specify `FUN=mean` in the call to the `aggregate` function.

```
aggregate(JohnsonJohnson, FUN=mean)
```

Aggregation of this type smoothes seasonal data by replacing all values within a sequence of periods by a single value obtained from the individual values in the periods. Thus, aggregated data contain fewer values than the original data covering the same time interval. It is not the same as a moving smoother, such as a moving average, because it yields only one value for the group of periods.

Seasonal adjustment should not be done just to smooth the data. Before attempting to adjust for seasonal effects, we should think about the nature of the data-generating process and our objectives in the analysis. In the case of JNJ, the seasonal variation in earnings is somewhat mysterious, but it has to do with bookings of expenses, and less to do with seasonal variations in sales.

Financial data relating to prices usually do not have seasonal effects that cannot be discounted out by using present values. If they did, traders would take advantage of them, and then they would no longer be present.

5.2   Analysis of Discrete Time Series Models

Models for discrete time series begin with a sequence of random variables:

$$\ldots, \; x_{-2}, \; x_{-1}, \; x_0, \; x_1, \; x_2, \; \ldots,$$

which we often denote as $\{X_t\}$. (Recall that in this book, I do not use the case of letters to distinguish random variables from their realizations.) The white noise, the moving average white noise, the random walk, and the other models we defined in the introduction to this chapter are all based on a random sequence such as this. The sequence may have a fixed starting point, such as in the random walk model, and in that case, we just ignore any values in the sequence before the starting point.

Means and Variances; Marginal and Conditional

As with any random variable, we are interested in properties such as the mean and variance, $\mu_t = \mathrm{E}(x_t)$ and $\sigma_t^2 = \mathrm{V}(x_t)$.

Our first concern is whether or not they are constant for all t. Lack of dependence on t relates to stationarity, which we have referred to often, and which we will discuss more formally in Section 5.2.1. In some of the simple models we have discussed so far, the mean, variance, and other measures are constant in t, but in some cases, such as a random walk or a random walk with drift, they may depend on t.

Another type of dependence is on previous values of the time series. We often refer to measures that may depend on other variables in the time series as *conditional*: that is, the *conditional mean*, the *conditional variance*, the *conditional covariance*, and so on. We refer to measures in which expectations have been taken over the full domain of all variables as *marginal* values. (See Section 3.1.5, page 258, for definitions and discussion of conditional and marginal probability measures.)

We worked out some marginal measures for the simple models in the introductory section; for example, in equations (5.10) and (5.11), we showed that the mean of a process $\{x_t\}$ that follows a random walk with a starting value of x_0 is x_0, and that the variance of x_t is $t\sigma_w^2$, where σ_w^2 is the variance of the underlying white noise.

Now consider the conditional mean and variance of a random walk, given all previous values $x_{t-1}, x_{t-2}, \dots$. From equation (5.8), we have

$$
\begin{aligned}
\mathrm{E}(x_t | x_{t-1}, x_{t-2}, \dots) &= \mathrm{E}(x_{t-1}) + \mathrm{E}(w_t) \\
&= x_{t-1} \qquad\qquad (5.48)
\end{aligned}
$$

and

$$
\begin{aligned}
\mathrm{V}(x_t | x_{t-1}, x_{t-2}, \dots) &= \mathrm{V}(x_{t-1}) + \mathrm{V}(w_t) \\
&= \sigma_w^2. \qquad\qquad (5.49)
\end{aligned}
$$

The property in equation (5.48) means that a random walk is a *martingale* (the essential property in the equation is essentially the definition of a martingale). The property is also related to the *Markovian property*, which is the property of a sequence of random variables in which the distribution of any one does not depend on any others in the sequence except possibly the previous one. (That property does not require the mean to be the same, however, as does the martingale property.) A *Markov chain*, in which the distribution of each successive random variable depends only on the immediately preceding one, is one of the most important general types of time series.

Autocovariances and Autocorrelations; Marginal and Conditional

The most important properties in a time series are often not just the means and variances, but how the terms in the sequence relate to each other. The basic measure of this relationship is the covariance of terms in the sequence.

Let us denote the covariance between the sth and the tth items in the sequence as $\gamma(s,t)$:

$$\gamma(s,t) = \text{Cov}(x_s, x_t) = \text{E}\left((x_s - \mu_s)(x_t - \mu_t)\right), \qquad (5.50)$$

and denote the correlation between the sth and the tth items in the sequence as $\rho(s,t)$:

$$\rho(s,t) = \text{Cor}(x_s, x_t) = \frac{\gamma(s,t)}{\sqrt{\gamma(s,s)\gamma(t,t)}}. \qquad (5.51)$$

Because these are relations among items within a single time series, we refer to them as *autocovariances* and *autocorrelations*.

Note, from the definitions,

$$\gamma(t,t) = \sigma_t^2,$$

and

$$\rho(t,t) = 1.$$

Means, variances, autocovariances, and so on, are possibly functions of time; hence, we often refer to the "mean function", or the "autocorrelation function", for example.

Note that if the autocovariance is 0, then the autocorrelation is also 0, and vice versa.

The autocorrelation of white noise is 0 if $s \neq t$. This just follows from the definition of white noise.

The autocorrelation $\rho(s,t)$ of a moving average of a white noise process depends on the difference in s and t, and on the number of terms in the average, k. The autocorrelation is 0 outside the window of the moving average; that is, for x_s and x_t in a k-moving average of a white noise, $\rho(s,t) = 0$ if $|s - t| \geq k$ (see Exercise 5.9a). Inside the window of the moving average it depends on $|s - t|$; for example, if $k = 3$, it is

$$\rho(s,t) = \begin{cases} 1 & \text{for } s = t, \\[2mm] \frac{2}{3} & \text{for } |s - t| = 1, \\[2mm] \frac{1}{3} & \text{for } |s - t| = 2, \\[2mm] 0 & \text{for } |s - t| \geq 3. \end{cases} \qquad (5.52)$$

(Exercise 5.9b.)

Note that the autocorrelation of a moving average of a white noise process does not depend on s and t individually, but only on $|s - t|$.

The autocovariance of a random walk is

$$\gamma(s,t) = \text{Cov}\left(X_0 + \sum_{i=1}^{s} w_i, \ X_0 + \sum_{i=1}^{t} w_i\right)$$

$$= \text{Cov}\left(\sum_{i=1}^{s} w_i, \ \sum_{i=1}^{t} w_i\right)$$

$$= \min(s,t)\sigma_w^2. \qquad (5.53)$$

Note that the variance of x_t, $\gamma(t,t)$, is $t\sigma_w^2$, as we saw above.

We can compute the autocorrelation of a random walk using equation (5.53). It is proportional to $\sqrt{s/t}$ or to $\sqrt{t/s}$. We see that it depends on the difference in s and t to some extent, but, more importantly, it depends on s and t individually.

Detrending or seasonally adjusting the data may of course change these properties.

These properties are marginal, or unconditional. Because in applications, at a time s, we may have access to $\{X_r : r \le s\}$ but not to X_t for $t > s$, often we are interested in the conditional distributions and conditional expectations given the observations up to that point in time. A model for the relationships among successive terms in a time series may yield conditional expectations, such as $V(X_s|\{X_r : r \le s\})$. Time series models are often formulated in terms of conditional expectations.

The specific nature of these properties distinguish one time series from another. One of the most important issues is how these quantities change over time. In order to make any statistical inferences, we must have a certain amount of constancy over time or else we must have a model that specifies how the properties change over time. Otherwise, without any constancy, we would have only one observation for each point in the model.

Multivariate Time Series; Cross-Covariances and Cross-Correlations

We are often interested in how multiple time series are related. As with most multivariate analyses, we concentrate on pairwise relationships. We consider two discrete time series observed at the same times,

$$\ldots, x_{-2}, x_{-1}, x_0, x_1, x_2, \ldots,$$
$$\ldots, y_{-2}, y_{-1}, y_0, y_1, y_2, \ldots,$$

The individual means, variances, and autocovariance are similar whether we have one time series or two, so long as the series themselves are not correlated. The important property to consider, therefore, is the covariance or correlation *between* the series. We will call this the "cross-covariance" or "cross-correlation". Of course, we still have the "auto" relationships to be concerned with.

We denote the *cross-covariance* of the time series $\{X_t\}$ and $\{Y_t\}$ at s and

t as $\gamma_{XY}(s,t)$, and define it as

$$\gamma_{XY}(s,t) = \mathrm{Cov}(x_s, y_t) = \mathrm{E}\left((x_s - \mu_{Xs})(y_t - \mu_{Yt})\right). \tag{5.54}$$

A single subscript, as in $\gamma_X(s,t)$, refers to the univariate time series. The subscripts may become somewhat unwieldy, but their meaning should be clear.

Note that while $\mathrm{Cov}(x_s, y_t) = \mathrm{Cov}(y_t, x_s)$,

$$\mathrm{Cov}(x_s, y_t) \neq \mathrm{Cov}(x_t, y_s), \tag{5.55}$$

in general. This gives rise to the concept of *leading* and *lagging* indicators.

We denote the *cross-correlation* of the time series as $\rho_{XY}(s,t)$, and define it in the obvious way as

$$\rho_{XY}(s,t) = \mathrm{Cor}(x_s, y_t) = \frac{\gamma_{XY}(s,t)}{\sqrt{\gamma_X(s,s)\gamma_Y(t,t)}}. \tag{5.56}$$

5.2.1 Stationarity

We have observed some important differences in a moving average of white noise and a random walk. In the case of a moving average, the mean and variance, μ_t and σ_t^2, are constant, and the autocorrelation, $\rho(s,t)$, does not depend on s and t, but only on $|s - t|$. On the other hand, the variance of a random walk grows without bound in t, and the autocorrelation of a random walk depends on both s and t.

Time series processes such as a moving average of white noise in which the mean and variance are constant for all values of t, and for which $\gamma(s,t)$ and $\rho(s,t)$ depend on s and t only through $|s - t|$, are easy to analyze. They serve as simple models for some financial processes, and they are useful starting points for more realistic financial time series models.

We call this condition of constant mean and variance and autocorrelations that depend only on the time difference *stationarity*.

(Actually, we sometimes call this *weak stationarity* as there is a stronger degree of constancy, which we call *strict stationarity*. In strict stationarity any ordered subset of variables in the time series has the same multivariate distribution as the ordered subset of corresponding variables, each a constant distance in time from the variable in the other ordered subset. We may refer to strict stationarity from time to time, but we will not put it to practical use. We will take the term "stationary" to mean "weakly stationary", although we may occasionally use the latter term for clarity or emphasis.)

Stationarity is one of the most important properties that time series may have because it allows us to make statistical inferences that we could not otherwise make.

The defining properties of a (weakly) stationary time series are

- the mean μ_t is finite and constant (that is, it does not depend on t),

- the variance at each point is finite, and

- the autocovariance $\gamma(s, t)$ is constant for any fixed difference $|t - s|$.

Note in the properties listed above defining stationarity, we did not state that the variance is constant. The constancy of the variance follows from the statement that the autocovariance is constant for any fixed difference $|t - s|$, including $|t - s| = 0$.

For a stationary time series we can omit the time subscripts on the symbols for the mean and variance, μ and σ^2.

From the properties that we have observed, we can say

- a white noise process is stationary

- a moving average of a white noise process is stationary

- a random walk is not stationary

Evolutionary Time Series

It is unlikely that even a weakly stationary time series model can be appropriate for a financial data-generating process over time intervals of any length. Economies develop, companies grow, and availability of resources change. The time series of almost any sequence of financial data are likely to be *evolutionary*.

A major effort in time series analysis often involves an attempt to transform an evolutionary time series into a stationary one. Often a simple evolutionary time series can be transformed into a stationary time series by differencing.

Autocovariances and Autocorrelations

Because the autocovariance function $\gamma(s, t)$ of a stationary time series depends only on the difference $|s - t|$, we can write $s = t + h$, and write the autocovariance function as $\gamma(h)$:

$$\gamma(h) = \text{Cov}(x_{t+h}, x_t) = \text{E}\left((x_{t+h} - \mu)(x_t - \mu)\right). \tag{5.57}$$

This yields

$$\gamma(0) = \sigma^2.$$

We define the autocorrelation function as before, and write it as $\rho(h)$:

$$\rho(h) = \text{Cor}(x_{t+h}, x_t) = \frac{\gamma(h)}{\gamma(0)}. \tag{5.58}$$

(Although this yields inconsistencies in the notation for γ and ρ, the meanings are usually clear from the context.)

We work with the autocorrelation function most often. We refer to it as the

ACF. Note the similarity of the ACF to the sample ACF, the SACF, defined
on page 100.

Note that
$$\gamma(h) = \gamma(-h) \quad \text{and} \quad \rho(h) = \rho(-h). \tag{5.59}$$

Partial Autocovariances and Autocorrelations

In addition to the autocovariance or autocorrelation at a lag of h, it may be of
interest to determine the covariance or correlation of x_{t+h} and x_t conditional
on $x_{t+h-1}, \ldots, x_{t+1}$; that is, conditional on (or "adjusted for") the values in
the time series between x_{t+h} and x_t. This adjusted measure is called the *partial
autocovariance* or *partial autocorrelation*. The partial autocorrelation function
is called the *PACF*. This is similar to partial regression and the partial sums
of squares and the partial F value that we discussed beginning on page 424.

Because there are $h-1$ values between x_t and x_{t+h}, the notation for partial
autocovariances and partial autocorrelations becomes rather cumbersome.

We will discuss the PACF in Section 5.3.3.

Integration

Consider again the random walk process, as in equation (5.8),

$$x_t = x_{t-1} + w_t,$$

where $\{w_t\}$ is a white noise. This process is not stationary. Now consider the
process $\triangle(\{x_t\})$:

$$\triangle(x_t) = w_t.$$

This differenced process is stationary, since it is white noise. See the lower
panel in Figure 5.3.

Differencing a process can change properties of the process in fundamental
ways. In this case, differencing forms a stationary white noise from a nonstationary random walk. Because integration is the inverse of differencing (see
equation (5.27)), we say that the random walk is an *integrated white noise*.

When one process is differenced to form a second process, we say that the
original process is an "integrated" process of the second process.

Now consider the random walk with drift process, as in equation (5.13),

$$x_t = \delta + x_{t-1} + w_t,$$

where again $\{w_t\}$ is a white noise. This process is not stationary, and first
order differencing yields a white noise process plus a constant. Differencing a
second time yields

$$\triangle^2(x_t) = \widetilde{w}_t,$$

which is white noise.

If a white noise process results from d differences, we say the original
process is *d-integrated white noise*, and we denote the process by I(d); thus,

a random walk is an I(1) process and a random walk with drift is an I(2) process. Note that a white noise process is itself an I(1) process.

We also use a similar notation for other types of processes, such as IMA(d, q) and ARIMA(p, d, q), which we will encounter later.

Joint Stationarity

A bivariate process $\{(X_t, Y_t)\}$ is said to be *jointly stationary* if each process is stationary and the cross-covariance function $\gamma_{XY}(s, t)$ is constant for fixed values of $s - t$; that is, for fixed $h = s - t$. In that case, we write the cross-covariance as $\gamma_{XY}(h)$, and we write the cross-correlation as $\rho_{XY}(h)$.

With this notation, we have the *cross-correlation function (CCF)* of a jointly stationary process as

$$\rho_{XY}(h) = \frac{\gamma_{XY}(h)}{\sqrt{\gamma_X(0)\gamma_Y(0)}}. \tag{5.60}$$

Note that $\gamma_{XY}(0)$ is just the ordinary covariance between X and Y at the same times, and $\rho_{XY}(0)$ is the ordinary correlation.

We do not have the symmetries of the autocovariance function and the ACF as in equation (5.59). This is because of the property in equation (5.55). In general,

$$\gamma_{XY}(h) \neq \gamma_{XY}(-h) \quad \text{and} \quad \rho_{XY}(h) \neq \rho_{XY}(-h). \tag{5.61}$$

This means that we have the property of "leading" or "lagging" indicators.

It is the case, however, that

$$\gamma_{XY}(h) = \gamma_{YX}(-h) \quad \text{and} \quad \rho_{XY}(h) = \rho_{YX}(-h), \tag{5.62}$$

which can be shown using the definition and the fact that $\text{Cov}(X, Y) = \text{Cov}(Y, X)$. (You are asked to go through the steps in Exercise 5.14.)

Stationary Multivariate Time Series

We can extend the ideas of a bivariate time series to a multivariate time series. Most measures for multivariate data are actually bivariate measures, such as covariances or correlations. Of course, with multivariate data, we may make adjustments of some variables on others before computing partial bivariate measures among the variables.

First, we extend the notation. Instead of using different letters to distinguish the variables, we use subscripts. The underlying variable of the time series is a vector, and x_t and x_{t+h} represent two vectors at different times. The means are vectors, and the variances are variance-covariance matrices. We will use similar notation as in the univariate case, sometimes with upper-case letters to emphasize that the quantities are matrices. (Recall that I do not make any notational distinction between vectors and scalars.)

For a stationary multivariate time series, the mean μ is a constant vector.

A multivariate time series is stationary if the mean is finite and constant and the autocovariance matrices between x_s and x_t are finite and depend only on $|s - t|$.

The autocovariances in a stationary multivariate time series are a matrix function:

$$\Gamma(h) = \mathrm{E}\left((x_{t+h} - \mu)(x_t - \mu)^{\mathrm{T}}\right). \qquad (5.63)$$

The individual elements of the matrix $\Gamma(h)$ are

$$\gamma_{ij}(h) = \mathrm{E}\left((x_{t+h,i} - \mu_i)(x_{tj} - \mu_j)\right). \qquad (5.64)$$

Notice that the $\gamma_{ii}(h)$ are just the autocovariances of the time series of the individual elements of the x_t vectors. (Hence, notice in our statement that defines stationarity of the multivariate time series, we did not need to say the individual series are stationary as we did in defining joint stationarity in a bivariate time series. The independence of $\Gamma(h)$ from t implies that.)

Note that $\gamma_{ij}(h) = \gamma_{ji}(-h)$, and so

$$\Gamma(h) = \Gamma^{\mathrm{T}}(-h) \qquad (5.65)$$

Most of the statistical methods used in multivariate time series can be developed in a bivariate time series.

5.2.2   Sample Autocovariance and Autocorrelation Functions; Stationarity and Estimation

All of the measures we have described above, μ_t, $\gamma_x(s,t)$, $\gamma_x(h)$, and so on, are expectations of random variables. For stationary series, the means are constant, and all autocovariances and autocorrelations are just functions of h.

In the stationary case, these measures all have finite-sample analogues. We assume that we have samples $x_1, \ldots, x_n$ and $y_1, \ldots, y_n$ from the time series, where the time indexes have the same meaning, that is, x_t and y_t correspond to the same time t.

These sample analogues can be used as estimators of the population quantities that we have defined, so we will denote them as $\hat{\mu}$, $\hat{\gamma}_x(h)$, and so on. The basic principle is that we replace an expected value by an average over the sample.

The sample mean, $\hat{\mu}$, is just the average over the sample $\bar{x}$. Note that in the general case, for μ_t dependent on t, there is no simple useful estimator, and neither the sample autocovariance function nor the sample autocorrelation function are appropriate in an analysis.

For the stationary case, we have sample quantities that can serve as estimators. It is important to recognize the difference between the autocovariance or autocorrelation of nonstationary process, and the corresponding autocovariance or autocorrelation from a sample of the nonstationary process (see Exercise 5.12).

- sample autocovariance function:

$$\hat{\gamma}(h) = \frac{1}{n} \sum_{t=1}^{n-h} (x_{t+h} - \bar{x})(x_t - \bar{x}) \tag{5.66}$$

- sample cross-covariance function:

$$\hat{\gamma}_{xy}(h) = \frac{1}{n} \sum_{t=1}^{n-h} (x_{t+h} - \bar{x})(y_t - \bar{y}) \tag{5.67}$$

Notice that in $\hat{\gamma}_x(h)$ and $\hat{\gamma}_{xy}(h)$ the divisor is n instead of $n-1$, or the number of terms $n-h$. (Compare this with $\hat{\sigma}^2$, where the divisor is $n-1$, which makes $\hat{\sigma}^2$ unbiased for σ^2. In the case of $\hat{\gamma}(h)$, neither a divisor of n nor a divisor of $n-1$ makes it unbiased. Using a divisor of n, however, guarantees that the expression for various values of h yields a non-negative definite matrix. As we will see in the following, we generally only use asymptotic properties of these statistics, so unbiasedness is not an important issue.)

From the covariances, we form the corresponding correlations.

- sample autocorrelation function (SACF):

$$\hat{\rho}(h) = \frac{\hat{\gamma}(h)}{\hat{\gamma}(0)}. \tag{5.68}$$

- sample cross-correlation function (CCF):

$$\hat{\rho}_{xy}(h) = \frac{\hat{\gamma}_{xy}(h)}{\sqrt{\hat{\gamma}_x(0)\hat{\gamma}_y(0)}}. \tag{5.69}$$

Since $\hat{\gamma}(-h) = \hat{\gamma}(h)$, $\hat{\rho}(-h) = \hat{\rho}(h)$. For either, we only need to compute the values for nonnegative values of h. The cross-covariance and cross-correlation, however, depend on the sign of h.

Correlograms and Cross-Correlograms

A plot of $\hat{\rho}(h)$ is called a *correlogram*. Since $\hat{\rho}(-h) = \hat{\rho}(h)$ and $\hat{\rho}(0) = 1$, a correlogram need include only values of $h = 1, 2, \ldots$. An example of a correlogram is shown in Figure 1.30 in Chapter 1, and several are shown in Figure 5.5 below.

A plot of $\hat{\rho}_{XY}(h)$ is called a *cross-correlogram*. The cross-correlation is different depending on the sign of h, so a cross-correlogram includes both positive and negative values of h. Also, $\hat{\rho}_{XY}(0)$ is not just always 1. It is the correlation of X and Y at the same time; hence, it is an important point in the cross-correlogram. An example of a cross-correlogram is shown in Figure 5.6.

Examples

The R function `acf` computes, and by default plots, the sample autocovariance or autocorrelation function for lags from 0 up to a maximum lag specified. Because $\hat{\rho}(0) = 1$, if a correlogram includes $h = 0$, the other values in the correlogram may be overwhelmed. The R function `Acf` in the `forecast` package performs the same functions as `acf` but does not include lag 0. It is the one I use most often.

The autocorrelations in a series of prices of a stock, of a stock index, of the prices of a commodity, or of other economic series such as the numbers of airline passengers over any short lags are always positive. That is just because the prices move from one price to a nearby price, and then from that price to a nearby price and so on, and numbers of airline passengers during one period is likely to be similar to the numbers in the immediately preceding period. For such series, however, the autocorrelations of measures of changes, such as returns, may show interesting relationships. Although, as we have seen, the correlations are all relatively small, we sometimes do see a tendency for an increase to be followed by a decrease; that is a negative autocorrelation. After one or two lags, the autocorrelations generally appear to be random. We see this in the correlogram of the time series of the S&P 500 Index daily returns for 1987 through 2017 in Figure 1.30 on page 101. (That correlogram was produced by R function `Acf`, so that the autocorrelation at lag 0, which is always 1, is not displayed.)

Other series may show larger (absolute) correlations at a number of lags that correspond to a season, such as a year.

Figure 5.5 displays autocorrelations of various financial and economic time series. (All were produced by `Acf`, and the data came either from Yahoo Finance or from FRED, except for the airline passengers data, which is a built-in dataset in R. The price returns and the index returns data are log returns for the year 2017. The change data for the other datasets are simple returns. The data obtained from FRED are for varying periods of time. The autocorrelations shown in each case are for the full set available at FRED. We will describe the two horizontal dashed lines in the plots on page 525.)

The ACFs in the two panels at the top of Figure 5.5 are for monthly returns of stock prices or indices. All of the autocorrelations seem to be relatively small, and the variation could be attributed to inherent randomness. (Figure 1.30 on page 101, which is for daily returns, is similar.)

The monthly returns of the VIX display more interesting patterns. The negative autocorrelations at short lags may be related to its tendency for mean reversion (see Exercise 1.19 on page 138 and Exercise A1.18 on page 199).

The monthly changes in the CPI, the monthly Case-Shiller Index (of US national housing prices), and the monthly number of PanAm passengers (for

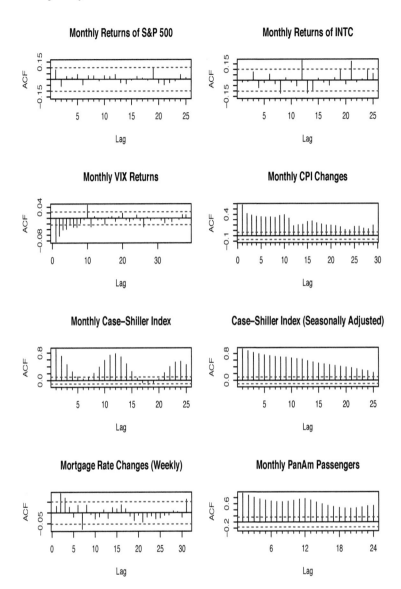

FIGURE 5.5
Autocorrelation Functions of Various Financial and Economic Time Series

the years 1949 through 1960) all show evidence of seasonal effects at a twelve-month lag. The seasonally-adjusted Case-Shiller Index, in which the data are aggregated and further smoothed, do not show the seasonal effects.

The Cross-Correlation Function, CCF

The R function `ccf` computes, and by default plots, the sample cross-covariance function or the cross-correlation function, CCF.

A cross-correlogram is different depending on the sign of h, so it includes both positive and negative lags. The lag h value returned by the R function invocation `ccf(x, y)` is the sample correlation between x_{t+h} and y_t.

The CCF of the S&P 500 Index daily log returns and the daily log returns of INTC for the period 1987 through 2017 is shown in Figure 5.6. (A scatterplot of these two series is shown in a scatterplot matrix in Figure 1.22 on page 86. Note that a bivariate scatterplot does not convey time series information, but it does give an indication of the CCF at $h = 0$.)

Intel and S&P 500 Returns

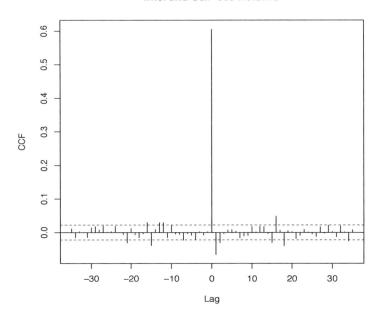

FIGURE 5.6
Cross-Correlations of S&P 500 Index and INTC Daily Log Returns for 1987 through 2017

Note the strong ordinary correlation between the S&P 500 Index and INTC daily returns shown by $\hat{\rho}_{xy}(0)$. (This is the same as the sample correlation given in Table 1.9 on page 84.)

Another pair of series that we may expect to have some interesting cross-correlations are the S&P 500 Index returns and the VIX, either the VIX itself or the changes in the VIX. Figure 5.7 shows the CCF for the daily returns of the S&P 500 Index versus the daily returns of the VIX computed in such a way that positive lag corresponds to VIX leading.

S&P 500 Daily Returns versus VIX Daily Returns

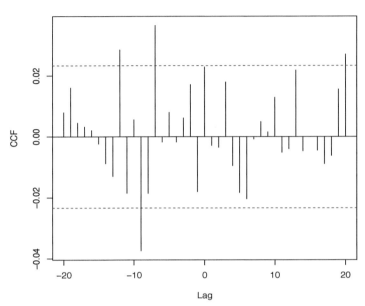

FIGURE 5.7
Cross-Correlations of S&P 500 Index Daily Log Returns and the VIX for 1987 through 2017

The cross-correlations are not large, in general. The largest ones (in absolute value) are for S&P 500 Index returns leading the VIX returns within a two-week period. There do not seem to be any strong conclusions that we can draw from the CCF, however.

5.2.3 Statistical Inference in Stationary Time Series

Statistical inference involves formulation of a model for describing a data-generating process and then using data from that process to estimate various parameters in the model or to test hypotheses concerning them.

Given a time series model such as equation (5.2),

$$x_t = f(x_{t-1}, x_{t-2}, \ldots; \theta) + w_t,$$

statistical inference is based on past observations. Use of past observations obviously implies some degree of stationarity. In the following, we assume weak stationarity, but in applications, mild departures from stationarity may not invalidate the methods. Statistical inference may involve estimation or tests of hypotheses concerning general properties of the time series, such as the mean, variance, and autocorrelation; it may involve estimation or tests concerning the parameter vector θ; or it may involve prediction of future values of x_t. Prediction of future values is called *forecasting*.

Given data from a time series, we estimate the mean, variance, autocovariance, and autocorrelation, which are all expected values in a probability model, by their sample analogues. For a pair of time series, we likewise estimate the cross-covariance and cross-correlation, by the sample analogues which are averages over a sample. The basic principle is that we replace an *expected value* by a *sample average*.

In Section 4.2.2, we discussed various criteria for statistical estimators, such as least squares and maximum likelihood. Use of sample averages as estimators of expected values is the method of moments (see page 349).

In statistical inference, we need to know something about the probability distribution of the estimators or test statistics so as to form confidence intervals or test statistical hypotheses. These distributions, of course, depend on the probability distributions in the underlying model.

Inference Concerning Autocorrelations

One of the most important properties of a time series is the autocorrelation function. The null case is that the autocorrelations are zero. An important model that we have described for the null case is white noise, in which $\rho(h) = 0$ for $h \neq 0$. Although the properties defining a white noise process are somewhat restrictive, the white noise model is rather weak, because no distribution is specified. Nevertheless, we can test the hypothesis that $\rho(h) = 0$ for $h \neq 0$ in this model using the asymptotic normal distribution implied by the central limit theorem. The tests that we develop can also be used in the stronger models that we discuss in Section 5.3.

We first consider the null hypothesis for a single value of h, say $h = k$, and we assume that the mean and variance are constant and finite. A null hypothesis for given $k \geq 1$ is

$$H_0 \ : \ \rho(k) = 0. \tag{5.70}$$

This hypothesis is composite, as it does not specify a specific distribution.

The obvious sample statistic to use is $\hat{\rho}(h)$, but we need its probability distribution under the null hypothesis. We use its probability distribution from a process that is white noise, but even so, the probability distribution of $\hat{\rho}(h)$ is very complicated. As in most complicated cases, we seek a general asymptotic distribution.

For a single lag, say $h = k$, the asymptotic distribution of $\hat{\rho}(k)$ for a sample of size n from a white noise process is $N(0, 1/n)$. Proof of this result is beyond the scope of this book. The interested reader is referred to Fuller (1995), Chapter 6.

Using the asymptotic distribution of $\hat{\rho}(k)$, the hypothesis is simple to test. For an asymptotic test at the α significance level, the hypothesis is rejected if $|\hat{\rho}(k)|$ exceeds the two-sided $(1 - \alpha)$-quantile from a $N(0, 1/n)$ distribution, which for $\alpha = 0.05$, is approximately $2/\sqrt{n}$.

The R functions `acf` and `Acf` plot two lines (blue and dashed by default) at these critical values. Note, for example, in Figure 1.30 on page 101, the two dashed horizontal lines are at ± 0.023. There are 7,813 observations in the dataset and the lines are at $\pm 2/\sqrt{n}$. In the ACF plots in Figure 5.5, the critical lines are at different heights, depending on the differing numbers of observations in the series.

In the bivariate case, as in the example of the cross-correlogram of the S&P and INTC returns in Figure 5.6, the asymptotic distribution of $\hat{\rho}_{xy}(k)$, if either x_t or y_t is a white noise process, so is also $N(0, 1/n)$. This asymptotic distribution provides a basis for a test of the null hypothesis that the processes do not "move together".

The asymptotic distribution provides a basis for a test of the null hypothesis that all autocorrelations at nonzero lag are zero:

$$H_0 \,:\, \rho(h) = 0 \quad \text{for each } h \geq 1. \tag{5.71}$$

This hypothesis is a set of simultaneous hypotheses, as we discussed on page 384. There are various approaches to testing it, including use of simultaneous tests, sometimes called "portmanteau tests", for the null hypotheses for $1 \leq h \leq m$, for some positive integer m, against the alternative hypothesis, $H_1 \,:\, \rho(k) \neq 0$, for some $k = 1, \ldots, m$.

The *Box-Pierce portmanteau test* is based on the statistic

$$Q^*(m) = n \sum_{k=1}^{m} \hat{\rho}(k)^2, \tag{5.72}$$

which under the null hypothesis (and some other technical assumptions that we will ignore here) has an asymptotic chi-squared distribution with m degrees of freedom. (This is not surprising, as we recall that the sum of m squared independent standard normal variables is chi-squared, and $\sqrt{n}\hat{\rho}(k)$ has an asymptotic $N(0, 1)$ distribution.)

A modified version of this test, called the *Ljung-Box test*, or sometimes the *Box-Ljung test*, is based on the statistic

$$Q(m) = n(n + 2) \sum_{k=1}^{m} \frac{\hat{\rho}(k)^2}{n - k}. \tag{5.73}$$

The Ljung-Box test also has an asymptotic chi-squared distribution with m

degrees of freedom under the null hypothesis, but it has better power, so it is more commonly used.

We note that the test statistics $Q^*(m)$ and $Q(m)$ involve the same sample statistics $\hat{\rho}(k)$; the only difference is how those statistics are weighted. This is also essentially the same sample statistic as in the Durbin-Watson test. The Durbin-Watson statistic in equation (4.138) on page 434 just involves $\hat{\rho}(1)$. The Durbin-Watson statistics for larger lags correspond to values of $\hat{\rho}(k)$ with larger values of k.

The R function Box.test in the basic stats package implements these tests. As an illustration, we perform the Box-Ljung test for the first 12 lags of the daily returns of the S&P 500 Index for the period shown in Figure 1.30.

```
> Box.test(GSPCdReturns, lag=12, type="Ljung")

        Box-Ljung test

data:  GSPCdReturns
X-squared = 97.328, df = 12, p-value = 1.907397e-08
```

We reject the null hypothesis (based on the very small computed p-value), and conclude that the autocorrelations are not 0, for some lags between 1 and 12. This confirms our visual assessment from Figure 1.30.

We may notice that the size of the sample from which the ACF is computed does not enter into the distribution of the test statistic. The asymptotic distribution depends on a large sample, but a sample size in the hundreds is generally sufficient. (The sample size in examples is 7,813 returns.) The statistics in these tests are highly variable; tests on different samples may yield quite different results.

Autocorrelations at any lag among returns over longer periods of time are likely to be small. Recall the "stylized facts" in Section 1.6, and compare the ACF of monthly returns in Figure 1.32 on page 103 with the ACF of daily returns in Figure 1.30. The Box-Ljung test for the first 12 lags of the monthly returns of the S&P 500 Index for the same period (shown in the bottom panel of Figure 1.33) yields the following.

```
> Box.test(GSPCmReturns, lag=12, type="Ljung")

        Box-Ljung test

data:  GSPCmReturns
X-squared = 7.442, df = 12, p-value = 0.8270675
```

We do not reject the null hypothesis of 0 autocorrelations for monthly returns, and this confirms our visual assessment from Figure 1.32.

We will discuss other types of statistical inference in the context of specific time series models in Section 5.3.6 and in other sections. We will also discuss tests of autocorrelations in the context of the distributions of the errors in regression models in Section 5.3.12.

Forecasting

In economic and financial applications, time series models are often used to forecast or predict future values in the series.

In a time series model as above,

$$x_{t+1} = f(x_t, x_{t-1}, \ldots; \theta) + w_{t+1},$$

with $\mathrm{E}(w_{t+1}) = 0$, given $x_t, x_{t-1}, \ldots$, the best predictor of x_t in the sense of smallest conditional mean squared error, is $\mathrm{E}(x_{t+1} | x_t, x_{t-1}, \ldots)$, that is, $f(x_t, x_{t-1}, \ldots; \theta)$. We observed this fact in Chapter 4.

Of course we do not know θ.

We fit the model based on available information. We denote "available information" at time t by $\mathcal{F}_t$, which includes all observed data and assumptions we have made about the data-generating process up through time t. The best predicted value $\hat{x}_t^{(1)}$ is just the conditional expected value, so we take

$$\hat{x}_t^{(1)} = \mathrm{E}(x_{t+1} | \mathcal{F}_t). \tag{5.74}$$

In time series applications, the fitted model is updated as new observations come in. In forecasting, we call the point in time at which the model is fitted, the *forecast origin*.

Letting $\hat{f}_n(x_n, x_{n-2}, \ldots; \widehat{\theta}_n)$ represent the fitted model at time $t = n$, the predicted value of x_{n+1}

$$\hat{x}_n^{(1)} = \hat{f}_n(x_n, x_{n-1}, \ldots; \widehat{\theta}_n). \tag{5.75}$$

is called the 1-step-ahead forecast. We can iterate on this to obtain the 2-step-ahead forecast, $\hat{x}_{n+2}$,

$$\hat{x}_n^{(2)} = \hat{f}_n(\hat{x}_n^{(1)}, x_n, x_{n-1}, \ldots; \widehat{\theta}_n), \tag{5.76}$$

and so on.

Given the information at the forecast origin n, the h-step-ahead forecast is $\hat{x}_n^{(h)}$ formed in this way. The number of time units ahead, h, is called the *forecast horizon*. The term "forecast horizon" may also refer to the time for which the forecast is made, in this expression, $n + h$. We will illustrate forecasting after we have discussed some specific models and how to fit them.

5.3   Autoregressive and Moving Average Models

The basic moving average of white noise model (5.5) and the random walk model (5.8) can be extended in various ways. An extension of the moving average model allows general linear combinations of past white noise realizations, and an extension to the random walk allows the current step to be taken from a linear combination of past values. The latter models are called autoregressive (AR) and the former are called moving average (MA) models, and together the autoregressive and moving average models are the most commonly used linear models for time series analysis.

These models relate a present value x_t in the data-generating process to a finite number of values at previous times, $t-1, t-2, \ldots, t_n$. An important thing to note is that there is no starting point; the model for x_1 holds for x_{-1}, x_{-2}, and so on. This fact yields useful infinite power series for these models, as in equations (5.92) and (5.103) below.

ARMA models and the integrated versions, called ARIMA models, are some of the most fundamental and widely-used time series models in economic and financial applications. The models and their extensions (such as ARMA+GARCH, discussed in a later section) do not necessarily fit financial data very well, but they often serve as useful approximations. In this section, we will describe their general properties. The AR(1) model, which we emphasize, is very useful as a general descriptor of economic and financial time series.

General Properties and Alternate Representations

Because there is no starting or ending point in autoregressive and moving average time series models, there are some interesting transformations of the models. For example, a moving average model that relates a present observable value x_t to a finite number of noise values at previous times possibly can be inverted so that one of those previous noise values can be represented as an infinite weighted sum of previous values ("invertibility", page 533). Alternatively, an autoregressive model that relates a present observable value x_t to a finite number of past observables values may be rewritten so that it can be represented as an infinite weighted sum of noise values ("causality", below and page 537).

Causality

Depending on the linear relationships among a present value and past values, it may be possible to express a time series model as a one-sided linear process (page 489):

$$x_t = \mu + \sum_{j=0}^{\infty} \psi_j w_{t-j}, \tag{5.77}$$

where $\{w_t\}$ is white noise and $\sum_{j=0}^{\infty} |\psi_j| < \infty$. This is an interesting representation. Heuristically, it means that the present, except for one white noise variate, is determined by the past. We call a process that can be represented as the one-sided linear process (5.77) *causal*. We discuss an example of this on page 536.

Another interesting transformation may change a model that relates a present value x_t to a finite number of values at previous times to a model that relates the present value x_t to values at *future* times (page 538). Models that relate a present value to future values are not "causal". (We should note that "causality" in this sense is not the same as "causality" in statistical inference.)

In the first few sections below, we will consider general properties of autoregressive and moving average models. In these sections we treat the parameters as known. Of course to use the models on real data, the parameters must be estimated from the data. In Section 5.3.6, we will discuss problems of fitting such model to data, and then consider some applications in finance.

5.3.1 Moving Average Models; MA(q)

The moving average of white noise (5.5) can be extended to be a general linear combination,

$$x_t = w_t + \theta_1 w_{t-1} + \cdots + \theta_q w_{t-q}, \tag{5.78}$$

where $\theta_1, \ldots, \theta_q$ are constants and $\{w_t\}$ is a white noise process. If $\theta_q \neq 0$, this is called a *moving average model* of order q, or an MA(q) model.

We often write this model as $x_t = \sum_{j=0}^{q} \theta_j w_{t-j}$, in which $\theta_0 = 1$.

We see immediately from the representation of the model in (5.78) that

- a moving average process is causal.

The properties of an MA(q) model are similar to those of a moving average of white noise that we worked out in Section 5.2; it is merely a multiple of a weighted moving average of white noise. The unconditional or marginal mean, which we denote as μ_x, is a constant 0:

$$\begin{aligned} \mathrm{E}(x_t) &= \mathrm{E}(w_t) + \theta_1 \mathrm{E}(w_{t-1}) + \cdots + \theta_q \mathrm{E}(w_{t-q}) \\ &= 0. \end{aligned} \tag{5.79}$$

The conditional mean of x_t, given $x_{t-1}, \ldots, x_q$, is

$$\begin{aligned} \mathrm{E}(x_t | x_{t-1}, x_{t-2}, \ldots) &= \mathrm{E}(w_t) + \theta_1 x_{t-1} + \cdots + \theta_q x_{t-q} \\ &= \theta_1 x_{t-1} + \cdots + \theta_q x_{t-q}. \end{aligned} \tag{5.80}$$

The marginal variance is also a constant, in this case, similar to before,

$$\sigma_x^2 = (1 + \theta_1^2 + \cdots + \theta_q^2)\sigma_w^2, \tag{5.81}$$

where σ_w^2 is the variance of the white noise. The conditional variance, given $x_{t-1}, x_{t-2}, \ldots$, is just σ_w^2, because all terms in the sum (5.78) are constants except w_t.

Multivariate MA(q) Models

A simple generalization of the moving average model is the *vector moving average model*, in which the elements x_t and w_t are d-vectors and the θs are replaced by constant $d \times d$ matrices.

Specifically, a *d-variate moving average model* of order q is

$$x_t = w_t + \Theta_1 w_{t-1} + \cdots + \Theta_q w_{t-q}, \tag{5.82}$$

where x_t and w_t are d-vectors, $\Theta_1, \dots, \Theta_q$ are constant $d \times d$ matrices with $\Theta_q \neq 0$, and $\{w_t\}$ is a d-variate white noise process.

Autocorrelations in MA(q) Models; Stationarity

The autocorrelation $\rho(s,t)$ of a MA(q) process is somewhat more complicated than that of a simple moving average of a white noise given in equation (5.52). As before, for the moving average $\{x_t\}$, we begin with the autocovariance $\gamma(s,t)$ (omitting the subscript) and, writing $s = t + h$, we have

$$\gamma(s,t) = \text{Cov}\left(\sum_{j=0}^{q} \theta_j w_{t+h-j}, \ \sum_{i=0}^{q} \theta_i w_{t-i} \right), \tag{5.83}$$

where, for convenience, we have let $\theta_0 = 1$.

From equation (5.83), we see that $\gamma(s,t)$ is constant for a fixed difference $|s-t|$, and is 0 for $|s-t| > q$. (Showing this latter property requires some algebra; it is essentially the same as you are asked to work through in Exercise 5.9 for the autocorrelations in a moving average of white noise.)

The fact that the autocorrelation depends only on the length of the lag $|s - t|$, together with the fact that the mean is constant, implies that

- a moving average process is stationary.

Because of the stationarity, we can now write the autocovariance function with just one argument as $\gamma(h)$, and the autocorrelation function as $\rho(h)$.

Now, simplifying the sums in expression (5.83), and using $\rho(h) = \gamma(h)/\gamma(0)$, for nonnegative values of h, we have the ACF for an MA(q) model,

$$\rho(h) = \begin{cases} \dfrac{\sum_{j=0}^{q-h} \theta_j \theta_{j+h}}{\sum_{j=0}^{q} \theta_j^2} & 0 \le h \le q \\ 0 & q < h. \end{cases} \tag{5.84}$$

The R function `ARMAacf` computes the values of the theoretical or population ACF for given values of the coefficients, as in expression (5.84).

The usefulness of a statistical model for analyzing data or understanding a data-generating process depends on how well the model fits the data. To use an MA(q) for real data, the first consideration is the choice of q.

The dependence of the ACF on the order q of an MA model suggests use of the sample ACF to determine an appropriate value of q. For example, consider the ACF of the S&P daily returns in Figure 1.30 on page 101. As we have emphasized, a sample ACF may be rather noisy, and the large magnitudes of the correlations at lags 15 through 20 or lags in the mid-30s may not be informative. The large magnitudes of the correlations at lags 1 and 2 would suggest, however, that *if an MA model is appropriate*, its order should be 2. You are asked to explore this model in Exercise 5.18.

The Moving Average Operator

Notice that if we define the function $\theta(B)$ of the backshift operator as

$$\theta(B) = 1 + \theta_1 B + \cdots + \theta_q B^q, \tag{5.85}$$

we can write the MA(q) model as

$$x_t = \theta(B)(w_t). \tag{5.86}$$

We call $\theta(B)$ in equation (5.85) the *moving average operator* of order q, for given coefficients $\theta_1, \ldots, \theta_q$.

A simple generalization of the moving average operator is the multivariate moving average operator in which the θs are square matrices, as in equation (5.82), and the backshift operator is replaced with the multivariate backshift operator.

The Moving Average Characteristic Polynomial

As on page 503, we identify the characteristic polynomial for an MA(q) model as

$$\theta(z) = 1 + \theta_1 z + \cdots + \theta_q z^q. \tag{5.87}$$

(Note that the signs on the θs are reversed from those on page 503. That does not change the properties.)

Notation

Because the basic symbols are the same, to distinguish the operator $\theta(B)$ from the coefficients $\theta_1, \ldots, \theta_q$, I will write the coefficients with subscripts, even if there is only one of them. For example, in the MA(1) model, while it may be more natural to write the coefficient without a subscript, I will write it with one:

$$x_t = w_t + \theta_1 w_{t-1}.$$

In this case, $\theta(B) = 1 + \theta_1 B$.

It should also be noted that some authors use the same notation for MA(q) models as above, but reverse the signs on the θs.

The notation I use is in common usage, and it corresponds to the notational model used in the stats package of R. Some other R packages for time series use different notation, however.

The MA(1) Model; Autocorrelations and Partial Autocorrelations

Consider the simple MA(1) model. From equation (5.84) or just by working it out directly, we see that the autocorrelation is

$$\rho(1) = \frac{\theta_1}{1 + \theta_1^2} \tag{5.88}$$

and $\rho(h) = 0$ for $h > 1$, as we noted above.

Although the ACF is zero for $h > q$ in an MA(q) model, the partial autocorrelation is not.

We can see this easily in the MA(1) model. Consider the autocorrelation between x_t and x_{t-2} adjusted for x_{t-1}. Let $\rho_1 = \theta_1/(1 + \theta_1^2)$. The adjusted value of x_t is $x_t - \rho_1 x_{t-1}$ and likewise the adjusted value of x_{t-1} is $x_{t-1} - \rho_1 x_{t-2}$, so the partial autocorrelation of x_t and x_{t-2} is

$$\frac{\mathrm{Cov}(x_t - \rho_1 x_{t-1}, \; x_{t-1} - \rho_1 x_{t-2})}{\mathrm{V}(x_t - \rho_1 x_{t-1})} = \frac{\theta_1^2}{1 + \theta_1^2 + \theta_1^4}, \tag{5.89}$$

which is not zero.

This same kind of steps can be applied to partial autocorrelations at greater lags and to MA models of higher order.

The MA(1) Model; Identifiability

We note an interesting property in equation (5.88): the autocorrelation for the MA(1) model with $\theta_1 = c$ is the same as the autocorrelation when $\theta_1 = 1/c$. Because of this, the MA model is not identifiable based on its autocorrelation function.

The concept of *identifiability* in statistical models generally refers to the uniqueness of parameters in the model.

If the variance σ_w^2 of the white noise $\{w_t\}$ in the MA model is specified, the indeterminacy in θ is resolved, as we see from $\gamma(t, t)$ in equation (5.83). The two parameters θ_1 and $\sigma_w^2 = 1$ yield essentially the same model as the two parameters $1/\theta_1$ and $\sigma_w^2 = \theta_1^2$.

The MA(1) Model; Invertibility

We now note another interesting relationship. From

$$x_t = w_t + \theta_1 w_{t-1}, \tag{5.90}$$

we have

$$
\begin{aligned}
w_t &= -\theta_1 w_{t-1} + x_t \\
&= \theta_1^2 w_{t-2} - \theta_1 x_{t-1} + x_t \\
&= -\theta_1^3 w_{t-3} + \theta_1^2 x_{t-2} - \theta_1 x_{t-1} + x_t \\
&= (-\theta_1)^k w_{t-k} + \sum_{j=0}^{k-1} (-\theta_1)^j x_{t-j}.
\end{aligned} \tag{5.91}
$$

The implications of equation (5.91) for the process $\{x_t\}$ is that previous realizations $x_{t-1}, x_{t-2}, \ldots$ affect x_t. In most applications, we would assume that this effect decreases with the elapsed time; hence, we assume that $|\theta_1^j|$ decreases. This is the case if $|\theta_1| < 1$, and in this case (see equation (5.31 on page 501), we have the limit

$$w_t = \sum_{j=0}^{\infty} (-\theta_1)^j x_{t-j}. \tag{5.92}$$

This means that we have "inverted" a model in which the observable x_t was represented as a linear combination of past noise values (the MA model of equation (5.90)), and written an equivalent model in which the noise w_t is represented as a sum of past observable values x_{t-j}.

A process in which the white noise variable can be expressed in terms of past observable values is said to be *invertible*. The MA(1) process with $|\theta_1| < 1$ is invertible.

Under the condition $|\theta_1| < 1$, the weights decrease farther back in time.

The expression (5.92) represents an autoregressive process of infinite order; that is, an invertible MA(1) process is an AR(∞) process, in the notation we will develop in Section 5.3.2 below.

Note a relationship between invertibility and identifiability. If to achieve invertibility we require $|\theta_1| < 1$, then only one of the possibilities θ_1 or $1/\theta_1$ is possible.

Roots of the Characteristic Polynomial

We note that the presence of the properties of invertibility and identifiability in the MA(1) is equivalent to the one root, z_1, of its characteristic polynomial $1 + \theta_1 z$ satisfying $|z_1| > 1$.

The relationship between invertibility and the root of the characteristic polynomial in the MA(1) model can be extended to similar results in general MA(q) models:

An MA(q) model is invertible if and only if all roots of the characteristic polynomial lie outside of the unit circle.

The proof of this is beyond the scope of this book. The interested reader is referred to Fuller (1995) for a proof.

Integrated MA Models, IMA(d, q)

We discussed integrated processes on page 516. An I(d) process, for example, is a d-integrated white noise process; that is, it is a white noise process after being differenced d times.

Suppose we have a process $\{y_t\}$ that when differenced d times becomes an MA(q) process. We call the $\{y_t\}$ process a d-integrated MA(q) process, and denote it as IMA(d, q).

Exponentially Weighted Moving Average Models, EWMA

The IMA(1, 1) process is often used as a model for financial time series. It also leads to a common model used in economic forecasting.

Suppose the MA(1) process resulting from differencing is invertible. Then, for some λ where $|\lambda| < 1$, we can write such a model as

$$
\begin{aligned}
x_t &= x_{t-1} + w_t - \lambda w_{t-1} \\
&= x_{t-1} + y_t,
\end{aligned}
\tag{5.93}
$$

where

$$
y_t = w_t - \lambda w_{t-1}.
$$

Now because $\{y_t\}$ is invertible, using equation (5.92) and substituting, we have

$$
x_t = \sum_{k=1}^{\infty} (1 - \lambda)\lambda^{k-1} x_{t-k} + w_t.
\tag{5.94}
$$

This makes some sense as a model, because it weights the past in an exponentially decreasing way. With the smoothing parameter λ chosen appropriately, for $k = 1$, this is the exponentially weighted moving average of equation (1.30) on page 40, and for $k = 2, 3, \ldots$, it corresponds to higher order exponential smooths. The Holt-Winters smoothing method, which is used in Exercise A1.5 on page 194, is based on this model.

5.3.2 Autoregressive Models; AR(p)

One of the simplest models of the general form of equation (5.2) is the autoregressive linear model

$$
x_t = \phi_0 + \phi_1 x_{t-1} + \cdots + \phi_p x_{t-p} + w_t,
\tag{5.95}
$$

where $\phi_0, \phi_1, \ldots, \phi_p$ are constants and w_t is a white noise process. If $\phi_p \neq 0$, the model is of order p. The constant term ϕ_0 is often taken as 0. In this model, which is somewhat similar to the linear regression models of Chapter 4 that had "error terms", we often call the error terms w_t, *innovations*.

We see immediately from the representation of the model in (5.95) that

- an autoregressive process is invertible.

Other properties of time series that we have discussed may or may not obtain. The important property of stationarity determines some divergence in the terminology of autoregressive processes among different authors.

The simple random walk model is of this form with $p = 1$ and $\phi_1 = 1$. As we recall, the random walk is not stationary; hence, immediately we see that unlike moving average models, not all autoregressive processes are stationary.

The terminology regarding autoregressive models varies. Many authors use the term *autoregressive model of order p*, and denote it as AR(p), only if the model is stationary, meaning that there are restrictions on the ϕs, other than just the requirement that $\phi_p \neq 0$. Although usually I will use the term only in the case of stationarity, I will often prepend "stationary" to indicate that condition. I will, however, often refer to any model of the form of equation (5.95) as autoregressive or AR, even if it is not stationary.

The Autoregressive Operator

Notice that the model of equation (5.95) can be written as

$$(1 - \phi_1 B - \cdots - \phi_p B^p)(x_t) = \phi_0 + w_t, \tag{5.96}$$

or

$$\phi(B)(x_t) = \phi_0 + w_t, \tag{5.97}$$

where

$$\phi(B) = 1 - \phi_1 B - \cdots - \phi_p B^p. \tag{5.98}$$

We call the operator $\phi(B)$ the *autoregressive operator* of order p. Note the similarity between the representation of the AR model in equation (5.97) and that of the MA model in equation (5.86).

As before, because of the ambiguity of the notation for the operator $\phi(B)$ and for the associated coefficients $\phi_1, \ldots, \phi_p$, I will generally use a subscript for the coefficients even if there is only one. Also, we note that some authors reverse the signs on the ϕs in the AR model. The notation above is in common usage, and it corresponds to the notational model used in the stats package of R.

The Autoregressive Characteristic Polynomial

Note from the form of equation (5.97) that the AR(p) model is a pth order difference equation, as equation (5.33), with an additional random component w_t.

The autoregressive characteristic polynomial, denoted as $\phi(z)$, is the polynomial of degree p similar to the operator of order p with the backshift operator replaced by a complex number z, as on page 503.

$$\phi(z) = 1 - \phi_1 z - \cdots - \phi_p z^p. \tag{5.99}$$

Vector Autoregressive Processes

Before continuing to examine properties of the autoregressive model, we will note a simple generalization to multivariate processes, that is, to multiple AR series that may be correlated with each other. All of the univariate properties generalize to the multivariate case.

A multivariate autoregressive model is similar to (5.95), except that the variables are vectors and the coefficients are square matrices. For the d-variate model we have

$$x_t = \Phi_1 x_{t-1} + \cdots + \Phi_p x_{t-p} + w_t, \tag{5.100}$$

where $x_t, x_{t-1}, \ldots, x_{t-p}$ and w_t are a d-vectors, $\Phi_1, \ldots, \Phi_p$ are constant $d \times d$ matrices and $\Phi_p \neq 0$, and $\{w_t\}$ is a multivariate white noise process. The multivariate random walk model is of this form with $p = 1$ and $\Phi_1 = I$, the identity matrix. (Notice that I do not make a distinction in the symbols that represent scalars and those that represent vectors, but I usually denote matrices with upper-case letters.)

The multivariate autoregressive operator is the same as in equation (5.98) except that the scalar ϕs are replaced with the square matrices Φs and the backshift operator is the multivariate backshift operator.

This multivariate version in equation (5.100) is sometimes called a *vector autoregressive model* of order p or VAR(p). Despite the fact that VAR models may seem to be simple generalizations of AR models, depending on $\Phi_1, \ldots, \Phi_p$, even if the individual white noise processes are independent of each other, there may be some surprising properties of the x_t, as we will see on page 590.

The AR(1) Model and Random Walks

Let us consider the simple AR(1) model,

$$x_t = \phi_0 + \phi_1 x_{t-1} + w_t,$$

and observe some interesting properties that depend on ϕ_1. Note that for $\phi_1 = 1$, this model is similar to a random walk model without a starting point (t can take on any value).

The AR(1) model, while very simple, serves as a good approximate model for many time series of economic and financial time series. The AR(1) model in time series applications is somewhat similar to the linear regression model in other settings.

We see that the conditional mean is given by

$$\mathrm{E}(x_t|x_{t-1}) = \phi_0 + \phi_1 \mathrm{E}(x_{t-1}). \tag{5.101}$$

In the evolution of x_t over time, we have

$$
\begin{aligned}
x_t &= \phi_0 + \phi_1 x_{t-1} + w_t \\
&= \phi_0 + \phi_1(\phi_1 x_{t-2} + w_{t-1}) + w_t \\
&\;\;\vdots \\
&= \phi_0 + \phi_1^k x_{t-k} + \sum_{j=0}^{k-1} \phi_1^j w_{t-j}. \tag{5.102}
\end{aligned}
$$

The AR(1) Model; Stationarity and Causality

The properties of the AR(1) model depend on the absolute value of the autoregressive parameter. The characteristic polynomial has only one root, the reciprocal of that parameter, which of course is real.

If $|\phi_1| = 1$, the process is a random walk for any given starting point.

We recall that a random walk is not stationary, but it is an I(1) process. A simulated random walk is shown in Figure 5.1.

We now consider some properties of AR(1) models that depend on whether $|\phi_1| < 1$ or $|\phi_1| > 1$.

The AR(1) Model with $|\phi_1| < 1$

Now, if $|\phi_1| < 1$, we can continue this evolution, to get the representation

$$x_t = \phi_0 + \sum_{j=0}^{\infty} \phi_1^j w_{t-j}. \tag{5.103}$$

This is a one-sided linear process (see equation (5.3)) because $\sum_{j=0}^{\infty} \phi_1^j < \infty$, and hence, the AR(1) model with $|\phi_1| < 1$ is causal.

The ACF of an AR(1) Model with $|\phi_1| < 1$

The autocovariances, without further assumptions on the model (5.95), are complicated, and are not very useful in practice. Let us make the simplifying assumption that for the mean in equation (5.101), we have $E(x_t) = E(x_{t-1})$. This would be the case, for example, if the process is stationary, or just if the process is a martingale. Under this assumption and with $\phi_1 < 1$, we have the unconditional mean,

$$E(x_t) = \frac{\phi_0}{1 - \phi_1}. \tag{5.104}$$

(You are asked to show this in Exercise 5.19.)

Now, under the assumption that the mean is constant, the autocorrelations are the same for any value of the mean, so instead of the expectation given in equation (5.104), let us just take $E(x_t) = 0$ in order to simplify the notation. The autocovariance function, therefore, is

$$
\begin{aligned}
\gamma(s,t) &= \text{cov}(x_s, x_t) \\
&= E(x_s, x_t) \\
&= E\left(\left(\sum_{j=0}^{\infty} \phi_1^j w_{s-j} \right) \left(\sum_{j=0}^{\infty} \phi_1^j w_{t-j} \right) \right).
\end{aligned}
\tag{5.105}
$$

Now, with $s = t + h$, we see that we can write $\gamma(s,t)$ as $\gamma(h)$.

If $E(w_i w_j) = 0$ for $i \neq j$ and $E(w_i w_j) = \sigma_w^2$ for $i = j$, we have

$$\gamma(h) = \sigma_w^2 \sum_{j=0}^{\infty} \phi_1^j \phi_1^{j+h}. \tag{5.106}$$

For $h \geq 0$, we factor out ϕ_1^h to get

$$\gamma(h) = \sigma_w^2 \phi_1^h \sum_{j=0}^{\infty} \phi_1^{2j};$$

and if $|\phi_1| < 1$, from equation (5.31) on page 501, we see that the series is finite and converges to

$$\gamma(h) = \frac{\phi_1^h}{1 - \phi_1^2}\sigma_w^2; \tag{5.107}$$

that is, the variance (when $h = 0$) and the autocovariance (when $h \neq 0$) are finite if $|\phi_1| < 1$ (which is the condition that allowed us to get equation (5.103) in the first place).

Since the mean is constant and $\gamma(s, t)$ is finite and depends on s and t only through $h = |s - t|$, if $|\phi_1| < 1$, the AR(1) process is stationary.

From equation (5.107) we have the ACF for an AR(1) process with $|\phi_1| < 1$:

$$\rho(h) = \phi_1^h, \quad \text{for } h \geq 0. \tag{5.108}$$

It is interesting to note from this expression that the ACF is positive at all lags if ϕ_1 is positive, and the ACF alternates in sign if ϕ_1 is negative.

Note that, while the autocorrelation decreases in absolute value as the lag h increases, it never becomes 0 for any finite h, unlike the ACF of an MA(q) process.

The AR(1) Model with $|\phi_1| > 1$

If $|\phi_1| \geq 1$, the process is *explosive*; that is, the values increase in magnitude without bound. Because of the very different properties in this case, some authors do not consider it an AR(p) model. Also, some authors refer to the model as explosive only for the positive case, that is, only if $\phi_1 \geq 1$.

With $|\phi_1| \geq 1$, consider the reverse of the steps in equation (5.102):

$$\begin{aligned}
x_t &= \phi_1^{-1}x_{t+1} - \phi_1^{-1}w_{t+1} \\
&= \phi_1^{-1}(\phi_1^{-1}x_{t+2} - \phi_1^{-1}w_{t+2}) - \phi_1^{-1}w_{t+1} \\
&\vdots \\
&= \phi_1^{-k}x_{t+k} - \sum_{j=1}^{k-1}\phi_1^{-j}w_{t+j}. \tag{5.109}
\end{aligned}$$

Now, if $|\phi_1| > 1$, then $|\phi_1|^{-1} < 1$, so we can continue this evolution, to get the representation

$$x_t = -\sum_{j=0}^{\infty}\phi_1^{-j}w_{t+j}, \tag{5.110}$$

which is similar to equation (5.103), resulting from an AR(1) process with $|\phi_1| < 1$. This process is stationary, as you are asked to show formally in Exercise 5.8.

There is an important difference in the two AR(1) processes, however. The process in equation (5.110), that is, AR(1) with $|\phi_1| > 1$, depends on the future. As a model of a data-generating process, it is not useful.

The AR(1) model with $|\phi| < 1$ is causal, as we have seen, because it depends only on the past, The AR(1) model with $|\phi| > 1$, however, is not causal; the linear process (5.110) involves future values. (Note that some authors reserve the term "autoregressive model" or "AR" only for causal models.)

Note that the model is causal if and only if the root of the autoregressive polynomial is outside the unit circle.

Figure 5.8 shows some simulated AR(1) with various values of ϕ_1. In order to emphasize differences in the processes, all are based on the same sequence of white noise, which, in fact is Gaussian white noise with $\sigma_w^2 = 1$. Note, in particular, the two explosive processes in the figure.

For AR(1) processes with $|\phi_1| < 1$, we observed in equation (5.108) that the ACF is positive at all lags if ϕ_1 is positive, and if ϕ_1 is negative, the ACF alternates in sign. In Figure 5.9, we show the sample ACFs of the simulated data in Figure 5.8 from AR(1) processes with various values of ϕ_1. When $|\phi_1|$ is small, the noise dominates the process, and so the ACFs for $\phi_1 = 0.05$ and $\phi_1 = -0.05$ do not exhibit these properties (and, in fact, they are almost identical because the same sequence of white noise was used in the simulations). The behavior of the population ACF is very evident in the sample ACFs for $\phi_1 = 0.95$ and $\phi_1 = -0.95$.

The properties of stationarity and causality in the AR(1) model depend on ϕ_1; therefore, equivalently, the properties depend on the one root of the characteristic polynomial $1 - \phi_1 z$. For AR models of higher order, we cannot determine these properties directly from the ϕs, but we can determine them for the roots of the characteristic polynomial, which of course depends on the ϕs.

The Partial Autocorrelation Function in the AR(1) Model

As with the MA(1) model on page 532, we now consider the partial autocovariance of x_t and x_{t-2} in the AR(1) model. This is the autocovariance between x_t and x_{t-2} adjusted for x_{t-1}. Let ρ_1 be the correlation between x_t and x_{t-1}. This is also the correlation between x_{t-1} and x_{t-2}. The adjusted value of x_t is $x_t - \rho_1 x_{t-1}$ and likewise the adjusted value of x_{t-2} is $x_{t-2} - \rho_1 x_{t-1}$, so the partial autocovariance is

$$\text{Cov}(x_t - \rho_1 x_{t-1},\ x_{t-2} - \rho_1 x_{t-1}) = (\rho_1^2 - \rho_1^2 - \rho_1^2 + \rho_1^2)V(x_t) = 0. \quad (5.111)$$

The partial autocovariance and, hence, the PACF, for lag $h = 2$ in a stationary AR(1) model is zero. This fact could help us identify an AR(p) model as indeed AR(1). In Section 5.3.3, we see that this property extends to lags greater than p in AR(p) models.

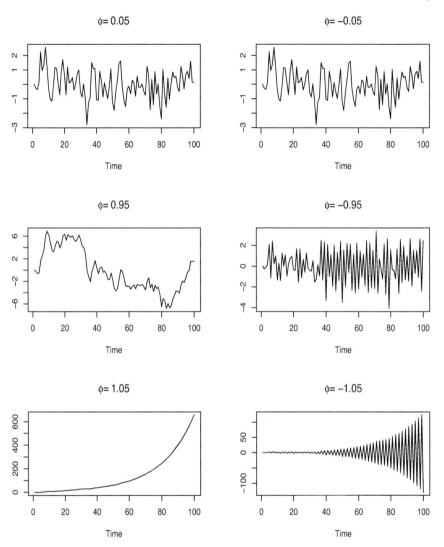

FIGURE 5.8
Simulated AR(1) Paths with Various Values of ϕ, Conditional on $x_0 = 0$

The AR(2) Model

The discussion above concerning the AR(1) model illustrates some general principles, and identifies a critical property of the root of the characteristic polynomial that determines stationary, explosive, and causal properties of the autoregressive process. Rather than attempt to generalize these properties to the AR(p) model immediately, it is instructive first to consider the AR(2)

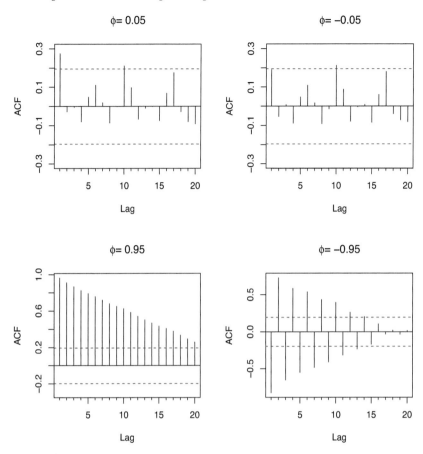

FIGURE 5.9
Sample ACFs of Simulated AR(1) Paths with Various Values of ϕ Shown in
Figure 5.8

model,
$$x_t = \phi_0 + \phi_1 x_{t-1} + \phi_2 x_{t-2} + w_t,$$
or
$$(1 - \phi_1 B - \phi_2 B^2)(x_t) = w_t. \qquad (5.112)$$

Let us assume the model to be stationary. Even so, we cannot get a simple
expression for the ACF or the autocovariance function. We will, hence, work
out a recursive relationship in h.

Under the assumption of stationarity, the mean is constant, and so, as in
equation (5.104) for AR(1), we have for AR(2)

$$\mathrm{E}(x_t) = \frac{\phi_0}{1 - \phi_1 - \phi_2}, \qquad (5.113)$$

so long as $\phi_1 + \phi_2 \neq 1$.

Letting $\mu = \mathrm{E}(x_t)$, we have

$$x_t - \mu = \phi_1(x_{t-1} - \mu) + \phi_2(x_{t-2} - \mu) + w_t.$$

Now consider $\gamma(h)$, for $h > 0$:

$$
\begin{aligned}
\gamma(h) &= \mathrm{E}((x_t - \mu)(x_{t-h} - \mu)) \\
&= \mathrm{E}(\phi_1(x_{t-1} - \mu)(x_{t-h} - \mu)) + \mathrm{E}(\phi_2(x_{t-2} - \mu)(x_{t-h} - \mu)) \\
&\qquad + \mathrm{E}((x_{t-h} - \mu)w_t) \\
&= \phi_1\gamma(h-1) + \phi_2\gamma(h-2).
\end{aligned}
\tag{5.114}
$$

The recursive equation (5.114) is sometimes called the *moment equation* for AR(2).

For the variance of x_t, $\gamma(0)$, we have

$$\gamma(0) = \phi_1\gamma(h-1) + \phi_2\gamma(h-2) + \sigma_w. \tag{5.115}$$

The Yule-Walker Equations

Equation (5.114) can be expanded to an AR(p) model. Following the same steps as above, we get

$$\gamma(h) = \phi_1\gamma(h-1) + \cdots + \phi_p\gamma(h-p). \tag{5.116}$$

Recalling that $\gamma(-j) = \gamma(j)$, and letting ϕ be the p-vector $(\phi_1, \ldots, \phi_p)$, γ_p be the p-vector $(\gamma(1), \ldots, \gamma(p))$, and Γ_p be the $p \times p$ autocovariance matrix

$$
\Gamma_p = \begin{bmatrix}
\gamma(0) & \gamma(1) & \cdots & \gamma(p-1) \\
\gamma(1) & \gamma(0) & \cdots & \gamma(p-2) \\
\vdots & \vdots & \ddots & \vdots \\
\gamma(p-1) & \gamma(p-2) & \cdots & \gamma(0)
\end{bmatrix},
$$

we can write the first p equations as

$$\Gamma_p \phi = \gamma_p. \tag{5.117}$$

And with $h = 0$ in equation (5.116), we have an equation for the variance,

$$\gamma_p^{\mathrm{T}} \phi + \sigma_w = \gamma(0). \tag{5.118}$$

(Note the slight differences in the elements of the matrix Γ_p and the vector γ_p.)

These $p + 1$ equations are called the *Yule-Walker* equations.

The AR(2) Model; The Yule-Walker Equations

The Yule-Walker equations allow us to express the coefficients ϕ in terms of the autocovariance function or vice versa. For the AR(2) model, we can solve for ϕ_1 and ϕ_2 and get, in terms of the autocovariances,

$$\phi_1 = \frac{\gamma(1)\gamma(2) - \gamma(0)\gamma(1)}{(\gamma(0))^2 - (\gamma(1))^2}, \tag{5.119}$$

and

$$\phi_2 = \frac{(\gamma(1))^2 - \gamma(0)\gamma(2)}{(\gamma(0))^2 - (\gamma(1))^2}. \tag{5.120}$$

These equations provide estimators of ϕ_1 and ϕ_2 based on the sample autocovariance function. These are method-of-moments estimators. We discuss estimation of the model parameters in general beginning on page 557.

The AR(2) Model; The ACF

Dividing the autocovariances in the moment equation by the constant variance, we have a recursive expression, that is, a difference equation, for the ACF:

$$\rho(h) = \phi_1\rho(h-1) + \phi_2\rho(h-2). \tag{5.121}$$

For any given ϕ_1 and ϕ_2, there are many functions $\rho(h)$ that would satisfy equation (5.121). If however, we fix $\rho(0)$ and $\rho(1)$, equation (5.121) yields $\rho(h)$ for $h \geq 0$. From the properties of the ACF (see page 515), we have

$$\rho(0) = 1 \quad \text{and} \quad \rho(1) = \rho(-1),$$

and so a recursive expression for ACF for an AR(2) process is

$$\rho(h) = \begin{cases} \dfrac{\phi_1}{1 - \phi_2} & \text{for } h = 1 \\[2mm] \phi_1\rho(h-1) + \phi_2\rho(h-2) & \text{for } h \geq 2. \end{cases} \tag{5.122}$$

Equation (5.122) allows us to compute the ACF for any h, but it does not tell us anything about the properties of the AR(2) process, whether or not it is explosive or causal, for example.

The AR(2) Model; The Characteristic Polynomial and Causality

Equation (5.121), which we can write as

$$(1 - \phi_1 B - \phi_2 B^2)\rho(h) = 0, \tag{5.123}$$

is a second-order deterministic finite difference equation. Following the analysis on page 504, we know that the solution to the difference equation depends on the roots of the characteristic polynomial, z_1 and z_2, which we can obtain easily using the quadratic formula (5.42).

For constants c_1 and c_2, determined by initial conditions from equations (5.44) and (5.45), the solution is either

$$\rho(h) = (c_1 + c_2 h) z_1^{-h}, \tag{5.124}$$

or

$$\rho(h) = c_1 z_1^{-h} + c_2 z_2^{-h} \tag{5.125}$$

depending on whether or not $z_1 = z_2$. If z_1 has an imaginary component, then $z_2 = \bar{z}_1$ and also $c_2 = \bar{c}_1$. Whether the roots are equal (and hence real), or unequal and real, or unequal and complex depends on the discriminant, $\sqrt{\phi_1^2 + 4\phi_2}$.

If and only if both roots (equal or not) are greater than 1 in modulus, the ACF converges to 0, as we see in equations (5.124) and (5.125). If additionally the roots have imaginary components, then by equation (5.46) on page 505, we see that the convergence is sinusoidal. If either root lies within the unit circle, the ACF grows toward one, and the series itself is explosive.

Figure 5.10 shows three simulated AR(2) processes and their ACFs with various values of ϕ_1 and ϕ_2. Each sequence in Figure 5.10 was generated by the `arima.sim` R function, and the `Acf` function was used to produce the ACF. In each case, all roots of the characteristic polynomial are greater than 1 in modulus. The first has a single root, the second has two distinct real roots and the third has roots with imaginary components. In Exercise 5.20a, you are asked to determine the characteristic polynomials and evaluate the roots for these processes.

Notice the sinusoidal decay in the ACF for the series whose roots have imaginary components. (We can see three cycles in this sample.) This may suggest that a time series process following such a model could exhibit cycles.

The roots of the characteristic polynomial corresponding to the AR(2) processes in Figure 5.10 all lie outside the unit circle. Figure 5.11 shows a simulated path of an AR(2) process whose characteristic polynomial has a unit root. Notice that after the first few lags, the ACF does not continue to go to zero, even up to 100 lags.

While it is rather simple to determine conditions on the coefficient in the model to ensure causality in an AR(1) model, it is not so straightforward in an AR(2) model. It turns out, however, that the relationship between causality and the root of the characteristic polynomial in the AR(1) model is the same for the AR(2) models. The AR(2) model is causal if and only if all roots of the characteristic polynomial lie outside of the unit circle.

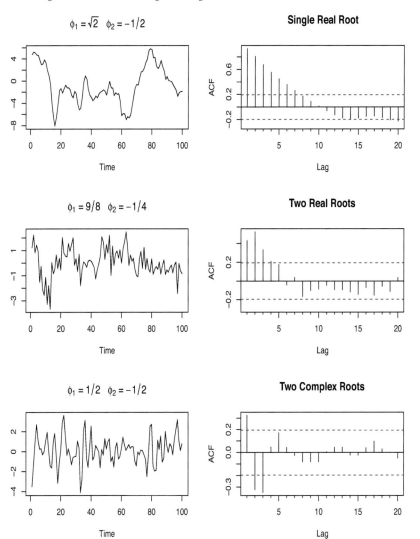

FIGURE 5.10
Simulated AR(2) Paths and the Corresponding Sample ACFs

These roots depend on ϕ_1 and ϕ_2 in the quadratic formula, of course, so we can summarize the conditions that ensure causality in terms of the coefficients of an AR(2) model:

$$\phi_1 + \phi_2 < 1;$$
$$\phi_2 - \phi_1 < 1; \qquad (5.126)$$
$$|\phi_2| < 1.$$

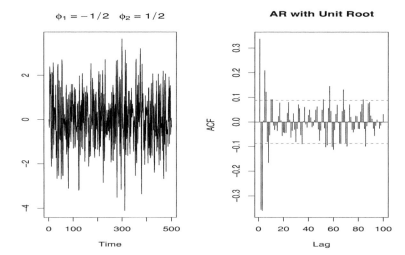

FIGURE 5.11
An AR(2) Process with a Unit Root; Simulated Path and the Corresponding Sample ACF

Properties of AR(p) Models

We obtained the ACF of an AR(2) model by first developing a difference equation in h for the ACF $\rho(h)$. The ACF, given as equation (5.122), is the solution of the difference equation (5.121) or (5.123) with the conditions $\rho(0) = 1$ and $\rho(h) = \rho(-h)$.

A difference equation for the ACF of an AR(p) model can be developed in the same way as equation (5.121) or (5.123). It is merely

$$\phi(\mathrm{B})\rho(h) = 0, \tag{5.127}$$

in operator form. If and only if all roots (equal or not) are greater than 1 in modulus, the ACF converges to 0. If additionally some roots have imaginary components, then by equation (5.46) on page 505, we see that the convergence is sinusoidal. If any root lies on or within the unit circle, the ACF does not converge to 0.

The relationship between causality and the roots of the characteristic polynomials in AR(1) and AR(2) models can be extended to similar results in general AR(p) models:

An AR(p) model is causal if and only if all roots of the characteristic polynomial lie outside of the unit circle.

The proof of this is beyond the scope of this book. The interested reader is referred to Fuller (1995) for a proof.

The R function `ARMAacf` computes the values of the theoretical, or population ACF for given values of the coefficients if the coefficients satisfy the conditions for convergence.

5.3.3   The Partial Autocorrelation Function, PACF

As we have seen, the ACF for an $MA(q)$ model relates strongly to the order q, but for an AR model, the ACF does not tell us much about the order. Any difference between the ACF of an $AR(1)$ model, equation (5.108), and that of an $AR(2)$ model, equation (5.122), is not immediately apparent.

We would like to have a function that will be zero for lags greater than or equal to p in an $AR(p)$ model, similar to the ACF in an $MA(q)$ model.

Because in an AR model the relationships are essentially linear regressions, when working out the relationship of one variable to another, we can adjust each of the two variables for all other variables between them in the sequence. Thus we can determine a *partial correlation*, just as we did in the $MA(1)$ and $AR(1)$ models on pages 532 and 539. The basic ideas of adjustments in linear models to yield "partial" statistics were developed in pages 424 through 426.

As before, we build a function based on a sequence of models. The function is defined so that it is the correlation that results after removing the effect of any correlation due to the terms of shorter lags. Because there are $h-1$ values between x_t and x_{t+h}, the notation becomes rather cumbersome, however.

As in the sequence of regression models (4.121) on page 424, we form the sequence of AR models,

$$x_t = \phi_{0,1} + \phi_{1,1}x_{t-1} + w_{1t}$$

$$x_t = \phi_{0,2} + \phi_{1,2}x_{t-1} + \phi_{2,2}x_{t-2} + w_{2t} \tag{5.128}$$

$$\vdots$$

$$x_t = \phi_{0,h} + \phi_{1,h}x_{t-1} + \cdots + \phi_{h,h}x_{t-h} + w_{ht}.$$

(In the following, we will take $\phi_{0,j} = 0$ for each j, just because the development proceeds more smoothly, but it could be modified to accommodate a nonzero mean.) Note that $\phi_{1,1}$ is the correlation of x_t and x_{t-1}. Next, $\phi_{2,2}$ is the correlation of $x_t - \phi_{1,1}x_{t-1}$ and x_{t-2}; that is, it is the correlation of x_t adjusted for x_{t-1} and x_{t-2}. Continuing, for a given h, we have the *partial autocorrelations* or *PACF*:

$$\phi_{1,1}, \phi_{2,2}, \ldots, \phi_{h,h}.$$

Consider a stationary $AR(p)$ model with mean 0,

$$x_t = \phi_1 x_{t-1} + \cdots + \phi_p x_{t-p} + w_t.$$

Using the AR model equation, we can solve for the ϕ_{ij} in equations (5.128).

If $p = 2$, for example, we have

$$\phi_{1,1} = \frac{\phi_1}{1 - \phi_2} = \rho(1)$$

$$\phi_{2,2} = \frac{\dfrac{\phi_1^2}{1 - \phi_2} + \phi_2 - \left(\dfrac{\phi_1}{1 - \phi_2}\right)^2}{1 - \left(\dfrac{\phi_1}{1 - \phi_2}\right)^2} = \frac{\rho(2) - (\rho(1))^2}{1 - (\rho(1))^2} \qquad (5.129)$$

$$\phi_{2,1} = \phi_1 = \rho(1)(1 - \phi_{2,2})$$

$$\phi_{3,3} = \frac{\rho(3) - \rho(2)\rho(1)(1 - \phi_{2,2}) - \rho(1)\phi_{2,2}}{1 - (\rho(1))^2(1 - \phi_{2,2}) - \rho(2)\phi_{2,2}} = 0.$$

We obtain the last equation, $\phi_{3,3} = 0$, using the recursive equation (5.122) for the ACF (see Exercise 5.21).

This same development would apply to an AR(p) model. We see from equation (5.128) that, in general, $\phi_{1,1} \neq 0$, ..., $\phi_{p,p} \neq 0$, but $\phi_{p+1,p+1} = 0$. The partial autocorrelation of x_{t-p-1} and x_t adjusted for $x_{t-1}, \ldots, x_{t-p}$, that is, $\phi_{p+1,p+1}$, is 0. Hence, if we have the PACF for an AR model, we immediately know a lower bound on the order of the model.

Recall from page 532, however, that the partial autocorrelations in an MA model do not go to 0.

The R function `ARMAacf` computes the values of the theoretical, or population PACF for given values of the coefficients.

Because the recursive equations (5.129) for the PACF only involve the ACF, we immediately have a sample version, using the sample version of the ACF. The sample version is an estimator, and from equations (5.129),

$$\widehat{\phi}_{k,j} = f(\widehat{\rho}(1), \ldots, \widehat{\rho}(k)),$$

for some function f of the sample ACFs. Based on this estimator, we estimate the AR order.

The R function `pacf` computes and optionally plots the sample PACF. Figure 5.12 shows the sample PACFs for the samples from the AR(2) models shown in Figures 5.10 and 5.11.

We see that for each of these samples, the PACF correctly indicates that the order of the AR process is 2.

Comparison of the PACF with the ACF of the same data can sometimes be useful. The PACF for data arising from a data-generating process with no autoregressive component is generally quite similar to the ACF for those data.

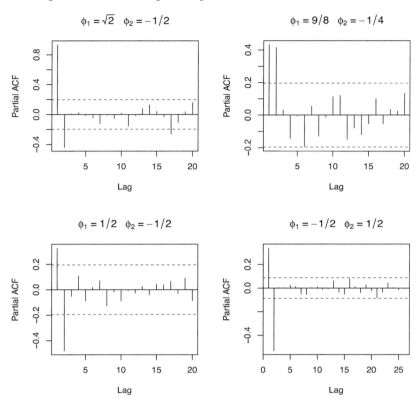

FIGURE 5.12
PACFs for the Simulated AR(2) Data in Figures 5.10 and 5.11

5.3.4   ARMA and ARIMA Models

A time series model with both autoregressive and moving average components is called an autoregressive moving average or ARMA model; specifically, an *ARMA model of orders p and q*, written ARMA(p,q), is the time series model, for $t = 0, \pm 1, \pm 2, \ldots,$

$$x_t = \phi_1 x_{t-1} + \cdots + \phi_p x_{t-p} + w_t + \theta_1 w_{t-1} + \cdots + \theta_q w_{t-q}, \qquad (5.130)$$

where $\theta_1, \ldots, \theta_q$, $\phi_1, \ldots, \phi_p$ are constants and $\theta_q \neq 0$ and $\phi_p \neq 0$, and w_t is a white noise process.

In terms of the autoregressive operator of order p and the moving average operator of order q, we can write the ARMA(p,q) model as

$$\phi(B)(x_t) = \theta(B)(w_t). \qquad (5.131)$$

The roots of the AR and MA polynomials formed from the AR and MA operators with a complex number in place of B are important tools in determining the properties of the time series.

The analysis of ARMA models involves the simultaneous application of the analyses we have developed for AR and MA models separately. The characteristics of one aspect of the model may mask those of the other component. In an ARMA model, the ACF may not go to zero, as it does in an MA model, and the PACF may not go to zero, as it does in an AR model. Nevertheless both of these functions are useful in the analysis.

Properties such as invertibility and causality in MA and AR models have the same meanings in ARMA models.

Autocovariances and Autocorrelations

The autocorrelation function for ARMA models depends on the parameters p and q.

Autocorrelation Function for ARMA(1,1)

It is a simple exercise to work out the ACF for a ARMA(1,1).

$$x_t = \phi x_{t-1} + \theta w_{t-1} + w_t.$$

We can write the autocovariance function recursively as

$$\gamma(h) - \phi\gamma(h-1) = 0, \quad \text{for } h = 2, 3, \ldots,$$

with initial conditions

$$
\begin{aligned}
\gamma(0) &= \phi\gamma(1) + \sigma_w^2(1 + \theta\phi + \theta^2) \\
\gamma(1) &= \phi\gamma(0) + \sigma_w^2\theta\phi.
\end{aligned}
$$

From this, with some algebraic manipulations we get the autocovariance function

$$\gamma(h) = \sigma_w^2 \frac{(1 + \theta\phi)(\phi + \theta)}{1 - \phi^2} \phi^{h-1}, \tag{5.132}$$

and so

$$\rho(h) = \frac{(1 + \theta\phi)(\phi + \theta)}{1 + 2\theta\phi + \theta^2} \phi^{h-1}.$$

Autocovariance Function for ARMA(p, q)

In a causal model, for $h \geq 0$, we get

$$
\begin{aligned}
\gamma(h) &= \text{Cov}(x_{t+h}, x_t) \\
&= \text{E}\left(\left(\sum_{j=1}^p \phi_j x_{t+h-j} + \sum_{j=0}^q \theta_j w_{t+h-j}\right) x_t\right) \\
&= \sum_{j=1}^p \phi_j \gamma(h-j) + \sigma_w^2 \sum_{j=h}^q \theta_j \psi_{j-h}, \tag{5.133}
\end{aligned}
$$

for some $\psi_0, \ldots, \psi_{j-h}$. (This last results from causality, because $x_t = \sum_{k=0}^{\infty} \psi_k w_{t-k}$; see also equation (5.134) below.)

The R function `ARMAacf` computes the values of the ACF for given values of the coefficients.

Properties of ARMA Models

The properties that we have discussed for MA and AR models or for time series models in general are applicable to ARMA models, in some cases with no modification. Again, I remark that some authors require a model to be stationary or causal to be called an "ARMA(p, q)" model. I generally will use the phrase "stationary ARMA" model when I assume it to be stationary.

As individually with MA and AR models, many of the important properties are determined by the roots of the characteristic polynomials $\theta(z)$ and $\phi(z)$. In the next two sections, we show relationships of the model parameters to the coefficients of the characteristic polynomial. We state two important relationships of properties of the model to properties of the roots of the characteristic polynomial, but the derivation of these properties are beyond the scope of this book.

Causal ARMA Models

We have seen that all MA(q) models are causal, because they are one-sided linear processes, but an AR(p) model is causal if and only if the roots of the characteristic polynomial lie outside the unit circle, in which case the AR model can be expressed in terms of a one-sided infinite series, as in equation (5.110). The concept of causality immediately applies to ARMA(p, q) models, and it depends only on the AR portion.

The basic idea in causality is that the present may depend on the past but does not depend on the future.

An ARMA(p, q) model, $\phi(B)(x_t) = \theta(B)(w_t)$, is said to be *causal* if the time series x_t for $t = 0, \pm 1, \pm 2, \ldots$ can be written as a one-sided linear process,

$$x_t = \sum_{j=0}^{\infty} \psi_j w_{t-j} = \psi(B)(w_t), \tag{5.134}$$

where $\psi(B) = \sum_{j=0}^{\infty} \psi_j B^j$ and $\sum_{j=0}^{\infty} |\psi_j| < \infty$ and $\psi_0 = 1$.

Using the representation $\phi(B)(x_t) = \theta(B)(w_t)$, we have

$$x_t = \psi(B)(w_t) = \frac{\theta(B)}{\phi(B)}(w_t),$$

in operator form. This yields a characteristic polynomial of the same form

$$\psi(z) = \sum_{j=0}^{\infty} \psi_j z^j \tag{5.135}$$

$$= \frac{\theta(z)}{\phi(z)}, \tag{5.136}$$

where we assume $\theta(z)$ and $\phi(z)$ have no roots in common. (If they have a common root, the two polynomials have a common factor and the fraction can be reduced.)

Again, we state a result without proof:

An ARMA(p,q) model is causal if and only if all roots of $\phi(z)$ in equation (5.136) lie outside the unit circle; that is, $\phi(z) \neq 0$ for $|z| \leq 1$.

The proof of this result can be found in Fuller (1995), among other places.

The R function ARMAtoMA performs the computations implied in equation (5.136), for given values of the coefficients ϕ and θ.

The ARMA(2,2) model

$$x_t = \frac{1}{4}x_{t-1} + \frac{1}{2}x_{t-2} + w_t - \frac{1}{5}w_{t-1} - \frac{2}{5}w_{t-2},$$

for example, is causal, but the model

$$x_t = \frac{3}{4}x_{t-1} + \frac{1}{2}x_{t-2} + w_t - \frac{1}{5}w_{t-1} - \frac{2}{5}w_{t-2}$$

is not. In either case, however, ARMAtoMA computes the specified number of coefficients of the polynomial resulting from the division:

```
> ARMAtoMA(c(1,1/4,1/2), c(-1/5,-2/5), 5)
[1] 0.8000 0.6500 1.3500 1.9125 2.5750
> ARMAtoMA(c(1,3/4,1/2), c(-1/5,-2/5), 5)
[1] 0.8000 1.1500 2.2500 3.5125 5.7750
```

Invertible ARMA Models

We have seen that all AR(p) models are invertible, but an MA(q) model is invertible if and only if the roots of the characteristic polynomial lie outside the unit circle. The concept of invertibility immediately applies to ARMA(p, q) models, and it depends only on the MA portion. Invertibility in the ARMA model is related to the identifiability of the MA portion.

An ARMA(p, q) model, $\phi(B)(x_t) = \theta(B)(w_t)$ is said to be *invertible* if the time series can be written as

$$\pi(B)(x_t) = \sum_{j=0}^{\infty} \pi_j x_{t-j} = w_t, \tag{5.137}$$

where $\pi(B) = \sum_{j=0}^{\infty} \pi_j B^j$ and $\sum_{j=0}^{\infty} |\pi_j| < \infty$ and $\pi_0 = 1$. Formally, we may write

$$w_t = \pi(B)(x_t) = \frac{\phi(z)}{\theta(z)}(x_t), \tag{5.138}$$

and identify the characteristic polynomial $\pi(x)$ as

$$\pi(z) = \sum_{j=0}^{\infty} \pi_j z^j \qquad (5.139)$$

$$= \frac{\phi(z)}{\theta(z)}, \qquad (5.140)$$

where again we assume $\phi(z)$ and $\theta(z)$ have no roots in common.

In this form we see the primal role of the MA polynomial, but we will not prove the main result:

An ARMA(p, q) model is invertible if and only if all roots of $\theta(z)$ in equation (5.140) lie outside the unit circle; that is, $\theta(z) \neq 0$ for $|z| \leq 1$.

The proof of this result can be found in Fuller (1995), among other places.

Just as a way to determine if an ARMA model is causal is to use the AR polynomial, a way to determine if an ARMA model is invertible is to use the MA polynomial. The ARMA(2,2) model

$$x_t = \frac{3}{4}x_{t-1} + \frac{1}{2}x_{t-2} + w_t - \frac{1}{5}w_{t-1} - \frac{2}{5}w_{t-2},$$

for example, is invertible, but the model

$$x_t = \frac{3}{4}x_{t-1} + \frac{1}{2}x_{t-2} + w_t - \frac{3}{5}w_{t-1} - \frac{2}{5}w_{t-2}$$

is not, and neither model is causal.

The R function ARMAtoMA can also be used to perform the computations implied in equation (5.140), for given values of the coefficients ϕ and θ, but the AR and MA coefficients must be interchanged.

Integration; ARIMA Models

In the general terminology used in time series, "integration" refers to the inverse of differencing. Simple properties of a time series can be changed by differencing. The original process before differencing is an integrated process.

As we have seen, some nonstationary models can be transformed into a stationary process by differencing. The original nonstationary process is an integrated stationary process. The problems in analysis that may be due to the nonstationarity of the original model, therefore, can possibly be addressed through the stationary model.

We may be able to make a nonstationary process into a stationary ARMA process by differencing. The original process in this case is an *ARIMA process*.

A time series $\{x_t\}$ is an ARIMA(p, d, q) process if $\{\triangle^d x_t\}$ is an ARMA(p, q) process.

Since the operator $\triangle^d(\cdot)$ is equivalent to $(1 - B)^d(\cdot)$, we can write an ARIMA(p, d, q) model as

$$\phi(B)(1 - B)^d(x_t) = \theta(B)(w_t). \qquad (5.141)$$

If there is a nonzero mean of the stationary differenced process, $\mathrm{E}(\triangle^d x_t) = \mu$, then we can take it out and write the model as

$$\phi(\mathrm{B})(1 - \mathrm{B})^d(x_t) = \alpha + \theta(\mathrm{B})(w_t), \tag{5.142}$$

where $\alpha = \mu(1 - \phi_1 - \cdots - \phi_p)$.

Fractional ARIMA Models

If the order of differencing d in an ARIMA(p, d, q) model satisfies $-0.5 < d < 0.5$ and $d \neq 0$, then the model is a fractional ARIMA model, sometimes called a FARIMA model or an ARFIMA model.

Such models are useful in studying long-memory processes, which are processes in which the decay in autocorrelations is slower than an exponential decrease. We will not pursue these topics in this book, however.

Seasonal ARMA Models

Both AR and MA models are based on linear combinations of values at points $t, t \pm 1, t \pm 2, \ldots$. A *seasonal* AR or MA process is one based on linear combinations of values at points $t, t \pm s, t \pm 2s, \ldots$, where $s \geq 1$ is the length of the *season*.

The commonly used notation for the seasonal differences uses upper-case letters that correspond to the lower-case letters in ordinary differences (not to be confused with our earlier use of upper-case letters for the multivariate processes).

The *seasonal AR operator* of order P with seasonal period s is:

$$\Phi_P(\mathrm{B}^s) = 1 - \Phi_1\mathrm{B}^s - \Phi_2\mathrm{B}^{2s} - \cdots - \Phi_P\mathrm{B}^{Ps} \tag{5.143}$$

The *seasonal MA operator* of order Q with seasonal period s is:

$$\Theta_Q(\mathrm{B}^s) = 1 + \Theta_1\mathrm{B}^s + \Theta_2\mathrm{B}^{2s} + \cdots + \Theta_Q\mathrm{B}^{Qs} \tag{5.144}$$

The seasonal effects can be handled as lower-frequency AR or MA effects as before.

The lower-frequency seasonal effect can be superimposed on the ordinary AR and MA factors:

$$\Phi_P(\mathrm{B}^s)\phi(\mathrm{B})(x_t) = \Theta_Q(\mathrm{B}^s)\theta(\mathrm{B})(w_t). \tag{5.145}$$

The model in equation (5.145) is denoted as ARMA$(p, q) \times (P, Q)_s$, in obvious notation.

A "pure seasonal ARMA model" is one with no dependencies at the higher frequency:

$$\Phi_P(\mathrm{B}^s)(x_t) = \Theta_Q(\mathrm{B}^s)(w_t). \tag{5.146}$$

It is denoted as ARMA$(P, Q)_s$. In this case, there are no model effects within seasons.

The ACF and PACF in pure seasonal ARMA models are similar to those in ARMA models, except at lags of Ps and Qs.

The PACF goes to 0 after lag Ps in an $AR(P)_s$ process, and the ACF goes to 0 after lag Qs in an $MA(Q)_s$ process.

In an $ARMA(P, Q)_s$ process, they both tail off, but over the length of the season.

Seasonal ARIMA Models

Using the concepts and notation above we can form a seasonal ARIMA, or SARIMA model:

$$\Phi_P(B^s)\phi(B)\triangle_s^D\triangle^d(x_t) = \alpha + \Theta_Q(B^s)\theta(B)(w_t), \qquad (5.147)$$

which we denote as $ARIMA(p, d, q) \times (P, D, Q)_s$.

5.3.5   Simulation of ARMA and ARIMA Models

One of the best ways of studying a class of data-generating models is to simulate data from the models. In Figures 5.1, 5.2, and 5.3 we showed simulated datasets from random walks. The random walk model is rather simple because usually we assume a starting point. The method of simulation is simple; as we mentioned, all that is required is a starting point and a source of white noise. Other time series models that we have described, including ARMA and ARIMA models, are for variables x_i with $i = 0, \pm1, \pm2, \ldots$; that is, there is no starting point. Whatever datum is generated is conditional on the ones before it.

One way of addressing this problem is to choose an arbitrary starting point and to generate a "burn-in" sequence that is discarded. The idea is that in moving far away from the arbitrary starting point, any conditional dependence on that point will be washed out.

Simulation of ARMA and ARIMA Models in R

The R function `arima.sim` generates simulated data from an ARIMA model. It first generates a burn-in sequence that is discarded.

The model parameters are specified in the `model` argument, which is a list that has three components, `ar`, which is a numeric vector specifying the AR coefficients; `ma`, which is a numeric vector specifying the MA coefficients; and `order`, which is optional and if supplied is a numeric vector with three elements, the order of the AR component, the order of differencing for an ARIMA model, and the order of the MA component. The orders of the AR and MA components must be the same as the length of the `ar` and `ma` components. The AR and MA coefficients are in the same form that we have used. (As we have mentioned, some authors reverse the signs of one or both sets of these values.) The AR coefficients must yield a causal model. (If they do not, the error message from R states that the AR component must be "stationary".)

The n argument to `arima.sim` specifies the number of observations to simulate. Unlike the other random number generators in R (see Section 3.3.3 beginning on page 321), the n argument follows the `model` argument, as a positional argument. It is best to specify the arguments with keywords.

Other arguments to `arima.sim` allow specification of the distribution of the innovations w_t, with the **rand.gen** argument, and the nature of the burn-in, with the **start.innov** argument.

The function produces an object of class `ts`.

The R code below generates 500 simulated observations from each of four different ARIMA time series. In each case, the innovations are generated from a standard normal distribution, which is the default. The seed was reset using `set.seed(12345)` prior to generation of each series. Plots of the time series are shown in Figure 5.13.

```
n <- 500
x1 <- arima.sim(n, model=list(ar=c(1/4,-1/2),order=c(2,0,0)))
x2 <- arima.sim(n, model=list(ma=c(1/4,1/2),order=c(0,0,2)))
x3 <- arima.sim(n, model=list(ar=c(1/4,-1/2),ma=c(1/4,1/2),
                order=c(2,0,2)))
x4 <- arima.sim(n, model=list(ar=c(1/4,-1/2),ma=c(1/4,1/2),
                order=c(2,1,2)))
```

One thing to note from Figure 5.13 is that there is very little visual difference in these different types of series, except of course for the integrated series, which is clearly nonstationary (see Exercise 5.25). In Section 5.3.10 we will return to this example, but with models whose innovations follow a t distribution.

5.3.6   Statistical Inference in ARMA and ARIMA Models

The first step in statistical inference is to formulate a preliminary model based on whatever information is available about the data-generating process and knowledge of the properties of the various classes of tractable models. Exploratory analysis can help to identify relevant properties of the data-generating process. Time series plots and plots of the ACF and PACF may reveal interesting aspects of time series data, and may suggest an appropriate model. The series may be differenced, and then the same exploratory analyses applied.

If an ARMA type model appears to be appropriate, the next considerations are the orders p and q of the MA and AR components. Choice of larger values of p and q than are appropriate results in overfitting. As we discussed in Chapter 4, overfitting not only results in a model that is more complex than appropriate, but it also increases the variances of the estimators.

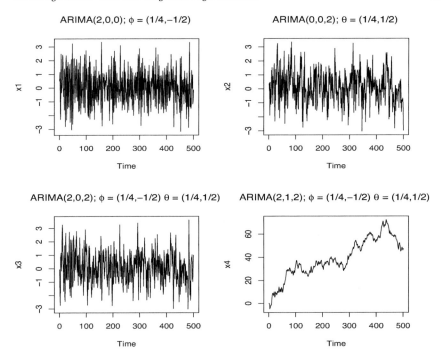

FIGURE 5.13
Simulated ARIMA Data with Gaussian Innovations

Determining the AR and MA Orders

In a pure AR process, the PACF indicates the order, and in a pure MA process, the ACF indicates the MA order. In an ARMA process, neither the ACF nor the PACF gives clear signals of the order.

Another approach is to fit ARMA models of different orders and choose the orders as the ones that yield the "best" fit. This is a problem in model selection, which we discussed in general in Section 4.6.3 on page 467. We will briefly discuss selection of ARIMA models in Section 5.3.7.

In the next few sections we will assume that the orders p and q are known and fixed.

Fitting ARMA Models

Even for fixed p and q, the problem of fitting an ARMA model is computationally difficult, and we will not address the computational details here.

There are several approaches to estimation based on different criteria, as we discussed in Section 4.2.2 on page 349. The three most common criteria for statistical inference in time series are methods of moments, maximum likelihood, and least squares. Each criterion has advantages and limitations.

Although the statistical properties of the various estimators depend on the underlying probability distribution (the distribution of w_t), the only one of these criteria that is based on a probability distribution is maximum likelihood, and for that, we often assume a normal distribution. More realistic distributions that have heavier tails present problems in formulation of the model and in computation of the MLEs.

Before fitting a model to the data, we should plot the data and compute some preliminary statistics from it, such as the ACF and PACF. A visual assessment may indicate some properties of the process that must be dealt with, such as a trend or seasonality.

One approach to estimation of the parameters of an ARMA model is to consider the two parts separately. This may be computationally simpler, and it may also help in deciding on an appropriate ARMA model. We have already indicated one method to estimate the AR parameters, based on the Yule-Walker equations. Substitution of the estimated AR parameters into the model yields a pure MA model, and then the MA parameters can be estimated conditionally.

Method of Moments Estimation in AR Models

The estimation of the autocovariance function or the ACF and the PACF is based on the method of moments; the estimators are just the sample versions. In the case of a pure AR model, there are direct linear relations among the autocovariance function and the coefficients of the AR model. The relations are the Yule-Walker equations, which we developed on page 542, and illustrated for an AR(2) model in equations (5.119) and (5.120).

For an AR(p) model, there are p equations that we can write in matrix notation as $\Gamma_p \phi = \gamma_p$ (equation (5.117)), plus the equation $\sigma_w^2 = \gamma(0) - \phi^{\mathrm{T}} \gamma_p$ (from equation (5.118)).

Given the sample autocovariance function $\hat{\gamma}_p$ for a given set of observations on the time series, we can form $\widehat{\Gamma}_p$ and substitute into the Yule-Walker equations. We then solve the Yule-Walker equations to get our estimators of the coefficients,

$$\hat{\phi} = \widehat{\Gamma}_p^{-1} \hat{\gamma}_p, \tag{5.148}$$

and of the variance of the white noise,

$$\hat{\sigma}_w^2 = \hat{\gamma}(0) - \hat{\phi}^{\mathrm{T}} \hat{\gamma}_p. \tag{5.149}$$

Instead of using the sample autocovariance function in the Yule-Walker equations, we usually use the sample ACF $\hat{\rho}_p$ and the sample autocorrelation matrix, which is equivalent.

These Yule-Walker estimators have useful properties. They are consistent estimators (see equation (4.18)). Not only that, but $\hat{\phi}$ has a simple asymptotic normal distribution:

$$\sqrt{n}(\hat{\phi} - \phi) \overset{\mathrm{d}}{\to} \mathrm{N}_p(0, \sigma_w^2 \Gamma_p^{-1}). \tag{5.150}$$

Maximum Likelihood Estimation

We discussed the likelihood function and maximum likelihood estimation (MLE) beginning on page 352. The first step in maximum likelihood estimation is formation of a likelihood function, which, as we have emphasized, requires a probability distribution. The selection of a probability distribution is often overlooked in MLE in time series analysis because the analyst immediately proceeds with formulas that result from the normal distribution.

The definition of the models requires white noise, which allows almost any distribution so long as the mean is 0. The sequential realizations are not even required to be independent so long as their correlations are 0.

Realistically, we can assume the distribution to be symmetric and unimodal. The normal distribution is symmetric and unimodal. Furthermore, under some simple assumptions, aggregated realizations from other distributions have an asymptotic normal distribution. Also the normal distribution has the advantage of being easy to work with, both mathematically and computationally.

The main disadvantage of the normal distribution is the fact that its tails are lighter than those of many frequency distributions encountered in financial applications. The problem of light tails is compounded by the fact that MLE in a normal distribution generally involves minimizing a sum of squares. The squared elements magnify large deviations resulting from observations in the tails of the distribution.

Despite the obvious problems with a normal distribution, in many ways it is a satisfactory approximation, and it is the one used in the usual approaches to MLE in ARMA models. We therefore assume that the white noise process $\{w_t\}$ is Gaussian.

In a simple model with $Y \sim \mathrm{N}(\mu, \sigma^2)$ and data $y_1, \ldots, y_n$ iid as $\mathrm{N}(\mu, \sigma^2)$, recall that the likelihood function is

$$L(\mu, \sigma^2 \; ; \; y_1, \ldots, y_n) = \frac{1}{(\sqrt{2\pi\sigma^2})^n} \mathrm{e}^{-\sum_{i=1}^{n}(y_i-\mu)^2/2\sigma^2}. \tag{5.151}$$

(This is equation (4.29) on page 353. Compare also the log-likelihood function for a linear regression model with errors normally distributed, equation (4.68) on page 374.)

The likelihood function for an ARMA has the general form as in equation (5.151), but it is quite a bit more complicated, and we will not write it out here. The parameters in the ARMA model, which are the variables in the likelihood function are the p ϕs, the q θs, and σ_w.

Maximizing the likelihood for the ARMA model with normal innovations is computationally difficult, but good algorithms to do so are available. The values at the maximum are the MLEs, and the second derivatives at the maximum provide estimates of the variances of the MLEs. The algorithm for optimizing the likelihood is generally a quasi-Newton method (see page 361). It is important to have a good starting point for Newton's method in this case, and there are various methods for getting preliminary estimates to begin the iterative optimization process. We will not consider the computational details here.

Inference in ARIMA Models

An ARIMA model is addressed in two stages. The first stage is a differencing to get to an ARMA model. The number of differencing operations is usually only one or two, and is often decided upon based on how well the differenced series seems to fit an ARMA model. We discuss the issue of model assessment and selection in Section 5.3.7 below.

Residuals

After fitting any statistical model, we should examine the residuals. The residuals do not have the same probability distribution as the errors, but they do correspond to the errors and the frequency distribution should be similar to that of the error probability distribution. Histograms or q-q plots may be useful to assess their distribution. While the distribution of the errors is likely to have heavier tails than a normal distribution, much of the theory underlying inference, including often the MLE fitting, is based on the assumption of a normal distribution of the errors. A q-q plot of the residuals with a normal reference distribution may indicate whether these approximations are suitable.

Another major issue in residuals in a fitted time series model is whether there are patterns in the residuals. An assumption in ARMA models is that the errors have 0 autocorrelations at all lags; hence, an ACF or PACF plot of the residuals may be useful to assess how well the model fits the data.

5.3.7   Selection of Orders in ARIMA Models

The selection of the orders in ARIMA models is difficult. There is very little obvious difference in time series following different ARMA models, as we see in the first three plots in Figure 5.13. A general practical rule is to choose small values for the orders. One reason for this is the general assumption that effects of observations at greater lags are less than those at shorter lags. Another reason is that models of greater complexity introduce more noise in the process of fitting the models to data, as discussed in Section 4.6.3.

One way of choosing a statistical model is to fit several different models to the given dataset, compute a measure of goodness-of-fit for each (say, based on residuals), determine some measure of the complexity of each, and then choose the model that yields the best combined measure. The AIC and BIC measures, for example, are useful criteria, among others that we discussed in Chapter 4.

5.3.8 Forecasting in ARIMA Models

One of the most important uses of time series analysis is to forecast, or make predictions of future values. We introduced some notation and concepts, such as forecast origin and horizon, on page 527.

For a model fitted using data up to time t, we denote the estimates at subsequent times as $\hat{x}_t^{(1)}, \hat{x}_t^{(2)}, \ldots$. The *h-step ahead forecast* is $\hat{x}_t^{(h)}$. (Other authors may use different notation.)

There are two aspects of a forecast: the time at which it is made, the forecast origin, and the point in future time that the forecast applies to, the forecast horizon. At time t, the "best conditional predictor" for x_{t+h}, in the sense of minimum mean squared error, is the conditional expected value of x_{t+h}, given $x_t, x_{t-1}, x_{t-2}, \ldots$, that is,

$$\hat{x}_t^{(h)} = \mathrm{E}(x_{t+h}|x_t, x_{t-1}, \ldots). \tag{5.152}$$

It is based on the available information at time t, which we sometimes denote as $\mathcal{F}_t$, as in $\mathrm{E}(x_{t+h}|\mathcal{F}_t)$.

Given a model and linearly-unbiased estimators of its parameters, the forecast $\hat{x}_t^{(h)}$ can be computed by substitution into the fitted model, as in equations (5.75) and (5.76) on page 527. These substitutions generally must be done recursively.

The forecasts are random variables. Their distributions are rather complicated because they depend on the distributions of the parameter estimators. We use normal approximations to estimate their standard deviations. The standard deviations increase as the time horizon moves out, because the forecasts (after the one-step forecast) depend on previous forecasts.

We will not go into the details of the ARIMA forecasts here; rather, we will illustrate them in an examples using R below.

5.3.9 Analysis of ARMA and ARIMA Models in R

There are many R packages for time series analysis. The packages are frequently updated, and more packages are being developed. Some R packages may use slightly different definitions of the models (for example, in the signs of the coefficients, as we have mentioned).

In this chapter, we will generally use functions from the basic stats package or else just from a small set of widely-used packages for time series, primarily tseries and forecast. The forecast package includes several functions that duplicate the functionality of functions in the stats package, but provide a better interface or cleaner output, such as Acf for acf, Arima for arima and forecast for the generic predict. The forecast package also includes some useful functions not in the regular stats package, such as auto.arima (see Hyndman and Khandakar 2008). There are also some other R packages that we will use in later sections.

Fitting of ARMA and ARIMA Models in R

The R function `arima` in the `stats` package computes estimates of the parameters in a seasonal or non-seasonal ARMA or ARIMA model. The model is specified in the `order` argument, which is a 3-vector of the form of (p, d, q) in the standard notation ARIMA(p, d, q). The model can include a mean or not by the use of the keyword argument `include.mean`. Various computational methods can be chosen.

The function produces a list that includes the coefficient estimates, estimates of their variance-covariance matrix, residuals, and other computed statistics. The list is of class `Arima` and the names of the list components are similar to the names of the components of the list (of class `lm`) produced by the R functions for analysis of linear models discussed in Chapter 4. The functions in the `forecast` package produce a list of class `ARIMA`, which inherits from `Arima`.

As an example, consider the data in the four simulated time series plotted in Figure 5.13. Figure 5.14 shows the code for fitting ARIMA models to those data sets. The output is shown with some lines omitted. The parameter values used in generating the data are shown in comments labeled "true".

First, we note that in each case, we specified the "correct" values of p, d, and q. We also note that, except for the differenced model, the estimates include an intercept term. Although, we know the data were generated without an intercept, it would not be appropriate to include the argument `include.mean=FALSE` because we are analyzing data with "unknown" properties.

Comparing the estimates with the "true" values, we see that the estimates are fairly good. Most differ from the value used in generating the data by less than one standard deviation.

In Section 5.3.10 we will return to this example, but with models whose innovations follow a t distribution.

Residuals

After fitting a statistical model, we should inspect the residuals. The residuals from fitting an ARIMA model using the R function `arima` are available in the output component `$resid`. We may inspect the residuals for outliers or their frequency distribution, either in a histogram of a q-q plot. As an example, for the x4 time series in Figures 5.13 and 5.14, we execute

```
x4fit <- arima(x4, order=c(2,1,2))
qqnorm(x4fit$residuals)
```

and get the plot shown in Figure 5.15.

```
> arima(x1, order=c(2,0,0))
          ar1        ar2  intercept
       0.2504    -0.5264     0.0537
s.e.   0.0380     0.0379     0.0354
# true 0.2500    -0.5000     0

> arima(x2, order=c(0,0,2))
          ma1       ma2   intercept
       0.2701    0.4727      0.1279
s.e.   0.0405    0.0401      0.0775
# true 0.2500    0.5000      0

> arima(x3, order=c(2,0,2))
          ar1        ar2       ma1       ma2  intercept
       0.1458    -0.5253    0.3516    0.5729     0.0967
s.e.   0.1115     0.0672    0.1076    0.0721     0.0629
# true 0.2500    -0.5000    0.2500    0.5000     0

> arima(x4, order=c(2,1,2))
          ar1        ar2       ma1       ma2
       0.1430    -0.5200    0.3573    0.5731
s.e.   0.1107     0.0677    0.1065    0.0717
# true 0.2500    -0.5000    0.2500    0.5000
```

FIGURE 5.14
Fitted ARIMA Models for the Time Series Shown in Figure 5.13

The residuals seem to follow a normal distribution model, indicating that the fit is probably good.

Selection of Orders in ARIMA Models

The auto.arima function in the forecast package fits ARIMA models of various orders and chooses the "best" one based on an AIC or the BIC. The function applies to both seasonal and nonseasonal models, and allows differencing both seasonally and nonseasonally. It also has several control parameters. A simple application of the function using the x4 time series in Figure 5.13 is shown below. We know the "true" model was ARIMA(2,1,2) (at least, that was the model used to generate the data). We limit the search to maximum orders of 5 for each of the AR and MA components.

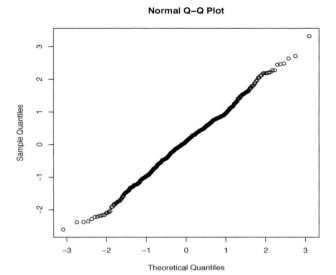

Normal Q–Q Plot

FIGURE 5.15
Residuals from Fitting an ARIMA Model

```
> library(forecast)
> auto.arima(x4, max.p = 5, max.q = 5, ic="bic")
Series: x4
ARIMA(2,1,2)
Coefficients:
            ar1       ar2       ma1       ma2
        0.1430   -0.5200    0.3573    0.5731
s.e.    0.1107    0.0677    0.1065    0.0717
sigma^2 estimated as 1.031:  log likelihood=-715.47
AIC=1440.93    AICc=1441.05    BIC=1462.01
```

In this case, the auto.arima function use of the BIC "correctly" identified
the model as ARIMA(2,1,2), and so of course it yields the same estimates as
computed previously.

In the example above, if the AIC or the AICc is used on the x4 dataset
with the same restriction on the orders, an ARIMA(3,1,5) model is chosen.
Recall that BIC occasionally selects a smaller or simpler model than AIC
(Section 4.6.3; also see Exercise 5.25b).

There are other methods of selecting the orders in ARIMA models, such as
bootstrap methods or cross-validation, as discussed in Section 4.6. The caveats

in that section of course apply to the selection of the orders in an ARIMA model as well.

Forecasting in ARIMA Models

Forecasts in ARIMA models can be computed in R using the object produced by the function used to fit the model. For example, the `arima` function produces an object of class `Arima`. If an `Arima` object is passed to the generic R function `predict`, it will produce forecast values for the time series model. (Recall that on page 448, we used the generic `predict` function to estimate observations in a linear regression model using an `lm` object.)

The `forecast` package also includes a function `forecast`, which can accept an `Arima` object from the usual functions or an `ARIMA` object from the functions in the `forecast` package. The `forecast` function computes 80% and 95% confidence intervals for the predictions, based on a normal approximation. Notice that the forecast value itself does not depend on an assumed probability distribution, but a confidence interval for the forecast does.

As an example of forecasting, consider the simulated data from the ARIMA(2,1,2) model shown in the lower right panel of Figure 5.13. (The data were generated by the R code shown on page 556. It is the x4 series.)

We will use the first 400 observations in the series to fit an ARIMA model, with the "correct" orders p, d, and q. (We recall for this series that `arima.auto` selected the "correct" orders using the BIC.) Then we will use the fit object in the R `predict` function to forecast 5 observations ahead. We then compare the forecast values to the next 5 actual values in the simulated series. The `round` function is used to facilitate comparisons. Some lines of the output are omitted. (We note that the estimates for the ARMA parameters are slightly different from the estimates based on 500 observations on page 564, and both sets of estimates differ slightly from the "true" values, $(0.25, -0.50, 0.25, 0.50)$.)

```
> x4fit400 <- arima(x4[1:400], order=c(2,1,2))
> x4fit400$coef
       ar1          ar2          ma1          ma2
 0.1815580 -0.5447515  0.3077949  0.5449485
> x4pred <- predict(x4fit400, n.ahead=5)
> round(x4pred$pred, 4)   # forecasts
[1] 54.7942 54.6500 54.2055 54.2033 54.4451
> round(x4pred$se, 5)   # standard errors
[1] 1.01363 1.81838 2.42203 2.77096 3.04020
> round(x4[401:405], 4)   # "true" values
[1] 55.1134 54.6791 54.2400 55.3459 55.8225
```

These predictions seem fairly good. The "true" values are all within one standard error of the predicted values.

Notice that the standard errors increase as the forecast horizon moves farther out.

Before we convince ourselves that forecasting in ARIMA models works well, we might consider another one of the series generated on page 556, the ARMA(2,2) series, x3. This series was also "correctly" identified by arima.auto using the BIC, and the MLEs for the coefficients were the same as those for x4 above. (Again, the round function is used to facilitate comparisons, and some lines of the output are omitted.)

```
> x3fit400 <- arima(x3[1:400], order=c(2,0,2))
> x3pred <- predict(x3fit400, n.ahead=5)
> round(x3pred$pred, 4)   # forecasts
[1]   0.1167 -0.2761   0.0695   0.3573   0.2212
> round(x3pred$se, 5)   # standard errors
[1] 1.00724 1.11842 1.12122 1.15005 1.15395
> round(x3[401:405], 4)   # "true" values
[1] -0.4343 -0.4391   1.1059   0.4766   1.4046
```

In a relative sense, these predictions do not seem nearly as good; they miss the "true" value by more than 100%. The problem, however, is the relatively large variance of the estimators and consequently of the forecasts. The standard errors are so large that in most cases, the forecast values are within one standard error of the true values.

Compared to our general experiences with statistical estimators, however, the variances of the coefficient estimators and predictors in ARIMA models often seem unusually large.

5.3.10   Robustness of ARMA Procedures; Innovations with Heavy Tails

In ARIMA models, the basic random variables are the innovations, the w_t. All randomness in the models derive from their randomness. The only restrictions on the distributions of the innovations have been that they are white noise. Some procedures of statistical inference, however, have assumed a normal distribution for the innovations. On the other hand, throughout our exploration of financial data, we have noticed heavy tails in the probability distributions. Notice that the simple assumption of white noise includes not only the normal distribution but also heavy-tailed distributions.

Let us consider, for example, ARIMA models whose innovations have a t distribution, still with constant mean and variance and with 0 autocorrelations. Specifically, let us consider a variation on the simulated data from ARIMA models on page 556. We will use an AR(2) model, an MA(2) model, an ARMA(2,2) model, and an ARIMA(2,1,2) model as before, with the same coefficients, which made the models causal and invertible. In this case, however, the innovations are generated from a t distribution with 5 degrees of freedom. As before, we use `arima.sim` and reset using `set.seed(12345)` prior to generation of each series.

```
n <- 500
xt1 <- arima.sim(n, model=list(ar=c(1/4,-1/2),
                 order=c(2,0,0)), rand.gen=rt,df=5)
xt2 <- arima.sim(n, model=list(ma=c(1/4,1/2),
                 order=c(0,0,2)), rand.gen=rt,df=5)
xt3 <- arima.sim(n, model=list(ar=c(1/4,-1/2),ma=c(1/4,1/2),
                 order=c(2,0,2)), rand.gen=rt,df=5)
xt4 <- arima.sim(n, model=list(ar=c(1/4,-1/2),ma=c(1/4,1/2),
                 order=c(2,1,2)), rand.gen=rt,df=5)
```

Although we do not show them here, plots of these series would look somewhat similar to the plots in Figure 5.13, except that they would show more variability, including, perhaps, some outliers. (You are asked to produce the plots in Exercise 5.26a.)

Now, as on page 563, we use the R function `arima` to fit models to these data sets. Figure 5.16 shows the code and the output for computing estimates of the coefficients of the four simulated time series. In each case, we have chosen the "correct" values of p, d, and q. The parameter values used in generating the data are shown in comments labeled "true". Some lines in the output are omitted.

The estimates computed by `arima` are maximum likelihood estimates, which means among other things, that a specific family of probability distributions is assumed at the outset. (Recall that, while the properties of the resulting estimators depend on the underlying distribution, in the case of method of moments or least squares, no assumed underlying distribution is required.) The distribution assumed by default for MLE in `arima` is the normal distribution.

Standard Errors

Comparing the estimates with the "true" values or comparing these estimates with the estimates of the same model parameters with a normal distribution as on page 563, we see that these estimates are not as good. Several are not within one standard error of the true value. In this case, the standard error is not a good estimator of the standard deviation of the parameter estimators. The distribution of these estimators depend on the underlying t distribution, and they are affected by the heavier tails of the t distribution. (See page 377 for general discussions of "standard errors".)

```
> arima(xt1, order=c(2,0,0))
          ar1       ar2  intercept
       0.2324   -0.4821     0.0745
s.e.   0.0392    0.0391     0.0451
# true 0.2500   -0.5000          0

> arima(xt2, order=c(0,0,2))
          ma1       ma2  intercept
       0.2170    0.5147     0.1752
s.e.   0.0376    0.0398     0.0977
# true 0.2500    0.5000          0

> arima(xt3, order=c(2,0,2))
          ar1       ar2       ma1       ma2  intercept
       0.5241   -0.6115   -0.0710    0.4546     0.1267
s.e.   0.1184    0.0639    0.1278    0.0774     0.0713
# true 0.2500   -0.5000    0.2500    0.5000          0

> arima(xt4, order=c(2,1,2))
          ar1       ar2       ma1       ma2
       0.5112   -0.6039   -0.0533    0.4586
s.e.   0.1194    0.0652    0.1282    0.0765
# true 0.2500   -0.5000    0.2500    0.5000
```

FIGURE 5.16
Fitted ARIMA Models for Time Series Whose Innovations Follow a t Distribution

Lack of robustness means that many standard procedures in time series analysis may not be very useful in financial applications.

In Exercise 5.26d, you are asked to analyze residuals from a fitted ARIMA(2,1,2) model in which the innovations follow a t with 5 df as above.

5.3.11   Financial Data

For the type of financial data that we have studied most often in this book, namely stock prices and returns, ARMA and ARIMA models do not work very well, although they are often used for that kind of data. Daily data is just too noisy, and data at lower frequencies that are based on longer time intervals are subject to structural changes in the individual security or in the economy that occur over time. Other types of financial data that are less subject to noise, such as GDP data for example, sometimes can be usefully modeled as ARMA or ARIMA processes.

The numeraire values in much economic data naturally become larger over time. GDP increases over time because of, among other things, population increases. GDP per capita increases over time because of, among other things, inflation. Inflation-adjusted GDP per capita increases over time because of, among other things, efficiency of production and inefficiency (in the statistical sense) of inflation adjustments. Processes with trends such as these do not follow an ARMA model, although sometimes trends can be accounted for in an ARIMA model.

Despite the obvious concerns about stationarity, let us consider ARIMA modeling of the logs of the daily closes of the S&P 500 Index. The differences in the logs at any two times is just the log return over that interval.

The daily closes of the S&P 500 Index is an interesting time series. We displayed it for the thirty-one-year period from 1987 through 2017 in Figure 1.10 on page 51, and again in a log scale in Figure 1.13. Over that period the Index generally increased, and so clearly it does not follow a stationary time series.

Using the `auto.arima` function in the `forecast` package to select an ARIMA model for the logs of the daily closes in the dataset `GSPCdlog`, we get an ARIMA(1,1,1) model.

```
> auto.arima(GSPCdlog, ic="bic")
ARIMA(1,1,1)
Coefficients:
          ar1       ma1
       0.6506   -0.7037
s.e.   0.0763    0.0712
```

Note that the first order differencing in the ARIMA(1,1,1) model yields an ARMA(1,1) model.

Figure 5.17 is a time series plot of the residuals of the actual closes of the log of the Index from the fitted ARIMA(1,1,1) model.

The residuals in Figure 5.17 show that the trend in the log series has been removed by the differences. The plot, however, clearly indicates that the process is not stationary. The volatility of the residuals shows clusters of high volatility and others of relatively low volatility. This clustering of volatility is similar to that of the volatility of the S&P 500 log returns evident in Figure 1.29 on page 99. (The graphs appear almost identical, but they are not.) The patterns in Figure 5.17 indicate that the ARMA(1,1) portion of the model does not fit well.

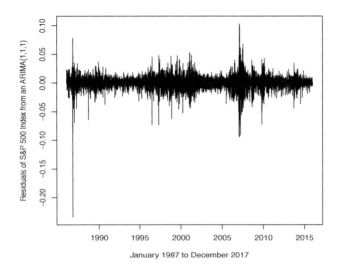

FIGURE 5.17
Residuals of Daily Log Closes of S&P 500 Index from a Fitted ARIMA(1,1,1) Model

Use of the fitted ARIMA model to forecast the S&P 500 closes for the first few days following December 31, 2017, yields useless results. For example, using the `forecast` function for the log of the close on the first trading day of 2018, we get a forecast value of 7.891725 with an 80% confidence interval of (7.88, 7.91):

```
> forecast(GSPCdfit,h=1)
     Point Forecast    Lo 80     Hi 80     Lo 95     Hi 95
7815       7.891725 7.877063 7.906386 7.869301 7.914148
```

The actual close on January 2, 2018, was 2696, compared to the forecast value of 2675 ($e^{7.891725}$). The mismatch of forecasts and actual values becomes even worse during the next 13 days of January. We might mention, however, that the market was on a bull run until January 22, 2018, at which point it took a precipitous drop until February 5. During the drop it began to match the forecast values, and continued on until on February 5, it closed below the lower 80% confidence limit.

The point of this example is not just to emphasize that the distribution of the actual S&P 500 log closes has a much heavier tail than a normal distribution (although that is the case). While ARIMA models may be useful in studying the structure of the process, they are not very useful for forecasts of stock prices or of stock indexes.

5.3.12 Linear Regression with ARMA Errors

The linear regression model, equation (4.106),

$$y_i = \beta_0 + \beta_1 x_{i1} + \cdots + \beta_m x_{im} + \epsilon_i, \tag{5.153}$$

discussed extensively in Chapter 4, is one of the most useful models for statistical analysis. It can be helpful in understanding relationships among various financial variables. It can also be used to predict values of the dependent variable y for specified values of the regressors or predictors, the xs.

We assume that the "errors", the ϵs, are uncorrelated random variables with $E(\epsilon_i) = 0$. Under this assumption, ordinary least squares, OLS, is generally the best way to fit the model.

In economic and financial applications, each observation $(y_i, x_{i1}, \ldots, x_{im})$ is just the observed or computed values of the different variables at a specific time. The errors in linear regression models involving economic data often fail to satisfy the common assumption of 0 correlation, however, because the changes in one variable from one time to the next is related to the changes in another variable between those two times.

We can use the Durbin-Watson test (see page 434) to decide if the assumption that $\text{Cor}(\epsilon_i, \epsilon_{i-h}) = 0$ for small values of h is reasonable. If the Durbin-Watson test suggests rejection of the null hypothesis of zero serial correlations, we conclude that the regression model errors are serially correlated, and so the variance-covariance matrix of the random vector of errors, $V(\epsilon)$, is not the diagonal matrix $\sigma^2 I$ (as in equation (4.108)). Instead, we have the more general matrix $V(\epsilon) = \Sigma$, as on page 368.

Generalized Least Squares

If the variance-covariance Σ is known, we use the weighted least squares methodology (GLS) as described in Chapter 4 instead of OLS, and all results remain valid. The problem here is that we do not know Σ.

There are many possible structures of the correlations among the ϵs. We may reasonably assume, however, that the only nonzero correlations among the ϵs are for nearby observations; that is, $\text{Cor}(\epsilon_i, \epsilon_{i-h})$ may be nonzero for h relatively small, but it is zero for large h. If the ϵs are stationary and we can estimate the correlations, then we can estimate the autocovariances and from those estimates form an estimate of Σ.

We could estimate Σ directly, which may obscure the relationships, or we may assume that the errors follow an ARMA process, fit that process, and use the resulting correlation estimates from the coefficients of the ARMA process.

Modeling the Dependent Variable as an ARMA Process

Another approach to accounting for serial correlations in the errors is to include ARMA terms in the dependent variable directly in the model. We write the model as

$$y_i = \beta_0 + \phi_1 y_{i-1} + \cdots + \phi_p y_{i-p} + \theta_1 \epsilon_{i-1} + \cdots + \theta_q \epsilon_{i-q} + \beta_1 x_{i1} + \cdots + \beta_m x_{im} + \epsilon_i.$$

(5.154)

This model includes the ordinary regressors as in (5.153), and accounts for serial correlations through the lagged variables.

Example

In Section 4.5.6 beginning on page 435, we discussed an example of multiple linear regression and illustrated many of the relevant techniques in analysis. We plotted the residuals in Figure 4.12 and, aside from some apparent clustering of large values, there were no visual anomalies. A Durbin-Watson test on the residuals of the fitted model, however, indicated presence of autocorrelations. The results (from page 443) were

```
lag Autocorrelation D-W Statistic p-value
 1       0.52982469      0.9385733    0.000
 2       0.27848495      1.4297074    0.000
 3       0.09229081      1.7996013    0.272
Alternative hypothesis: rho[lag] != 0
```

indicating that both the lag 1 and lag 2 serial correlations are nonzero. We compute the ACF of the residuals and it also indicates significant autocorrelations; see Figure 5.18.

Example: Estimating the ARIMA Error Structure

Let us now see what kind of low-order ARIMA model fits the residuals well. We use auto.arima and the BIC criterion. (The original OLS fit of the regression model was stored in **regr**, and so the residuals are in **regr$residuals**. Also see page 435 for a description of the dataset, variable names, and so on.)

```
> library(forecast)
> arfit <- auto.arima(regr$residuals, ic="bic")
> arfit
Series: regr$residuals
ARIMA(2,0,1) with zero mean
```

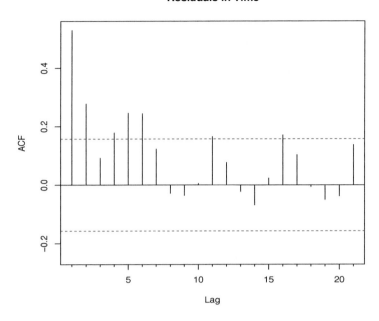

FIGURE 5.18
ACF of Residuals of Corporate Bonds and Treasuries

```
Coefficients:
          ar1      ar2      ma1
      -0.1857   0.4197   0.6848
s.e.   0.2398   0.1309   0.2391

sigma^2 estimated as 0.0007436:  log likelihood=339.71
AIC=-671.43    AICc=-671.16   BIC=-659.26
```

Using this fitted model, we have an estimate of the variance-covariance of the errors in the regression model.

Example: Fitting by GLS

We can now redo the regression using generalized least squares. The R function gls in the nlme package referred to earlier allows us to specify the variance-covariance (actually, the correlations) and then perform the GLS computations. The gls function has keyword arguments for various correlation structures. The correlations in this example have an ARMA structure, corARMA.

```
> library(nlme)
> glsarfit <- gls(dBaa ~ dfed + d3m + d2y + d30y, data=YX,
+            correlation = corARMA(arfit$coef[1:3],form=~1,2,1))
> glsarfit$coef
(Intercept)           dfed             d3m               d2y             d30y
-0.00258232 -0.01919563   0.04414342 -0.13200955   0.93614909
```

The coefficients in the OLS fit (see page 439) are

```
> regr$coef
 (Intercept)             dfed               d3m               d2y              d30y
-0.003300489   0.001432205   0.128557641 -0.146626730   0.903769166
```

There are major differences in the coefficients of dfed and d3m, but only slight differences in the coefficients of d2y and d30y. As we saw in the OLS analysis, however, neither dfed nor d3m were significantly different from 0. Likewise, they are not significantly different from 0 in the GLS fit:

```
> summary(glsarfit)
Coefficients:
                  Value  Std.Error    t-value p-value
(Intercept) -0.0025823 0.00509587 -0.506748  0.6131
dfed           -0.0191956 0.03874977 -0.495374  0.6211
d3m             0.0441434 0.08001335  0.551701  0.5820
d2y            -0.1320095 0.05757529 -2.292816  0.0232
d30y            0.9361491 0.03713014 25.212646  0.0000
```

(The output was edited for compactness.)

The GLS fit is to be preferred, however, since the data come closer to satisfying the assumptions of the model with correlated errors.

Example: Modeling the Dependent Variable as an ARMA Process with Covariates

For a given order (p, d, q), the R function arima implements the method suggested in equation (5.154). The dependent variable is specified as the time series and the regressors are specified in the xreg argument. Of course arima requires specification of the form of the model, and for the example above, we used ARIMA(2,0,1) because that is what was chosen by the BIC criterion as the ARIMA model for the residuals from an OLS fit. We get the following.

```
> arima(dBaa, order=c(2,0,1), xreg=cbind(dfed,d3m,d2y,d30y))

Call:
arima(x=dBaa, order=c(2,0,1), xreg = cbind(dfed,d3m,d2y,d30y))

Coefficients:
         ar1     ar2    ma1  intercept    dfed    d3m      d2y    d30y
     -0.1473  0.4186  0.6689   -0.0026  -0.0193  0.0448  -0.1323  0.9362
s.e.  0.2629  0.1474  0.2621    0.0050   0.0388  0.0795   0.0574  0.0372

sigma^2 estimated as 0.0007158: loglikelihood=341.14, aic=-664.29
```

Note that the regression coefficient estimates are the same (except for rounding) as in the GLS fit using the estimated variances from the fitted ARIMA(2,0,1) model.

In this latter approach, a linear regression model with ARMA errors is formulated as an ARMA process in the dependent variable with the regressors as covariate time series processes ("exogenous" variables, in economics parlance). The ARMA time series model with the covariates is sometimes called an ARMAX process (the "X" representing regressors).

5.4   Conditional Heteroscedasticity

We have observed that ARMA (or ARIMA) models do not work very well on some important financial time series. For the log S&P 500 Index closes analyzed in Section 5.3.11, for example, the values forecast by a fitted ARIMA model were generally not close to the actual values. One problem, of course, is the fact that the distributions have heavy tails. This affects estimates of variances of estimated or fitted parameters. Usually the actual variance is larger than the estimated variance. This means, among other things, that computed confidence intervals are not as wide as they should be. Forecast values, while perhaps in the right direction, and the confidence intervals associated with the forecasts, do not reflect the underlying variability in the process.

A major problem with fitting ARIMA models is that the variances of the distributions in financial models vary with time. A key property of stationarity is not present. The classic time series plot of the S&P 500 returns between 1987 and 2017 shown in Figure 1.29 on page 99 illustrates this vividly for volatility of returns; the volatility itself is volatile, or stochastic. This is also obvious in Figure 5.17 showing the residuals of the actual log S&P 500 closes from an inappropriate stationary model fitted to a time series of the closes.

Figure 1.29 indicates a pattern in the randomness of the volatility. Its variation occurs in clusters. The volatility at one point seems to be correlated with the volatility at nearby points in time. We call this *conditional heteroscedasticity*.

In this section, we will discuss ways of incorporating conditional heteroscedasticity into the time series model. As we have emphasized, we must assume some level of constancy in order to apply statistical methods. Even in the presence of conditional heteroscedasticity, we will assume some aspects are unconditionally constant.

5.4.1   ARCH Models

The general time series model with additive error term, as in equation (5.2), is

$$x_t = f(x_{t-1}, x_{t-2}, \dots ; \theta) + w_t, \tag{5.155}$$

where f is some function, θ is some vector of constant, possibly unknown, parameters, and w_t is a white noise random error term. Since $\mathrm{E}(w_t) = 0$, this is a model for the conditional mean of x_t.

A key aspect of the model is the assumption that w_t is a white noise, and so the variances $\mathrm{V}(w_t)$ are equal for all t (in the time domain of consideration).

The violation of this assumption is the focus of this section.

Models with Heteroscedastic Innovations

The time series models we have considered so far have all been variations on (5.155), that is, they are models of conditional means, with simple additive error terms. To preserve our notation of w_t to represent white noise, we now write the basic model as

$$x_t = f(x_{t-1}, x_{t-2}, \dots ; \theta) + a_t, \tag{5.156}$$

in which the random error or innovation, a_t, has a distribution that is not necessarily constant, as in the distribution of w_t in equation (5.155).

Our objective is to build an adequate model for x_t. This also involves the distribution of a_t. Our main interest in a_t is its conditional variance, $\mathrm{V}(a_t | a_{t-1}, a_{t-2}, \dots)$, because the conditional variance of x_t is directly related to it.

The model for a_t, as with most statistical models, has a systematic and a random component, and since we are interested in the variance of a_t rather than its mean, a model with a multiplicative random component is appropriate. A general form is

$$a_t = \epsilon_t g(a_{t-1}, a_{t-2}, \dots ; \psi), \tag{5.157}$$

where g is some function with parameter ψ, and ϵ_t is a random error.

The error term a_t is an additive error in the model (5.156) for x_t, and the error term ϵ_t is a multiplicative error in the model (5.157) for a_t. (This form of statistical models was discussed in Section 4.1, specifically in equation (4.4), which had the general form,

$$x_t = f(x_{t-1}, x_{t-2}, \ldots ; \theta) + \epsilon_t g(x_{t-1}, x_{t-2}, \ldots ; \psi). \qquad (5.158)$$

The additive error itself is composed of a systematic component and a purely random factor, ϵ_t.)

In the following, we will assume ϵ_t is strict white noise.

The conditional expectation of a_t is 0,

$$E(a_t | a_{t-1}, a_{t-2}, \ldots) = 0, \qquad (5.159)$$

and so the systematic portion of the model (5.156), $f(x_{t-1}, x_{t-2}, \ldots ; \theta)$, is the conditional expected value of x_t, just as in the time series models with white noise innovations.

Modeling the Heteroscedasticity

The variance of a_t is the expected value of a_t^2. We model a_t^2 as a function of past values and random noise. This is the basic time series problem, for which a general model without an explicit error term is

$$a_t^2 \approx h(a_{t-1}^2, a_{t-2}^2, \ldots ; \psi), \qquad (5.160)$$

for some function h and some vector of constant parameters ψ. (This is in the general form of the time series models we have used throughout this chapter.) In an AR(1) model for the general form (5.160), for example, we would have

$$a_t^2 \approx \psi_0 + \psi_1 a_{t-1}^2.$$

In the general time series models we have used earlier, we added a random error term, usually assumed to be white noise. Here, because of our interest in the variance of a_t, we use a multiplicative error. If we use a model similar to an AR(1) model, but with a multiplicative error, we have

$$a_t^2 = \epsilon_t^2 (\psi_0 + \psi_1 a_{t-1}^2), \qquad (5.161)$$

where we assume a_t and ϵ_t are independent. We also require $\psi_0 > 0$ and $\psi_1 \geq 0$ in order to ensure that the squared values are positive.

Recall that we have restricted ϵ_t to be strict white noise; that is, the ϵ_t are iid and $E(\epsilon_t) = 0$. Furthermore, for simplicity but without loss of generality, let us assume $V(\epsilon_t) = 1$. From equations (5.159) and (5.161) and these assumptions, we have

$$
\begin{aligned}
V(a_t | a_{t-1}, a_{t-2}, \ldots) &= E((\psi_0 + \psi_1 a_{t-1}^2)\epsilon_t^2 | a_{t-1}, a_{t-2}, \ldots) \\
&= (\psi_0 + \psi_1 a_{t-1}^2) E(\epsilon_t^2 | a_{t-1}, a_{t-2}, \ldots) \\
&= \psi_0 + \psi_1 a_{t-1}^2. \qquad (5.162)
\end{aligned}
$$

We note that this expression for conditional heteroscedasticity can account for the volatility clustering in a general time series following a model of the form (5.156). Conditional on the size of a_{t-1}, the variance of a_t may be larger or smaller.

Although the conditional variance is not constant, we assume that the underlying unconditional variance $V(a_t)$ is constant. Because the conditional mean is constant (equation (5.159)), we can take expectations on both sides of equation (5.162) to get the unconditional variance as

$$V(a_t) = \psi_0 + \psi_1 V(a_{t-1}), \tag{5.163}$$

with $V(a_{t-1}) = V(a_t)$, yielding

$$V(a_t) = \psi_0/(1 - \psi_1). \tag{5.164}$$

In order for this to be valid, we impose another restriction on the model; we require $\psi_1 < 1$.

A process following equation (5.161) and the conditions stated is called an *autoregressive conditionally heteroscedastic model of order 1*, or ARCH(1) model.

This kind of heteroscedasticity in the general time series model (5.156) is called an *ARCH effect*.

ARCH(p) Models

The ARCH(1) model can easily be extended to higher order, which for order p, we call ARCH(p). Equation (5.161) is extended as

$$a_t^2 = \epsilon_t^2 \left(\psi_0 + \sum_{i=1}^{p} \psi_i a_{t-i}^2 \right). \tag{5.165}$$

Given $a_{t-i}, \ldots, a_{t-p}$, the term $\psi_0 + \sum_{i=1}^{p} \psi_i a_{t-i}^2$ is the conditional variance of a_t. Denoting the conditional variance as σ_t^2, we can write

$$a_t = \epsilon_t \sigma_t. \tag{5.166}$$

We refer to a model consisting of a part for the underlying time series (the means) and an ARCH model for the variances of the innovations by the descriptors of the two parts. For example, a p_M-order autoregressive process with p_V-order ARCH innovations may is called an AR(p_M)+ARCH(p_V) model. (The subscripts obviously stand for "mean" and "variance".)

An alternative approach to modeling the heteroscedasticity is to transform the time axis; that is, instead of time measured in the usual units t, we measure time in the units of $\tilde{t}$, where $\tilde{t}$ is some nonlinear function of t. The idea is that heteroscedasticity in an ordinary time scale can be made homoscedastic in a transformed time scale. This is the same idea as any of the variance-stabilizing transformations referred to in various places in Chapter 4. One possibility for transforming t is the Box-Cox family of power transformations with the tuning parameter λ (see page 345). The BoxCox.Arima function in the fitAR R package accepts a fitted ARIMA model in an Arima object and computes values of the likelihood corresponding to a range of values of λ. (This is similar to the boxcox function in the MASS package for lm objects we referred to in Chapter 4.) We will not pursue this approach further here.

Neither of these approaches to modeling conditionally heteroscedastic processes adds any meaningful insight to the behavior of the process. They are merely mechanical devices to model the observed behavior of a system with varying conditional variance.

Testing for an ARCH Effect

Nonzero autocorrelations among squared deviations from the mean in a time series is a general indication that the process is heteroscedastic. This is not an exact implication. It is possible to construct a systematic model with nonzero autocorrelations among squared deviations from the mean, but such that the differences in the process from the systematic model have zero autocorrelations. Nevertheless, autocorrelations among the centered and squared x_t in the model (5.156) are useful surrogates for autocorrelations among the a_t.

We can, therefore, test for an ARCH effect in a time series using a Box-Ljung test for autocorrelations in the centered and squared data.

For example, using the daily returns of the S&P 500 for the period 1987 through 2017 (the same data as used in Section 5.3.11), we have the following.

```
> xsqd <- (GSPCdReturns-mean(GSPCdReturns))^2
> Box.test(xsqd, lag=12, type="Ljung")

        Box-Ljung test

data: xsqd
X-squared = 1390.4, df = 12, p-value < 2.2e-16
```

This test strongly indicates nonzero autocorrelations among the squared deviations from the mean, and these in turn indicate nonconstant variances; that is, an ARCH effect.

As pointed out in Chapter 1, the ACF of the absolute values of the returns, as shown in Figure 1.31, is also an indication that the volatility has positive autocorrelations.

Following the same idea as above of using residuals from a simple mean fit, it might be preferable to use residuals from an ARIMA fit. Using the daily returns data above, we get the same fitted ARIMA(1,1,1) model as on page 569. We then compute the residuals, and perform the Box-Ljung test for autocorrelations in the squared residuals.

```
> resd<-auto.arima(GSPCdReturns)$residuals
> Box.test(resd^2, lag=12, type="Ljung")

        Box-Ljung test

data:  resd^2
X-squared = 1181.7, df = 12, p-value < 2.2e-16
```

This test is similar to the one above on the squared centered data, and it also strongly indicates nonzero autocorrelations among the squared residuals from a fitted ARIMA model. Again, the inference is that an ARCH effect exists in the daily log returns.

   We should note that the Box-Ljung tests are based on an assumption of stationarity, so confidence levels associated with the test here are only approximate. Very large values of the test statistic, nevertheless, indicate nonconstant variances. There are other ways of formally testing for ARCH effects, but we will not describe them here. Formal statistical hypothesis tests rarely add to our understanding.

5.4.2   GARCH Models and Extensions

An ARCH(p) model often requires a large value of p to model the volatility process in asset returns adequately. ("Large" here means 9 or 10.) More effective models can be formed by augmenting the AR components of the a_{t-i}^2 in the model (5.165) with MA-like components.

   In the same basic model as in the ARCH(p) process,

$$a_t = \epsilon_t \sigma_t, \tag{5.167}$$

where ϵ_t is strict white noise, we express σ_t^2 as

$$\sigma_t^2 = \psi_0 + \sum_{i=1}^{p} \psi_i a_{t-i}^2 + \sum_{i=1}^{q} \beta_i \sigma_{t-i}^2. \tag{5.168}$$

This is called a *generalized* ARCH process, or GARCH(p, q) process.

   This model for the innovations a_t in a basic model such as equation (5.156), allows the variance of x_t, the variable of interest, to vary.

   We also note that the restriction on ϵ_t to be strict white noise allows both normal distributions and heavy-tailed distributions. (There are technical details regarding the assumption of iid ϵ_t that we omit here.)

$\mathbf{ARIMA}(p_\mathrm{M}, d, q_\mathrm{M}) + \mathbf{GARCH}(p_\mathrm{V}, q_\mathrm{V})$ Models

We now return to consideration of the time series model for the variable of interest, whose general form is equation (5.156). The GARCH(p, q) process is a model for the random component. The systematic component can be the same systematic component as any of the time series models we have discussed so far. Just as we combined an ARCH model with an AR model, we combine a GARCH model with an ARIMA model to form an ARIMA($p_\mathrm{M}, d, q_\mathrm{M}$)+GARCH($p_\mathrm{V}, q_\mathrm{V}$) model. Again, the subscripts stand for "mean" and "variance".

Aside from the order of differencing in the ARIMA portion of the model, there are $p_\mathrm{M} + q_\mathrm{M} + p_\mathrm{V} + q_\mathrm{V} + 3$ unknown parameters in this model. (The 3 additional parameters are two means, often assumed to be 0, and the overall variance.) There is, however, only one underlying random variable, ϵ_t; the other random variables are function of ϵ_t. Hence, there is only one probability distribution to model. If we assume the form of the probability distribution, we can form a likelihood function and obtain MLEs for all parameters.

Residuals in Fitted $\mathbf{ARIMA}(p_\mathrm{M}, d, q_\mathrm{M}) + \mathbf{GARCH}(p_\mathrm{V}, q_\mathrm{V})$ Models

The ordinary residuals from a fitted ARIMA + GARCH model are very similar to the residuals from the same ARIMA model fitted in the same way. They should mimic the a_t. They would show a similar correlational structure, which would be detected by a Box-Ljung tests on their squared values.

When standardized by their standard deviations, the ordinary residuals are homoscedastic, and these standardized residuals correspond to the ϵ_ts in equation (5.168). The standardized residuals are not "estimates" of the ϵ_ts (see the discussion regarding residuals and errors on page 415), but if the fitted model accounts for the stochastic volatility adequately, they should have insignificant autocorrelations.

It is appropriate to use the ordinary residuals as a measure of the variance of the observations, and the ordinary residuals should be used in setting confidence intervals for forecasts, for example.

R Software for GARCH Models

GARCH modeling is implemented in the `tseries` R package, which we have used for various tasks in time series. GARCH modeling is also implemented in the `rugarch` package and in the `fGarch` package, which we have used in Chapter 3 for its functions for probability distributions.

The `rugarch` package, developed by Alexios Ghalanos, contains a function for fitting ARMA + GARCH models as well as number of additional functions for analysis of ARMA + GARCH models. (The related `rmgarch` package contains many similar functions for computations for multivariate GARCH models.) The functions in `rugarch` accept a number of R data objects (in particular, `xts` objects).

The `ugarchspec` function in the `rugarch` package is used to specify the order of the ARMA + GARCH models. This function has an argument `mean.model`, which is a list to specify the model for the conditional mean, and an argument `variance.model`, which is a list to specify the model for the conditional variance. Although other possibilities are allowed, the function is oriented toward ARMA + GARCH models, so a component of the `mean.model` list is a vector `armaOrder` and a component of the `variance.model` list is a vector `garchOrder`.

The `ugarchspec` function also allows different probability distributions to be specified. The `distribution.model` parameter can be set to `norm` for the normal distribution, which is the default; `snorm` for the skewed normal distribution; `std` for the t distribution; `sstd` for the skewed t; `ged` for the generalized error distribution; and `sged` for the skew-generalized error distribution.

For a specified ARMA + GARCH model, the `ugarchfit` function in the `rugarch` package performs the computations for the fit.

An Example

We will illustrate with the S&P 500 Index daily log returns data used previously. We used the log of the Index daily closes in Section 5.3.11 to illustrate an ARIMA model. The time series of daily log returns of the S&P 500 is the differenced series of log closes. We observed several aspects of this series in Chapter 1, including importantly that it is heavy-tailed, as we see in the histogram in Figure 1.25 and the q-q plot in Figure 1.27. The relatively small values of the ACF of the returns shown in Figure 1.29 on page 99 indicate that the innovations in the time series may have the important 0 correlation of white noise. The characteristic that obviously violates the assumptions of an ARMA model, however, is the apparent heteroscedasticity of the series. This stochastic volatility is visually obvious in the time series plot in Figure 1.29.

In the example in Section 5.3.11, the `auto.arima` function, based on the BIC, selected an ARIMA(1,1,1) fit for the daily closes. We therefore choose an ARMA(1,1) as a possible model for the means of the daily returns. We choose a GARCH(1,1) model for the variance. There is no particular reason to choose this model, other than the general rule of choosing small values for the orders, just as in fitting ARIMA models.

```
library(rugarch)
agmodel <- ugarchspec(mean.model=list(armaOrder=c(1,1)),
                      variance.model=list(garchOrder=c(1,1)))
agfit <- ugarchfit(data=GSPCdReturns, spec=agmodel)
```

The object returned by `ugarchfit` is of class uGARCHfit, which is an S3 type of R class (see page 143). All components, such as coefficients or residuals, are extracted by functions that apply methods for the uGARCHfit class. For example, the coefficients in the fit performed by ugarchfit are given by

```
> round(coef(agfit),3)
   mu    ar1    ma1  omega alpha1  beta1
 0.001  0.881 -0.908  0.000  0.098  0.890
```

The residuals are obtained by the residuals function.

The uGARCHfit object in this example contains many more statistics, including the AIC, the BIC, Shibata's information criterion, and the Hannan-Quinn information criterion. These criteria mean nothing in isolation, but if different models are tried, they can be useful in evaluating the relative merits of the models.

The ugarchfit function also performs a Box-Ljung test on both the standardized residuals and the squared standardized residuals. The results of the tests are shown below.

```
Weighted Ljung-Box Test on Standardized Residuals
-----------------------------------
                           statistic p-value
Lag[1]                         2.648  0.1037
Lag[2*(p+q)+(p+q)-1][5]        4.319  0.0282
Lag[4*(p+q)+(p+q)-1][9]        5.488  0.3494
d.o.f=2
H0 : No serial correlation

Weighted Ljung-Box Test on Standardized Squared Residuals
-----------------------------------
                           statistic p-value
Lag[1]                         0.4355  0.5093
Lag[2*(p+q)+(p+q)-1][5]        2.4446  0.5177
Lag[4*(p+q)+(p+q)-1][9]        3.3823  0.6945
d.o.f=2
```

These results seem to indicate no remaining autocorrelations.

Although the autocorrelations among the residuals from the fitted ARMA(1,1)+GARCH(1,1) model are closer to 0 than those among the residuals from the fitted ARMA(1,1) model, it is not clear what additional insight the ARMA(1,1)+GARCH(1,1) model brings to the time series of returns on the S&P 500 Index.

5.5 Unit Roots and Cointegration

An objective in financial analysis is to find relationships among time series of prices, rates, indexes, and other observable variables. The relationships aid our understanding of the various underlying processes, and if the relationships involve lagged variables, it may be possible to predict the course of a process based on observations of another process. Unfortunately, the noise in the system renders most predictive models almost useless.

Some financial variables move together in such a way that a linear combination of them is almost stationary. Such variables are said to be cointegrated. (We will define this concept more precisely below.) Identification of cointegrated variables is useful both in general financial analysis and in formulating investment strategies, such as pairs trading.

Relationships *within* each of the time series of two variables may obscure relationships *between* the two variables. Relationships within each of two independent time series may cause the time series to appear to be cointegrated.

5.5.1 Spurious Correlations; The Distribution of the Correlation Coefficient

The correlation coefficient is the primary statistic used to decide if two variables have a linear relationship. Nonlinear relationships between two variables may also be identified by transforming the variables prior to computing the correlation coefficient. If more than two variables are involved, partial correlations or regression coefficients are the basis for deciding if variables have a relationship, again possibly computed on transformations of the variables.

Many financial processes seem to be correlated because they move in unison. The production rate of electrical vehicles and the number of airline passenger-miles in the United States have both increased steadily in recent years, for example, so the cross-correlations of the two time series may be significantly nonzero. Another example is the number of housing starts and the price of MSFT stock. From 2006 until mid-2008 both the number of housing starts and the price of MSFT stock increased, and then in the next 18 months they declined together, but it is doubtful that these two series are directly related to each other.

The apparently statistically significant correlations of unrelated financial time series that move in unison are called "'spurious correlations", and regression of one unrelated series on another with apparently significant estimated coefficients and large R-squared values is called "spurious regression". Spurious "significance" is a result of using critical values from an incorrect distribution for the computed statistic.

Variables with linear drifts in time have significant correlations, but the phenomenon of spurious correlation goes much beyond that. Two unrelated random walks, for example, will often appear to have nonzero correlation.

Independent Random Walks

Random walks are conceptually simple: the price of an asset closes at a certain level today; it begins at that price tomorrow and takes a step from there to close at tomorrow's price; and the process continues. (These are basic ingredients of a random walk, but a random walk includes other properties in the definition (page 490), and the Random Walk Hypothesis (page 121) includes other components.)

Unrelated random walks often appear to have nonzero correlation. We illustrate this by simulating two independent random walks, each of size 1,000 as shown in Figure 5.19.

```
> n <- 1000
> set.seed(128)
> white1 <- rnorm(n)
> white2 <- rnorm(n)
> walk1 <- cumsum(white1)
> walk2 <- cumsum(white2)
> cor(white1, white2)
[1] 0.03261427
> cor(walk1, walk2)
[1] 0.9257856
```

FIGURE 5.19
Example of Large Positive Correlation Between Independent Random Walks

The large sample correlation coefficient 0.93 between the two independent random walks `walk1` and `walk2` in Figure 5.19 is "statistically significant". (See the t test described on page 409, which is based on the assumption of independent normal samples. This value is so large that no calculations of significance are necessary under the assumptions.) We note that the large correlation in Figure 5.19 does not result from relations between the two underlying Gaussian white noise processes, `white1` and `white2`, which has a sample correlation of 0.03.

The simulated data in Figure 5.19 just represent one "random sample". Now suppose we do this again with a different random sample. We choose a different seed, and again simulate two random walks, each of size 1,000, as shown in Figure 5.20.

Again, we have a sample correlation coefficient, -0.92, that is "statistically

```
> set.seed(145)
> white3 <- rnorm(n)
> white4 <- rnorm(n)
> walk3 <- cumsum(white3)
> walk4 <- cumsum(white4)
> cor(white3, white4)
[1] 0.02377493
> cor(walk3, walk4)
[1] -0.9192005
```

FIGURE 5.20
Example of Large Negative Correlation Between Independent Random Walks

significantly different from zero", yet with an opposite sign from that of the other bivariate sample above. Again, the sample correlation between the two underlying Gaussian white noise processes is insignificant at 0.02.

Although the two random walks are not related in either case, the sample correlation coefficients are highly statistically significant based on the asymptotic null distribution of the correlation coefficient described on page 409. (The fact that the distribution is asymptotic is not relevant; the sample size of 1,000 is adequate.)

The problem arises from the fact that the distribution of the sample correlation coefficient between two variables that follow unrelated random walks is different from that of the sample correlation coefficient between two unrelated variables that do not follow random walks, such as white noise variables. To illustrate this, we will simulate a sample of 5,000 sample correlation coefficients from two white noise processes each of length 1,000, and a sample of 5,000 sample correlation coefficients from two random walks each of length 1,000.

Figure 5.21 shows histograms of the two samples of 5,000 correlation coefficients. The total area covered by each histogram is 1. The top row shows histograms that are scaled separately to show the general shape best. The histograms in the bottom row are the same as in the top row, but they have the same scales as each other (again, with total area 1) for easier comparisons between the two distributions. A plot of the approximate t distribution corresponding to correlation coefficients between two independent normal samples is superimposed on each histogram. Most of that curve is not visible over the unscaled histogram of the correlation coefficients between the random walks (in the top right side of the figure).

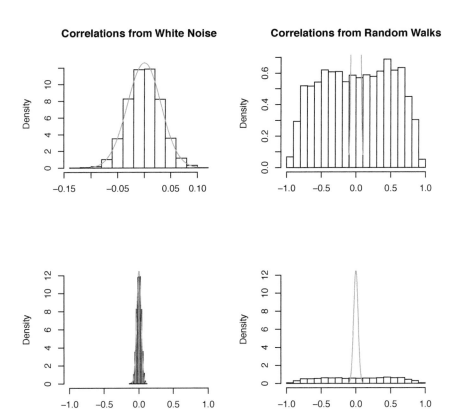

FIGURE 5.21
Frequency Distributions of Correlation Coefficients Between Simulated Independent White Noise Processes and Between Simulated Independent Random Walks

These simulations indicate why random walk processes are likely to exhibit spurious correlation. The null distribution of the correlation coefficient between them is different from the null distribution of the correlation coefficient between white noise processes.

If the data from either independent random walk process is regressed linearly on the data from the other, a significant linear regression may be obtained. The relevance of this in financial applications is that some time series may appear to be related to another time series, even accounting for any drift in either series, when in fact they are not related.

Correlations among AR(1) Processes

The main point in the illustrations above is that the null distribution of the correlation coefficient between two independent random walks is not the same as the distribution of the correlation coefficient between two white noise processes. This point is important because, if ignored, it can lead to misinterpretations in data analysis.

We now investigate the distribution of the correlation in another context, related to a random walk, that is, in AR(1) processes.

The distribution of the correlation coefficient between two AR(1) processes with coefficients ϕ_{11} and ϕ_{12} depends on $|\phi_{11}|$ and $|\phi_{12}|$.

Figure 5.22 shows some histograms of simulated samples of correlation coefficients between AR(1) processes each with 0 mean but with various values of the first-order coefficients. Each histogram is based on 5,000 simulations. The lengths of all of the processes is 1,000. (This is the value of n in equation (4.96) that shows the relation of the correlation coefficient between two independent Gaussian white noise samples to a t random variable.) The scales in the histograms are the same for easier comparisons. The total area covered by each histogram is 1. A plot of the approximate t distribution corresponding to correlation coefficients between two independent samples is superimposed on each histogram.

The distribution of the correlation coefficient between two AR(1) processes with AR coefficients of 0.99 in Figure 5.22 (whose roots of the characteristic polynomial are 1.0101) indicate that random samples from these processes will often exhibit correlations that are large with respect to correlations computed from independent normal samples, hence, "spurious".

Differenced AR(1) Processes; Residuals

As we have noted, a differenced random walk is white noise. The distribution of the correlations between two differenced random walks would not correspond to the distribution shown in the histograms on the right-hand side of Figure 5.21, but rather, would correspond to the histograms of correlations between white noise processes shown on left-hand side.

Likewise, for other AR(1) processes, the distribution of the correlation coefficients of the corresponding differenced processes is similar to that of the distribution of the correlation coefficient between white noise processes. This is illustrated in the histograms of the correlations of the differenced series, in Figure 5.23. The two histograms in Figure 5.23 correspond respectively to the histogram in the first row and second column of Figure 5.22 and to that in the second row and first column.

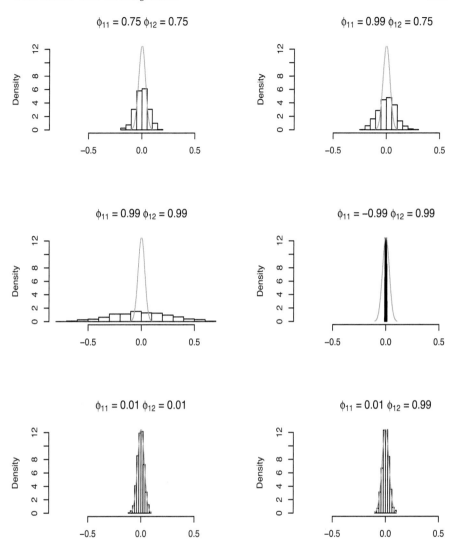

FIGURE 5.22

Frequency Distributions of Correlation Coefficients Between Independent Simulated AR(1) Processes

There are actually slight differences in the distributions of the correlation coefficients between different AR(1) processes that are differenced. The differences evident in the two histogram in Figure 5.23, however, are primarily due to sampling variations. They both correspond well to null t distributions, but the distribution is closer to a null t distribution when both $\phi_1 = 0.99$.

Although the distributions of the correlation coefficients between two dif-

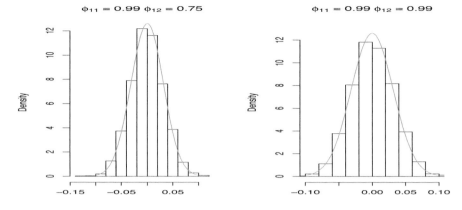

FIGURE 5.23
Frequency Distributions of Correlation Coefficients Between Independent Differenced Simulated AR(1) Processes

ferenced AR(1) processes are similar to the distribution between two white noise processes, AR(1) processes are not I(1), unless $\phi_1 = 1$.

Correlations among AR(2) Processes

The spurious correlations that occur between AR(1) processes also may occur in AR(p) processes.

Consider, for example, two AR(2) processes, each following the model

$$x_t = \phi_0 + \phi_1 x_{t-1} + \phi_2 x_{t-2} + w_t.$$

We consider two processes each with $\phi_1 = -0.25$ and $\phi_2 = 0.25$, and two processes each with $\phi_1 = -0.7499$ and $\phi_2 = 0.25$. We simulated 5,000 pairs of 1,000 observations for each process, and for each of the two pairs of processes, computed the ordinary correlation coefficient. Figure 5.24 shows histograms of the simulated samples. The histograms have the same scale for ease of comparison.

The unusual shapes of the distributions of the correlation coefficient of the AR(2) processes evident in the histograms of Figure 5.24 will not become more like that of a t distribution by simple differencing, as was the case with the AR(1) processes considered above.

Vector Autoregressive Processes

The pairs of unrelated AR(1) processes and the pairs of unrelated AR(2) processes considered above are both vector autoregressive (VAR) processes with diagonal coefficient matrices.

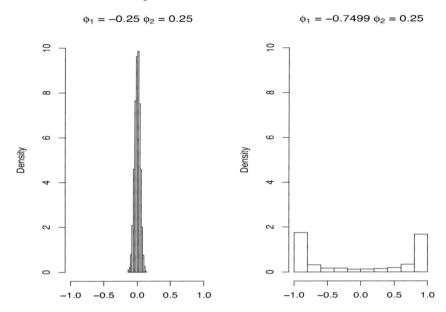

FIGURE 5.24
Frequency Distributions of Correlation Coefficients Between Independent
AR(2) Processes

We briefly discussed vector autoregressive models beginning on page 535.
The d-vector autoregressive model has the form

$$x_t = \Phi_1 x_{t-1} + \cdots + \Phi_p x_{t-p} + w_t,$$

where $x_t, x_{t-1}, \ldots, x_{t-p}$ and w_t are a d-vectors, $\Phi_1, \ldots, \Phi_p$ are constant $d \times d$
matrices and $\Phi_p \neq 0$, and $\{w_t\}$ is a d-variate white noise process.
    The multivariate distribution of the x_t vectors obviously depends on the
multivariate distribution of the white noise, but as we have seen in the examples above, even if the w_t are uncorrelated, the distribution of the x_t may
have an interesting correlational structure.
    In the AR(1) examples above, $d = 2$, $p = 1$, and

$$\Phi_1 = \begin{bmatrix} \phi_1 & 0 \\ 0 & \phi_1 \end{bmatrix} \quad \text{and} \quad \mathrm{V}(w_t) = \begin{bmatrix} 1 & 0 \\ 0 & 1 \end{bmatrix} \sigma_w^2.$$

This is a very simple VAR, yet we have observed some interesting distributional properties of $\{x_t\}$; in particular, of the off-diagonal element of the 2×2
matrix $\mathrm{Cor}(x_t)$. These are the quantities whose simulated distributions are
shown in Figure 5.22.

5.5.2   Unit Roots

Random walks, which tend to generate spurious regressions are AR(1) processes, with $\phi_1 = 1$. They are processes with unit roots.

It turns out that the roots of the characteristic polynomial are related to the distribution of the correlation coefficients. Independent AR processes in which a root is close to unity have unusual distributions of the correlation coefficients. This is what we observed for AR(1) processes with near-unit roots in Figure 5.22. (We also noted that relative values of the AR coefficients in pairs of processes affect the distribution of the correlation coefficient, but we did not explore that.)

This extends to AR(p) processes with unit roots, or near-unit roots. An important difference between the two pairs of AR(2) processes in Figure 5.24 is the moduli of the roots of the characteristic polynomials. The processes in the second pair have a root very close to the unit circle. The moduli of the roots of each of the first pair of processes are

```
> Mod(polyroot(c(1, 0.25, -0.25)))
[1] 2.561553 1.561553
```

and the moduli of the roots of each of the second pair of process are

```
> Mod(polyroot(c(1, 0.7499, -0.25)))
[1] 1.00008 3.99968
```

one of which is quite close to unity.

The unusual distribution of the correlation coefficient is also reflected in the distribution of the cross-correlation coefficients. Figure 5.25 shows the CCF of a single pair of independent AR(2) processes each with a root with modulus 1.00008. They both have the same coefficients, $\phi_1 = -0.7499$ and $\phi_2 = 0.25$, as in right panel of Figure 5.24. The negative value of ϕ_1 that is relatively large in absolute value is what causes the CCF to alternate in sign. The value at lag 0, which happens to be negative, approximately -0.5, of course is the ordinary correlation coefficient between the two series. Note that while Figure 5.24 displays results from a Monte Carlo simulation of 5,000 pairs of samples, Figure 5.25 represents a single sample from the two processes; a different sample may exhibit very different characteristics.

The foregoing examples illustrate unusual behavior of correlation and cross-correlation coefficients for two independent time series that have characteristic roots close to unity. Different values of the parameters in the models

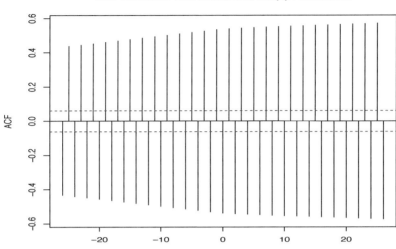

CCF Between Two Simulated AR(2) Processes

FIGURE 5.25

CCF Between Two Simulated Independent AR(2) Processes Each with $\phi_1 = -0.7499$ and $\phi_2 = 0.25$

would of course yield different results, most less extreme than the examples shown. See the simulated distributions in Figure 5.22 for examples in which the two processes follow different AR(1) models.

As we have seen, in many cases, especially processes that are similar to random walks, differencing the time series may affect the distribution of the correlation coefficient. In the case of random walks, the differenced series are white noise, and if the processes are independent, the distribution of the correlation coefficient between the differenced process follows an asymptotic t distribution.

The main point of this section is that processes with unit roots (or near-unit roots) yield statistics such as correlation coefficients whose distributions are quite different from the distributions of the same statistics computed from other time series. Engle and Granger (1987) list several of these differences (see also Choi, 2015, particularly page 4, for a discussion of the differences). In Exercise 5.30, you are invited to explore some of the differences via Monte Carlo simulation.

Unit roots are important in identifying processes that may exhibit what appears to be anomalous behavior. It is not immediately obvious from plots or simple statistics computed from a time series that the characteristic polynomial of the process has a unit root. In the next section, we consider some statistical tests for the presence of unit roots.

Tests for Unit Roots

In simple cases, we may visually distinguish unit-root processes from ones with no roots near unity either in the time series plots or in the ACF; for example, see Figures 5.8 and 5.9 for AR(1) processes, and Figures 5.10 and 5.11 for AR(2) processes. In general, however, the presence of a unit root may not be so obvious.

There are various statistical tests for unit roots. We will describe the basic ideas leading up to a test for a unit root, but we will not describe all of the considerations underlying the test. The development below leads to the *Dickey-Fuller* tests.

Consider an AR(1) model,

$$x_t = \phi_1 x_{t-1} + w_t. \tag{5.169}$$

If the characteristic polynomial has a unit root, then $|\phi_1| = 1$. Let us focus just on the case $\phi_1 = 1$; that is, on the case of a random walk. Although the random walk is not stationary, it is I(1).

We formulate statistical hypotheses of interest as

$$H_0 : \phi_1 = 1$$

$$vs$$

$$H_a : \phi_1 < 1$$

(see Section 4.4.2). The equivalent hypotheses in terms of roots of the characteristic polynomial are $H_0 : z_1 = 1$ and $H_a : z_1 > 1$. Note that the hypotheses exclude the explosive case of $\phi_1 > 1$. We also generally exclude the case of $\phi_1 < -1$, although often that is not formally stated.

The formulation of the null hypothesis as the existence of a unit root may not seem to be the natural way to set up a statistical hypothesis to be tested. Failure to reject this null hypothesis indicates that there is a unit root. (*Failure to reject* a null hypothesis does not yield the same kind of confidence level as a *rejection* of the hypothesis; see the discussion in Section 4.4.2.)

In terms of the root of the characteristic polynomial, say z_1, the null hypothesis is the same, $z_1 = 1$, but the alternative hypothesis is that the root is outside the unit circle, or simply that $z_1 > 1$, since it is real.

We rewrite the model (5.169) as

$$\triangle(x_t) = (\phi_1 - 1)x_{t-1} + w_t,$$

or

$$\triangle(x_t) = \pi x_{t-1} + w_t, \tag{5.170}$$

which is of the form of a simple linear regression model for the data pairs $(\triangle(x_2), x_1)$, $(\triangle(x_3), x_2)$, ...

The regression coefficient π can be estimated by ordinary least squares. The model (5.170), however, does not satisfy the assumptions of the regression models in Section 4.5.2 on page 413, which led to statistical tests concerning the regression coefficients. In particular, the OLS estimator $\hat{\pi}$ of π may not have a t distribution, not even an asymptotic t distribution. This is because of serial correlations among the pairs $(\triangle(x_t), x_{t-1})$.

A statistic in the same form as the t statistic in equation (4.115), that is, $t = \hat{\pi}/\sqrt{\text{MSE}}$ in this case, may, nevertheless, be useful in testing $H_0 : \pi = 0$ versus $H_a : \pi < 0$. Its distribution must be worked out so as to determine the critical values. Fuller (1995) and others have calculated critical values of the test statistic using Monte Carlo methods. The test based on this approach is called the *Dickey-Fuller test*. Notice that extreme values of the t-type test statistic indicate rejection of the null hypothesis; that is, small (negative) values indicate that there is no unit root.

There are many additional considerations that may affect this testing procedure. One is the possibility that the regression errors may be correlated, as we discussed in Section 5.3.12. The method to address this possibility that is commonly employed is the same one suggested by equation (5.154). Usually only AR terms are included. We augment the model (5.170) to form the model

$$\triangle(x_t) = \pi x_{t-1} + \sum_{j=1}^{k} \gamma_j \triangle(x_{t-j}) + w_t. \qquad (5.171)$$

Various methods have been suggested to determine the maximum lag order k in the AR model. Some possibilities are to use information criteria, such as AIC or BIC (as is done in `auto.arima`, for example), or to perform Box-Ljung tests on residuals of AR models of different orders.

Another consideration is the existence of deterministic trends in the data. The simple AR(1) model does not account for a trend, but economic data such as GDP and other measures of production or of wealth or financial data such as stock prices often have an upward trend.

A simple linear trend in the data may be accounted for by including a constant drift term α in the model for $\triangle(x_t)$, and a quadratic trend can be modeled by including a linear term in the model for $\triangle(x_t)$. Hence, the model (5.171) is augmented to form the model

$$\triangle(x_t) = \alpha + \beta t + \pi x_{t-1} + \sum_{j=1}^{k} \gamma_j \triangle(x_{t-j}) + w_t. \qquad (5.172)$$

The *augmented Dickey-Fuller test* for $\pi = 0$ in the model (5.172) is based on a test statistic that is in the same form as a t statistic, as in the Dickey-Fuller test based on the model (5.170). It does not have the same distribution as the simpler test statistic, however. Critical values of the t-type test statistic have been calculated by Monte Carlo methods. They are available in published tables and are incorporated in the R testing functions.

If the test indicates rejection of the null hypothesis, the decision is that there is no unit root. Although this is not a sufficient condition for stationarity, we often interpret rejection as indicating that the series is stationary.

If the null hypothesis is not rejected, we conditionally decide that $\pi = 0$, but the failure to reject may be due to a trend in time. If the null hypothesis is not rejected, we may proceed to test for a trend; that is, test that $\beta = 0$ in the model (5.172).

If, given $\pi = 0$, we cannot reject the null hypothesis that $\beta = 0$, we may proceed to test for a drift; that is, test that $\alpha = 0$ conditional on $\beta = 0$.

Finally, if given $\pi = 0$ and $\beta = 0$, we cannot reject the null hypothesis that $\alpha = 0$ from the model (5.172), we then return to the model in equation (5.171), and in that model test a null hypothesis that $\pi = 0$. In that case, selection of the AR order k becomes more important. Failure to reject the null hypothesis in this case also indicates the presence of a unit root.

The conditional nature of this process means that the concept of a significance level is meaningless; the process is not formal statistical inference. Rather, the analysis is exploratory and merely descriptive.

Example

We now consider an example of the augmented Dickey-Fuller unit root test on simulated data from a process we know not to have a unit root and from a process we know to have a unit root.

The augmented Dickey-Fuller test is performed by several different R functions, for example, ur.df in urca, adf.test in tseries, and adfTest in dUnitRoots.

We will use adf.test in tseries to illustrate the augmented Dickey-Fuller test on data from a white noise process and on data from a random walk that were generated in Figure 5.19. The R code is shown below.

```
library(tseries)
n <- 1000
set.seed(128)
white1 <- rnorm(n)
walk1 <- cumsum(white1)
adf.test(white1)
adf.test(walk1)
```

The results of the tests are shown in Figure 5.26. In each case neither a drift term, corresponding to α in equation (5.172), nor a linear trend, corresponding to β, was identified.

We see, as expected, for the white noise the null hypothesis of a unit root is rejected at a p-value less than 0.01; but the p-value of the test in the case of a random walk is not rejected.

```
        Augmented Dickey-Fuller Test
data:  white1
Dickey-Fuller = -10.254, Lag order = 9, p-value = 0.01
alternative hypothesis: stationary
Warning message:
In adf.test(white1) : p-value smaller than printed p-value

        Augmented Dickey-Fuller Test
data:  walk1
Dickey-Fuller = -2.2177, Lag order = 9, p-value = 0.4861
alternative hypothesis: stationary
```

FIGURE 5.26
Results of Augmented Dickey-Fuller Test on White Noise and on a Random
Walk

Other Unit Root Tests

The steps we followed from equations (5.169) to (5.172) in developing the ideas
behind the augmented Dickey-Fuller test could be modified in various ways at
each step. For example, in equation (5.171), instead of accounting for serial
correlations in the basic regression model of equation (5.170) by incorporating
an AR component, we could estimate the autocorrelations directly, using some
technique such as the kernel methods described in Section 2.3.3. This is what
is done in the *Phillips-Perron test*.

There are several other tests for unit roots, and there are multiple R functions that implement them. The tests differ from each other and from the
augmented Dickey-Fuller test in various ways. We will not describe the tests
here, but rather list some of the more common ones in Table 5.1. (Recall also
that we used the Kwiatkowski-Phillips-Schmidt-Shin (KPSS) test on page 507
to test for a trend.)

In Table 5.1, we list the R functions in the tseries and urca packages that
perform the tests. Some tests are not included in the tseries package. Also,
some of the functions are not designed specifically for testing the null hypothesis of a unit root, and so the output of the R function may not be structured
similarly to the results shown in Figure 5.26 for the augmented Dickey-Fuller
test as produced by adf.test. Most functions in the urca package require
use of the summary function to display the relevant statistics.

The number of tests is somewhat overwhelming, and there is no one test
that is better than the others. We list them here just as a matter of information

TABLE 5.1

R Functions for Unit Root Tests in the `tseries` and `urca` Packages

augmented Dickey-Fuller test	`adf.test`	`ur.df`
Elliott-Rothenberg-Stock test		`ur.ers`
Kwiatkowski-Phillips-Schmidt-Shin (KPSS) test	`kpss.test`	`ur.kpss`
Phillips-Perron test	`pp.test`	`ur.pp`
Schmidt-Phillips test		`ur.sp`
Zivot-Andrews test		`ur.za`

and we will make an effort to describe their advantages and disadvantages. In Exercise 5.32, however, you are asked to use all six of the tests shown in Table 5.1 on the same set of data to observe similarities and differences.

Other Approaches and General Considerations in Testing for Unit Roots

Tests for unit roots depend on the basic model assumptions. Variations in the model, such as seasonality, general structural shifts in time, heavy-tailed distributions, and so on, may lead to different tests, even based on the same general approach. The tests we considered above are Wald-type tests, but there are other general approaches to constructing tests, such as tests based on likelihood ratios, and Lagrange multipliers or "scores". See Gentle (2019) for definitions, but the distinctions are not important here, and we will not discuss these methods. Other approaches are based on Bayesian models, in which parameters in the model are random variables. Choi (2015) provides descriptions of a variety of tests for unit roots and the general approaches underlying them.

Unit root tests are somewhat unusual in that the null hypothesis is that a unit root exists, and hence, the measures of significance are oriented towards the nonexistence of a unit root. Statistical hypothesis testing treats the null and alternative hypotheses asymmetrically. The null hypothesis that there is a unit root is accepted as the default, and rejection of that hypothesis is associated with a level of confidence (the p-value). There is no stated level of confidence in the decision that there is a unit root. This decision is made when the null hypothesis is not rejected.

A quandary results from the basic paradigm of statistical hypothesis testing. In general, "statistical significance" may play a larger role in decision making than is appropriate. Certainly the significance of a computed statistic is important, but the empirical evidence should be examined more broadly. These issues in statistical hypothesis testing are discussed in Section 4.4.2, in particular, on page 385.

There are other issues in testing for unit roots. We might consider what is special about a root with modulus exactly equal to 1, instead of slightly less than one. A model in which a real parameter being exactly a specific value determines important behavioral characteristics of the model is ill-conditioned. To illustrate the point in the case of an AR(1) model, the relevance of a unit root is equivalent to discontinuities at $\phi_1 = 1$. This is the case mathematically, as we see for the mean and variance expressions (equations (5.104) and (5.107)), but finite samples of data from AR(1) models with $\phi_1 = 1$ or $\phi_1 > 1$ do not exhibit discontinuities as ϕ_1 varies from $1 - \epsilon$ to $1 + \epsilon$.

5.5.3 Cointegrated Processes

Linear combinations of financial time series are of interest. As discussed beginning on page 493, linear combinations of separate processes of some types retain the properties of the individual processes. Linear combinations of independent white noise processes are white noise. Linear combinations of independent random walks are random walks. It may be of interest to determine if some processes that are not white noise can be combined to form a white noise process or at least a stationary process.

An important type of linear combination of time series is one that yields a stationary process, especially if the combined process is white noise. If such a combination exists, the time series are cointegrated.

Formally, if $\{x_t\}$ is an m-variate time series each of which is I(d) and there exists a nonzero constant m-vector c such that $\{c^T x_t\}$ is I($d-1$), then we say that $\{x_t\}$ is a *cointegrated* time series with *cointegrating vector* c. If there are linearly independent vectors $c_1, \ldots, c_r$ such that each is a cointegrating vector of $\{x_t\}$, then we say $\{x_t\}$ is *cointegrated with rank r*.

Our interest is often in the simpler case of two I(1) time series $\{x_t\}$ and $\{y_t\}$, for which there exists a nonzero constant c such that

$$\{cx_t + y_t\} \tag{5.173}$$

is stationary. In this simple case, clearly $\{x_t\}$ and $\{y_t\}$ are cointegrated.

A simple case of two time series being cointegrated is when each follows the same (or similar) underlying process. For example, if the series $\{x_t\}$ and $\{y_t\}$ are such that $x_t = z_t + w_{x,t}$ and $y_t = z_t + w_{y,t}$. With $c = -1$, the process $\{cx_t + y_t\}$ is white noise, which of course is stationary.

Although a model for two time series each of which is the same underlying random walk plus independent white noise errors may seem very simple, in fact, this is a good approximate model for many financial time series.

An obvious approach for determining if two series are related is to regress one series on the other and examine the residuals for stationarity; that is, we form the regression model

$$y_t = \alpha + \beta x_t + u_t, \tag{5.174}$$

fit the model, and inspect the residuals $r_t = y_t - \hat{y}_t$. If, indeed, the two series are cointegrated, the process $\{r_t\}$ should be stationary (subject to the properties of $\hat{y}_t$ that result from the properties of the estimators $\widehat{\alpha}$ and $\widehat{\beta}$).

Tests for Cointegration

Three tests for cointegration based on these ideas, are the Engle-Granger test, the Phillips-Ouliaris test, and the Johansen test. These tests differ in how they do the regression, and also how they test for stationarity of the residuals.

The null hypothesis for each of these tests is that the series are not cointegrated. Rejection of the null hypothesis, therefore, leads to the decision that the series are cointegrated. This is a more natural formulation of the hypotheses than that for unit root tests on page 594.

Each of these tests assesses the question of stationarity by testing for a unit root. They each conclude that the residuals are stationary if the unit root test is not rejected. (Recall the nature of the standard hypotheses in unit root tests, pages 594 and 598.)

There are three main issues in the regression step of these tests: which variable to choose as the dependent variable; what assumptions to make on the distribution of the residuals, u_t in equation (5.174); and how to model the residuals (recall the differences between residuals and errors in regression modeling; see page 415). In the Phillips-Ouliaris test, the residuals are tested for the presence of a unit root, for which, as we have seen, there are various approaches. The Phillips-Ouliaris test for cointegration uses the Phillips-Perron test for unit roots, for example.

The Johansen test is designed for multiple series and treats them symmetrically. (The difference in this approach and other approaches is similar to the difference in a regression model (asymmetric) and a principal components model (symmetric).) The question is whether a cointegrating vector exists; that is, whether or not the time series are contained in a lower rank space. This kind of question can be addressed either by use of the trace of the cross-products matrix or by an eigen-decomposition of that matrix (similar to identifying the principal component). Obviously, if there exists a lower dimension subspace, the time series are cointegrated. The Johansen test is based on these ideas. Of course, in a common application, there are only two times series (rank 2), and the lower-rank space is one-dimensional.

The Phillips-Ouliaris test is implemented in the R function po.test in the tseries package and uses the pp.test function in that package for the Phillips-Perron test for unit roots. The ca.po function in the urca package also performs the Phillips-Ouliaris test, but its output is organized completely differently.

The ca.jo function in the urca package performs the Johansen test.

Example

Because stationary series exhibit mean reversion, identification of cointegrated financial time series allows determination of points at which one series is larger than usual with respect to another series. In trading of financial assets this allows for statistical arbitrage and it provides guidance for pairs trading.

Unfortunately, noise, exogenous events, and structural change often render the cointegrated models useless.

Cointegration is often used in modeling forex rates with some success. In Figure 5.27, we illustrate the use of the Phillips-Ouliaris cointegration test for the series of daily US dollar to the euro rates and the Canadian dollar to the US dollar rates.

```
> library(quantmod)
> library(tseries)
> dollareuro <- getSymbols("DEXUSEU", env=NULL, src="FRED")
> candollar <- getSymbols("DEXCAUS", env=NULL, src="FRED")
> rates <- na.omit(merge(dollareuro, candollar, all=FALSE))
> po.test(rates)
        Phillips-Ouliaris Cointegration Test

data:  rates
Phillips-Ouliaris demeaned = -27.67, Truncation lag parameter = 52,
p-value = 0.01216
```

FIGURE 5.27
Phillips-Ouliaris Cointegration Test on US Dollar / Euro and Canadian Dollar / US Dollar Exchange Daily Rates for January 1999 through August 2019

This test suggests rejection of the null hypothesis that the series are not cointegrated. The relationships between the two series could be explored further by regression one series on the other. The relationships could be examined over time. Other variables, such as measures of GDP could be incorporated. In this example, we have just shown the simple test itself.

The R functions implementing the Phillips-Ouliaris test and the Johansen test allow for more than just two series. (The po.test function restricts the number of series for the Phillips-Ouliaris test to 6, and the ca.jo function restricts the number of series for the Johansen test to 11.)

Error Correction Models

The regression step in the general approach to the analysis of cointegration described above requires selection of a dependent variable. Which one is chosen as the dependent variable can make a difference in the results. The differences are generally small. For example, using the data in Figure 5.27 but switching the order of the variables so that the other one is chosen as the dependent variable, we have the results shown in Figure 5.28.

To eliminate the effect of the choice of which variable to treat as the de-

```
> rates1 <- na.omit(merge(candollar, dollareuro, all=FALSE))
> po.test(rates1)
        Phillips-Ouliaris Cointegration Test

data:  rates1
Phillips-Ouliaris demeaned = -25.856, Truncation lag parameter = 51,
p-value = 0.01819
```

FIGURE 5.28

Phillips-Ouliaris Cointegration Test on Canadian Dollar / US Dollar and US Dollar / Euro Exchange Daily Rates for January 1999 through August 2019; Compare Figure 5.27

pendent variable, instead of using the vertical residuals in equation (5.174), we could use the distances from the observed points to the line, as in orthogonal regression which we mentioned in Chapter 4.

Instead of orthogonal regression, however, the methods generally used in cointegration models are called "error correction" methods. For the $m + 1$ univariate series $\{y_t\}, \{x_{1,t}\}, \dots, \{x_{m,t}\}$, in an error correction method, a particular series is chosen as the dependent variable in a regression, but the regression model is for the changes $\{\triangle y_t\}, \{\triangle x_{1,t}\}, \dots, \{\triangle x_{m,t}\}$. The error correction regression model includes a term that works sequentially to favor a change in sign of the residual at each time point:

$$\triangle y_t = \beta_1 \triangle x_{1,t} + \cdots + \beta_m \triangle x_{m,t} + \lambda(y_{t-1} - \hat{y}_{t-1}) + \epsilon_t. \tag{5.175}$$

This model is called an *error correction model*, or ECM. It is also called a vector error correction model, or VECM.

To complete the cointegration analysis, the residuals from the ECM fitted model are tested for stationarity, perhaps using a unit root test, as in the Phillips-Ouliaris cointegration test described above.

The R function ecm in the ecm package performs the fit of the model. The result of the ecm function is an object of class lm, the same as the result of the lm function for linear regression.

Notes and Further Reading

The subject of this chapter is very broad. I have attempted to cover the fundamental concepts carefully and in detail. The level of detail, however, decreases as the topics become more advanced.

Because time is a continuous process, it may be appropriate to model a time series in continuous time. This approach is common in the valuation of derivatives and is covered in many texts, including the classic text by Hull (2017).

Since all observational data are taken at discrete times, a discrete-time model for time-series is often appropriate, as it maps directly to real data. The analysis of discrete time series models usually takes one of two approaches, called the "spectral domain" and the "time domain". The spectral approach begins with a transform of the data using periodic orthogonal functions. This approach is appropriate for physical wave processes, but it has limited usefulness in financial applications. It has been used in studying financial "cycles".

The field of discrete time domain analysis is large and there are many texts in the area. The one I used when I first studied time series was Box and Jenkins (first edition, 1970), the current edition of which is Box, Jenkins, Reinsel, and Ljung (2015). The one I have used most often recently is Fuller (1995), although my notation is not always consistent with Fuller's.

Long memory processes and fractionally integrated ARMA models useful in their analysis were developed and described by Granger and Joyeux (1980).

A failure of many time series models arises from assumptions about stationarity or assumptions about constancy of volatility. The GARCH family of models addresses this issue. Tsay (2010) covers GARCH models and their applications in finance. Duan (1995) developed a GARCH approach to option pricing. Lütkepohl (2007) covers the standard ARIMA models as well as GARCH models in a multivariate setting.

Unit roots and cointegration have received much attention in recent years. The discussion beginning on page 594 should indicate that there are many open issues in the subject, and the material in Sections 5.5.2 and 5.5.3 have only touched the surface. Maddala and Kim (1998) provide background coverage of the topics. Unit root tests and the properties of series with unit roots are discussed at a nontechnical level by Choi (2015). Sims (1988) discusses general issues in testing for unit roots, both from a Bayesian and a frequentist perspective. Herranz (2017) reports on simulation studies to compare different unit root tests under various scenarios, including structural shifts.

Bootstrap methods can be used in time series analysis, such as for confidence intervals for estimated parameters, or for testing for unit roots. Resampling methods such as the bootstrap, however, must be used with some care because of the structure of the data, which would be obscured under the usual independent resampling. Various methods, some ad hoc, have been suggested. A general approach is called a *block bootstrap* in which subsequences are sampled together. Lahiri (2003) describes the block bootstrap and other methods of resampling in time series data.

R Packages

The `base` and `stats` packages in R have many functions for time series analysis. Other R packages that we have used or referred to in this chapter include `tseries`, `forecast`, `FitAR`, `rugarch`, `urca`, `car`, `lmtest`, `nlme`, `ecm`, and `KFAS`. These packages may also be useful in the exercises.

- `tseries` contains many functions for analysis of time series, including goodness-of-fit tests, unit root tests, and tests for stationarity. It is especially oriented toward financial time series, and includes many interesting old financial datasets.

- `forecast` (see Hyndman and Khandakar, 2008) contains many utility functions (such as `Acf`, which I prefer to `acf`) and it also contains the very useful `auto.ariam` function for order selection in ARIMA models.

- `FitAR` also contains functions for order selection.

- `rugarch` (and `rmgarch`) contains functions for specifying, fitting, forecasting with, and simulation of, univariate (multivariate) GARCH models.

- `urca` contains functions for unit root test and cointegration analysis.

- `car`, `lmtest`, and `nlme` contain several functions for inference in linear models with serial correlations.

- `ecm` contains the basic fitting function for ECM models.

- `KFAS` contains functions and data objects for a state space approach to time series analysis.

Pfaff (2008) and Berlinger et al. (2015) discuss R functions for unit root tests and analysis of cointegrated time series.

Exercises

5.1. **Linear combinations of random walks.**

Let $\{x_t\}$ and $\{y_t\}$ be univariate random walk processes with $\text{Cor}(x_s, y_t) = 0$ for all s and t. Now for a constant a, let $z_t = ax_t + y_t$.

Show that $\{z_t\}$ is a random walk process.

5.2. Linear operators.

Show that $h(x) = x_1^2 + x_2$ is not linear. (Here, $x = (x_1, x_2)$ is a real 2-vector.)

5.3. The backshift and difference operators.

(a) Show that $(1 - B^2)(x_t) = (1 - B)(1 + B)(x_t)$.

(b) Show that $\triangle^k(x_t) = \sum_{h=0}^{k}(-1)^h \binom{k}{h} x_{t-h}$ (equation (5.26)).

(c) Show that for any integer k, $(1 - B^k)(x_t) = \triangle_k(x_t)$.

5.4. Difference equations.

Consider a specific instance of the difference equation (5.33):

$$\left(1 - 3B + 2B^2\right)(x_t) = 0, \quad \text{for } t = 2, 3, \dots.$$

(a) What is the characteristic polynomial for this difference equation, in the form of equation (5.35) (that is, with $x_t = z^{-t}$)? What are its roots?

(b) What are the general solutions to this difference equation?
Show that your solutions are indeed solutions to the difference equation.

5.5. Random walk; detrending.

Generate 100 observations in a random walk process starting at $x_0 = 10$ with a drift term of $\delta = 2$.

Use a white noise process with variance 16, as in Figure 5.2, but distributed as a multiple of a t random variable with 4 df. (This is a $(4, 0, 4)$ location-scale t distribution; see Exercise 3.13.)

(a) Plot the observations and fit a least squares trend line.
Generate the observations efficiently.

(b) Detrend the data in the random walk with drift by subtracting the fitted values on the trend line, and plot the observations that result.

(c) Now detrend the data in the random walk with drift by use of the difference operator, and plot the observations that result.

(d) At this point, you have three series. Use the KPSS test to assess whether each has a linear trend.

5.6. White noise.

(a) Suppose $\{w_t\}$ is a white noise with constant variance σ_w^2. What is the variance of the process $\{\triangle(w_t)\}$?
Why is this so?

(b) Generate 100 realizations of a white noise sequence with a variance of 1.
Compute the variance of your sample `w` and the variance of `diff(w)`.

(c) Generate three bivariate white noise sequences of 100 realizations each.

Let the variance of the first of the bivariate variables be 1 and let the variance of the second be 2 in all three processes. For the first series, let the correlation between the two elements be 0.5; for the second, let the correlation be 0.0; and for the third, let the correlation be -0.5.

Plot each bivariate time series on a separate graph (three graphs in all). In each graph use different line types or colors for the two terms in each time series.

5.7. **d-integrated processes.**

Suppose that $\{x_t\}$ is an I(d) process for $d > k$, for some integer $k > 1$.

(a) Show that $\{x_t\}$ is an I($d + k$) process
(b) Show that $\triangle^k(\{x_t\})$ is an I($d - k$) process.

5.8. **Non-causal processes.**

Show that the process in equation (5.110), $x_t = -\sum_{j=0}^{\infty} \phi_1^{-j} w_{t+j}$ where $|\phi_1| < 1$, is stationary.

Note an important property: in this representation x_t depends on future values of the white noise. (It is non-causal.)

5.9. **ACF of moving averages of white noise.**

(a) Show that the ACF of a moving average of a white noise process $x_t = \frac{1}{k}\sum_{j=1}^{k} w_{t-j}$ is 0 outside the window of the moving average; that is, for x_s and x_t in a k-moving average of a white noise, $\rho(s,t) = 0$ if $|s - t| \geq k$.
(b) Show that the ACF of a moving average of a white noise process with $k = 3$ is as given in equation (5.52) on page 512.

5.10. **Moving average of white noise.**

Consider a moving average of white noise, $x_t = \frac{1}{k}\sum_{j=1}^{k} w_{t-j}$, where the w_i have independent t distributions with 4 df.

Generate random samples of length 1,000 of this process with $k = 1$ (which is just white noise) and with $k = 2$, $k = 4$, and $k = 8$.

In each case compute and plot the ACF. How does the ACF relate to the window width, k?

5.11. **Lognormal process.**

Let
$$P_t = P_0 \exp(r_t + \cdots + r_1),$$
where the r_i are distributed iid as N(μ, σ^2). Show that P_t is distributed as lognormal with parameters $t\mu + \log(P_0)$ and $t\sigma$. (If $\mu = 0$, this is a lognormal geometric random walk with parameter σ.)

5.12. Trends and stationarity.

Suppose x_t is a time series with a linear and a quadratic trend, with white noise superimposed:

$$x_t = \beta_0 + \beta_1 t + \beta_2 t^2 + w_t, \quad \text{for } t = 1, 2, \ldots,$$

where β_0, β_1, and β_2 are constants, and w_t is a white noise with variance σ^2.

(a) Is $\{x_t\}$ (weakly) stationary? Why or why not?

(b) What is the ACF of $\{x_t\}$?

(c) Simulate 200 observations for $t = 1, 2, \ldots, 500$ following the $\{x_t\}$ with $\beta_0 = 1$, $\beta_1 = 0.1$, and $\beta_2 = 0.01$, and compute the sample ACF.

Plot the series and the ACF.

How does this compare with your work in Exercise 5.13b?

5.13. Trends and stationarity.

Let x_t be a time series with linear and quadratic trend, as in Exercise 5.12,

$$x_t = \beta_0 + \beta_1 t + \beta_2 t^2 + w_t, \quad \text{for } t = 1, 2, \ldots,$$

where β_0, β_1, and β_2 are constants, and w_t is a white noise with variance σ^2.

Let $\{y_t\}$ be a first-order difference process on $\{x_t\}$:

$$y_t = \triangle(x_t) = x_t - x_{t-1}, \quad \text{for } t = 2, 3, \ldots,$$

and let $\{z_t\}$ be a second-order difference process on $\{x_t\}$:

$$z_t = \triangle^2(x_t) = y_t - y_{t-1}, \quad \text{for } t = 3, 4, \ldots,$$

In answering the following questions, be sure you note which series the question concerns. "x_t", "y_t", and "z_t" can look very much alike!

(a) Is $\{y_t\}$ (weakly) stationary? Why or why not?

(b) Is $\{z_t\}$ stationary? Why or why not?

5.14. Bivariate processes.

For the jointly stationary bivariate process $\{(x_t, y_t)\}$, show that

$$\gamma_{xy}(h) = \gamma_{yx}(-h)$$

and

$$\rho_{xy}(h) = \rho_{yx}(-h).$$

5.15. **ACFs and CCFs of returns.**

Obtain the closing prices of MSFT and the S&P 500 Index for the period January 1, 2017, through December 31, 2017, and compute the daily log returns for this period.

(a) Compute and plot the ACFs of the returns of MSFT and of the S&P 500 Index for this period.

(b) Test the hypothesis that the ACF of MSFT daily returns is 0 at all lags. (A statistical test does not consist of just computing a p-value; state and interpret your results.)

Test the hypothesis that the ACF of S&P 500 returns is 0 at all lags. State and interpret your results.

You are given very few guidelines in either case. It is a part of the exercise to make some decisions.

(c) Compute and plot the CCFs of the returns of MSFT and of the S&P 500 Index for this period.

5.16. **CCFs of changes in economic time series.**

Obtain the Consumer Price Index for All Urban Consumers: All Items for the period January 1, 2000 to December 31, 2018. These monthly data, reported as of the first of the month, are available as `CPIAUCSL` from FRED. Compute the simple monthly returns for this period. (The first return is as of February 1, 2000; see Exercise A1.23.)

Now obtain the 15-Year Fixed Rate Mortgage Average in the United States for each month in the same period, January 1, 2000 to December 31, 2018.

Weekly data, reported on Thursdays, are available as `MORTGAGE15US` from FRED. This series contains four values. The "close" is the fourth variable (column). To convert these data to correspond approximately to the same monthly dates as those of the CPI data, we take the data for last Thursday in each month as the data for the first day of the following month; hence, the first "monthly" data is for the last Thursday in December 1999. Compute the simple monthly differences for this period. (The first difference corresponds to February 1, 2000.) Also, note that these differences are percentages, so to make them comparable to the returns of the CPI, we divide by 100.

Now compute and plot the CCFs of these two series of monthly returns and differences (from February 1, 2000 through December 1, 2018).

5.17. **Autocovariance for MA(2) model.**

Work out the autocovariance $\gamma(h)$ for the MA(2) model.

5.18. **MA(2) models.**

Obtain the S&P 500 Index daily closes for the period from January 1987 to December 2017 and compute the log returns. (These are the data shown in Figure 1.30.) Fit an MA(2) model to the returns.

Comment on the fit and this method of analysis generally. Can ARMA models add meaningful insight into stock returns?

5.19. **AR(1) and AR(2) models.**

(a) In the AR(1) model $x_t = \phi_0 + \phi_1 x_{t-1} + w_t$, with $\phi_1 < 1$ and $E(x_t) = E(x_{t-1})$, show that

$$E(x_t) = \frac{\phi_0}{1 - \phi_1}.$$

(b) In the AR(2) model $x_t = \phi_0 + \phi_1 x_{t-1} + \phi_2 x_{t-2} + w_t$, with a constant mean, $E(x_t) = E(x_{t-1}) = E(x_{t-2})$, show that

$$E(x_t) = \frac{\phi_0}{1 - \phi_1 - \phi_2},$$

so long as $\phi_1 + \phi_2 \neq 1$.

5.20. **Characteristic roots of an AR(2) model.**

(a) For each pair of coefficients, write the characteristic polynomial, evaluate the discriminant, determine the characteristic roots, and compute their moduli.
   i. $\phi_1 = \sqrt{2}$; $\phi_2 = -1/2$
   ii. $\phi_1 = 9/8$; $\phi_2 = -1/4$
   iii. $\phi_1 = 1/4$; $\phi_2 = 1/2$
   iv. $\phi_1 = -1/2$; $\phi_2 = 3/4$
   v. $\phi_1 = -1/2$; $\phi_2 = 1/2$

(b) For each pair of roots, write a characteristic polynomial that has those roots, and write the coefficients of an AR(2) process that has a characteristic polynomial with the given roots.
   i. Roots $z_1 = 2$ and $z_2 = 4$.
   ii. Roots $z_1 = 1 + 2i$ and $z_2 = 1 - 2i$.

5.21. **ACF of AR(2) models.**

For each pair of coefficients, determine the ACF up to the lag specified and comment on the results. You may want to use the R function `ARMAacf`.

(a) $\phi_1 = \sqrt{2}$; $\phi_2 = -1/2$; to lag 15
(b) $\phi_1 = 9/8$; $\phi_2 = -1/4$; to lag 15
(c) $\phi_1 = 1/4$; $\phi_2 = 1/2$; to lag 15
(d) $\phi_1 = -1/2$; $\phi_2 = 3/4$; to lag 100

5.22. **PACF of an AR(2) model.**

    (a) For an AR(2) model, show that $\phi(3,3) = 0$ in equation (5.129).

    (b) For an AR(2) model with $\phi_1 = \sqrt{2}$; $\phi_2 = -1/2$, compute the PACF. You may want to use the R function `ARMAacf`.

5.23. **Yule-Walker equations in an AR(2) model.**

    (a) Show that the denominator in the solutions to the Yule-Walker equations (5.119) and (5.120), $(\gamma(0))^2 - (\gamma(1))^2$, is positive and less than 1.

    (b) Suppose that for a certain process, we have $\gamma(0) = 2$, $\gamma(1) = 0.5$, and $\gamma(2) = 0.9$.

        If the process is an AR(2) model, what are ϕ_1 and ϕ_2?

    (c) Given an AR(2) model with ϕ_1 and ϕ_2 as in Exercise 5.23b, what is the ACF for lags 1 through 5? You may want to use the R function `ARMAacf`.

5.24. **Data generated by an AR(2) model.**

Some previous exercises have involved model (theoretical) properties of an AR(2) process. In this exercise, you are to use simulated data from an AR(2) process. Data can be simulated as discussed in Section 5.3.5.

For each part of this exercise, generate 1,000 observations from the causal AR(2) model

$$x_t = \sqrt{2}x_{t-1} - \frac{1}{2}x_{t-2} + w_t,$$

where w_t has a standardized t distribution with 5 degrees of freedom, and variance of 1 (equation (3.96)).

    (a) Write the code to generate the data as specified, and plot the time series.

    (b) Compute the ACF and the PACF for 15 lags; compare with Exercises 5.21a and 5.22b.

    (c) Use the Yule-Walker equations and the sample ACF to compute estimates of ϕ_1 and ϕ_2.

    (d) Estimate ϕ_1 and ϕ_2 using maximum likelihood (assuming a normal distribution).

5.25. **ARMA and ARIMA models.**

As seen from Figure 5.13, there may be very little visual difference in ARMA models with different values of the coefficients. ARIMA models with $d > 0$ are generally nonstationary, however.

   (a) Generate 500 values of an ARIMA(2,1,2) process with Gaussian innovations and produce a time series plot of the differenced values.

       Does the differenced series appear to be stationary?

   (b) Now fit both the original series and the differenced series using `auto.arima` with both the AIC and the BIC. What models are chosen?

5.26. ARIMA models with innovations with heavy tails.

   (a) Generate four sets of ARIMA data using the same models as used to produce the series shown in Figure 5.13, except for the innovations, use a t distribution with 5 degrees of freedom. (These are the same models used to produce Figure 5.16 in Section 5.3.10.)

       Plot the four series in a manner similar to Figure 5.13, in which a normal distribution was used.

       Comment on the differences in your plots and the plots in Figure 5.13.

   (b) As in Exercise 5.25a, generate 500 values of an ARIMA(2,1,2) process, except with innovations following a t with 5 df. (This can be one of the sets generated in Exercise 5.26a.)

       Produce a time series plot of the differenced values.

       Does the differenced series appear to be stationary?

   (c) Now fit both the original series and the differenced series using `auto.arima` with both the AIC and the BIC. What models are chosen?

   (d) Fit the data generated in Exercise 5.25b using the "correct" model, and compute the residuals from the fitted model.

       Make two q-q plots of the residuals, one using a normal reference distribution and one using a t with 5 df.

5.27. Linear regression with ARMA errors.

This exercise is a continuation of Exercise 4.28 in Chapter 4, and uses the same data, the weekly yields of Moody's seasoned Aaa corporate bonds, the yields of 3-month Treasury bills, the yields of 2-year Treasury notes, the yields of 10-year Treasury notes, and the yields of 30-year Treasury bonds for the years 2008 through 2018.

Using OLS, fit the linear regression of the change in weekly yields of Moody's seasoned Aaa corporate bonds on the weekly changes in yields of 3-month treasury bills, of 2-year treasury notes, of 10-year treasury notes, and of 30-year treasury bonds.

Perform a Durbin-Watson test of the null hypothesis that serial correlations of the errors in the linear regression model up to lag 3 are 0. (This is Exercise 4.28c.)

Discuss the results of the Durbin-Watson test.

Now, based on those results, use GLS to refit the linear regression model.

Discuss the results.

5.28. **Linear regression with ARMA errors.**

Consider the regression of the US gross domestic product (GDP) on some other economic measures relating to the labor force.

There are various measure of the GDP. We will use the dataset GDP at FRED, which is a quarterly, seasonally-adjusted series begun in 1947 and maintained by the US Bureau of Economic Analysis. For regressors, use

- Unemployment Rate: Aged 15-64: All Persons in the United States LRUN64TTUSQ156S

- Average Weekly Hours of Production and Nonsupervisory Employees: Manufacturing AWHMAN

- Civilian Employment-Population Ratio EMRATIO

- Natural Rate of Unemployment (Long-Term) NROU

Information on exactly what these series are can be obtained at the FRED website.

Use all quarterly data from the first quarter of 1987 through the last quarter of 2018. The data obtained from FRED are quarterly except for the monthly data in AWHMAN and EMRATIO, which you must convert to quarterly data.

(a) Fit a linear regression model for GDP on all four regressors listed above.

(b) Analyze the time series of the residuals from the regression fit in Exercise 5.28a.

(c) Based on your analysis of the residuals in Exercise 5.28b, determine an autocorrelation structure, and fit a linear regression model using generalized least squares.

(d) Fit an ARIMA model for GDP with the four regressors as covariates using maximum likelihood, assuming a Gaussian white noise in the ARMA process.

5.29. **GARCH models.**

Obtain the weekly log returns of the S&P 500 index for the period January 1, 1978 through December 31, 2017, and compute the weekly log returns. (The daily data for the same period were used in Section 5.3.11.)

(a) Test for an ARCH effect in this time series.

(b) Fit an ARIMA(1,1,1)+GARCH(1,1) model to the weekly log returns of the S&P 500 index for the period January 1, 1978 through December 31, 2017. (See example on page 582.)

5.30.**Unit roots.**

(a) i. Generate a random walk of length 10 and perform an augmented Dickey-Fuller test on the series for a unit root.
   ii. Generate a random walk of length 100 and perform an augmented Dickey-Fuller test on the series for a unit root.
   iii. Generate a random walk of length 1,000 and perform an augmented Dickey-Fuller test on the series for a unit root.

(b) Determine the coefficients for an AR(2) process with roots 1.00008 and 3.9968 (see Exercise 5.20b and the example on page 592), and simulate a series of length 500 with this model.

Perform an augmented Dickey-Fuller test for unit root.

5.31.**Unit roots.**

(a) Obtain the daily adjusted closing prices of MSFT for the period January 1, 2007, through December 31, 2007.
   i. Perform an augmented Dickey-Fuller test on the series of adjusted closing prices for a unit root.
   ii. Now compute the returns for that period and perform an augmented Dickey-Fuller test on the series of returns for a unit root.

(b) Obtain the daily adjusted closing prices of MSFT for the period January 1, 2018, through December 31, 2018.
   i. Perform an augmented Dickey-Fuller test on the series of adjusted closing prices for a unit root.
   ii. Now compute the returns for that period and perform an augmented Dickey-Fuller test on the series of returns for a unit root.

5.32.**Unit roots; other tests.**

Obtain the quarterly, seasonally-adjusted US gross domestic product (GDP) in the FRED dataset GDP (beginning in 1947). Test for a unit root in the log CDF.

Use six different tests: augmented Dickey-Fuller, Elliott-Rothenberg-Stock, Kwiatkowski-Phillips-Schmidt-Shin (KPSS), Phillips-Perron, Schmidt-Phillips, and Zivot-Andrews. See Table 5.1.

5.33.**Cointegration.**

Simulate a random walk $\{z_t\}$ of length 500. Now generate two series, $\{x_t\}$ and $\{y_t\}$, each of which follows the random walk with an additive white noise, one with variance $\sigma_1^2 = 1$ and the other with variance $\sigma_2^2 = 9$.

Perform a Phillips-Ouliaris cointegration test on the two simulated series.

5.34. **Cointegration.**

Obtain the exchange rates for the US dollar versus the euro from 1999-01-04 to the present and the exchange rates for the Japanese yen versus the US dollar for the same period, and test for cointegration.

What are your conclusions? Discuss.

References

Achelis, Steven B. (2001), *Technical Analysis from A to Z*, McGraw-Hill, New York.

Adler, Robert J.; Raisa Feldman; and Murad S. Taqqu (Editors) (1998), *A Practical Guide to Heavy Tails*, Birkhäuser, Boston.

Arnold, Barry C. (2008), Pareto and generalized Pareto distributions, in *Modeling Income Distributions and Lorenz Curves* (edited by Duangkamon Chotikapanich), Springer, New York, 119–146.

Azzalini, A. (1985), A class of distributions which includes the normal ones, *Scandinavian Journal of Statistics* **12**, 171-178.

Azzalini, Adelchi (2005). The skew-normal distribution and related multivariate families, *Scandinavian Journal of Statistics* **32**, 159-188.

Bacro, J. N., and M. Brito (1998), A tail bootstrap procedure for estimating the tail Pareto-index, *Journal of Statistical Planning and Inference* **71**, 245–260.

Belsley, David A.; Edwin Kuh; and Roy E. Welsch (1980), *Regression Diagnostics: Identifying Influential Data and Sources of Collinearity*, John Wiley & Sons, New York.

Berger, James O. (1993), *Statistical Decision Theory and Bayesian Analysis*, second edition, Springer, New York.

Berlinger, Edina; Ferenc Illés; Milán Badics; Ádám Banai; Gergely Daróczi; Barbara Dömötör; Gergely Gabler; Dániel Havran; Péter Juhász; István Margitai; Balázs Márkus; Péter Medvegyev; Julia Molnár; Balázs Árpád Szücs; and Ágnes Tuza; Tamás Vadász; Kata Váradi; Ágnes Vidovics-Dancs (2015), *Mastering R for Quantitative Finance*, Packt Publishing, Birmingham.

Box, George E. P.; Gwilym M. Jenkins; Gregory C. Reinsel; and Greta M. Ljung (2015), *Time Series Analysis: Forecasting and Control*, fifth edition, John Wiley & Sons, New York.

Burnham, Kenneth P., and David R. Anderson (2002), *Model Selection and Multimodel Inference: A Practical Information-Theoretic Approach*, second edition, Springer, New York.

Buthkhunthong, P.; A. Junchuay; I. Ongeera; T. Santiwipanont; and S. Sumetkijakan (2015), Local Kendalls tau, in *Econometrics of Risk. Studies in Computational Intelligence, volume 583* (edited by V. N. Huynh, V. Kreinovich, S. Sriboonchitta, and K. Suriya), Springer, Cham, 161–169.

CBOE (2019), "Cboe SKEW Index (SKEW)", URL www.cboe.com/SKEW (accessed December 1, 2019).

Chambers, John M. (2008), *Software for Data Analysis. Programming with R*, Springer, New York.

Chambers, John M. (2016), *Extending R*, Chapman & Hall / CRC Press, Boca Raton.

Chen, Ying, and Jun Lu (2012), Value at risk estimation, in *Handbook of Computational Finance* (edited by Jin-Chuan Duan, Wolfgang Härdle, and James E. Gentle), Springer, Berlin, 307–333.

Cherubini, Umberto; Sabrina Mulinacci; Fabio Gobbi; and Silvia Romagnoli (2012), *Dynamic Copula Methods in Finance*, John Wiley & Sons, New York.

Choi, In (2015), *Almost All About Unit Roots*, Cambridge University Press, Cambridge.

Chotikapanich, Duangkamon (Editor) (2008), *Modeling Income Distributions and Lorenz Curves*, Springer, New York.

Choudhry, Moorad (2004), *Analysing & Interpreting the Yield Curve*, John Wiley & Sons, New York.

Claeskens, Gerda, and Nils Lid Hjort (2008), *Model Selection and Model Averaging*, Cambridge University Press, Cambridge.

Cotton, Richard (2017), *Testing R Code*, Chapman & Hall / CRC Press, Boca Raton.

Cox, John C.; Stephen A.Ross; and Mark Rubinstein (1979), Option pricing: A simplified approach, *Journal of Financial Economics* **7**, 229–263.

Davidson, Russell, and James G. MacKinnon (2004), *Econometric Theory and Methods*, Oxford University Press, Oxford.

Drees, Holger; Laurens de Haan; and Sidney Resnick (2000), How to make a Hill plot, *Annals of Statistics* **28**, 254–274.

Duan, Jin-Chuan (1995), The garch option pricing model, *Mathematical Finance* **5**, 13–32.

Duan, Jin-Chuan; Wolfgang Härdle; and James E. Gentle (Editors) (2012), *Handbook of Computational Finance*, Springer, Berlin.

Durbin, J., and S. J. Koopman (2012), *Time Series Analysis by State Space Methods*, second edition, Oxford University Press, Oxford.

Eddelbuettel, Dirk (2013), *Seamless R and C++ Integration with Rcpp*, Springer, New York.

Efron, Bradley, and Robert J. Tibshirani (1993), *An Introduction to the Bootstrap*, Chapman & Hall / CRC Press, Boca Raton.

Ehrentreich, Norman (2008), *Agent-Based Modeling: The Santa Fe Institute Artificial Stock Market Model Revisited*, Springer, New York.

Engle, Robert F., and C. W. J. Granger (1987), Co-integration and error correction: representation, estimation, and testing, *Econometrica* **55**, 251–276.

Fama, Eugene F., and Richard Roll (1968), Some properties of symmetric stable distributions, *Journal of the American Statistical Association* **63**, 817-836.

Fama, Eugene F., and Richard Roll (1971), Parameter estimates for symmetric stable distributions, *Journal of the American Statistical Association* **66**, 331–338.

Fernandez, Carmen, and Mark F. J. Steel (1998). On Bayesian modeling of fat tails and skewness, *Journal of the American Statistical Association* **93**, 359-371.

Fischer, David Hackett (1996), *The Great Wave. Price Revolutions and the Rhythm of History*, Oxford University Press, New York.

Fisher, Mark; Douglas Nychka; and David Zervos (1995), Fitting the term structure of interest rates with smoothing splines, Working Paper No. 95-1, *Finance and Economics Discussion Series*, Federal Reserve Board.

Fouque, Jean-Pierre; George Papanicolaou; and Ronnie Sircar (2000), *Derivatives in Financial Markets with Stochastic Volatility*, Cambridge University Press, Cambridge.

Fouque, Jean-Pierre; George Papanicolaou; Ronnie Sircar; and Knut Sølna (2011), *Multiscale Stochastic Volatility for Equity, Interest Rate, and Credit Derivatives*, Cambridge University Press, Cambridge.

Fox, John, and Sanford Weisberg (2018), *An R Companion to Applied Regression*, third edition, Sage Publications, Inc., Washington.

Fraser, Steve (2005), *Every Man a Speculator. A History of Wall Street in American Life*, HarperCollins Publishers, New York.

Fridson, Martin S. (1998), *It Was a Very Good Year. Extraordinary Moments in Stock Market History*, John Wiley & Sons, New York.

Friedman, Jerome H.; Trevor Hastie; and Rob Tibshirani (2010), Regularization paths for generalized linear models via coordinate descent, *Journal of Statistical Software* **33**(1), doi: 10.18637/jss.v033.i01.

Fuller, Wayne A. (1995), *Introduction to Statistical Time Series*, second edition, John Wiley & Sons, New York.

Galbraith, John Kenneth (1955) *The Great Crash 1929*, Houghton Mifflin, Boston. (Revised edition, 2007, Mariner Books, Boston.)

Gentle, James E. (2003), *Random Number Generation and Monte Carlo Methods*, second edition, Springer, New York.

Gentle, James E. (2009), *Computational Statistics*, Springer, New York.

Gentle, James E. (2017), *Matrix Algebra. Theory, Computations, and Applications in Statistics*, Springer, New York.

Gentle, James E. (2019), *Theory of Statistics*,
`mason.gmu.edu/~jgentle/books/MathStat.pdf`

Gilleland, Eric; Mathieu Ribatet; and Alec G. Stephenson (2013), A software review for extreme value analysis, *Extremes* **16**, 103-119.

Glasserman, Paul (2004), *Monte Carlo Methods in Financial Engineering*, Springer, New York.

Glosten, Lawrence R.; Ravi Jagannathan; and David E. Runkle (1993), On the relation between the expected value and the volatility of the nominal excess return on stocks, *The Journal of Finance* **48**, 1779–1801.

Gordon, John Steele (1999), *The Great Game. The Emergence of Wall Street as a World Power 1653–2000*, Scribner, New York.

Granger, C. W. J., and Roselyne Joyeux (1980), An introduction to long-memory time series models and fractional differencing, *Journal of Time Series Analysis* **1**, 15-30.

Griva, Igor; Stephen G. Nash; and Ariela Sofer (2009), *Linear and Nonlinear Optimization*, second edition, Society for Industrial and Applied Mathematics, Philadelphia.

Harrell, Frank E., Jr. (2015), *Regression Modeling Strategies: With Applications to Linear Models, Logistic and Ordinal Regression, and Survival Analysis*, second edition, Springer, New York.

Haug, Espen Gaarder (2007), *The Complete Guide to Option Pricing Formulas*, second edition, McGraw-Hill Education, New York.

Hastie, Trevor; Robert Tibshirani; and Jerome Friedman (2009), *The Elements of Statistical Learning. Data Mining, Inference, and Prediction*, second edition, Springer, New York.

Helske, Jouni (2017), KFAS: Exponential family state space models in R, *Journal of Statistical Software* **78**(10), doi: 10.18637/jss.v078.i10.

Herranz, Edward (2017), Unit root tests, *WIREs Computational Statistics*, **9**:e1396. doi: 10.1002/WICS.1396

Hill, Bruce M. (1975), A simple general approach to inference about the tail of a distribution, *Annals of Statistics* **3**, 1163–1174.

Hull, John C. (2015) *Risk Management and Financial Institutions*, fourth edition, John Wiley & Sons, Inc., Hoboken.

Hull, John C. (2017) *Options, Futures, and Other Derivatives*, tenth edition, Pearson, New York.

Hyndman, Rob J., and Yeasmin Khandakar (2008), Automatic time series forecasting: The forecast package for R, *Journal of Statistical Software* **26**(3), doi 10.18637/jss.v027.i03.

James, Gareth; Daniela Witten; Trevor Hastie; and Robert Tibshirani (2013), *An Introduction to Statistical Learning with Applications in R*, Springer, New York.

Joanes, D. N., and C. A. Gill (1998), Comparing measures of sample skewness and kurtosis, *Journal of the Royal Statistical Society (Series D): The Statistician* **47**, 183–189.

Jobson, J. D., and B. Korkie (1980), Estimation for Markowitz efficient portfolios, *Journal of the American Statistical Association* **75**, 544–554.

Joe, Harry (2015), *Dependence Modeling with Copulas*, Chapman & Hall / CRC Press, Boca Raton.

Jondeau, Eric; Ser-Huang Poon; and Michael Rockinger (2007), *Financial Modeling Under Non-Gaussian Distributions*, Springer, New York.

Jones, M. C. (1996), The local dependence function, *Biometrika* **83**, 899-904.

Jorion, Philippe (2006), *Value at Risk: The New Benchmark for Managing Financial Risk*, third edition, McGraw-Hill Education, New York.

Karian, Zaven A., and Edward J. Dudewicz (2000), *Fitting Statistical Distributions: The Generalized Lambda Distribution and Generalized Bootstrap Methods*, Chapman & Hall / CRC Press, Boca Raton.

Keen, Kevin J. (2018), *Graphics for Statistics and Data Analysis with R*, second edition, Chapman & Hall / CRC Press, Boca Raton.

Kindleberger, Charles P., and Robert Aliber (2011), *Manias, Panics, and Crashes. A History of Financial Crashes*, sixth edition, John Wiley & Sons, Inc., Hoboken.

Kleiber, Christian (2008), A guide to the Dagum distributions, in *Modeling Income Distributions and Lorenz Curves* (edited by Duangkamon Chotikapanich), Springer, New York, 97–118.

Knight, Frank (1921), *Risk, Uncertainty, and Profit*, Hart, Schaffner & Marx, Boston.

Krishnamoorthy, K. (2015), *Handbook of Statistical Distributions with Applications*, second edition, Chapman & Hall / CRC Press, Boca Raton.

Kruschke, John K. (2015), *Doing Bayesian Data Analysis. A Tutorial with R, JAGS, and Stan*, second edition, Academic Press / Elsevier, Cambridge, Massachusetts.

Kutner, Michael; Christopher Nachtsheim; John Neter; and William Li (2004), *Applied Linear Statistical Models*, fifth edition, McGraw-Hill/Irwin, New York.

Lahiri, S. N. (2003), *Resampling Methods for Dependent Data*, Springer, New York.

Leemis, Lawrence M., and Jacquelyn T. McQueston (2008), Univariate distribution relationships, *The American Statistician*, **62**, 45–53.

Lehalle, Charles-Albert, and Sophie Laruelle (2013), *Market Microstructure in Practice*, World Scientific Publishing Company, Singapore.

Lo, Andrew W., and A. Craig MacKinlay (1999), *A Non-Random Walk Down Wall Street*, Princeton University Press, Princeton.

Lütkepohl, Helmut (2007), *New Introduction to Multiple Time Series Analysis*, Springer, New York.

Maddala, G. S., and In-Moo Kim (1998), *Unit Roots, Cointegration, and Structural Change*, Cambridge University Press, Cambridge.

Malkiel, Burton G. (1999), *A Random Walk Down Wall Street*, revised edition, W. W. Norton and Company Inc., New York.

Mandelbrot, Benoit B., and Richard L. Hudson (2004), *The (mis)Behavior of Markets. A Fractal view of Risk, Ruin, and Reward*, Basic Books, New York.

Markowitz, H. M. (1952), Portfolio selection, *Journal of Finance* **7**, 77–91.

McMillan, Lawrence G. (2012), *Options as a Strategic Investment*, fifth edition, Prentice Hall Press, Upper Saddle River, NJ.

McNeil, Alexander; Rüdiger Frey; and Paul Embrechts (2015), *Quantitative Risk Management: Concepts, Techniques and Tools*, revised edition, Princeton University Press, Princeton.

Michaud, Richard O. (1989), The Markowitz optimization enigma: Is "optimized" optimal? *Financial Analysts Journal* **45**, 31–42.

Murrell, Paul (2018), *R Graphics*, third edition, Chapman & Hall / CRC Press, Boca Raton.

Nelson, Charles R., and Andrew F. Siegel (1987), Parsimonious modeling of yield curves, *Journal of Business* **60**, 473-489.

Niederreiter, Harald (2012), Low-Discrepancy Simulation, in *Handbook of Computational Finance* (edited by Jin-Chuan Duan, Wolfgang Härdle, and James E. Gentle), Springer, Berlin, 703–730.

Nolan, Deborah, and Duncan Temple Lang (2014), *XML and Web Technologies for Data Science with R*, Springer, New York.

Nolan, John P. (1997), Numerical calculation of stable densities and distribution functions, *Communications in Statistics: Stochastic Models* **13**, 759-774.

Novak, Serguei Y. (2012), *Extreme Value Methods with Applications to Finance*, CRC Press, Boca Raton.

O'Hagan, A., and Tom Leonard (1976), Bayes estimation subject to uncertainty about parameter constraints, *Biometrika* **63**, 201–203.

Okhrin, Ostap (2012), Fitting high-dimensional copulae to data, in *Handbook of Computational Finance* (edited by Jin-Chuan Duan, Wolfgang Härdle, and James E. Gentle), Springer, Berlin, 469–501.

Pinheiro, Jose, and Douglas M. Bates (2009), *Mixed-effects Models in S and S-PLUS*, Springer, New York.

Perlin, Marcelo S. (2017), *Processing and Analyzing Financial Data with R*, Independently Published, ISBN: 978-85-922435-5-5.

Petitt, Barbara S.; Jerald E. Pinto; and Wendy L. Pirie (2015), *Fixed Income Analysis*, third edition, John Wiley & Sons, Inc., Hoboken.

Pfaff, Bernhard (2008), *Analysis of Integrated and Cointegrated Time Series with R*, Springer, New York.

Pictet, Olivier V.; Michel M. Dacorogna; and Ulrich A. Müller (1998), Hill, bootstrap and jackknife estimators for heavy tails, in *A Practical Guide to Heavy Tails* (edited by Robert J. Adler, Raisa Feldman, and Murad S. Taqqu), Birkhäuser, Boston, 283–310.

Politis, Dimitris N. (2002), A new approach on estimation of the tail index, *Comptes Rendus Mathematique* **335**, 279–282.

R Core Team (2019), *R: A language and environment for statistical computing*, R Foundation for Statistical Computing, Vienna, Austria. URL www.R-project.org/.

Ramberg, John S., and Bruce W. Schmeiser (1974), An approximate method for generating asymmetric random variables, *Communications of the ACM* **17**, 78–82.

Reiss, Rolf-Dieter, and Michael Thomas (2007), *Statistical Analysis of Extreme Values with Applications to Insurance, Finance, Hydrology and Other Fields*, third edition, Birkhäuser, Boston.

Royall, Richard (1997), *Statistical Evidence: A Likelihood Paradigm*, Chapman & Hall / CRC Press, Boca Raton.

Scott, David W. (2015), *Multivariate Density Estimation: Theory, Practice, and Visualization*, second edition, John Wiley & Sons, Inc., Hoboken.

Sharpe, W. F. (1966), Mutual fund performance, *Journal of Business* **39**, 119–138.

Sharpe, William F.; Gordon J. Alexander; and Jeffrey W. Bailey (1999), *Investments*, sixth edition, Prentice-Hall, Upper Saddle River, NJ.

Sheather, Simon J. (2009), *A Modern Approach to Regression with R*, Springer, New York.

Sims, Christopher A. (1988), Bayesian skepticism on unit root econometrics, *Journal of Economic Dynamics and Control* **12**, 463–474.

Smithers, Andrew (2009), *Wall Street Revalued: Imperfect Markets and Inept Central Bankers*, John Wiley & Sons, Inc., Hoboken.

Stuart, Alan, and J. Keith Ord (1994), *Kendall's Advanced Theory of Statistics: Volume 1: Distribution Theory*, sixth edition, Hodder Education Publishers, London.

Svensson, Lars (1994), *Estimating and Interpreting Forward Interest Rates: Sweden 1992 – 1994*, Working Paper No. 4871, National Bureau of Economic Research

Taleb, Nassim Nicholas (2010), *The Black Swan: The Impact of the Highly Improbable*, second edition, Random House, New York.

Taveras, John L. (2016), *R for Excel Users: An Introduction to R for Excel Analysts*, CreateSpace Independent Publishing Platform.

Thaler, Richard H. (2015), *Misbehaving. The Making of Behavioral Economics*, W. W. Norton and Company Inc., New York.

Tsay, Ruey S. (2010), *Analysis of Financial Time Series*, third edition, John Wiley & Sons, Inc., Hoboken.

Tukey, John W. (1962), The future of data analysis, *Annals of Mathematical Statistics* **33**, 1–67.

Vidyamurthy, Ganapathy (2004), *Pairs Trading. Quantitative Methods and Analysis*, John Wiley & Sons, Inc., Hoboken.

Wang, Yuedong (2012), Model Selection, in *Handbook of Computational Statistics: Concepts and Methods*, second revised and updated edition, (edited by James E. Gentle, Wolfgang Härdle, and Yuichi Mori), Springer, Berlin, 469–497.

Wasserstein, Ronald L.; Allen L. Schirm; and Nicole A. Lazar (2019), Moving to a world beyond "$p < 0.05$", *The American Statistician* **73**, 1–19.

Wickham, Hadley (2019), *Advanced R*, second edition, Chapman & Hall / CRC Press, Boca Raton.

Wickham, Hadley (2016), *ggplot2. Elegant Graphics for Data Analysis*, second edition, Springer, New York.

Wilder, J. Welles, Jr. (1978), *New Concepts in Technical Trading Systems* Trend Research.

Wiley, Matt, and Joshua F. Wiley (2016), *Advanced R: Data Programming and the Cloud*, Apress, New York.

Williams, John Burr (1938), *The Theory of Investment Value*, North Holland, Amsterdam.

Index

$\nabla f(\theta, \gamma)$, 359
$\triangle(x_t)$, 497
$\triangle^{-1}(x_t)$, 500

absolute return, 101
acceptance/rejection method, 317
accredited investor, 18
ACF (autocorrelation function), 8,
    100, 515
    AR model, 541–546
    AR(1), 538
    difference equation, 544
    MA model, 531
    partial (PACF), 516, 547
    sample (SACF), 100, 519
adjusted price, 45–48
    dividend, 46, 134
    effect on option strike, 64
    return of capital, 48
    stock split, 45
adjusted R^2, 423, 468
after hours, 10, 37
agent-based model, 357
aggregate (R function), 510
AIC (Akaike information criterion),
    445, 468
AICc, 469
Akaike information criterion (AIC),
    445, 468
algo (programmed trading), 126
alpha, 63, 126
alternative hypothesis, 382
alternative trading system (ATS),
    16, 35, 129
analysis of variance (ANOVA),
    415–421, 440
Anderson-Darling test, 461

annualized log return, 30
annualized simple return, 29
annualized volatility, 107, 491
ANOVA (analysis of variance),
    415–421, 440
apply (R function), 150
APT (arbitrage pricing theory), 44
AR model, 534–571
    AR(1), 536–539
        ACF, 538
        PACF, 539
    AR(2)
        ACF, 543
    AR(p), 534–548
        ACF difference equation, 546
        order determination, 548
    autocorrelation, 538, 543, 547
    causality, 539
    explosive, 538
    partial autocorrelation function
        (PACF), 547
    stationarity, 535
arbitrage, 38
    statistical, 600
arbitrage pricing theory (APT), 44
ARCH (GARCH *see*), 576
    ARCH effect, 578
        tests for, 579
Archimedean copula, 266
ARFIMA model, 554
arima (R function), 561
ARIMA model, 553–571
    order determination, 560,
        563–565
arima.sim (R function), 555
ARMA model, 549–571
    causality, 551

invertibility, 552
order determination, 557, 560,
    563–565
ARMAX model, 575
asymptotic distribution, 270, 271,
    308
asymptotic inference, 378, 393
ATS (alternative trading system),
    16, 35, 129
augmented Dickey-Fuller test,
    595–596, 598
autocorrelation, 7, 511–513, 518
    returns, 99–101
    tests, 526, 580
autocorrelation function (ACF), 8,
    100, 511–513, 515
    AR model, 541–546
    AR(1), 538
    difference equation, 544
    MA model, 531
    partial (PACF), 516, 547
    sample (SACF), 100, 519
autocovariance function, 515
    partial, 516
    sample, 519
autoregressive model, 338, 492,
    534–571
    autocorrelation, 538, 543, 547
    stationarity, 535
    vector autoregressive model,
        535, 590
autoregressive operator, 535
    multivariate, 536

backshift operator, B, 495, 531, 535
    inverse, B^{-1}, 496
    multivariate, 497, 531, 536
bagplot, 227, 236
Bank of Japan (BOJ), 22
Basel Committee, 19, 119
    Basel Accords, 119
    Basel II, 119
basis point, 22
Bayes information criterion (BIC),
    468

Bayesian model, 346, 386–393
    conjugate prior, 388
    MCMC, 389
Bayesian network, 284
bear market, 126
Bernoulli distribution, 283, 450
beta, 61–63, 126
beta distribution, 291
bias, 347
    bootstrap estimate, 395
bias-variance tradeoff, 218, 220, 370,
    379, 402, 463, 465
BIC (Bayes information criterion),
    468
bid/ask, 35
    bid-ask spread, 37
big data, 139
bill, Treasury, 22
binary regression, 451
binary regressor, 415, 419
binary tree, 281
binomial distribution, 283, 450
binomial tree, 281, 283
Black Monday (October 19, 1987),
    99, 159
black swan, 132
Black Thursday (October 24, 1929),
    159
Black Tuesday (October 29, 1929),
    159
Black-Scholes-Merton differential
    equation, 280
BlackRock iShares, 56
BOJ (Bank of Japan), 22
bond, 13, 20
    coupon rate, 26
    coupon stripping, 28
    coupon yield to maturity, 28
    par value, 21
    price, 28
    yield to maturity, 28
    zero-coupon, 28
bond, Treasury, 22
book value, 38
bootstrap, 393–396

bias estimate, 395
boot (R package), 396
bootstrap (R package), 396
confidence interval, 395
variance estimate, 394
Bowley's skewness, 96
box-and-whisker plot, 224, 225
Box-Cox transformation, 345, 422, 579
Box-Ljung test, 526, 580
Box-Pierce test, 526
boxplot, 110, 224, 225
bivariate, 227, 236
Brownian motion, 279
BUGS (software), 389
Bulletin Board (OTC), 16
burn-in for a stochastic process, 357
Burr distributions, 324
buyback, stock, 33, 38, 43, 44

C_p, 424, 468
CAC-40 (index), 60, 167
calendar spread, 74
call option, *see* option
call spread, 74
candlestick graphic, 33, 182
CAPE (Shiller P/E ratio), 43
capital asset pricing model (CAPM), 44
capital market line, 72
capitalization, market, 32
CAPM (capital asset pricing model), 44
Case-Shiller Index, 168
cash settlement, 64
Cauchy distribution, 293
causality, 528, 537, 539, 551
CBOE (Chicago Board Options Exchange), 17, 108, 113
CBOE SKEW Index, 113
CBOE Volatility Index, 108–111, 166
CCF (cross-correlation function), 517
sample, 519
CDF (cumulative distribution function), 211, 249

complementary, 258
CDF-skewing, 302
CDO (collateralized debt obligation), 20
Center for Research in Security Prices (CRSP), 53, 171
centered data, 208–210, 408
central limit theorem, 272
Chambers, John, 140
change of variables technique, 255
characteristic polynomial, 503, 531, 533, 535, 539, 543, 545, 546
AR model, 535, 543
discriminant, 504
MA model, 531, 533
chi-squared distribution, 289, 292, 313, 383
chi-squared test, 457
Chicago Board Options Exchange (CBOE), 17, 108, 113
Chicago Mercantile Exchange (CME), 17
Cholesky factor, 269, 310, 320
class (R object), 143
classification model, 337, 449
logistic regression, 449
Clayton copula, 266
cleaning data, 183–186
missing data, 185
CME (Chicago Mercantile Exchange), 17
CME Group, 17
coefficient of multiple correlation, 417
coefficient of multiple determination, 417
coherent measure, 118, 257
cointegration, 599–602
Engle-Granger test, 600
Johansen test, 600
Phillips-Ouliaris test, 600
collateralized debt obligation (CDO), 20
COMEX, 17
comma separated file (CSV), 162–163

header, 163
Commodity Futures Trading
        Commission (CFTC), 18
complementary cumulative
        distribution function
        (CCDF), 258
composite hypothesis, 381
composition of functions, 495
compound distribution, 307
compounding, 26
        continuous, 28
computational inference, 378, 393
conditional distribution, 6, 260, 310,
        513
conditional heteroscedasticity,
        575–583
conditional value at risk, 119, 404
confidence interval, 379, 448
        basic bootstrap, 395
conjugate prior, 388
consistency, 348
consistent estimator, 348
constrained optimization, 359, 362
        feasible set, 362
Consumer Price Index (CPI), 168
continuous compounding, 28
continuous data, 206
continuous distribution, 252, 285
contour plot, 235
Cook's distance, 433, 440, 443
copula, 264–267, 310, 321, 334, 350
        Archimedean, 266
        copula (R package), 350
        definition, 265
        fitting, 350
        Gaussian, 266
        normal, 266
Cor(·, ·) (correlation, *see*), 4
Cornish-Fisher expansion, 406
corporate bond, 25
        seasoned, 26
correction, 25
correlation, 83, 85, 208, 210,
        261–263, 584–591

among assets in financial crises,
        127, 130, 276
autocorrelation, 100
distribution, 408–409, 584–591
Kendall's tau, 98, 399
Pearson correlation coefficient,
        83
sample, 83, 408
Spearman's rank correlation, 97,
        399
spurious, 584
tail dependence, 127, 130, 276,
        402
test, 409
correlogram, 519
coupon bond, 21
coupon rate, 26
coupon stripping, 28
Cov(·, ·) (covariance, *see*), 4
covariance, 82, 208, 210, 261–263
covered call, 74
Cox-Ross-Rubinstein (CRR) model,
        282
Cramér-von Mises test, 461
CRAN (Comprehensive R Archive
        Network), 146
credit rating agency, 19
cross-correlation function (CCF),
        513, 517
        sample, 519
cross-correlogram, 519, 522
cross-covariance function, 513
        sample, 519
cross-validation, 463–467
        k-fold, 465
        leave out one (LOOCV), 464
crossover, 122
CRSP (Center for Research in
        Security Prices), 53, 171
        database, 53, 171
cryptocurrency, 14, 38
CSV file (comma separated file),
        162–163
        header, 163
cumsum (R function), 154

cumulative distribution function
(CDF), 211, 249
complementary, 258
currency exchange rates, 167, 168
CVaR, 119, 404

D'Agostino-Pearson test, 461
DAG (directed acyclic graph), 388
daily log return, 30
dark pool, 16, 35, 129
data, 2
continuous, 206
copyright, 164
derived, 158
discrete, 206
observed, 158
repository, 163
data cleansing, 183–186
data quality, 197, 481
missing data, 185
data frame, 152–153, 161
stringsAsFactors, 152, 163
converting to xts object, 174
data-generating process, 2, 311
model, 311
data reduction, 341
data structure, 160
data wrangling and munging, 183
datasets
CanadaDollar/USDollar daily,
601
Dow Jones daily, 50, 55, 85, 98,
127
Dow Jones daily 1987-2017, 83
GLD daily, 85
INTC daily, 9, 34, 39, 51, 70, 76,
77, 79, 80, 85, 86, 88, 98,
124, 127
INTC daily 2017, 61
INTC monthly, 9
INTC price adjustments, 47, 48
INTC weekly, 9
Nasdaq daily, 50, 55, 85, 86, 98,
127
Nasdaq daily 1987-2017, 83

S&P 500 daily, 50, 53, 55, 61,
83, 85, 89, 90, 92, 98–101,
108, 127, 569, 579
S&P 500 monthly, 93, 102, 105
S&P 500 Total Return daily, 53,
56
S&P 500 weekly, 93, 105
SKEW daily, 108
SPY daily, 56
T-Bill, 3-month monthly, 22
USDollar/Euro daily, 14, 601
VIX daily, 108, 110
yield curves, 24, 25
date data, 156, 174
in data frames, 161
in xts objects, 174
ISO 8601, 157, 176
POSIX, 157, 176
DAX (index), 60, 167
death cross, 123
deleted residual, 429, 464
delta (Greek), 65, 75, 137
density plot, 79, 226
bivariate, 227
derivative, 13, 63–66
descriptive statistics, 94, 205
detrending, 505, 508
deviance, 373, 451, 468
deviance residual, 374, 451
df (degrees of freedom), 289
DFBETAS, 433, 440, 443
DFFITS, 433, 440, 443
Dickey-Fuller test, 595
augmented, 595–596
diff (R function), 154, 174, 176, 499
cumsum, 154
diffinv, 154
on xts object, 155, 174
difference equation, 501–505, 545,
546
difference operator, $\triangle$ or $(1-B)$, 497
fractional differencing, 498
differential scaling, 302
diffusion process, 279, 280, 490
dimension reduction, 341

directed acyclic graph (DAG), 388
discount rate, 22
discounted cash flow, 42
discounted value, 28
discrete data, 206
discrete distribution, 248, 283
discrete uniform distribution, 213,
　　323, 393
dispersion, 207
distribution family
　　Bernoulli, 283, 450
　　beta, 286, 313
　　binomial, 283, 284, 313, 450
　　Burr distributions, 324
　　Cauchy, 286, 293, 313
　　chi-squared, 289, 292, 313, 383
　　Dagum, 301
　　Dirichlet, 287, 294, 313
　　discrete uniform, 213, 323, 393
　　double exponential, 286, 292,
　　　　299, 313
　　elliptical, 263, 295
　　exponential, 286, 292, 313
　　extreme value, 309
　　F, 289, 313, 383
　　gamma, 286, 313
　　Gaussian, 87, 286, 287, 312, 322
　　generalized error distribution,
　　　　299–300
　　generalized extreme value, 309
　　generalized lambda, 299, 317
　　generalized Laplace, 300
　　generalized Pareto, 301
　　geometric, 284, 313
　　Gumbel, 309
　　heavy-tailed, 273–277
　　hypergeometric, 284, 313
　　Johnson distributions, 324
　　Laplace, 286, 292, 299, 313
　　location-scale, 229, 298
　　logistic, 286, 313
　　lognormal, 286, 288, 313
　　Lorentz, 286
　　multinomial, 284, 294, 313

　　multivariate normal, 287, 295,
　　　　313
　　multivariate t, 295, 313
　　negative binomial, 284, 313
　　normal, 87, 286, 287, 312, 313,
　　　　322
　　　skewed, 302
　　outlier-generating, 277
　　Pareto, 286, 292, 313
　　Pearson distributions, 324
　　Poisson, 284, 285, 312, 313, 322,
　　　　352
　　power law, 292
　　R functions for, 144, 313
　　skewed, 302
　　skewed generalized error
　　　　distribution, 300
　　stable, 298, 313
　　t, 289, 296–298, 313, 383
　　　location-scale t, 297, 298
　　　noncentral t, 290
　　　skewed, 302
　　　"standardized" t, 298
　　uniform, 252, 286, 313, 315
　　variance gamma, 300
　　Weibull, 286, 292, 313
diversification, 66, 114
dividend, 45, 171
　　adjustment to prices, 46, 134
　　`getDividends` (R function), 171
　　reinvestment program, 45
　　stock, 45
DJIA (Dow Jones Industrial
　　　Average), 49, 166, 479
　　divisor, 49
　　historical values, 51, 55
double exponential distribution, 286,
　　292, 299, 313
Dow Jones Industrial Average
　　　(DJIA), 49, 166, 479
　　divisor, 49
　　historical values, 51, 55
Dow Jones Transportation Average,
　　50, 166
Dow Jones Utility Average, 50, 166

`dplyr` (R package), 153, 497
Durbin-Watson test, 434, 443, 526, 571
DVaR, 119, 276

E(·) (expectation, *see*), 4
earnings, 41
    earnings per share (EPS), 41
    forward earnings, 41
    trailing earnings, 41
EBITDA, 43
ECB (European Central Bank), 22
ECDF (empirical cumulative distribution function), 211–217, 276, 350
    folded, 214
EDA (exploratory data analysis), 1, 205
EDGAR, 169
effective fed fund rate, 22, 25, 436, 438
efficient frontier, 71
efficient market, 120
Efficient Market Hypothesis, 120
80/20 rule, 293
elastic net, 370, 376
Elliott-Rothenberg-Stock test, 598
elliptical distribution, 263, 295
emerging markets, 60
empirical cumulative distribution function (ECDF), 211–217, 276, 350
    folded, 214
empirical quantile, 95, 215–217
endogenous/exogenous variable, 337
Engle-Granger test, 600
enterprise value (EV), 33
equity asset, 13
error (in a model), 246, 534
    compared to residual, 415
error correction model (ECM), 601
estimation, 311, 349–354
    computational inference, 393
    least squares, 351, 363–370

    maximum likelihood, 352–354, 371–375
    method of moments, 349
    Monte Carlo, 355–357
    tuned inference, 354, 401, 402
ETF (exchange-traded fund), 18, 56–59
    index, 56
    inverse, 57–59
    leveraged, 57–59
    sector, 56
    short, 59
    tracking, 56
    ultra, 57–59
ETN (exchange-traded note), 18, 111
Euromarket, 19, 22
European Central Bank (ECB), 22
EV (enterprise value), 33
EV/EBITDA, 43
evolutionary time series, 515
EWI (S&P 500 Equal Weight Index), 50
EWMA (exponentially weighted moving average), 39, 123
    model, 534
exact inference, 393
exceedance distribution, 232, 275
    q-q plot, 232
exceedance threshold, 232, 275
excess kurtosis, 82, 88, 251, 254
excess return, 32, 60–63, 73
    alpha, 63
    mixing log returns and interest rates, 32
exchange, 16
exchange-traded fund (ETF), 18, 56–59
    index, 56
    inverse, 57–59
    leveraged, 57–59
    sector, 56
    short, 59
    ultra, 57–59
exchange-traded note (ETN), 18, 111
expectation, 250, 254

conditional, 6, 260, 513
  linear operator, 256
expected shortfall, 119, 404
exploratory data analysis, 1, 205
exponential distribution, 292
exponential smoothing, 41, 534
exponential tail, 274
exponentially weighted moving
        average (EWMA), 39, 123
  model, 534
externally studentized residual, 432,
        433
extreme value theory, 308

F distribution, 289, 313, 383
F test, 421
face value, 21
factor model, 44, 413
fair market value, 32, 120
FARIMA model, 554
fat-tailed distribution, 273
fed discount rate, 22
fed fund rate, 22, 25, 438
  effective, 22, 25, 436, 438
Federal Open Market Committee
        (FOMC), 21, 25, 438
filter, 40, 172, 495–500
financial instrument, 13
FINRA (Financial Industry
        Regulatory Authority), 18
fiscal policy, 19
Fisher scoring, 375
Fitch Ratings, 19
fitting matrix, 420
fixed-income asset, 13
FOMC (Federal Open Market
        Committee), 21, 25, 438
foreach (R function), 396, 465
forecasting in time series, 527, 561,
        565–566
  forecast horizon, 527
  forecast origin, 527
foreign exchange, 13, 167, 168
forex, 13, 170
formula in R, 454

forward earnings, 41
forward rate, 27
forward selection, 425, 444
forward shift operator, B^{-1}, 496
fractional differencing, 498
Frank copula, 267
FRED (Federal Reserve Economic
        Data), 167–168
  missing data, 189
front-running, 129
FTSE-100 (index), 60, 167
fundamental analysis, 35
future value, 27, 28
futures, 13
FX, 13, 170

GAAP (Generally Accepted
        Accounting Principles), 43
gamma distribution, 291
GARCH model, 308, 580–583
Gauss-Markov theorem, 419
Gauss-Newton method, 367
Gaussian copula, 266
Gaussian distribution, 87, 287, 312,
        322
Gaussian white noise, 488
GED (generalized error distribution),
        299–300
generalized error distribution,
        299–300
  skewed, 300
generalized extreme value
        distribution, 309
generalized inverse (of a function),
        250
generalized lambda distribution, 299,
        317
generalized least squares, 369, 376
generalized linear model, 376, 451
generalized Pareto distribution, 301
Generally Accepted Accounting
        Principles (GAAP), 43
generating random numbers, 314–321
  quasirandom numbers, 316, 321
Gentleman, Robert, 140

geometric Brownian motion, 279
geometric distribution, 284
geometric random walk, 492
geometric series, convergence of, 500
`getSymbols` (R function), 165, 168,
        171–172
    missing data, 165, 186
`ggfortify` (R package), 180
`ggplot2` (R package), 146, 156, 187
GICS (Global Industry Classification
        Standard), 56
Gini index, 214, 293
GLD (ETF), 56, 86, 403
Global Industry Classification
        Standard (GICS), 56
GLS (generalized least squares), 369,
        376
golden cross, 123
Gompertz model, 366
goodness-of-fit test, 456
    Anderson-Darling, 461
    chi-squared, 457
    Cramér-von Mises, 461
    D'Agostino-Pearson, 461
    Jarque-Bera, 461
    Kolmogorov-Smirnov, 460
    Lilliefors, 461
    Shapiro-Francia, 461
    Shapiro-Wilk, 461
    test for normality, 461
Google, 164
government statistics, 164
gradient, $\nabla f(\theta, \gamma)$, 359
graphics, 221–238
    bagplot, 227, 236
    bivariate boxplot, 227, 236
    bivariate density plot, 87, 227,
        236
    boxplot, 224
    color, 156
    density plot, 79, 226
    ECDF plot, 212
    folded ECDF plot, 214
    half-normal plot, 233
    histogram, 76, 115, 222

layer, 235, 237
log density plot, 90, 91, 227
mountain plot, 214
OHLC dataset, 34, 182
parametric line plot, 236
q-q plot, 91–93, 229–233, 276,
        306
quantile-quantile plot, 91–93,
        229–233, 276, 306
R, 234–238
scatterplot, 235
scatterplot matrix, 86, 235
starburst plot, 227
three-dimensional, 235
`xts` object, 179, 182
Greeks, 65
Group of Ten, 19
Gumbel copula, 266
Gumbel distribution, 309

half-normal plot, 233
Halton sequence, 316
Hang Seng (index), 60, 167
Hannan-Quinn information criterion,
        471, 583
hard-to-borrow, 66
hat matrix, 420
hazard function, 257
head and shoulders, 124
heavy-tailed distribution, 89,
        273–277
hedge, 73–75
    cross hedge, 73
    dynamic, 75
    hedge ratio, 74
hedge fund, 17, 75
Hessian, $H_f(\theta, x)$, 359
heteroscedasticity, 99, 414
    conditional, 576
high-frequency trading, 12, 132
Hill estimator, 400–402
    Hill plot, 402, 471
    interval of stability, 402
histogram, 76, 115, 217–218, 222
    Sturges's rule, 218

historical volatility, 103
    compared with implied
        volatility, 108
Hmisc (R package), 146
Holt-Winters method, 41, 534
hypergeometric distribution, 284
hypothesis testing, 311, 381–384
    multiple hypotheses, 384, 525
    one-sided test, 384
    power, 382
    simultaneous tests, 384, 525
    type I and II errors, 382

I(d) process, 517, 533
ICO (initial coin offering), 38
identifiability, 532
Ihaka, Ross, 140
iid (independently and identically
        distributed), 5, 260
IMA(d, q) model, 533
implied volatility, 107
    compared with historical
        volatility, 108
income, 45
independent random variables, 259
index mutual fund, 18
index, market, *see also* stock index,
        49–55, 413, 479
inference, 347, 523
    asymptotic inference, 378, 525
    computational inference, 378,
        393
    exact inference, 378
    tuned inference, 218, 220, 354,
        370, 379, 401, 402, 463
information criteria, 471, 583
initial coin offering (ICO), 38
initial public offering (IPO), 37
innovation, 534
inputting data from the internet,
        163, 165
    getSymbols, 171
insider information, 128
institutional investor, 15, 125
instrumental variable, 337

integration (in time series), 500, 516
    integrated ARMA model
        (ARIMA), 553
    integrated moving average
        model, 533
    integrated white noise, 516
    integration operator, $S(x_t)$, 500
interest, 20–28
    compounded, 26
        continuously, 28
interest rate, 21–28
    annualized, 21, 29
    daily, 29
    fed discount rate, 22
    LIBOR, 22
    prime rate, 22
    risk-free interest rate, 22
    structure, 24
internally standardized residual, 433
international markets, 59, 166
interquartile range (IQR), 96, 208,
        216, 225, 397
    relationship to standard
        deviation, 96, 208
intrinsic value, 64
inverse CDF method, 317
inverse ETF, 59, 111
invertibility, 532, 534, 552
investor, 15
    accredited, 18
IPO (initial public offering), 37
IQR (interquartile range), 96, 208,
        216, 225, 397
    relationship to standard
        deviation, 96, 208
iShares, 56
ISLR (R package), 451
ISO 8601 date format, 157, 176

Jacobian (of a transformation), 268
JAGS (software), 389
Jarque-Bera test, 461
Joe copula, 267
Johansen test, 600
Johnson S_U distributions, 324

joint distribution, 258
joint stationarity, 517–518
Journal of Statistical Software, 187
jump, 77, 285

Kaggle, 164
Kendall's tau, 98, 399
kernel density estimation, 79,
219–221, 226, 236
multivariate, 220, 227
Knightian uncertainty, 127
Kolmogorov-Smirnov test, 460
KPSS (Kwiatkowski-Phillips-
Schmidt-Shin) test, 507,
598
KS test, 460
kurtosis, 81, 251, 254
excess, 82, 88
Kwiatkowski-Phillips-Schmidt-Shin
(KPSS) test, 507, 598

L_1 and L_2 norms, 351
L_1 estimation, 351, 376, 397, 429
lag, 495, 497
backshift operator, B, 495
lagging indicator, 514, 517
Lagrange multiplier, 362
Lagrangian function, 362
Laplace distribution, 286, 292, 299,
313
lasso regression, 369, 427, 447
law of one price, 38
leading indicator, 25, 514, 517
least absolute values (L_1) estimation,
351, 376, 397, 429
least squares estimation, 351,
363–370, 376
Gauss-Markov theorem, 419
Gauss-Newton method, 367
generalized least squares, 369,
376
nonlinear least squares, 366–368,
376
nonnegative least squares, 370,
376

penalized, 369, 427, 447, 466
prediction, 448
properties (algebraic), 364
properties (statistical), 415, 418,
430
regularized, 369, 427, 447, 466
shrunken, 369, 427, 447, 466
weighted least squares, 369
leave-out-one deletion, 433, 440
cross-validation, 464
leptokurtic, 88
leverage in regression, 429
Cook's distance, 433, 440
DFBETAS, 433, 440
DFFITS, 433, 440
leveraged instrument, 13, 57–59
LIBOR, 22
likelihood function, 352–354,
371–375, 398
likelihood ratio, 385
Lilliefors test, 461
limit order, 35
linear function, 495
linear model, 376, 413–448, 493, 599
intercept (or not), 365, 419
linear operator, 256, 495–505
expectation, 256
linear process, 489
one-sided, 489, 529, 537
linear transformation, 209, 229, 320,
495
link function, 450
Ljung-Box test, 526, 580
location, 207
location-scale family of distributions,
229, 298
log density plot, 91, 227
log return, 29–31, 491, 500
annualized, 30
log-likelihood function, 373
logarithmic scale, 54
logistic function, 450
logistic regression, 449
prediction, 453
logit function, 450

lognormal distribution, 288, 492
lognormal geometric random walk, 492
London Stock Exchange (LSE), 16
long position, 65
Lorenz curve, 214, 293
LSE (London Stock Exchange), 16

MA model, 529–571
    ACF, 530
    autocorrelation, 530
    identifiability, 532
    invertibility, 532
    MA(1) invertibility, 533
    order determination, 530
    partial autocorrelation function (PACF), 532
MACD, 123
MAD (median absolute deviation), 96, 208, 397
    relationship to standard deviation, 96, 208
Mallows's C_p, 424, 468
margin, 17
margin call, 17
marginal distribution, 259
market, 16, 38
    secondary, 38
market capitalization, 32
market index, *see also* stock index, 49–55
market microstructure, 12
market model, 60
market order, 36
market portfolio, 60
market price, 34
market risk, 113
market timing, 121, 200
market value, 32
Markov chain, 279, 511
Markov chain Monte Carlo (MCMC), 389
Markovian property, 511
married put, 74
martingale, 279, 511

MASS (R package), 146
master limited partnership (MLP), 13
MATLAB®, 146, 172
maximum likelihood estimator (MLE), 352–354, 371–376
    Fisher scoring, 375
MCMC, 389
mean, 3, 80, 250, 254, 272
    of a random variable, 250
    trimmed mean, 95, 397
    Winsorized mean, 95, 397
mean reversion, 122
mean squared error (MSE), 348, 369, 463
median, 95, 216
    median($\cdot$), 95
median absolute deviation (MAD), 96, 208, 397
    relationship to standard deviation, 96, 208
mesokurtic, 88
method of moments, 349
methods (R package), 143
Microsoft Excel®, 147, 162
minimum variance portfolio, 69
missing data, 142, 185–186, 189
mixture distribution, 303, 307
MLE (maximum likelihood estimator), 352–354, 371–376
    Fisher scoring, 375
MLP (master limited partnership), 13
model, 114, 336–346
    additive error, 339
    autoregressive, 338, 492
    Bayesian, 346, 386–393
    classification, 337, 449
    error, 246, 534
        compared to residual, 415
    factor, 44, 413
    linear, 376, 413–448
    model averaging, 472
    multimodel inference, 472

multiplicative error, 339
of relationship, 408
probability distribution, 245–323
regression, 337, 413
selection, 463–469, 472
time series, 560
smoothing, 346
time series, 346, 487–604
state-space, 494
model averaging, 472
model specification in R, 454
ARIMA, 562
moment, 80, 81, 207, 208, 254
method of moments, 349
population, 254
sample, 80, 81, 207, 208
momentum, 44
monetary policy, 19
moneyness, 64
Monte Carlo estimate, 355–357
Monte Carlo simulation, 314–321,
355–357
Moody's Investors Service, 19, 26
Moody's Seasoned Corporate
Bonds, 26, 168, 436
mortgage rates, 168
mountain plot, 214
moving average, 39–41, 123, 134, 172
convergence-divergence, 123
death cross, 123
golden cross, 123
MACD, 123
moving average model, *see*, 529
of white noise, 489
autocorrelation, 512
R functions, 172
rolling operation, 172
moving average model (MA),
529–571
autocorrelation, 530
identifiability, 532
invertibility, 532
moving average operator, 531
multivariate, 531

MSE, mean squared error, 348, 369,
463
MSE, mean squared residual (MSR),
364, 417, 463
multicollinearity, 426
multimodel inference, 472
multivariate backshift operator, B,
497, 531, 536
multivariate data, 82–87, 218,
258–269, 309
time series, 517
multivariate distribution, 258–269,
294–295, 309
variance-covariance matrix, 5,
260, 264
multivariate regression, 339, 344
multivariate time series, 517
munging data, 183
municipal bond, 26
mutual fund, 17
index, 18

$N(\mu, \sigma^2)$, 283
NA ("Not Available"), 142, 185
naked option, 66
Nasdaq Composite Index, 50, 166
historical values, 51, 55
Nasdaq National Market, 16
national best bid and offer (NBBO),
35, 129
NBBO (national best bid and offer),
35, 129
Nelder-Mead method, 361, 375
New York Mercantile Exchange
(NYMEX), 17
New York Stock Exchange (NYSE),
16
Newton's method, 360, 375
Nikkei-225 (index), 60, 167
noncentral t distribution, 290
nonlinear least squares, 366–368, 376
nonnegative definite matrix, 262
nonnegative least squares, 370, 376
nonparametric probability density
estimation, 217–221, 226

multivariate, 220, 227
nonsynchronous trading, 11
norm, 351, 363, 370
normal copula, 266
normal distribution, 87, 287, 312, 322
    multivariate, 295
    skewed, 302
    test for, 461
    visual assessment, 231, 457
normal equations, 364
note, Treasury, 22
null hypothesis, 381
numeraire, 2
NYMEX (New York Mercantile
    Exchange), 17
NYSE (New York Stock Exchange),
    16
NYSE Composite, 53

Oanda, 170
object, R, 143–144, 173–182
    class, 143
    mode, 144
    type, 144
    `xts`, 173–181
OHLC dataset, 33, 166, 180–182
OLS (ordinary least squares), 369
omnibus hypothesis test, 383
optimization, 358–376, 413
    constrained, 359, 362, 413
    direct method, 360
    Gauss-Newton method, 367
    iterative method, 360–362
    least squares, 363–370
    Levenberg-Marquardt method,
      368
    maximum likelihood, 352–354,
      371–375
    Nelder-Mead method, 361
    Newton's method, 360
    quasi-Newton method, 360, 361,
      368
      Gauss-Newton method, 367
      Levenberg-Marquardt
        method, 368

simulated annealing, 362
steepest descent, 360
option, 63–66, 74–75, 171
    adjustment of strike, 64
    American, 64
    calendar spread, 74
    call spread, 74
    cash settlement, 64
    chain, 171
    European, 64
    `getOptionChain` (R function),
      171
    intrinsic value, 64
    married put, 74
    moneyness, 64
    pricing, 65, 279, 280
    put spread, 74
    straddle, 75
    strangle, 75
    strap, 75
    strike, 64
      adjustment, 64
    strip, 75
    time value, 64
order statistic, 95, 211, 269–270, 308
    ECDF, 211
    notation, 95
    PDF, 270
    quantile, 95
orthogonal regression, 414, 471, 602
oscillator, 124
OTC Bulletin Board, 16
outlier, 77, 93, 210, 277, 428
outlier-generating distribution, 277
overfitting, 417, 423, 428, 468, 556
    regression, 423, 428
    time series, 556

p-value, 225, 239, 381, 385, 470
PACF (partial autocorrelation
    function), 425, 516, 532,
    547–548
    AR model, 547
    MA model, 532
    sample, 548

pairs trading, 75–76, 600
par value, 21
parameter space, 248
parametric line plot, 236
Pareto distribution, 292
    generalized, 301
Pareto tail, 275, 399
partial autocorrelation function
    (PACF), 425, 516, 532,
    547–548
  AR model, 547
  MA model, 532
  sample, 548
partial autocovariance function, 516
partial F test, 425, 440
partial regression, 424–426
partitioning the variation, 306, 340,
    415, 425
pattern (technical)
  crossover, 122
  death cross, 123
  golden cross, 123
  head and shoulders, 124
  MACD, 123
  moving average
    convergence-divergence, 123
    death cross, 123
    golden cross, 123
  oscillator, 124
  relative strength index, 124
  resistance line, 126
  stochastic oscillator, 124
  support line, 123, 126
payout ratio, 42
PCA (principal components
    analysis), 409–413, 479
PDF (probability density function),
    217, 248, 253
  estimation, 217–221, 399
  kernel, 253
PE, 41–43
  CAPE (Shiller P/E ratio), 43
  enterprise value to EBITDA, 43
  forward, 42
  trailing, 42

Pearson correlation coefficient, 83
Pearson distributions, 324
percentile, 92
$\Phi(\cdot)$, 288
$\phi(\cdot)$, 288
Phillips-Ouliaris test, 600
Phillips-Perron test, 597, 598
pivotal quantity, 380
platykurtic, 88
point estimate, 347
  consistent, 348
  unbiased, 347
Poisson distribution, 285, 312, 322
  jump, 285
polynomial regression, 449
polynomial tail, 275, 399
portfolio, 60, 66–76, 407
  efficient frontier, 71
  efficient portfolio, 71
  expected shortfall, 407
  minimum variance portfolio, 69
  optimal portfolio, 69, 371
  rebalancing, 67, 407
  return, 68, 407
  risk measure, 407
  Sharpe ratio, 72
  tangency portfolio, 72
  Treynor ratio, 72
  VaR, 407
portmanteau test, 526
positive definite matrix, 262
POSIX, 156, 174
power curve, 383
power law, 301
  distribution, 292, 301
power of test, 382
power series, convergence of, 500
pracma (R package), 146
precision, 207, 388
predict (R function), 448, 453, 565
prediction, 348, 385, 448, 453
prediction interval, 385, 448
present value, 26–28
price discovery, 37
price-to-earnings ratio, 41–43

CAPE (Shiller P/E ratio), 43
enterprise value to EBITDA, 43
forward, 42
trailing, 42
price-to-sales ratio (PSR), 41
prime rate, 22
principal components, 409–413, 479
constrained, 413
probability density function (PDF),
217, 248, 253
estimation, 217–221, 399
kernel, 253
probability distribution, 247–314
nondegenerate, 247
R functions for, 144, 313
support, 248, 253
product distribution, 310
programmed trading, 126
projection matrix, 420
ProShares, 56
pseudorandom number, 315, 321
PSR (price-to-sales ratio), 41
pump-and-dump, 128
put option, *see* option
put spread, 74
PVaR, 115

q-q plot, 91–93, 229–233, 276, 306
half-normal plot, 233
of exceedance distribution, 232
of mixtures, 306
two samples, 233
quadratic programming, 69, 370, 413
qualitative regressor, 415, 419
Quandl, 169, 170
quantile, 92, 95, 144, 215, 249, 253,
259, 312, 404
Cornish-Fisher expansion, 406
estimation, 215, 356, 404–407
quantile function, 249, 250, 265,
277
sample or empirical, 95,
215–217, 405
quantile-quantile plot, 91–93,
229–233, 276, 306

of mixtures, 306
quantmod (R package), 165, 171, 187
quartile, 96, 216
quasirandom number, 316, 321

R (software), 139–190
$ (list extractor), 150, 152, 154,
174
Bayesian methods, 389
caveats, 187
comma separated file (CSV), 163
header, 163
conditional, 143
data frame, 152–153
stringsAsFactors, 152, 163
merging, 153, 178
tibble, 153
data structure, 160
date data, 156, 174
environment, 145, 146
formula, 454, 455
function, 144
generic function, 143
ggplot2, 146, 187
graphics, 155, 234–238
color, 156
layout, 238
three-dimensional, 235
GUI app, 147
help, 141
vignette, 142
inputting data, 161, 171
from the internet, 163
linear algebra, 150
list, 149
log return, 154, 174, 176
logical operator, 142
logistic regression, 451–453
long vector, 148
matrix, 150
merging datasets, 153, 178, 186
methods, 143
missing value, 142, 185
model formula, 454, 455
object, 143

optimization, 375–376
package, 140, 145–147, 187
    caveats, 147
    lowest common denominator,
        188
    parallel processing, 396, 465
    pracma, 146
    probability distributions, 311
    quadratic programming, 371
    quantmod, 171, 187
    random number generation, 314,
        321–323
        quasirandom number, 316,
            321
        randtoolbox, 316, 321
    regression, 376, 435–448
    ridge regression, 370
    rjags, 389
    RStudio, 140, 147
    Shiny, 147
    simulation, 314, 321–323, 555
        of time series, 555
    task views, 141
    three-dimensional graphics, 235
    tibble, 153
    time series, 172–181, 496, 499,
        561
    TTR, 172
    vector, 147
    vignette, 142
    web technologies, 172
R^2, 416
    adjusted, 423, 468
random number generation, 314–321
    quasirandom number, 316, 321
random variable, 2–5, 247–282
    notation, 3
    transformation, 251, 254, 267
random walk, 132, 490–492, 505
    autocovariance, 513
    cumsum (R), 491
    linear combination, 493, 599
    multivariate, 493
    with drift, 492, 505
Random Walk Hypothesis, 120, 491

rank transformation, 97, 398
real estate investment trust (REIT),
        13
recession, 25
regression, 351, 354, 363, 413–448,
        571–575
    ANOVA, 415
    ARMA errors, 435, 571–575
    assumptions, 414
    binary, 451
    Cook's distance, 433, 440
    deleted residual, 429, 464
    DFBETAS, 433, 440
    DFFITS, 433, 440
    F test, 421
    hat matrix, 420
    intercept, 365, 419
    leave-out-one deletion, 433, 440
    leverage, 429
        Cook's distance, 433, 440
        DFBETAS, 433, 440
        DFFITS, 433, 440
    logistic, 449
    outlier, 428
    partial F test, 425
    polynomial, 449
    qualitative regressor, 415, 419
    residual, 424
    residuals, 415, 417, 428–435,
        440, 571
        compared to errors, 415
        deleted, 429, 464
        serial correlation, 434,
            571–575
        studentized, 431–433
    seemingly unrelated regressions,
        339, 344
    spurious, 584
    t test, 421
    variable selection, 425, 444
        forward selection, 425, 444
    variance inflation factor (VIF),
        427, 446
regression model, 337, 413
    intercept (or not), 365, 419

regulatory agency, 43
REIT (real estate investment trust), 13
relative strength index, 124
resampling, 393–396
residual, 415, 417, 424, 428–435, 440, 571
   compared to error, 415
   deleted residual, 429, 464
   deviance, 374, 451
   externally studentized, 432, 433
   internally standardized, 433
   serial correlation, 434, 571
   standardized, 433
   studentized, 431–433
   studentized deleted, 433
resistance line, 126
return, 29–32, 67–73, 99–102
   absolute, 101
   aggregating over assets, 31
   aggregating over time, 30
   annualized, 29, 30
   autocorrelations, 99–101
   excess, 32, 60–63, 73
      mixing log returns and interest rates, 32
   expected, 68
   log return, 29–31, 491, 500
      annualized, 30
   portfolio, 68, 407
   R functions for, 155, 500
   simple, 29, 500
      annualized, 29
   variance, 69
return of capital, 45
   adjustment to prices, 48
revenue, 41
reverse split, 45
ridge regression, 369, 376, 427, 428, 447, 466
   ridge trace, 369, 428, 447
risk, 15, 66, 67, 113, 127
   diversifiable, 114
   market, 113
   measure, 114, 118, 257, 407

   coherent, 118, 257, 407
   DVaR, 119
   expected shortfall, 119
   PVaR, 115
   simple returns, 114, 404
   tail loss, 119
   VaR, 114
risk-free asset, 15, 22, 71
   interest rate, 22
risky asset, 15
   simple returns, 114, 404
   specific, 114
   systematic, 113
   uncertainty, 127
risk management, 113–120, 404
risk neutral, 44
robust method, 97, 396–399
R-squared, 416
   adjusted, 423, 468
RSS, 363, 416
RStudio, 140, 147
Russell 2000, 53, 166
Ryan, Jeffrey A., 164

$S(\cdot)$ (summation operator), 500
S&P 500 (Standard & Poor's 500 Index), 50, 166
   historical values, 51, 54, 55
   total return, 53, 54, 166
S&P 500 Equal Weight Index (EWI), 50
SACF (sample ACF), 100, 519
sample autocorrelation function (ACF or SACF), 100, 519
sample autocovariance function, 519
sample CCF (cross-correlation function), 519
sample correlation, 83, 85, 208, 210, 408
sample covariance, 82, 208, 210, 408
sample cross-correlation function (CCF), 519
sample cross-covariance function, 519
sample mean, 80, 272
sample moment, 80

sample quantile, 95, 215–217
sample standard deviation, 80
    historical volatility, 104
sample variance, 80, 289, 408
SARIMA model, 555
scale, 80, 207
scatterplot, 235
    three-dimensional, 235
scatterplot matrix, 86, 235
Schmidt-Phillips test, 598
Schwarz information criterion (SIC), 468
SDS (ETF), 58, 59
seasonal adjustment, 508
seasonal effects, 200
    stock prices, 200
seasonality, 41, 121, 508, 554, 555
    ARIMA model, 555
    ARMA model, 554
    seasonal adjustment, 508
    stock prices, 121
seasoned security, 26, 38
secondary market, 38
sector, market, 55, 56
    GICS, 56
Securities and Exchange Commission (SEC), 17, 170
seemingly unrelated regressions, 339, 344
SH (ETF), 58
shape, frequency distribution, 80–82, 207, 229–232
Shapiro-Francia test, 461
Shapiro-Wilk test, 461
Sharpe ratio, 72
Shibata's information criterion, 471, 583
Shiller P/E ratio (CAPE), 43
Shiny (R GUI apps), 145, 147
short ETF, 59
short interest, 44, 65
short position, 65–66, 120
short-and-distort, 128
shrunken estimator, 325, 369

SIC (Schwarz information criterion), 468
significance, statistical, 381
    p-value, 225, 239, 381, 470
simple hypothesis, 381
simple return, 29, 500
    annualized, 29
simulated annealing, 375
simulating random numbers, 314–321
    quasirandom number, 316, 321
simulation, 355–357
    in R, 321–323
    of time series, 555
simultaneous equations model, 339
skew, 113
SKEW Index, 113
SkewDex®, 113
skewed distribution
    CDF-skewing, 302
    differential scaling, 302
    Lorenz curve, 214
skewed generalized error distribution, 300
skewed normal distribution, 302
skewed t distribution, 302
skewness, 81, 251, 254
    Bowley's, 96
    Gini index, 214
    Pearson's, 96
    quantile-based measure, 96
smoothing, 41, 346, 534
    moving average, 41
smoothing matrix, 221, 420
smoothing parameter, 41, 218, 220, 346, 534
Sobol' sequence, 316
SPDR, 56
Spearman's rank correlation, 97, 399
specific risk, 114
split, 45
    adjustment to prices, 45
spread, 207
spreadsheet, 162
spurious regression, 584
SPY (ETF), 56, 57

SSE (sum of squares of residuals), 363, 416
SSE Composite (index), 60, 167
SSO (ETF), 57
stable distribution, 298, 313
stale price, 37
Standard & Poor's, 19
Standard & Poor's 500 Index (S&P 500), 50, 166
    historical values, 51, 54, 55
    total return, 53, 54, 166
standard deviation, 4, 80, 251, 254
    historical volatility, 104
standard error, 377, 567
standardized data, 208–210, 408, 409, 411
standardized residual, 433
"standardized" t distribution, 298
starburst plot, 227
state space model, 494
stationarity, 7, 100, 278, 514–518
    joint, 517–518
    strict, 514
    weak, 514
statistical arbitrage, 600
statistical inference, 347, 376–379, 381
    asymptotic inference, 378
    computational inference, 378
    confidence interval, 379
    consistency, 348
    exact inference, 378
    hypothesis testing, 381–384
    point estimate, 347
        consistent, 348
        unbiased, 347
    tuned inference, 218, 220, 354, 370, 379, 401, 402, 463
statistical model, 336–346
statistical significance, 381
    p-value, 225, 239, 381, 470
Statistical Software, Journal of, 187
steepest descent, 360
stochastic difference equation, 502

stochastic differential equation (SDE), 279, 502
stochastic oscillator, 124
stochastic process, 278–282, 487
stochastic volatility, 99, 414, 576
stock, 13
stock buyback, 33, 38, 43, 44
stock dividend, 45
stock index, 49–55, 413, 479
    adjusted value, 53
    capitalization weighted, 50
    equally weighted, 50
    market-value weighted, 50
    price weighted, 49, 50
    principal components, 413, 479
stock split, 45
    adjustment to indexes, 49
    adjustment to option strikes, 64
    adjustment to prices, 45
stop order, 36
stop-limit order, 36
straddle, 75
strangle, 75
strap, 75
strict stationarity, 514
strike price, 63
strip, 75
stripping of coupons, 28
STRIPS, 28
Student's t distribution, *see also* t distribution, 289
studentization, 289, 431
studentized deleted residual, 433
studentized residual, 431–433
Sturges's rule, 218
sum of squares, 341
summation operator, S, 500
support, 248, 253
    continuous distribution, 253
    discrete distribution, 248
support line, 123, 126
survival function, 257
SVXY (inverse VIX), 111
swap, 13
systematic risk, 113

t distribution, 289, 296–298, 313, 383
    location-scale, 295, 297, 298, 313
    multivariate, 295, 313
    noncentral, 290
    skewed, 290, 302
    "standardized", 298
    tail index, 296
t test, 382, 421
tail dependence, 127, 276, 402–403
    coefficient, 277, 403
    copula, 277
    tail dependence function, 276
tail index, 275, 399–402
    estimation, 399–402, 471
        Hill estimator, 400
    exponential, 274
    generalized Pareto, 402
    of t distribution, 296
    Pareto, 275, 399
    polynomial, 275, 399
tail loss, 119, 404
tail risk index, 113
tail threshold exceedance, 232, 275,
        301
tail weight, 274
TailDex®, 113
tangency portfolio, 72, 73
technical analysis, 35, 122
10-K form, 169
term structure, 24
test statistic, 381
three-dimensional graphics, 235
threshold exceedance, 232, 275, 301
tidyverse (suite of R packages), 146
time series, 5–12, 222, 346, 487–604
    bivariate, 519
    evolutionary, 515
    model
        AR, 534
        ARCH, 576
        ARIMA, 553
        ARMA, 549
        EWMA, 534
        GARCH, 580
        Gaussian white noise, 488

        geometric random walk, 492
        IMA, 533
        linear process, 489
        MA, 529
        moving average of white noise,
            489
        multivariate random walk, 493
        multivariate white noise, 493
        random walk, 490
        random walk with drift, 492
        state-space, 494
        white noise, 488
    multivariate, 517
    stationary, 514–518
    xts (R object), 173–181
time value, 64
Tokyo Stock Exchange (TSE), 16
trader, 15
trading friction, 37
trading strategy, 73–76
    crossoveer, 122
    dynamic hedge, 75
    hedge, 73–75
    options, 74
    pairs trading, 75–76, 600
trailing earnings, 41
tranch, 20
transformation, 227, 251, 254, 267,
        310, 315, 320, 344, 412, 421
    Box-Cox, 345, 422, 579
    linear, 209, 229, 268, 320, 421,
        495
    of a random variable, 251, 254,
        267, 315, 320
        change of variables technique,
            255
    of data, 208–210, 344, 412, 421
        centering, 208, 209, 421
        covariance transformation,
            210
        standardizing, 208, 209, 421
    one-to-one, 267
    power, 345
    variance-stabilizing, 344, 422,
        579

Treasuries, 22–25
    rates, 168
Treasury Bill, 22
trend (in time series), 505
    test, 507
Treynor ratio, 72
trimmed mean, 95, 397
TSE (Tokyo Stock Exchange), 16
TTR (R package), 172
tuned inference, 218, 220, 354–355,
    370, 379, 401, 402, 463
tuning parameter, 218, 220, 354, 370,
    379, 401, 402

$U(0,1)$, 252, 283, 286, 315
Ulrich, Joshua, 172
unbiased estimator, 347
uncertainty, 127
underlying, 63
uniform distribution, 252, 286, 315
unit circle, 504, 533, 544, 546
unit root, 503, 505, 592–599
    tests, 594–599
UPRO (ETF), 57, 58
US regulatory agency, 17–18
    Commodity Futures Trading
        Commission (CFTC), 18
    Securities and Exchange
        Commission (SEC), 17, 170
US Treasuries, 22–25

$V(\cdot)$ (variance, *see*), 4
value at risk (VaR), 114–120, 232,
    273, 276, 301, 309, 404–407
    coherence, 118, 257, 407
    CVaR, 119, 404
    DVaR, 119, 276
    estimation, 404–407
    portfolio, 407
    simple returns, 114, 404
VAR (see vector autoregressive
    model), 535
VAR(p), 535, 590
variable selection, 425, 444
variance, 4, 80, 251, 254, 289

bootstrap estimate, 394
    of a linear combination, 4,
        268–269
    of a random variable, 4, 251
    partitioning, 306, 340, 415
    sample, 80, 289
variance inflation factor (VIF), 427,
    446
variance-bias tradeoff, *see*
    bias-variance tradeoff
variance-covariance matrix, 4, 5, 83,
    260, 264, 269, 408
    of a linear transformation, 268
    sample, 83, 408
variance-stabilizing transformation,
    344, 422, 579
vector autoregressive model, 339,
    535, 590
vector error correction model
    (VECM), 601
VIX, 108–111, 166
    historical values, 109–111
volatility, 98–113
    annualized, 107, 491
    definition, 103
    effect of length of time intervals,
        93, 105
    effect of outliers, 104
    historical compared with
        implied, 108
    historical volatility, 103
    implied, 107
    square-root rule, 107
    stochastic, 99, 414
volatility index, 108–112
volatility smile, 112
VolDex®, 112

Watanabe-Akaike information
    criterion, 471
weekend effect, 10
Weibull distribution, 292
weighted least squares, 369
white noise, 488
    linear combination, 493, 599

multivariate, 493
    with drift, 488
Wickham, Hadley, 146
window size, 221
Winsorized mean, 95, 397
World Bank Open Data, 170
wrangling data, 183

XIV (inverse VIX), 111
xts object, 173–181
    changing time period, 177
    converting to data frame, 173
    graphics, 179
    merging, 178, 186
    subsetting, 176

Yahoo Finance, 164–167, 190
    dividends, 182
    international markets, 166
    missing data, 186, 189
yield curve, 24, 169
    Nelson-Siegel, 24
    spline, 24
    Svensson, 24
yield to maturity, 28
Yule-Walker equations, 542, 558

Zivot-Andrews test, 598